STUDENT'S SOLUTIONS MANUAL

CRAIG C. JOHNSON
Brigham Young University Idaho

STATISTICS:
INFORMED DECISIONS
USING DATA
FOURTH EDITION

Michael Sullivan, III
Joliet Junior College

PEARSON

Boston Columbus Indianapolis New York San Francisco Upper Saddle River
Amsterdam Cape Town Dubai London Madrid Milan Munich Paris Montreal Toronto
Delhi Mexico City Sao Paulo Sydney Hong Kong Seoul Singapore Taipei Tokyo

The author and publisher of this book have used their best efforts in preparing this book. These efforts include the development, research, and testing of the theories and programs to determine their effectiveness. The author and publisher make no warranty of any kind, expressed or implied, with regard to these programs or the documentation contained in this book. The author and publisher shall not be liable in any event for incidental or consequential damages in connection with, or arising out of, the furnishing, performance, or use of these programs.

Reproduced by Pearson from electronic files supplied by the author.

ISBN-13: 978-0-321-75747-0
ISBN-10: 0-321-75747-5

2 3 4 5 6 EBM 15 14 13 12

www.pearsonhighered.com

Table of Contents

Preface

This solutions manual accompanies *Statistics: Informed Decisions Using Data, 4e* by Michael Sullivan, III. The Student's Solutions Manual contains detailed solutions to all odd exercises in the text and all solutions to chapter reviews, chapter tests, Consumer Reports® projects and the case studies. A concerted effort has been made to make this manual as user-friendly and error free as possible. We offer a special note of thanks to Nathan Kidwell who carefully checked this manual for accuracy.

Chapter 1

Data Collection

Section 1.1

1. Statistics is the science of collecting, organizing, summarizing and analyzing information in order to draw conclusions and answer questions. In addition, statistics is about providing a measure of confidence in any conclusions.

3. Individual

5. Statistic; Parameter

7. 18% is a parameter because it describes a population (all of the governors).

9. 32% is a statistic because it describes a sample (the high school students surveyed).

11. 0.366 is a parameter because it describes a population (all of Ty Cobb's at-bats).

13. 23% is a statistic because it describes a sample (the 6,076 adults studied).

15. Qualitative

17. Quantitative

19. Quantitative

21. Qualitative

23. Discrete

25. Continuous

27. Continuous

29. Discrete

31. Nominal

33. Ratio

35. Ordinal

37. Ratio

39. The population consists of all teenagers 13 to 17 years old who live in the United States. The sample consists of the 1,028 teenagers 13 to 17 years old who were contacted by the Gallup Organization.

41. The population consists of all of the soybean plants in this farmer's crop. The sample consists of the 100 soybean plants that were selected by the farmer.

43. The population consists of all women 27 to 44 years of age with hypertension. The sample consists of the 7,373 women 27 to 44 years of age with hypertension who were included in the study.

45. Individuals: Motorola Droid X, Motorola Droid 2, Apple iPhone 4, Samsung Epic 4G, Samsung Captivate.
Variables: Weight (ounces), Service Provider, Depth (inches).
Data for *weight*: 5.47, 5.96, 4.8, 5.5, 4.5 (ounces);
Data for *service provider*: Verizon, Verizon, ATT, Sprint, ATT;
Data for *depth*: 0.39, 0.53, 0.37, 0.6, 0.39 (inches).
The variable *weight* is continuous; the variable *service provider* is qualitative; the variable *depth* is continuous.

47. Individuals: Alabama, Colorado, Indiana, North Carolina, Wisconsin.
Variables: Minimum age for Driver's License (unrestricted); mandatory belt-use seating positions, maximum allowable speed limit (rural interstate) in 2007.
Data for *minimum age for driver's license*: 17, 17, 18, 16, 18;
Data for *mandatory belt-use seating positions*: front, front, all, all, all;
Data for *maximum allowable speed limit (rural interstate) 2007*: 70, 75, 70, 70, 65 (mph.)
The variable *minimum age for driver's license* is continuous; the variable *mandatory belt-use seating positions* is qualitative; the variable *maximum allowable speed limit (rural interstate) 2007* is continuous (although only discrete values are typically chosen for speed limits.)

49. (a) The research objective is to determine if adolescents who smoke have a lower IQ than nonsmokers.

    (b) The population is all adolescents aged 18-21. The sample consisted of 20,211 18-year-old Israeli military recruits.

    (c) Descriptive statistics: The average IQ of the smokers was 94, and the average IQ of nonsmokers was 101.

    (d) The conclusion is that individuals with a lower IQ are more likely to choose to smoke.

51. (a) The research objective is to determine the proportion of adult Americans who believe the Federal government wastes 51 cents or more of every dollar.

(b) The population is all adult Americans aged 18 years or older.

(c) The sample is the 1026 American adults aged 18 years or older that were surveyed.

(d) Descriptive statistics: Of the 1026 individuals surveyed, 35% indicated that 51 cents or more is wasted.

(e) From this study, one can infer that many Americans believe the Federal government wastes much of the money collected in taxes.

53. *Jersey number* is nominal (the numbers generally indicate a type of position played). However, if the researcher feels that lower caliber players received higher numbers, then *jersey number* would be ordinal since players could be ranked by their number.

55. (a) The research question is to determine the role that TV watching by children younger than 3 plays in future attention problems for the children.

(b) The population of interest is all children under the age of 3 years.

(c) The sample consisted of the 967 children whose parents answered questions about TV habits and behavior issues.

(d) Descriptive statistic: The risk of attention problems five years later doubled for each hour per day kids under 3 watched violent child-oriented programs.

(e) Inference: Children under the age of 3 years should not watch television. If they do watch, it should be educational and not violent child-oriented entertainment. Shows that are violent double the risk of attention problems for each additional hour watched each day. Even educational programs can result in a substantial risk for attention problems.

57. The values of a discrete random variable result from counting. The values of a continuous random variable result from a measurement.

59. We say data vary, because when we draw a random sample from a population, we do not know which individuals will be included. If we were to take another random sample, we would have different individuals and therefore different data. This variability affects the results of a statistical analysis because the results would differ if a study is repeated.

61. Age could be considered a discrete random variable. A random variable can be "discretized" by allowing, for example, only whole numbers to be recorded.

Section 1.2

1. The response variable is the variable of interest in a research study. An explanatory variable is a variable that affects (or explains) the value of the response variable. In research, we want to see how changes in the value of the explanatory variable affect the value of the response variable.

3. Confounding exists in a study when the effects of two or more explanatory variables are not separated. So any relation that appears to exist between a certain explanatory variable and the response variable may be due to some other variable or variables not accounted for in the study. A lurking variable is a variable not accounted for in a study, but one that affects the value of the response variable.

5. Cross-sectional studies collect information at a specific point in time (or over a very short period of time). Case-control studies are retrospective (they look back in time). Also, individuals that have a certain characteristic (such as cancer) in a case-control study are matched with those that do not have the characteristic. Case-control studies are typically superior to cross-sectional studies. They are relatively inexpensive, provide individual level data, and give longitudinal information not available in a cross-sectional study.

7. There is a perceived benefit to obtaining a flu shot, so there are ethical issues in intentionally denying certain seniors access to the treatment.

9. This is an observational study because the researchers merely observed existing data. There was no attempt by the researchers to manipulate or influence the variable(s) of interest.

11. This is an experiment because the explanatory variable (teaching method) was intentionally varied to see how it affected the response variable (score on proficiency test).

13. This is an observational study because the survey only observed preference of Coke or Pepsi. No attempt was made to manipulate or influence the variable of interest.

15. This is an experiment because the explanatory variable (carpal tunnel treatment regimen) was intentionally manipulated in order to observe potential effects on the response variable (level of pain).

17. **(a)** This is a cohort study because the researchers observed a group of people over a period of time.

    (b) The response variable is whether the individual has heart disease or not. The explanatory variable is whether the individual is happy or not.

    (c) There may be confounding due to lurking variables. For example, happy people may be more likely to exercise, which could affect whether they will have heart disease or not.

19. **(a)** This is an observational study because the researchers simply administered a questionnaire to obtain their data. No attempt was made to manipulate or influence the variable(s) of interest. This is a cross-sectional study because the researchers are observing participants at a single point in time.

    (b) The response variable is body mass index. The explanatory variable is whether a TV is in the bedroom or not.

    (c) Answers will vary. Some lurking variables might be the amount of exercise per week and eating habits. Both of these variables can affect the body mass index of an individual.

    (d) The researchers attempted to avoid confounding due to lurking variables by taking into account such variables as 'socioeconomic status'.

    (e) No. Since this was an observational study, we can only say that a television in the bedroom is associated with a higher body mass index.

21. Answers will vary. This is a prospective, cohort observational study. The response variable is whether the worker had cancer or not and the explanatory variable is the amount of electromagnetic field exposure. Some possible lurking variables include eating habits, exercise habits, and other health related variables such as smoking habits. Genetics (family history) could also be a lurking variable. This was an observational study, and not an experiment, so the study only concludes that high electromagnetic field exposure is associated with higher cancer rates.

The author reminds us that this is an observational study, so there is no direct control over the variables that may affect cancer rates. He also points out that while we should not simply dismiss such reports, we should consider the results in conjunction with results from future studies. The author concludes by mentioning known ways (based on extensive study) of reducing cancer risks that can currently be done in our lives.

23. **(a)** The research objective is to determine whether lung cancer is associated with exposure to tobacco smoke within the household.

    (b) This is a case-controlled study because there is a group of individuals with a certain characteristic (lung cancer but never smoked) being compared to a similar group without the characteristic (no lung cancer and never smoked). The study is retrospective because lifetime residential histories were compiled and analyzed.

    (c) The response variable is whether the individual has lung cancer or not. This is a qualitative variable.

    (d) The explanatory variable is the number of "smoker years". This is a quantitative variable.

    (e) Answers will vary. Some possible lurking variables are household income, exercise routine, and exposure to tobacco smoke outside the home.

    (f) The conclusion of the study is that approximately 17% of lung cancer cases among nonsmokers can be attributed to high levels of exposure to tobacco smoke during childhood and adolescence. No, we cannot say that exposure to household tobacco smoke causes lung cancer since this is only an observational study. We can, however, conclude that lung cancer is associated with exposure to tobacco smoke in the home.

(g) An experiment involving human subjects is not possible for ethical reasons. Researchers would be able to conduct an experiment using laboratory animals, such as rats.

Section 1.3

1. The frame is a list of all the individuals in the population.

3. Sampling without replacement means that no individual may be selected more than once as a member of the sample.

5. Answers will vary. We will use one-digit labels and assign the labels across each row (i.e. *Pride and Prejudice* – 0, *The Sun Also Rises* – 1, and so on). Starting at row 5, column 11, and proceeding downward, we obtain the following labels: 8, 4, 3
In this case, the 3 books in the sample would be *As I Lay Dying*, *A Tale of Two Cities*, and *Crime and Punishment*. Different labeling order, different starting points in Table I in Appendix A, or use of technology will likely yield different samples.

7. (a) {616, 630}, {616, 631}, {616, 632}, {616, 645}, {616, 649}, {616, 650}, {630, 631}, {630, 632}, {630, 645}, {630, 649}, {630, 650}, {631, 632}, {631, 645}, {631, 649}, {631, 650}, {632, 645}, {632, 649}, {632, 650}, {645, 649}, {645, 650}, {649, 650}

(b) There is a 1 in 21 chance that the pair of courses will be EPR 630 and EPR 645.

9. (a) Starting at row 5, column 22, using two-digit numbers, and proceeding downward, we obtain the following values: 83, 94, 67, 84, 38, 22, 96, 24, 36, 36, 58, 34,.... We must disregard 94 and 96 because there are only 87 faculty members in the population. We must also disregard the second 36 because we are sampling without replacement. Thus,

the 9 faculty members included in the sample are those numbered 83, 67, 84, 38, 22, 24, 36, 58, and 34.

(b) Answers will vary depending on the type of technology used. If using a TI-84 Plus, the sample will be: 4, 20, 52, 5, 24, 87, 67, 86, and 39.

Note: We must disregard the second 20 because we are sampling without replacement.

11. (a) Answers will vary depending on the technology used (including a table of random digits). Using a TI-84 Plus graphing calculator with a seed of 17 and the labels provided, our sample would be North Dakota, Nevada, Tennessee, Wisconsin, Minnesota, Maine, New Hampshire, Florida, Missouri, and Mississippi.

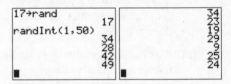

(b) Repeating part (a) with a seed of 18, our sample would be Michigan, Massachusetts, Arizona, Minnesota, Maine, Nebraska, Georgia, Iowa, Rhode Island, Indiana.

13. (a) The list provided by the administration serves as the frame. Number each student in the list of registered students, from 1 to 19,935. Generate 25 random numbers, without repetition, between 1 and 19,935 using a random number generator or table. Select the 25 students with these numbers.

(b) Answers will vary.

15. Answers will vary. Members should be numbered 1 – 32, though other numbering schemes are possible (e.g. 0 – 31). Using a table of random digits or a random-number generator, four different numbers (labels) should be selected. The names corresponding to these numbers form the sample.

Section 1.4

1. Stratified random sampling may be appropriate if the population of interest can be divided into groups (or strata) that are homogeneous and non-overlapping.

3. Convenience samples are typically selected in a nonrandom manner. This means the results are not likely to represent the population. Convenience samples may also be self-selected, which will frequently result in small portions of the population being overrepresented.

5. Stratified sample

7. False. In many cases, other sampling techniques may provide equivalent or more information about the population with less "cost" than simple random sampling.

9. True. Because the individuals in a convenience sample are not selected using chance, it is likely that the sample is not representative of the population.

11. Systematic sampling. The quality-control manager is sampling every 8th chip.

13. Cluster sampling. The airline surveys all passengers on selected flights (clusters).

15. Simple random sampling. Each known user of the product has the same chance of being included in the sample.

17. Cluster sampling. The farmer samples all trees within the selected subsections (clusters).

19. Convenience sampling. The research firm is relying on voluntary response to obtain the sample data.

21. Stratified sampling. Shawn takes a sample of measurements during each of the four time intervals (strata).

23. The numbers corresponding to the 20 clients selected are 16, $16 + 25 = 41$, $41 + 25 = 66$, $66 + 25 = 91$, $91 + 25 = 116$, 141, 166, 191, 216, 241, 266, 291, 316, 341, 366, 391, 416, 441, 466, 491.

25. Answers will vary. To obtain the sample, number the Democrats 1 to 16 and obtain a simple random sample of size 2. Then number the Republicans 1 to 16 and obtain a simple random sample of size 2. Be sure to use a different starting point in Table I or a different seed for each stratum.

For example, using a TI-84 Plus graphing calculator with a seed of 38 for the Democrats and 40 for the Republicans, the numbers selected would be 6, 9 for the Democrats and 14, 4 for the Republicans. If we had numbered the individuals down each column, the sample would consist of Haydra, Motola, Engler, and Thompson.

```
38→rand
            38
randInt(1,16)
             6
             9
■
```
```
40→rand
            40
randInt(1,16)
            14
             4
```

27. (a) $\dfrac{N}{n} = \dfrac{4502}{50} = 90.04 \to 90$; Thus, $k = 90$.

   (b) Randomly select a number between 1 and 90. Suppose that we select 15. Then the individuals to be surveyed will be the 15th, 105th, 195th, 285th, and so on up to the 4425th employee on the company list.

29. Simple Random Sample:
   Number the students from 1 to 1280. Use a table of random digits or a random-number generator to randomly select 128 students to survey.

   Stratified Sample:
   Since class sizes are similar, we would want to randomly select $\dfrac{128}{32} = 4$ students from each class to be included in the sample.

   Cluster Sample:
   Since classes are similar in size and makeup, we would want to randomly select $\dfrac{128}{32} = 4$ classes and include all the students from those classes in the sample.

31. Answers will vary. One design would be a stratified random sample, with two strata being commuters and noncommuters, as these two groups each might be fairly homogeneous in their reactions to the proposal.

33. Answers will vary. One design would be a cluster sample, with the clusters being city blocks. Randomly select city blocks and survey every household in the selected blocks.

35. Answers will vary. Since the company already has a list (frame) of 6,600 individuals with high cholesterol, a simple random sample would be an appropriate design.

37. **(a)** For a political poll, a good frame would be all registered voters who have voted in the past few elections since they are more likely to vote in upcoming elections.

(b) Because each individual from the frame has the same chance of being selected, there is a possibility that one group may be over- or underrepresented.

(c) By using a stratified sample, the strategist can obtain a simple random sample within each strata (political party) so that the number of individuals in the sample is proportionate to the number of individuals in the population.

39. Answers will vary.

Section 1.5

1. A closed question is one in which the respondent must choose from a list of prescribed responses. An open question is one in which the respondent is free to choose his or her own response. Closed questions are easier to analyze, but limit the responses. Open questions allow respondents to state exactly how they feel, but are harder to analyze due to the variety of answers and possible misinterpretation of answers.

3. Bias means that the results of the sample are not representative of the population. There are three types of bias: sampling bias, response bias, and nonresponse bias. Sampling bias is due to the use of a sample to describe a population. This includes bias due to convenience sampling. Response bias involves intentional or unintentional misinformation. This would include lying to a surveyor or entering responses incorrectly. Nonresponse bias results when individuals choose not to respond to questions or are unable to be reached. A census can suffer from response bias and nonresponse bias, but would not suffer from sampling bias.

5. **(a)** Sampling bias. The survey suffers from undercoverage because the first 60 customers are likely not representative of the entire customer population.

(b) Since a complete frame is not possible, systematic random sampling could be used to make the sample more representative of the customer population.

7. **(a)** Response bias. The survey suffers from response bias because the question is poorly worded.

(b) The survey should inform the respondent of the current penalty for selling a gun illegally and the question should be worded as: "Do you approve or disapprove of harsher penalties for individuals who sell guns illegally?" The order of "approve" and "disapprove" should be switched from one individual to the next.

9. **(a)** Nonresponse bias. Assuming the survey is written in English, non-English speaking homes will be unable to read the survey. This is likely the reason for the very low response rate.

(b) The survey can be improved by using face-to-face or phone interviews, particularly if the interviewers are multilingual.

11. **(a)** The survey suffers from sampling bias due to undercoverage and interviewer error. The readers of the magazine may not be representative of all Australian women, and advertisements and images in the magazine could affect the women's view of themselves.

(b) A well-designed sampling plan not in a magazine, such as a cluster sample, could make the sample more representative of the population.

13. **(a)** Response bias due to a poorly worded question.

(b) The question should be reworded in a more neutral manner. One possible phrasing might be: "Do you believe that a marriage can be maintained after an extramarital relation?"

15. **(a)** Response bias. Students are unlikely to give honest answers if their teacher is administering the survey.

(b) An impartial party should administer the survey in order to increase the rate of truthful responses.

17. No. The survey still suffers from sampling bias due to undercoverage, nonresponse bias, and potentially response bias.

19. It is very likely that the order of these two questions will affect the survey results. To alleviate the response bias, either question B could be asked first, or the order of the two questions could be rotated randomly.

21. The company is using a reward in the form of the $5.00 payment and an incentive by telling the reader that his or her input will make a difference.

23. For random digit dialing, the frame is anyone with a phone (whose number is not on a do-not-call registry). Even those with unlisted numbers can still be reached through this method.
Any household without a phone, households on the do-not-call registry, and homeless individuals are excluded. This could result in sampling bias due to undercoverage if the excluded individuals differ in some way than those included in the frame.

25. It is extremely likely, particularly if households on the do-not-call registry have a trait that is not part of those households that are not on the registry.

27. Some non-sampling errors presented in the article as leading to incorrect exit polls were poorly trained interviewers, interviewer bias, and over representation of female voters.

27. Answers will vary.

29. Answers will vary.

31. Answers will vary.

33. The *Literary Digest* made an incorrect prediction due to sampling bias (an incorrect frame led to undercoverage) and nonresponse bias (due to the low response rate).

35. **(a)** The population of interest is all vehicles that travel on the road (or portion of the road) in question.

(b) The variable of interest is the speed of the vehicles.

(c) The variable is quantitative.

(d) Because speed has a 'true zero', it is at the ratio level of measurement.

(e) A census is not feasible. It would be impossible to obtain a list of all the vehicles that travel on the road.

(f) A sample is feasible, but not a simple random sample (since a complete frame is impossible). A systematic random sample would be a feasible alternative.

(g) Answers will vary. One bias is sampling bias. If the city council wants to use the cars of residents who live in the neighborhood to gauge the prevailing speed, then individuals who are not part of the population were in the sample (likely a huge portion), so the sample is not representative of the intended population.

37. Nonresponse can be addressed by conducting callbacks or offering rewards.

39. Conducting a presurvey with open questions allows the researchers to use the most popular answers as choices on closed-question surveys.

41. Provided the survey was conducted properly and randomly, a high response rate will provide more representative results. When a survey has a low response rate, only those who are most willing to participate give responses. Their answers may not be representative of the whole population.

43. There is more than one type of CD. This can be interpreted as a medium used to store music or information electronically: a compact disk. It could also be understood as a special type of savings account: a certificate of deposit.

Section 1.6

1. **(a)** An experimental unit is a person, object, or some other well-defined item upon which a treatment is applied.

(b) A treatment is a condition applied to an experimental unit. It can be any combination of the levels of the explanatory variables.

   (c) A response variable is a quantitative or qualitative variable that measures a response of interest to the experimenter.

   (d) A factor is a variable whose effect on the response variable is of interest to the experimenter. Factors are also called explanatory variables.

   (e) A placebo is an innocuous treatment, such as a sugar pill, administered to a subject in a manner indistinguishable from an actual treatment.

   (f) Confounding occurs when the effect of two explanatory variables on a response variable cannot be distinguished.

3. In a single-blind experiment, subjects do not know which treatment they are receiving. In a double-blind experiment, neither the subject nor the researcher(s) in contact with the subjects knows which treatment is received.

5. Blocking

7. **(a)** The researchers used an innocuous treatment to account for effects that would result from any treatment being given (i.e. the placebo effect). The placebo is the flavored water that looks and tastes like the sports drink. It serves as the baseline for which to compare the results when the noncaffeinated and caffeinated sports drinks are administered.

   (b) Being double-blind means that neither the cyclists nor the researcher administering the treatments knew when the cyclists were given the caffeinated sports drink, the noncaffeinated sports drink, or the flavored-water placebo. Using a double-blind procedure is necessary to avoid any intentional or unintentional bias due to knowing which treatment is being given.

   (c) Randomization is used to determine the order of the treatments for each subject.

   (d) The population of interest is all athletes or individuals involved in prolonged exercise. The sample consists of the 16 highly trained cyclists studied.

   (e) There are three treatments in the study: caffeinated sports drink, noncaffeinated sports drink, and a flavored-water placebo.

   (f) The response variable is total work completed.

   (g) A repeated-measure design takes measurements on the same subject using multiple treatments. A matched-pairs design is a special case of the repeated-measures design that uses only two treatments.

9. **(a)** The response variable is the achievement test scores.

   (b) Some factors are teaching methods, grade level, intelligence, school district, and teacher.
Fixed: grade level, school district, teacher.
Set at predetermined levels: teaching method.

   (c) The treatments are the new teaching method and the traditional method. There are 2 levels of treatment.

   (d) The factors that are not controlled are dealt with by random assignment into the two treatment groups.

   (e) Group 2, using the traditional teaching method, serves as the control group.

   (f) This experiment has a completely randomized design.

   (g) The subjects are the 500 first-grade students from District 203 recruited for the study.

(h)

11. **(a)** This experiment has a matched-pairs design.

    (b) The response variable is the level of whiteness.

    (c) The explanatory variable is the whitening method. The treatments are Crest Whitestrips Premium in addition to brushing and flossing, and just brushing and flossing alone.

    (d) Answers will vary. One other possible factor is diet. Certain foods and tobacco products are more likely to stain teeth. This could impact the level of whiteness.

    (e) Answers will vary. One possibility is that using twins helps control for genetic factors such as weak teeth that may affect the results of the study.

13. **(a)** This experiment has a completely randomized design.

    (b) The population being studied is adults with insomnia.

    (c) The response variable is the terminal wake time after sleep onset (WASO).

    (d) The explanatory variable is the type of intervention. The treatments are cognitive behavioral therapy (CBT), muscle relaxation training (RT), and the placebo.

    (e) The experimental units are the 75 adults with insomnia.

    (f)

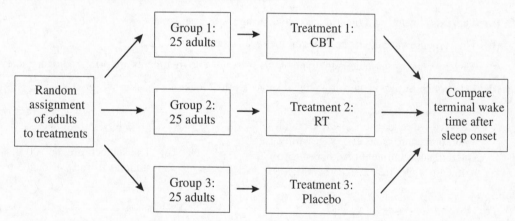

15. **(a)** This experiment has a completely randomized design.

    (b) The population being studied is adults over 60 years old and in good health.

    (c) The response variable is the standardized test of learning and memory.

    (d) The factor set to predetermined levels (explanatory variable) is the drug. The treatments are 40 milligrams of ginkgo 3 times per day and the matching placebo.

Copyright © 2013 Pearson Education, Inc.

9

(e) The experimental units are the 98 men and 132 women over 60 years old and in good health.

(f) The control group is the placebo group.

(g)

17. (a) This experiment has a matched-pairs design.

(b) The response variable is the distance the yardstick falls.

(c) The explanatory variable is hand dominance. The treatment is dominant versus non-dominant hand.

(d) The experimental units are the 15 students.

(e) Professor Neil used a coin flip to eliminate bias due to starting on the dominant or non-dominant hand first on each trial.

(f)

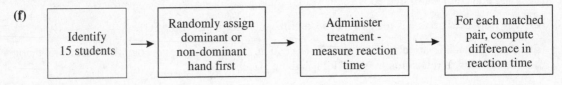

19. (a) This experiment has a randomized block design.

(b) The response variable is the score on the recall exam.

(c) The explanatory variable is the type of advertising. The treatments are print, radio, and television.

(d) Level of education is the variable that serves as the block.

(e)

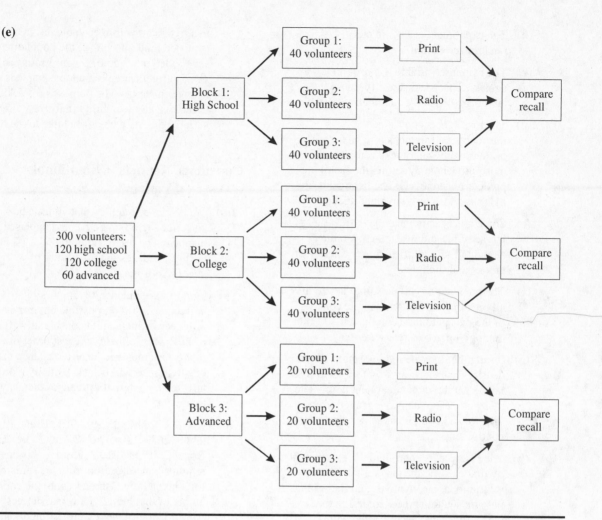

21. Answers will vary. Using a TI-84 Plus graphing calculator with a seed of 195, we would pick the volunteers numbered 8, 19, 10, 12, 13, 6, 17, 1, 4, and 7 to go into the experimental group. The rest would go into the control group. If the volunteers were numbered in the order listed, the experimental group would consist of Ann, Kevin, Christina, Eddie, Shannon, Randy, Tom, Wanda, Kim, and Colleen.

23. Answers will vary. A completely randomized design is likely the best.

25. Answers will vary. A matched-pairs design matched by type of exterior finish is likely the best.

27. Answers will vary. A randomized block design blocked by type of car is likely best.

29. (a) The response variable is blood pressure.

    (b) Three factors that have been identified are daily consumption of salt, daily consumption of fruits and vegetables, and the body's ability to process salt.

    (c) The daily consumption of salt and the daily consumption of fruits and vegetables can be controlled. The body's ability to process salt cannot be controlled. To deal with variability of the body's ability to process salt, randomize experimental units to each treatment group.

    (d) Answers will vary. Three levels of treatment might be a good choice – one level below the recommended daily allowance, one equal to the recommended daily allowance, and one above the recommended daily allowance.

31. Answers will vary.

33. (a) The research objective is to determine if nymphs of *Brachytron pretense* will control mosquitoes.

(b) The experiment is a completely randomized design.

(c) The response variable is mosquito larvae density. This is a discrete (because the larvae are counted), quantitative variable.

(d) The researchers controlled the introduction of nymphs of *Brachytron pretense* into water tanks containing mosquito larvae by setting the factor at predetermined levels. The treatments are nymphs added or no nymphs added.

(e) Some other factors that may affect the larvae of mosquitoes are temperature, amount of rainfall, fish, presence of other larvae, and sunlight.

(f) The population is all mosquito larvae in all breeding places. The sample consists of the mosquito larvae in the ten 300-liter outdoor, open concrete water tanks.

(g) During the study period, mosquito larvae density changed from 7.34 to 0.83 to 6.83 larvae per dip in the treatment tanks. The mosquito larvae density changed from 7.12 to 6.83 to 6.79 larvae per dip in the control tanks.

(h) The researchers controlled the experiment by first clearing the ten tanks of all non-mosquito larvae, nymphs, and fish so that only mosquito larvae were left in the tanks. Five tanks were treated with 10 nymphs of *Brachytron pretense*, while five tanks were left untreated, serving as a control group for baseline comparison. The researchers attempted to make the tanks as identical as possible, except for the treatment.

(i)

(j) The researchers concluded that nymphs of *Brachytron pretense* can be used effectively as a strong, ecologically friendly means of controlling mosquitoes, and ultimately mosquito-borne diseases.

35. In a randomized block design, experimental units are divided into homogeneous groups called blocks before being randomly assigned to a treatment within each block. In a stratified random sample, the population is subdivided into homogeneous groups called strata before a simple random sample is drawn from each strata. The purpose of blocking is to remove any variability in the response variable that may be attributable to the block.

Consumer Reports®: Emotional "Aspirin"

(a) Good science requires that all tests be performed in an objective and unbiased manner. The use of random codes ("blind testing") reduces the possibility of unconscious bias.

(b) In most experiments it is impossible to perfectly control all possible environmental influences which could conceivably affect test results. Randomizing the test order of replicate samples reduces the chance that any one treatment will be substantially more affected by external influences than any other treatment.

(c) Brand A: 231.2, 232.9, 261.1; Brand B: 224.6, 238.9, 263.7; Brand C: 207.8, 213.4, 227.1; Brand D: 192.8, 219.2, 251.1. This question requires generalization from the sample tested for a given brand (three lots) to the entire output of that brand. How one defines "meeting their label claim" is important. Certainly the brand average needs to be above the claimed value. In addition, it would seem necessary that few, if any of the lots fall below the claim. The following graph displays the observed amount of SAM-E for the various brands.

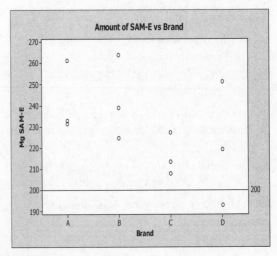

Brands A and B clearly appear to meet their label claim as all samples are well above the target of 200mg. The brand C average is the closest of the four to the target, but the three observations are the most consistent of any brand, suggesting that it too meets its label claim. Brand D is the most variable of the brands and has one observation below 200mg. Even though the sample mean is above 200, the lack of consistency suggests some chance that a larger sample could average below 200 or that a high proportion of lots might be below 200.

(d) **Step 1:** *Identify the problem to be solved.* *Consumer Reports®* would like to determine if the amount of SAM-E contained in several representative brands meets the label claims. Therefore, the response will be the amount of SAM-E (mg) in the pills.

Step 2: *Determine the explanatory variables that affect the response variable.* Some explanatory variables that affect the amount of SAM-E measured: Brand, lot from which pills are obtained, and testing mechanism.

Step 3: *Determine the number of experimental units.* In this experiment, we will measure the amount of SAM-E in 12 orders (4 brands) of mood-changing pills.

Step 4: *Determine the level of the explanatory variables*: We list the explanatory variables and their levels.

• **Brand** – We wish to determine the difference (if any) in the amount SAM-E as follows:
    Brand A: 3 orders
    Brand B: 3 orders
    Brand C: 3 orders
    Brand D: 3 orders

• **Lot from which pills are obtained** – Variability in the amount of SAM-E could possibly occur depending on the lot from which the pills are drawn. To account for this, we randomly select the samples from different lots.

• **Testing mechanism** – The measurement of SAM-E obtained could vary depending on the accuracy of the testing mechanism. We can control this somewhat by using the same testing approach for each order of pills. We also randomize the order in which the pills are tested. We also assign random codes to the various brands so that the

laboratory will not know the manufacturer it is testing, which helps to avoid bias.

Step 5: *Conduct the experiment.* (a) We determine the order in which the pills will be tested. (b) We determine the amount of SAM-E for each order of pills.

Step 6: *Test the claim.* We analyze the data to determine whether the manufacturers are meeting their label claims.

Chapter 1 Review Exercises

1. Statistics is the science of collecting, organizing, summarizing and analyzing information in order to draw conclusions.

2. The population is the group of individuals that is to be studied.

3. A sample is a subset of the population.

4. An observational study uses data obtained by studying individuals in a sample without trying to manipulate or influence the variable(s) of interest. Observational studies are often called *ex post facto* studies because the value of the response variable has already been determined.

5. In a designed experiment, a treatment is applied to the individuals in a sample in order to isolate the effects of the treatment on the response variable.

6. The three major types of observational studies are (1) Cross-sectional studies, (2) Case-control studies, and (3) Cohort studies.

   Cross-sectional studies collect data at a specific point in time or over a short period of time. Cohort studies are prospective and collect data over a period of time, sometimes over a long period of time. Case-controlled studies are retrospective, looking back in time to collect data either from historical records or from recollection by subjects in the study. Individuals possessing a certain characteristic are matched with those that do not.

7. The process of statistics refers to the approach used to collect, organize, analyze, and interpret data. The steps are
   (1) Identify the research objective
   (2) Collect the data needed to answer the research question.
   (3) Describe the data.
   (4) Perform inference.

8. The three types of bias are sampling bias, nonresponse bias, and response bias. Sampling bias occurs when the techniques used to select individuals to be in the sample favor one part of the population over another. Bias in sampling is reduced when a random process is to select the sample. Nonresponse bias occurs when the individuals selected to be in the sample that do not respond to the survey have different opinions from those that do respond. This can be minimized by using call-backs and follow-up visits to increase the response rate. Response bias occurs when the answers on a survey do not reflect the true feelings of the respondent. This can be minimized by using trained interviewers, using carefully worded questions, and rotating questions and answer selections.

9. Nonsampling errors are errors that result from undercoverage, nonresponse bias, response bias, and data-entry errors. These errors can occur even in a census. Sampling errors are errors that result from the use of a sample to estimate information about a population. These include random error and errors due to poor sampling plans, and result because samples contain incomplete information regarding a population.

10. The steps in conducting an experiment are:

   (1) *Identify the problem to be solved.*
      Gives direction and indicates the variables of interest (referred to as the claim).

   (2) *Determine the factors that affect the response variable.*
      List all variables that may affect the response, both controllable and uncontrollable.

   (3) *Determine the number of experimental units.*
      Determine the sample size. Use as many as time and money allow.

   (4) *Determine the level of each factor.*
      Factors can be controlled by fixing their level (e.g. only using men) or setting them at predetermined levels (e.g. different dosages of a new medicine). For factors that cannot be controlled, random assignment of units to treatments helps average out the effects of the uncontrolled factor over all treatments.

   (5) *Conduct the experiment.*
      Carry out the experiment using an equal number of units for each treatment. Collect and organize the data produced.

   (6) *Test the claim.*
      Analyze the collected data and draw conclusions.

11. 'Number of new automobiles sold at a dealership on a given day' is quantitative because its values are numerical measures on which addition and subtraction can be performed with meaningful results. The variable is discrete because its values result from a count.

12. 'Weight in carats of an uncut diamond' is quantitative because its values are numerical measures on which addition and subtraction can be performed with meaningful results. The variable is continuous because its values result from a measurement rather than a count.

13. 'Brand name of a pair of running shoes' is qualitative because its values serve only to classify individuals based on a certain characteristic.

14. 73% is a statistic because it describes a sample (the 1011 people age 50 or older who were surveyed).

15. 69% is a parameter because it describes a population (all the passes thrown by Chris Leak in the 2007 Championship Game).

16. Birth year has the *interval* level of measurement since differences between values have meaning, but it lacks a true zero.

17. Marital status has the *nominal* level of measurement since its values merely categorize individuals based on a certain characteristic.

18. Stock rating has the *ordinal* level of measurement because its values can be placed in rank order, but differences between values have no meaning.

19. Number of siblings has the *ratio* level of measurement because differences between values have meaning and there is a true zero.

20. This is an observational study because no attempt was made to influence the variable of interest. Sexual innuendos and curse words were merely observed.

21. This is an experiment because the researcher intentionally imposed treatments (experimental drug vs. placebo) on individuals in a controlled setting.

22. This was a cohort study because participants were identified to be included in the study and then followed over a period of time (over 26 years) with data being collected at regular intervals (every 2 years).

23. This is convenience sampling since the pollster simply asked the first 50 individuals she encountered.

24. This is a cluster sample since the ISP included all the households in the 15 randomly selected city blocks.

25. This is a stratified sample since individuals were randomly selected from each of the three grades.

26. This is a systematic sample since every 40th tractor trailer was tested.

27. (a) Sampling bias; undercoverage or nonrepresentative sample due to a poor sampling frame. Cluster sampling or stratified sampling are better alternatives.

    (b) Response bias due to interviewer error. A multi-lingual interviewer could reduce the bias.

    (c) Data-entry error due to the incorrect entries. Entries should be checked by a second reader.

28. Answers will vary. Using a TI-84 Plus graphing calculator with a seed of 1990, and numbering the individuals from 1 to 21, we would select individuals numbered 14, 6, 10, 17, and 11. If we numbered the businesses down each column, the businesses selected would be Jiffy Lube, Nancy's Flowers, Norm's Jewelry, Risky Business Security, and Solus, Maria, DDS.

29. Answers will vary. The first step is to select a random starting point among the first 9 bolts produced. Using row 9, column 17 from Table I in Appendix A, he will sample the 3rd bolt produced, then every 9th bolt after that until a sample size of 32 is obtained. In this case, he would sample bolts 3, 12, 21, 30, and so on until bolt 282.

30. Answers will vary. The goggles could be numbered 00 to 99, then a table of random digits could be used to select the numbers of the goggles to be inspected. Starting with row 12, column 1 of Table 1 in Appendix A and reading down, the selected labels would be 55, 96, 38, 85, 10, 67, 23, 39, 45, 57, 82, 90, and 76.

31. (a) To determine the ability of chewing gum to remove stains from teeth

    (b) Experimental design

    (c) Completely randomized design

    (d) Percentage of stain removed

    (e) Type of stain remover (gum or saliva); Qualitative

    (f) The 64 stained bovine incisors

    (g) The chewing simulator could impact the percentage of the stain removed

    (h) Gum A and B remove significantly more stain

32. (a) Matched-pairs

    (b) Reaction time; Quantitative

    (c) Alcohol consumption

    (d) Food consumption; caffeine intake

    (e) Weight, gender, etc.

    (f) To act as a placebo to control for the psychosomatic effects of alcohol

    (g) Alcohol delays the reaction time significantly in seniors for low levels of alcohol consumption; healthy seniors that are not regular drinkers.

33. (a) This experiment has a randomized block design.

    (b) The response variable is the exam grade.

    (c) The factor 'Notecard use' is set at predetermined levels. The treatments are 'with notecard' and 'without notecard'.

    (d) The experimental units are the instructor's statistics students.

(e)

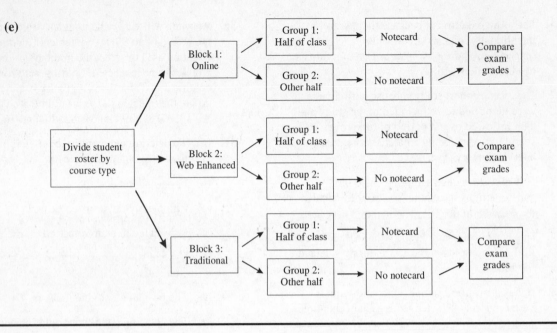

---

34. Answers will vary. Since there are ten digits (0 – 9), we will let a 0 or 1 indicate that (a) is to be the correct answer, 2 or 3 indicate that (b) is to be the correct answer, and so on. Beginning with row 1, column 8 of Table 1 in Appendix A, and reading downward, we obtain the following:

2, 6, 1, 4, 1, 4, 2, 9, 4, 3, 9, 0, 6, 4, 4, 8, 6, 5, 8, 5

Therefore, the sequence of correct answers would be:

b, d, a, c, a, c, b, e, c, b, e, a, d, c, c, e, d, c, e, c

35. **(a)** Answers will vary. One possible diagram is shown below.

(b) Answers will vary. One possible diagram is shown below.

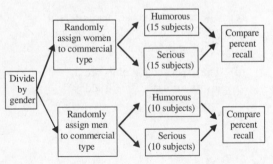

36. A matched-pairs design is an experimental design where experimental units are matched up so they are related in some way.

In a completely randomized design, the experimental units are randomly assigned to one of the treatments. The value of the response variable is compared for each treatment. In a matched-pairs design, experimental units are matched up on the basis of some common characteristic (such as husband-wife or twins). The differences between the matched units are analyzed.

37. Answers will vary.

38. Answers will vary.

Chapter 1 Test

1. Collect information, organize and summarize the information, analyze the information to draw conclusions, provide a measure of confidence in the conclusions drawn from the information collected.

2. The process of statistics refers to the approach used to collect, organize, analyze, and interpret data. The steps are
(1) Identify the research objective
(2) Collect the data needed to answer the research question.
(3) Describe the data.
(4) Perform inference.

3. The time to complete the 500-meter race in speed skating is quantitative because its values are numerical measurements on which addition and subtraction have meaningful results. The variable is continuous because its values result from a measurement rather than a count. The variable is at the *ratio* level of measurement because differences between values have meaning and there is a true zero.

4. Video game rating is qualitative because its values classify games based on certain characteristics but arithmetic operations have no meaningful results. The variable is at the *ordinal* level of measurement because its values can be placed in rank order, but differences between values have no meaning.

5. The number of surface imperfections is quantitative because its values are numerical measurements on which addition and subtraction have meaningful results. The variable is discrete because its values result from a count. The variable is at the *ratio* level of measurement because differences between values have meaning and there is a true zero.

6. This is an experiment because the researcher intentionally imposed treatments (brand-name battery versus plain-label battery) on individuals (cameras) in a controlled setting. The response variable is the battery life.

7. This is an observational study because no attempt was made to influence the variable of interest. Fan opinions about the asterisk were merely observed. The response variable is whether or not an asterisk should be placed on Barry Bonds' 756th homerun ball.

8. A *cross-sectional study* collects data at a specific point in time or over a short period of time; a *cohort study* collects data over a period of time, sometimes over a long period of time (prospective); a *case-controlled study* is retrospective, looking back in time to collect data.

9. An experiment involves the researcher actively imposing treatments on experimental units in order to observe any difference between the treatments in terms of effect on the response variable. In an observational study, the researcher observes the individuals in the study without attempting to influence the response variable in any way. Only an experiment will allow a researcher to establish causality.

10. A control group is necessary for a baseline comparison. This accounts for the placebo effect which says that some individuals will respond to any treatment. Comparing other treatments to the control group allows the researcher to identify which, if any, of the other treatments are superior to the current treatment (or no treatment at all). Blinding is important to eliminate bias due to the individual or experimenter knowing which treatment is being applied.

11. The steps in conducting an experiment are (1) Identify the problem to be solved, (2) Determine the factors that affect the response variable, (3) Determine the number of experimental units, (4) Determine the level of each factor, (5) Conduct the experiment, and (6) Test the claim.

12. Answers will vary. The franchise locations could be numbered 01 to 15 going across. Starting at row 7, column 14 of Table I in Appendix, and working downward, the selected numbers would be 08, 11, 03, and 02. The corresponding locations would be Ballwin, Chesterfield, Fenton, and O'Fallon.

13. Answers will vary. Using the available lists, obtain a simple random sample from each stratum and combine the results to form the stratified sample. Start at different points in Table I or use different seeds in a random number generator. Using a TI-84 Plus graphing calculator with a seed of 14 for Democrats, 28 for Republicans, and 42 for Independents, the selected numbers would be:
Democrats: 3946, 8856, 1398, 5130, 5531, 1703, 1090, and 6369
Republicans: 7271, 8014, 2575, 1150, 1888, 3138, and 2008
Independents: 945, 2855, and 1401

14. Answers will vary. Number the blocks from 1 to 2500 and obtain a simple random sample of size 10. The blocks corresponding to these numbers represent the blocks analyzed. All trees in the selected blocks are included in the sample. Using a TI-84 Plus graphing calculator with a seed of 12, the selected blocks would be numbered 2367, 678, 1761, 1577, 601, 48, 2402, 1158, 1317, and 440.

15. Answers will vary. $\dfrac{600}{14} \approx 42.86$, so we let

$k = 42$. Select a random number between 1 and 42 which represents the first slot machine inspected. Using a TI-84 Plus graphing calculator with a seed of 132, we select machine 18 as the first machine inspected. Starting with machine 18, every 42nd machine thereafter would also be inspected (60, 102, 144, 186, ..., 564).

16. In a completely randomized design, the experimental units are randomly assigned to one of the treatments. The value of the response variable is compared for each treatment. In a randomized block design, the experimental units are first divided according to some common characteristic (such as gender). Then each experimental unit within each block is randomly assigned to one treatment. Within each block the value of the response variable is compared for each treatment, but not between blocks. By blocking, we prevent the effect of the blocked variable from confounding with the treatment.

17. (a) Sampling bias due to voluntary response.

   (b) Nonresponse bias due to the low response rate.

   (c) Response bias due to poorly worded questions.

   (d) Sampling bias due to poor sampling plan (undercoverage).

18. (a) The response variable is the lymphocyte count.

   (b) The treatment is space flight.

   (c) This experiment has a matched-pairs design.

   (d) The experimental units are the 4 members of Skylab.

   (e)

19. (a) This experiment has a completely randomized design.

(b) The response variable is the level of dermatitis improvement.

(c) The factor set to predetermined levels is the topical cream concentration. The treatments are 0.5% cream, 1.0% cream, and a placebo (0% cream).

(d) The study is double-blind if neither the subjects, nor the person administering the treatments, are aware of which topical cream is being applied.

(e) The control group is the placebo (0% topical cream).

(f) The experimental units are the 225 patients with skin irritations.

(g)

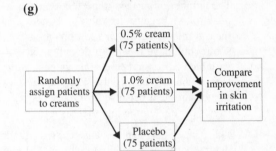

20. (a) This was a cohort study because participants were identified to be included in the study and then followed over a long period of time with data being collected at regular intervals (every 4 years).

(b) The response variable is bone mineral density. The explanatory variable is weekly cola consumption.

(c) The response variable is quantitative because its values are numerical measures on which addition and subtraction can be performed with meaningful results.

(d) The researchers observed values of variables that could potentially impact bone mineral density (besides cola consumption) so their effect could be isolated from the variable of interest.

(e) Answers will vary. Some possible lurking variables that should be accounted for are smoking status, alcohol consumption, physical activity, and calcium intake (form and quantity).

(f) The study concluded that women who consumed at least one cola per day (on average) had a bone mineral density that was significantly lower at the femoral neck than those who consumed less than one cola per day. The study cannot claim that increased cola consumption *causes* lower bone mineral density because it is only an observational study. The researchers can only say that increased cola consumption is *associated* with lower bone mineral density.

Case Study: Chrysalises for Cash

Reports will vary. The reports should include the following components:

Step 1: *Identify the problem to be solved.* The entrepreneur wants to determine if there are differences in the quality and emergence time of broods of the black swallowtail butterfly depending on the following factors: (a) early brood season versus late brood season; (b) carrot plants versus parsley plants; and (c) liquid fertilizer versus solid fertilizer.

Step 2: *Determine the explanatory variables that affect the response variable.* Some explanatory variables that may affect the quality and emergence time of broods are the brood season, the type of plant on which the chrysalis grows, fertilizer used for plants, soil mixture, weather, and the level of sun exposure.

Step 3: *Determine the number of experimental units.* In this experiment, a sample of 40 caterpillars/butterflies will be used.

Step 4: *Determine the level of the explanatory variables*:

- **Brood season** – We wish to determine the differences in the number of deformed butterflies and in the emergence times depending on whether the brood is from the early season or the late season. We use a total of 20 caterpillars/butterflies from the early brood season and 20 caterpillars/butterflies from the late brood season.

- **Type of plant** – We wish to determine the differences in the number of deformed butterflies and in the emergence times depending on the type plant on which the caterpillars are placed. A total of 20 caterpillars are placed on carrot plants and 20 are placed on parsley plants.

- **Fertilizer** – We wish to determine the differences in the number of deformed butterflies and in the emergence times depending on the type of fertilizer used on the plants. A total of 20 chrysalises grow on plants that are fed liquid fertilizer and 20 grow on plants that are fed solid fertilizer.

- **Soil mixture** – We control the effects of soil by growing all plants in the same mixture.

- **Weather** – We cannot control the weather, but the weather will be the same for each chrysalis grown within the same season. For chrysalises grown in different seasons, we expect the weather might be different and thus part of the reason for potential differences between seasons. Also, we can control the amount of watering that is done.

- **Sunlight exposure** – We cannot control this variable, but the sunlight exposure will be the same for each chrysalis grown within the same season. For chrysalises grown in different seasons, we expect the sunlight exposure might be different and thus part of the reason for potential differences between seasons.

Step 5: *Conduct the experiment.*

(a) We fill eight identical pots with equal amounts of the same soil mixture. We use four of the pots for the early brood season and four of the pots for the late brood season.

For the early brood season, two of the pots grow carrot plants and two grow parsley plants. One carrot plant is fertilized with a liquid fertilizer, one carrot plant is fertilized with a solid fertilizer, one parsley plant is fertilized with the liquid fertilizer, and one parsley plant is fertilized with the solid fertilizer. We place five black swallowtail caterpillars of similar age into each of the four pots.

Similarly, for the late brood season, two of the pots grow carrot plants and two grow parsley plants. One carrot plant is fertilized with a liquid fertilizer, one carrot plant is fertilized with a solid fertilizer, one parsley plant is fertilized with the liquid fertilizer, and one parsley plant is fertilized with the solid fertilizer. We place five black swallowtail caterpillars of similar age into each of the four pots.

(b) We determine the number of deformed butterflies and in the emergence times for the caterpillars/butterflies from each pot.

Step 6: *Test the claim.* We determine whether any differences exist depending on season, plant type, and fertilizer type.

Conclusions:

Early versus late brood season: From the data presented, more deformed butterflies occur in the late season than in the early season. Five deformed butterflies occurred in the late season while only one occurred in the early season. Also, the emergence time seems to be longer in the early season than in the late season. In the early season, all but one of the twenty emergence times were between 6 and 8 days. In the late season, all twenty of the emergence times were between 2 and 5 days.

Parsley versus carrot plants: From the data presented, the plant type does not seem to affect the number of deformed butterflies that occur. Altogether, three deformed butterflies occur from parsley plants and three deformed butterflies occur from the carrot plants. Likewise, the plant type does not seem to affect the emergence times of the butterflies.

Liquid versus solid fertilizer: From the data presented, the type of fertilizer seems to affect the number of deformed butterflies that occur. Five deformed butterflies occurred when the solid fertilizer was used while only one occurred when the liquid fertilizer was used. The type of fertilizer does not seem to affect emergence times.

Chapter 2
Organizing and Summarizing Data

Section 2.1

1. Raw data are the data as originally collected, before they have been organized or coded.

3. The relative frequencies should add to 1, although rounding may cause the answers to vary slightly.

5. (a) The largest segment in the pie chart is for "Washing your hands" so the most commonly used approach to beat the flu bug is washing your hands. 61% of respondents selected this as their primary method for beating the flu.

   (b) The smallest segment in the pie chart is for "Drinking Orange Juice" so the least used method is drinking orange juice. 2% of respondents selected this as their primary method for beating the flu.

   (c) 25% of respondents felt that flu shots were the best way to beat the flu.

7. (a) The largest bar corresponds to China, so China had the most internet users in 2010.

   (b) The bar for the United Kingdom appears to reach the line for 50. Thus, we estimate that there were 50 million internet users in the United Kingdom in 2007.

   (c) The bar for China appears to reach 420 on the vertical axis. The bar for Germany appears to reach 70. Since, 420-70=350, we estimate that there were about 350 million more internet users in China than in Germany during 2007.

   (d) This graph should use relative frequencies, rather than frequencies.

9. (a) 69% of the respondents believe divorce is morally acceptable.

   (b) 23% believe divorce is morally wrong. So, 240 million * 0.23 = 55.2 million adult Americans believe divorce is morally wrong.

(c) This statement is inferential, since it is a generalization based on the observed data.

11. (a) The proportion of 18-34 year old respondents who are more likely to buy when made in America is 0.42. For 34-44 year olds, the proportion is 0.61.

    (b) The 55+ age group has the greatest proportion of respondents who are more likely to buy when made in America.

    (c) The 18-34 age group has a majority of respondents who are less likely to buy when made in America.

    (d) As age increases, so does the likelihood that a respondent will be more likely to buy a product that is made in America.

13. (a) Total students surveyed = 125 + 324 + 552 + 1257 + 2518 = 4776
    Relative frequency of "Never"
    = 125 / 4776 ≈ 0.0262 and so on.

Response	Relative Frequency
Never	0.0262
Rarely	0.0678
Sometimes	0.1156
Most of the time	0.2632
Always	0.5272

   (b) 52.72%

   (c) 0.0262 + 0.0678 = 0.0940 or 9.40%

   (d)

"How Often Do You Wear Your Seat Belt?"

(e)

"How Often Do You Wear Your Seat Belt?"

(f)

"How Often Do You Wear Your Seat Belt?"

(g) The statement is inferential since it is inferring something about the entire population based on the results of a sample survey.

15. (a) Total adults surveyed = 377 + 192 + 132 + 81 + 243 = 1025
Relative frequency of "More than 1 hour a day" = $377/1025 \approx 0.3678$ and so on.

Response	Relative Frequency
More than 1 hr a day	0.3678
Up to 1 hr a day	0.1873
A few times a week	0.1288
A few times a month or less	0.0790
Never	0.2371

(b) 0.2371 (about 24%)

(c)

Time Spent Online

(d)

(e)

Time Spent Online

(f) The statement provides an estimate, but no level of confidence is given.

17. (a) Total adults = 1936
Relative frequency for "none" is:
173/1936=0.09, and so on.

Number of Texts	Rel. Freq. (Adults)
None	0.09
1 to 10	0.51
11 to 20	0.13
21 to 50	0.13
51 to 100	0.07
101+	0.08

(b) Total teens = 627
Relative frequency for "none" is:
13/627=0.02, and so on.

Number of Texts	Rel. Freq. (Teens)
None	0.02
1 to 10	0.22
11 to 20	0.11
21 to 50	0.18
51 to 100	0.18
101+	0.29

(c)

Number of Texts Each Day

(d) Answers will vary. Adults are much more likely to do fewer texts per day, while teens are much more likely to do more texting.

19. (a) Total males = 99; Relative frequency for "Professional Athlete" is 40/99 = 0.404, and so on.

Total number of females = 100; Relative frequency for "Professional Athlete" is 18/100 = 0.18, and so on.

Dream Job	Men	Women
Professional Athlete	0.404	0.180
Actor/Actress	0.263	0.370
President of the U.S.	0.131	0.130
Rock Star	0.131	0.130
Not Sure	0.071	0.190

(b)

Dream Job

(c) Answers will vary. Males are much more likely to want to be a professional athlete. Women are more likely to aspire to a career in acting than men. Men's desire to become athletes may be influenced by the prominence of male sporting figures in popular culture. Women may aspire to careers in acting due to the perceived glamour of famous female actresses.

21. (a), (b)

Total number of Winter Olympics = 22; relative frequency for Canada is 8/22=0.364.

Winner	Freq.	Rel. Freq.
Canada	8	0.364
Czech Republic	1	0.045
Great Britain	1	0.045
Soviet Union	7	0.318
Sweden	2	0.091
U.S.A.	2	0.091
Unified Team	1	0.045

(c)

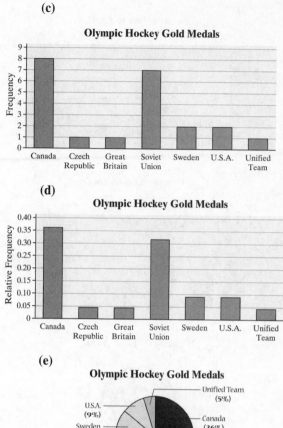

Olympic Hockey Gold Medals

(d)

Olympic Hockey Gold Medals

(e)

Olympic Hockey Gold Medals

23. (a), (b)

Total number of responses = 40; relative frequency for "Sunday" is 3/40=0.075.

Response	Freq.	Rel. Freq.
Sunday	3	0.075
Monday	2	0.05
Tuesday	5	0.125
Wednesday	6	0.15
Thursday	2	0.05
Friday	14	0.35
Saturday	8	0.2

(c) Answers will vary. If you own a restaurant, you will probably want to advertize on the days when people will be most likely to order takeout: Friday. You might consider avoiding placing an ad on Monday and Thursday, since the readers are least likely to choose to order takeout on these days.

(d)

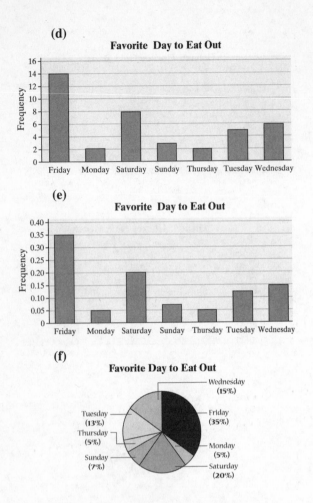

Favorite Day to Eat Out

(e)

Favorite Day to Eat Out

(f)

Favorite Day to Eat Out

(c)

**Foreign Languages
Studied by College Students**

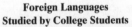

(d)

**Foreign Languages
Studied by College Students**

(e)

**Foreign Languages
Studied by College Students**

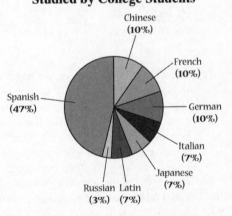

25. (a), (b)

Total number of students = 30
Relative frequency for "Chinese"

$= \dfrac{3}{30} = 0.100$ and so on.

Language	Freq.	Rel. Frequency
Chinese	3	0.100
French	3	0.100
German	3	0.100
Italian	2	0.067
Japanese	2	0.067
Latin	2	0.067
Russian	1	0.033
Spanish	14	0.467

27. (a) It would make sense to draw a pie chart for land area since the 7 continents contain all the land area on Earth. Total land area is 11,608,000 + 5,100,000 + ... + 9,449,000 + 6,879,000 = 57,217,000 square miles
The relative frequency (percentage) for

Africa is $\dfrac{11,608,000}{57,217,000} = 0.2029$.

Continent	Land Area (mi^2)	Rel. Freq.
Africa	11,608,000	0.2029
Antarctica	5,100,000	0.0891
Asia	17,212,000	0.3008
Australia	3,132,000	0.0547
Europe	3,837,000	0.0671
North America	9,449,000	0.1651
South America	6,879,000	0.1202

(d) The following bar charts were created in Microsoft® Excel:

(b) It does not make sense to draw a pie chart for the highest elevation because there is no whole to which to compare the parts.

29. Answers will vary.

31. Relative frequencies should be used when the size of two samples or populations differ.

33. A bar chart is preferred when trying to compare two specific values. Pie charts are helpful for comparing parts of a whole. A pie chart cannot be drawn if the data do not include all possible values of the qualitative variable.

Consumer Reports®: Consumer Reports Rates Treadmills

(a) A bar chart is used to display the overall scores. Because the bars are in decreasing order, this is an example of a Pareto chart.

(b) The Precor M9.33 has the highest construction score since it was the only model receiving an excellent rating. Two models, the Tunturi J6F and the ProForm 525E received a Fair rating, making them the models with the lowest ease of use score.

(c) 1 model was rated Excellent, 7 models were rated Very Good, 1 model was rated Good, and 2 models were rated Fair. No models were rated Poor for ease of use.

(e) The following scatterplot was obtained by eyeballing the value of the scores from the Overall Score Pareto chart. Although there is a great deal of scatter in the data, even within a similar price range, there appears to be a relationship between score and price. The more expensive models tested by Consumer Reports in March 2002 tended to score higher in overall performance. (One should be cautious about generalizing the conclusions to the universe of treadmills since only a small sample of treadmills have been tested here.)

Section 2.2

1. classes

3. class width

5. True

7. False

9. (a) 8

   (b) 2

   (c) 15

   (d) $11 - 7 = 4$

   (e) $\dfrac{15}{100} = 0.15$ or 15%

   (f) The distribution is bell shaped.

11. (a) Total frequency = 2 + 3 + 13 + 42 + 58 + 40 + 31 + 8 + 2 + 1 = 200

   (b) 10 (e.g. 70 – 60 = 10)

   (c)

IQ Score (class)	Frequency
60–69	2
70–79	3
80–89	13
90–99	42
100–109	58
110–119	40
120–129	31
130–139	8
140–149	2
150–159	1

   (d) The class '100 – 109' has the highest frequency.

   (e) The class '150 – 159' has the lowest frequency.

   (f) $\dfrac{8 + 2 + 1}{200} = 0.055 = 5.5\%$

   (g) No, there were no IQs above 159.

13. (a) Likely skewed right. Most household incomes will be to the left (perhaps in the $50,000 to $150,000 range), with fewer higher incomes to the right (in the millions).

   (b) Likely bell-shaped. Most scores will occur near the middle range, with scores tapering off equally in both directions.

   (c) Likely skewed right. Most households will have, say, 1 to 4 occupants, with fewer households having a higher number of occupants.

   (d) Likely skewed left. Most Alzheimer's patients will fall in older-aged categories, with fewer patients being younger.

15. (a) Total number of households = 16 + 18 + 12 + 3 + 1 = 50

   Relative frequency of 0 children = 16/50 = 0.32, and so on.

Number of Children Under Five	Relative Frequency
0	0.32
1	0.36
2	0.24
3	0.06
4	0.02

   (b) $\dfrac{12}{50} = 0.24$ or 24% of households have two children under the age of 5.

   (c) $\dfrac{18 + 12}{50} = \dfrac{30}{50} = 0.6$ or 60% of households have one or two children under the age of 5.

17. From the legend, 1|0 represents 10, so the original data set is:
10, 11, 14, 21, 24, 24, 27, 29, 33, 35, 35, 35, 37, 37, 38, 40, 40, 41, 42, 46, 46, 48, 49, 49, 53, 53, 55, 58, 61, 62

19. From the legend, 1|2 represents 1.2, so the original data set is:
1.2, 1.4, 1.6, 2.1, 2.4, 2.7, 2.7, 2.9, 3.3, 3.3, 3.3, 3.5, 3.7, 3.7, 3.8, 4.0, 4.1, 4.1, 4.3, 4.6, 4.6, 4.8, 4.8, 4.9, 5.3, 5.4, 5.5, 5.8, 6.2, 6.4

21. (a) 8 classes

  (b) Lower class limits: 775, 800, 825, 850, 875, 900, 925, 950
  Upper class limits: 799, 824, 849, 874, 899, 924, 949, 974

  (c) The class width is found by subtracting consecutive lower class limits. For example, $800 - 775 = 25$. Therefore, the class width is 25(dollars).

23. (a) 7 classes

  (b) Lower class limits: 15, 20, 25, 30, 35, 40, 45; Upper class limits: 19, 24, 29, 34, 39, 44, 49

  (c) The class width is found by subtracting consecutive lower class limits. For example, $20 - 15 = 5$. Therefore, the class width is 5 (years).

25. (a) Total frequency is:
    $22 + 68 + 15 + 5 + 0 + 0 + 0 + 1 = 111$
    Relative frequency for 775-779 is
    $22/111 = 0.1982$ and so on.

Tuition ($)	Relative Frequency
775-799	0.1982
800-825	0.6126
825-849	0.1351
850-874	0.0450
875-899	0.0000
900-924	0.0000
925-949	0.0000
950-974	0.0090

(b)

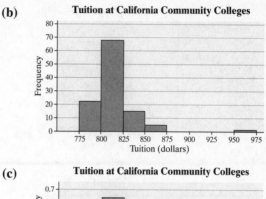

Tuition at California Community Colleges

(c)

Tuition at California Community Colleges

Total number of California community colleges with tuition less than $800 is 22.
$\frac{22}{111} \cdot 100\% \approx 19.82\%$ of California community colleges had tuition of less than $800.
Total number of colleges with tuition of $850 or more $= 5 + 1 = 6$
$\frac{6}{111} \cdot 100\% \approx 5.41\%$ of California community colleges had tuition of $850 or more.

27. (a) Total births $= 445,045 + 1,082,837 + 1,208,405 + 962,179 + 499,816 + 105,071 + 7,349 = 4,310,802$
Relative frequency for 15-19 $= 445,045/4,310,802 = 0.1032$ and so on.

Age	Relative Frequency
15-19	0.1032
20-24	0.2512
25-29	0.2803
30-34	0.2232
35-39	0.1160
40-44	0.0244
45-49	0.0017

(b)

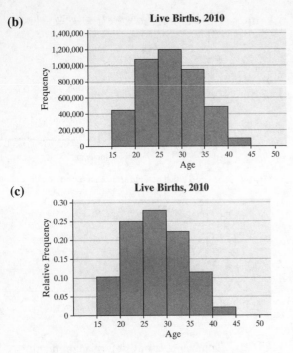

Live Births, 2010

(c)

Live Births, 2010

The relative frequency is 0.0244, so 2.44% of live births were to women 40-44 years of age.

0.1032 + 0.2512 = 0.3544 so 35.44% of live births were to women 24 years of age or younger.

29. (a) The data are discrete. The possible values for the number of color televisions in a household are countable.

(b), (c)

The relative frequency for 0 color televisions is 1/40 = 0.025, and so on.

Number of Color TVs	Frequency	Relative Frequency
0	1	0.025
1	14	0.350
2	14	0.350
3	8	0.200
4	2	0.050
5	1	0.025

(d) The relative frequency is 0.2 so 20% of the households surveyed had 3 color televisions.

(e) 0.05 + 0.025 = 0.075

7.5% of the households in the survey had 4 or more color televisions.

(f)

Televisions in Household

(g)

Televisions in Household

(h) The distribution is skewed right.

31. (a) and **(b)**

Relative frequency for 24,000-26,999 = 8/51 = 0.1569 and so on.

Disposable Income ($)	Freq.	Rel. Freq.
30,000 – 35,999	21	0.4118
36,000 – 41,999	18	0.3529
42,000 – 47,999	7	0.1373
48,000 – 53,999	3	0.0588
54,000 – 59,999	1	0.0196
60,000 – 65,999	0	0.0000
66,000 – 71,999	1	0.0196

(c)

Per Capita Disposable Income by State, 2009

(d)

Per Capita Disposable Income by State, 2009

(e) The distribution appears to be skewed right.

(f) Relative frequency for 30,000-32,999 = 10/51 = 0.1961 and so on.

Disposable Income ($)	Freq.	Rel. Freq.
30,000 – 32,999	9	0.1765
33,000 – 35,999	12	0.2353
36,000 – 38,999	10	0.1961
39,000 – 41,999	8	0.1569
42,000 – 44,999	5	0.0980
45,000 – 47,999	2	0.0392
48,000 – 50,999	3	0.0588
51,000 – 53,999	0	0.0000
54,000 – 56,999	1	0.0196
57,000 – 59,999	0	0.0000
60,000 – 62,999	0	0.0000
63,000 – 65,999	0	0.0000
66,000 – 68,999	1	0.0196

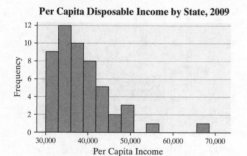

Per Capita Disposable Income by State, 2009

Per Capita Disposable Income by State, 2009

The distribution appears to be skewed right.

(g) Answers will vary. While both distributions indicate the data are skewed right, the first distribution provides a more detailed look at the data. The second distribution has a larger width of the bars, which can potential obscure details in the data.

33. **(a)** and **(b)**

Total number of data points = 51
Relative frequency of 0-0.49 is:
7/51 = 0.1373, and so on.

Cigarette Tax	Frequency	Relative Frequency
0.00-0.49	7	0.1373
0.50-0.99	14	0.2745
1.00-1.49	7	0.1373
1.50-1.99	8	0.1569
2.00-2.49	6	0.1176
2.50-2.99	4	0.0784
3.00-3.49	4	0.0784
3.50-3.99	0	0.0000
4.00-4.49	1	0.0196

(c)

Cigarette Tax Rates by State, 2010

(d)

Cigarette Tax Rates by State, 2010

(e) The distribution appears to be right skewed.

(f) Relative frequency of 0 – 0.99 is:
21/51 = 0.4118, and so on.

Cigarette Tax	Frequency	Relative Frequency
0.00-0.99	21	0.4118
1.00-1.99	15	0.2941
2.00-2.99	10	0.1961
3.00-3.99	4	0.0784
4.00-4.99	1	0.0196

Violent Crimes per 100,000	Freq.	Rel. Freq.
100 – 199	4	0.0784
200 – 299	17	0.3333
300 – 399	7	0.1373
400 – 499	11	0.2157
500 – 599	3	0.0588
600 – 699	7	0.1373
700 – 799	1	0.0196
800 – 899	0	0.0000
900 – 999	0	0.0000
1000 – 1099	0	0.0000
1100 – 1199	0	0.0000
1200 – 1299	0	0.0000
1300 – 1399	1	0.0196

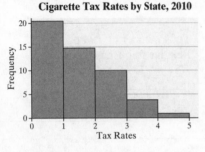

Cigarette Tax Rates by State, 2010

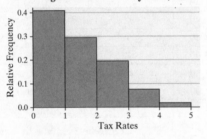

Cigarette Tax Rates by State, 2010

The distribution is right skewed.

(g) Answers will vary. The first distribution gives a more detailed pattern.

35. Answers will vary. One possibility follows.

(a) Lower class limit of first class: 100; We can determine a class width by subtracting the smallest value from the largest, dividing by the desired number of classes, then rounding up. For example,

$$\text{Class width} \approx \frac{1345.9 - 119.8}{13} = 94.3 \to 100$$

So, a class width 100 should suffice.

(b), (c)
Relative frequency for 100-199 is:
4/51 = 0.0784, and so on.

(d)

Violent Crimes per 100,000 Population by State, 2009

(e)

Violent Crimes per 100,000 Population by State, 2009

(f) The distribution is skewed right.

37. (a) President Ages at Inauguration

```
4 | 23
4 | 6677899
5 | 0011112244444
5 | 555566677778
6 | 0111244
6 | 589
```
Legend: 4 | 2 represents 42 years.

(b) The distribution appears to be roughly symmetric and bell shaped.

39. **(a)** **Fat in McDonald's Breakfast**

```
0 | 39
1 | 1266
2 | 1224577
3 | 0012267
4 | 6
5 | 159
```
Legend: 5 | 1 represents 51 grams of fat.

(b) The distribution appears to be roughly symmetric and bell shaped.

41. (a) Rounded data:

17.3	10.5	8.7	9.4	8.8	6.7	6.5
12.3	9.6	8.9	13.1	8.8	7.5	15.4
14.5	7.5	8.1	8.6	7.8	10.2	25.3
14.6	10.4	7.7	8.6	7.6	9.0	
14.6	9.4	8.1	8.8	8.1	7.6	
13.2	10.0	12.3	7.1	9.8	6.1	
15.7	7.8	13.9	9.2	10.5	14.4	
17.2	8.6	10.7	6.9	10.0	7.7	

Electric Rates by State, 2010

```
 6 | 1579
 7 | 155667788
 8 | 11166678889
 9 | 024468
10 | 0024557
11 |
12 | 33
13 | 129
14 | 4566
15 | 47
16 |
17 | 23
18 |
19 |
20 |
21 |
22 |
23 |
24 |
25 | 3
```
Legend: 6 | 1 represents 6.1 cents per kWh.

(b) The distribution is skewed right.

(c) Hawaii's average retail price is 25.33 cents/kWh. Hawaii's rate may be so much higher because it is an island far away from the mainland. Resources on the island are limited and importing resources increases the overall cost. (Answers will vary.)

43. (a) Violent crime rates rounded to the nearest tens:

450	1350	400	280	380	670	300
630	610	260	490	400	190	260
410	430	620	250	200	670	230
520	270	120	280	330	490	
470	230	590	700	500	210	
340	500	460	160	250	130	
300	330	500	310	380	230	
640	280	240	620	250	330	

(b) **Violent Crime Rates by State, 2009**

```
 1 | 2369
 2 | 013334555667888
 3 | 001333488
 4 | 001356799
 5 | 00029
 6 | 1223477
 7 | 0
 8 |
 9 |
10 |
11 |
12 |
13 | 5
```
Legend: 1 | 2 represents 120 violent crimes per 100,000 population

(c) **Violent Crime Rates by State, 2009**

```
 1 | 23
 1 | 69
 2 | 013334
 2 | 555667888
 3 | 0013334
 3 | 88
 4 | 0013
 4 | 56799
 5 | 0002
 5 | 9
 6 | 12234
 6 | 77
 7 | 0
 7 |
 8 |
 8 |
 9 |
 9 |
10 |
10 |
11 |
11 |
12 |
12 |
13 |
13 | 5
```
Legend: 1 | 2 represents 120 violent crimes per 100,000 population

(d) Answers will vary. The first display is decent. It clearly shows that the distribution is skewed right and has an outlier. The second display is not as good as the first. Splitting the stems did not reveal any additional information and has made the display more cluttered and cumbersome.

45. (a)

Home Run Distances

McGwire		Bonds
	32	0 0
	33	
1 0	34	7
0 0	35	0
9 0 0 0	36	0 0 0 1 5
7 0 0 0 0	37	0 0 5 5 5 5
8 5 5 0 0 0 0 0	38	0 0 0 0 0 5
8 0 0 0 0	39	0 0 1 4 6
9 0 0	40	0 0 0 0 4 5
0 0 0 0 0	41	0 0 0 0 0 0 0 0 0 0 1 5 5 6 7 7
5 3 0 0 0 0 0	42	0 0 0 0 0 0 0 9
0 0 0 0 0 0 0	43	0 0 0 0 0 5 5 6
0 0 0	44	0 0 0 0 2
8 2 0 0 0 0	45	0 4
1 0 0	46	
8 0 0 0	47	
0	48	8
	49	
0	50	
0 0	51	
7	52	
	53	
	54	
0	55	

Legend: 0 | 34 | 7 represents 340 feet
for McGwire and 347 feet for Bonds.

(b) Answers will vary. For both players, the distances of homeruns mainly fall from 360 to 450 feet. McGwire has quite a few extremely long distances.

47. Answers will vary.

49.

Televisions in Household

Number of Televisions

51. Because the data are quantitative, either a stem-and-leaf plot or a histogram would be appropriate. There were 20 people who spent less than 30 seconds, 7 people spent at least 30 seconds but less than 60 seconds, etc. One possible histogram is:

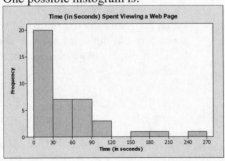

The data appear to be skewed right with a gap and one potential outlier. It seems as if the majority of surfers spent less than one minute viewing the page, while a few surfers spent several minutes viewing the page.

53. Classes should not overlap to avoid any confusion as to which class an observation belongs.

55. There is no such thing as *the* correct choice for a class width, however some choices are better than others. For example, if the class width is too small, the histogram will show many gaps between the bars.

57. Answers will vary.

Section 2.3

1. An ogive is a line graph of cumulative frequencies or cumulative relative frequencies against upper class limits.

3. True

5. (a) 10; The class width is the difference between successive class midpoints. e.g. $15 - 5 = 10$
There are 10 classes represented in the graph. The 12 plotted points are for the 10 class midpoints plus two additional points to connect the graph to the horizontal axis on the ends.

(b) The midpoint of the first class is 5. Assuming data values are given in whole numbers, the lower limit of the first class is 0 and the upper limit is 9.

(c) The midpoint of the last class is 95. Assuming data values are given in whole numbers, the lower limit of the last class is 90 and the upper limit is 99.

(d) The highest population occurs in the age group $40 - 49$; this population is about 44 million.

(e) The lowest population occurs in the age group $90 - 99$; this population is about 2 million.

(f) We estimate the total population by adding the estimates for the 10 classes.

$42 + 41 + 43 + 40 + 44 + 41 + 28 + 16 + 9$
$+ 2 = 306$
The approximate total number of U.S. residents below 100 years of age is 306 million.

(g) The population change is most extreme where the line segment is the longest. This occurs between the classes $50-59$ and $60-69$.
The population change is least extreme where the line segment is the shortest. This occurs between the classes $10-19$ and $20-29$.

7. (a) From the graph, it appears that about 20% of students had an ACT composite score of 16 or below.

(b) From the graph, it appears that about 60% of students had an ACT composite score of 19 or above.

(c) About $(0.20)(1,568,835) \approx 313,767$ students had ACT composite scores of 16 or below.
About 21% scored above 25.
About $(0.21)(1,568,835) \approx 329,455$ students had ACT composite scores above 25.

9. (a) The HOI in the first quarter of 1999 is about 70%.

(b) The lowest value of the HOI was about 40%. This occurred in 2006.

(c) The highest value of the HOI was about 73%. This occurred in 2009.

(d) $\dfrac{40-70}{70} = \dfrac{-30}{70} \approx -0.43$
The HOI decreased by about 43% from the first quarter of 1999 to the third quarter of 2006.

(e) There is an increase of about 82.5%.

11. (a) For 1992, the unemployment rate was about 7.5% and the inflation rate was about 3.0%.

(b) For 2009, the unemployment rate was about 9.2% and the inflation rate was about -0.4%.

(c) $7.5\% + 3.0\% = 10.5\%$
The misery index for 1992 was 10.5%.
$4.6\% + 3.4\% = 8.0\%$
The misery index for 2009 was 8.8%.

(d) Answers may vary. One possibility: An increase in the inflation rate seems to be followed by an increase in the unemployment rate. Likewise, a decrease in the inflation rate seems to be followed by a decrease in the unemployment rate.

13. (a), (b) Second class cumulative frequency = 22 + 58 = 90 and so on.
Total frequency is:
$22 + 68 + 15 + 5 + 0 + 0 + 0 + 1 = 111$
Second class cumulative relative
frequency $= \dfrac{90}{111} = 0.8108$ and so on.

Tuition ($)	Cumul. Freq.	Cumul. Rel. Freq.
775-799	22	0.1982
800-824	90	0.8108
825-849	105	0.9459
850-874	110	0.9910
875-899	110	0.9910
900-924	110	0.9910
925-949	110	0.9910
950-999	111	1

(c) The class midpoint is computed dividing the sum of successive lower class limits by 2. So, for example, the first class midpoint is: $\dfrac{775+800}{2} \approx 787.5$. The next class midpoint is: $\dfrac{800+825}{2} \approx 812.5$, and so on. We connect the graph with the x-axis at (762.5,0) and (987.5,0).

Tuition at Public Community Colleges in California

(d)

Tuition at Public Community Colleges in California

(e)

Tuition at Public Community Colleges in California

15. (a), (b)

Second class cumulative frequency is
21 + 18 = 39 and so on.
Second class cumulative relative
frequency

$$= \frac{39}{51} \approx 0.7647 \text{ and so on.}$$

Disposable Income ($)	Cumul. Freq.	Cumul. Rel. Freq.
$30,000 - 35,999$	21	0.4118
$36,000 - 41,999$	39	0.7647
$42,000 - 47,999$	46	0.9020
$48,000 - 53,999$	49	0.9608
$54,000 - 59,999$	50	0.9804
$60,000 - 65,999$	50	0.9804
$66,000 - 71,999$	51	1

(c) The class midpoint is computed dividing
the sum of successive lower class limits
by 2. So, for example, the first class

midpoint is: $\frac{30,000 + 36,000}{2} \approx 33,000$.

The next class midpoint is:

$\frac{36,000 + 42,000}{2} \approx 39,000$, and so on.

We connect the graph with the x-axis at
$(27000, 0)$ and $(75000, 0)$.

Average Income by State, 2009

(d)

Average Income by State, 2009

(e)

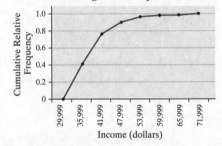

Average Income by State, 2009

17. The price of Disney stock over the year
seemed to fluctuate somewhat, with a general
upward trend.

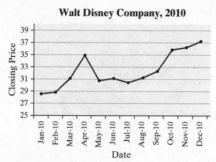

Walt Disney Company, 2010

The percent change in the stock price of Disney from January 2010 to December 1010 is 30.7%.

19. During the late 1990s debt as a percent of GDP was decreasing. It was increasing slightly during the early to mid 2000s. It has increased substantially from 2007 to 2010.

Debt as a percent of Gross Domestic Product

21. (a) Total frequency for financial stocks = 35

Total frequency for energy stocks = 35

Rate of Return	Frequency (Consumer Goods)	Frequency (Energy)
−20 to −10.01	1	0
−10 to −0.01	5	5
0 − 9.99	15	16
10 − 19.99	10	12
20 − 29.99	3	2
30 − 39.99	1	0

(b) Relative frequency of −20 to −10.01 for

consumer goods stocks = $\dfrac{1}{35} \approx 0.0286$,

and so on.

Relative frequency of −20 to −10.01 for

energy stocks = $\dfrac{0}{35} \approx 0.0000$, and so on.

Rate of Return	Rel. Freq. (Consumer Goods)	Rel. Freq. (Energy)
−20 to −10.01	0.0286	0.0000
−10 to −0.01	0.1429	0.1429
0 − 9.99	0.4286	0.4571
10 − 19.99	0.2857	0.3429
20 − 29.99	0.0857	0.0571
30 − 39.99	0.0286	0.0000

(c) The class midpoint is computed dividing the sum of successive lower class limits by 2. So, for example, the first class

midpoint is: $\dfrac{-20+(-10)}{2} \approx -15$. The

next class midpoint is: $\dfrac{-10+0}{2} \approx -5$, and

so on. We connect the graph with the x-axis at $(-25,0)$ and $(45,0)$.

5-Year Rate of Return on Consumer Goods and Energy Stocks

(d)

5-Year Rate Return on Consumer Goods and Energy Stocks

(e) Answers will vary.

23. Answers will vary. Reports should address the fact that the number of people going to the beach and participating in underwater activities (e.g. scuba diving, snorkeling) has also increased, so an increase in shark attacks is not unexpected. A better comparison would be the rate of attacks per 100,000 beach visitors. The number of fatalities could decrease due to better safety equipment (e.g. bite resistant suits) and better medical care.

25. Answers will vary. The cumulative relative frequency is the proportion of observations less than or equal to the current class. All the observations are less than or equal to the highest class, so the cumulative relative frequency of the last class must be 1, or 100%.

Section 2.4

1. The lengths of the bars are not proportional. For example, the bar representing the cost of Clinton's inauguration should be slightly more than 9 times as long as the one for Carter's cost, and twice as long as the bar representing Reagan's cost.

3. (a) The vertical axis starts at 34,500 instead of 0. This tends to indicate that the median earnings for females changed at a faster rate than actually occurred.

   (b) This graph indicates that the median earnings for females has decreased slightly over the given time period.

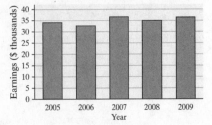

5. The bar for 12p-6p covers twice as many hours as the other bars. By combining two 3-hour periods, this bar looks larger compared to the others, making afternoon hours look more dangerous. If this bar were split into two periods, the graph may give a different impression. For example, the graph may show that daylight hours are safer.

7. (a) The vertical axis starts at 0.1 instead of 0. This might cause the reader to conclude, for example, that the proportion of people aged 25-34 years old who are not covered by any type of health insurance is more than 3 times the proportion for those aged 55-64 years old.

(b)

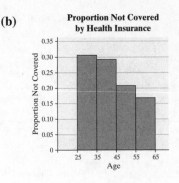

9. This graph is misleading because it does not take into account the size of the population of each state. Certainly, Vermont is going to pay less in total taxes than California simply because its population is so much lower. The moral of the story here is that many variables should be considered on a per capita (per person) bias. For example, this graph should be drawn with taxes paid per capita (per person).

11. (a) The graphic is misleading because the bars are not proportional. The bar for housing should be a little more than twice the length of the bar for transportation, but it is not.

   (b) The graphic could be improved by adjusting the bars so that their lengths are proportional.

13. (a)

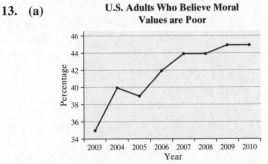

This graphic is misleading because the vertical scale starts at 34 instead of 0 without indicating a gap. This might cause the reader to think that the proportion of U.S. adults who believe moral values are poor is increasing more quickly than they really are.

(b)

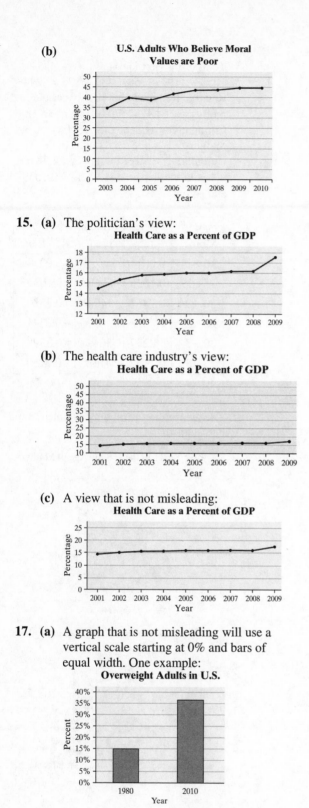

15. (a) The politician's view:

(b) The health care industry's view:

(c) A view that is not misleading:

17. (a) A graph that is not misleading will use a vertical scale starting at 0% and bars of equal width. One example:

(b) A graph that is misleading might use bars of unequal width or will use a vertical scale that does not start at 0%. One example, as follows, implies that the

percent of overweight adults in the U.S. has more than quadrupled between 1980 and 2010.

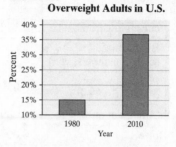

19. This is a histogram and the bars do not touch. In addition, there are no labels on the axes and there is not title on the graph.

Chapter 2 Review Exercises

1. (a) There are $614 + 154 + 1448 = 2216$ participants.

(b) The relative frequency of the respondents indicating that it makes no difference is

$$\frac{1448}{2216} \approx 0.653$$

(c) A Pareto chart is a bar chart where the bars are in descending order.

(d) Answers will vary.

2. (a) Total homicides = $9{,}146 + 1{,}825 + 611 + 801 + 121 + 99 + 895 = 13{,}498$
Relative frequency for Firearms =
$$\frac{9{,}146}{13{,}498} \approx 0.6776 \quad \text{and so on.}$$

Type of Weapon	Relative Frequency
Firearms	0.6776
Knives or cutting intstruments	0.1352
Blunt objects (clubs, hammers, etc.)	0.0453
Personal weapons (hands, fists, etc.)	0.0593
Strangulation	0.0090
Fire	0.0073
Other weapon or not stated	0.0663

(b) The relative frequency is 0.0453, so 4.5% of the homicides were due to blunt objects.

(c)

(d)

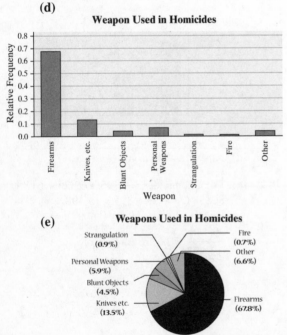

3. (a), (b), and **(c)**

Total births (in thousands) = 6 + 435 + 1053 + 1197 + 958 + 489 + 106 = 4244
Relative frequency for 10-14 year old mothers = $6 / 4244 \approx 0.0014$ and so on.
Cumulative frequency for 15-19 year old mothers = 6 + 435 = 441 and so on.
Cumulative relative frequency for 15-19 year old mothers = $441 / 4244 \approx 0.1039$ and so on.

Age of Mother	Rel. Freq.	Cumul. Freq.	Cumul. Rel. Freq.
10 – 14	0.0014	6	0.0014
15 – 19	0.1025	441	0.1039
20 – 24	0.2481	1494	0.3520
25 – 29	0.2820	2691	0.6341
30 – 34	0.2257	3649	0.8598
35 – 39	0.1152	4138	0.9750
40 – 44	0.0250	4244	1.0000

(d) The distribution is roughly symmetric and bell shaped.

(e)

(f)

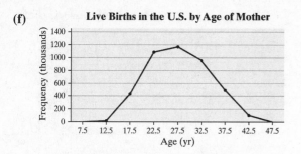

Live Births in the U.S. by Age of Mother

(g)

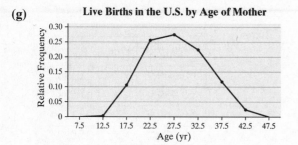

Live Births in the U.S. by Age of Mother

(h)

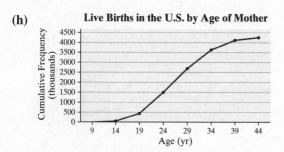

Live Births in the U.S. by Age of Mother

(i)

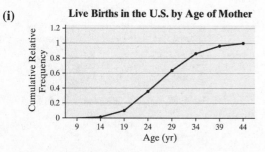

Live Births in the U.S. by Age of Mother

(j) From the relative frequency table, the relative frequency of 20-24 is 0.2481 and so the percentage is 24.81%.

(k) $\dfrac{958+489+106}{4244} = \dfrac{1553}{4244} \approx 0.3659$

36.59% of live births were to mothers aged 30 years or older.

4. (a) and **(b)**

Affiliation	Frequency	Relative Frequency
Democrat	46	0.46
Independent	16	0.16
Republican	38	0.38

(c)

Political Affiliation

(d)

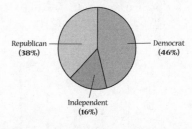

Political Affiliation

(e) Democrat appears to be the most common affiliation in Naperville.

5. (a), (b), (c), and **(d)**

Family Size	Freq.	Rel. Freq.	Cumul. Freq.	Cumul. Rel. Freq.
0	7	0.1167	7	0.1167
1	7	0.1167	14	0.2333
2	18	0.3000	32	0.5333
3	20	0.3333	52	0.8667
4	7	0.1167	59	0.9833
5	1	0.0167	60	1.0000

(e) The distribution is more or less symmetric.

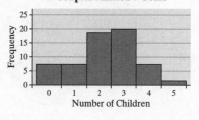

Number of Children for Couples Married 7 Years

(f)

Number of Children for
Couples Married 7 Years

(g) From the relative frequency table, the relative frequency of two children is 0.3000 so 30% of the couples have two children.

(h) From the frequency table, the relative frequency of at least two children (i.e. two or more) is
0.3000 + 0.3333 + 0.1167 + 0.0167 = 0.7667
or 76.67%. So, 76.67% of the couples have at least two children.

(i)

6. (a), (b)

Ownership Rate	Freq.	Rel. Freq.
40 – 44.9	1	0.0196
45 – 49.9	0	0.0000
50 – 54.9	1	0.0196
55 – 59.9	2	0.0392
60 – 64.9	2	0.0392
65 – 69.9	19	0.3725
70 – 74.9	21	0.4118
75 – 79.9	5	0.0980

(c)

Home Ownership Rates by State, 2009

(d)

Home Ownership Rates by State, 2009

(e) The distribution is skewed left.

(f)

Ownership Rate	Freq.	Rel. Freq.
40 – 49.9	1	0.0196
50 – 59.9	3	0.0588
60 – 69.9	21	0.4118
70 – 79.9	26	0.5098

Home Ownership Rates
by State, 2009

Home Ownership Rates
by State, 2009

The distribution is skewed left.

Answers will vary. Both class widths give a good overall picture of the distribution. The first class width provides a little more detail to the graph, but not necessarily enough to be worth the trouble. An intermediate value, say a width of 500, might be a reasonable compromise.

7. (a), (b), (c), and **(d)**
Answers will vary. Using 2.2000 as the lower class limit of the first class and 0.0200 as the class width, we obtain the following.

Class	Freq.	Rel. Freq.	Cumul. Freq.	Cumul. Rel.Freq.
2.2000 − 2.2199	2	0.0588	2	0.0588
2.2200 − 2.2399	3	0.0882	5	0.1471
2.2400 − 2.2599	5	0.1471	10	0.2941
2.2600 − 2.2799	6	0.1765	16	0.4706
2.2800 − 2.2999	4	0.1176	20	0.5882
2.3000 − 2.3199	7	0.2059	27	0.7941
2.3200 − 2.3399	5	0.1471	32	0.9412
2.3400 − 2.3599	1	0.0294	33	0.9706
2.3600 − 2.3799	1	0.0294	34	1

(e)

Diameter of Chocolate Chip Cookies

The distribution is roughly symmetric.

(f)

Diameter of Chocolate Chip Cookies

8.

Hours Spent Online

```
12 |
13 | 467
14 | 05578
15 | 1236
16 | 456
17 | 113449
18 | 066889
19 | 2
20 | 168
21 | 119
22 | 29
23 | 48
24 | 4
25 | 7
26 |
```
Legend: 13 | 4 = average 13.4 hours per week.

The distribution is slightly skewed right.

9. (a) Grade inflation seems to be happening in colleges. GPAs have increased every time period for all schools.

(b) GPAs increased 5.6% for public schools. GPAs increased 6.8% for private schools. Private schools have higher grade inflation because the GPAs are higher and they are increasing faster.

(c) The graph is misleading because it starts at 2.6 on the vertical axis.

10. (a) Answers will vary. The adjusted gross income share of the top 1% of earners shows steady increases overall, with a few minor exceptions. The adjusted gross income share of the bottom 50% of earners shows steady decreases overall, with a few minor exceptions.

(b) Answers will vary. The income tax share of the top 1% of earners shows steady increases overall, with few exceptions, including a notable decrease from 2007 to 2008. The income tax share of the bottom 50% of earners shows steady decreases over time.

2005 Average Earnings

11. (a) Graphs will vary. One way to mislead would be to start the vertical scale at a value other than 0. For example, starting the vertical scale at $30,000 might make the reader believe that college graduates earn more than three times what a high school graduate earns (on average).

(b) A graph that does not mislead would use equal widths for the bars and would start the vertical scale at $0. Here is an example of a graph that is not misleading:

2005 Average Earnings

12. (a) Flats are preferred the most (40%) and extra-high heels are preferred the least (1%).

(b) The graph is misleading because the bar heights and areas for each category are not proportional.

Chapter 2 Test

1. **(a)** The United States won the most men's singles championships between 1968 and 2010 with 15 wins.

   (b) $6 - 4 = 2$
   Representatives from Australia have won 2 more championships than representatives from Germany.

   (c) $15 + 8 + 6 + 5 + 4 + 2 + 1 + 1 + 1 = 43$
   championships between 1968 and 2010.
   $\frac{8}{43} = 0.186$
   Representatives of Sweden won 18.6% of the championships.

   (d) No, it is not appropriate to describe the shape of the distribution as skewed right. The data represented by the graph are qualitative so the bars in the graph could be placed in any order.

2. **(a)** There were 1005 responses. The relative frequency who indicated they preferred new tolls was $\frac{412}{1005} = 0.4100$ and so on.

Response	Freq.	Rel. Freq.
New Tolls	412	0.4100
Inc. Gas Tax	181	0.1801
No New Roads	412	0.4100

   (b) The relative frequency is 0.1801, so the percentage of respondents who would like to see an increase in gas taxes is 18.01%.

   (c)

   How Would You Prefer to Pay for New Road Construction?

(d)

How Would You Prefer to Pay for New Road Construction?

(e)

How Would You Prefer to Pay for New Road Construction?

3. **(a), (b)**

Education	Freq.	Rel. Freq.
No high school diploma	9	0.18
High school graduate	16	0.32
Some college	9	0.18
Associate's degree	4	0.08
Bachelor's degree	8	0.16
Advanced degree	4	0.08

   (c)

   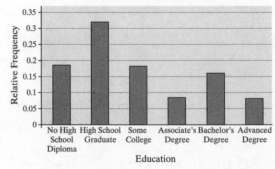

   Educational Attainment of Commuters

(d) **Educational Attainment of Commuters**

(e) The largest bar (and largest pie segment) corresponds to 'High School Graduate', so high school graduate is the most common educational level of a commuter.

4. **(a)**, **(b)**, **(c)**, and **(d)**

No. of Cars	Freq.	Rel. Freq.	Cumul. Freq.	Cumul. Rel. Freq.
1	5	0.10	5	0.10
2	7	0.14	12	0.24
3	12	0.24	24	0.48
4	6	0.12	30	0.60
5	8	0.16	38	0.76
6	5	0.10	43	0.86
7	2	0.04	45	0.90
8	4	0.08	49	0.98
9	1	0.02	50	1

(e)

Number of Cars Arriving at McDonald's

The distribution is skewed right.

(f)

Number of Cars Arriving at McDonald's

(g) The relative frequency of exactly 3 cars is 0.24. So, for 24% of the weeks, exactly three cars arrived between 11:50 am and 12:00 noon.

(h) The relative frequency of 3 or more cars
$$= 0.24 + 0.12 + 0.16 + 0.10$$
$$+ 0.04 + 0.08 + 0.02$$
$$= 0.76$$
So, for 76% of the weeks, three or more cars arrived between 11:50 am and 12:00 noon.

(i)

5. Answers may vary. One possibility follows:

(a), **(b)**
Using a lower class limit of the first class of 20 and a class width of 10:
Total number of data points = 40
Relative frequency of 20 – 29 = 1/40
= 0.025, and so on.

HDL Cholesterol	Frequency	Relative Frequency
20–29	1	0.025
30–39	6	0.150
40–49	10	0.250
50–59	14	0.350
60–69	6	0.150
70–79	3	0.075

(c) **Serum HDL of 20–29 Year Olds**

(d) **Serum HDL of 20–29 Year Olds**

(e) The distribution appears to be roughly bell-shaped.

6. (a), (b)

Closing Price	Cumul. Freq.	Cumul. Rel. Freq.
50,000 – 99,999	4	0.08
100,000 – 149,999	17	0.34
150,000 – 199,999	36	0.72
200,000 – 249,999	43	0.86
250,000 – 299,999	46	0.92
300,000 – 349,999	48	0.96
350,000 – 399,999	49	0.98
400,000 – 449,999	49	0.98
450,000 – 499,999	50	1

(c) The cumulative relative frequency for the class $150,000-$199,000 is 0.72. Therefore, 72% of the homes sold for less than $200,000.

(d)

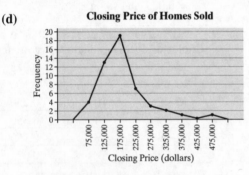

Closing Price of Homes Sold

The distribution is skewed right.

(e)

Closing Price of Homes Sold

(f)

Closing Price of Homes Sold

7. The stem-and-leaf diagram below shows a symmetric (uniform) distribution.

Time Spent on Homework

```
 4 | 0567
 5 | 26
 6 | 13
 7 | 01338
 8 | 59
 9 | 1369
10 | 3899
11 | 0018
12 | 556
```

Legend: 4 | 0 represents 40 minutes.

8. The curves in the figure below appear to follow the same trend. Birth rate increases as per capita income increases.

Fertility Rates (births per 1000 women aged 15–44) and per Capita Income (thousands of 2008 dollars)

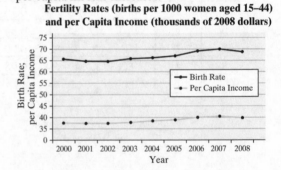

9. Answers may vary. It is difficult to interpret this graph because it is not clear whether the scale is represented by the height of the steps, the width of the steps, or by the graphics above the steps. The graphics are misleading because they must be increased in size both vertically and horizontally to avoid distorting the image. Thus, the resulting areas are not proportionally correct. The graph could be redrawn using bars whose widths are the same and whose heights are proportional based on the given percentages. The use of graphics should be avoided, or a standard size graphic representing a fixed value could be used and repeated as necessary to illustrate the given percentages.

Case Study: The Day the Sky Roared

1.

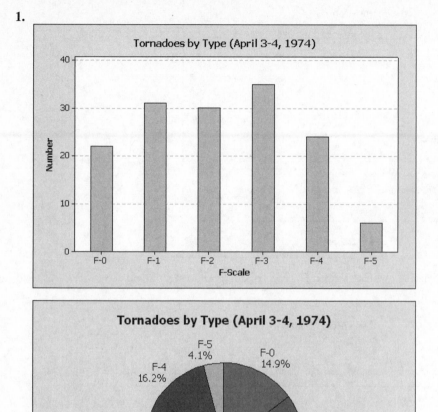

During the April 3-4, 1974 outbreak, 20% of the tornadoes exceeded F-3 on the Fujita Wind Damage Scale. This was much greater than the 1% that typically occurs.

2. The histogram will vary depending on the class width.

3. Histograms may vary depending on class widths. For comparison purposes, the same class width was used for each histogram.

The distributions all appear to be skewed right, though the distribution for F-5 tornadoes is difficult to see due to the low sample size. There is an obvious shift in the distributions. As the strength of the tornado increases, the duration of the tornado increases.

4. There were 305 deaths during the outbreak. Of these, 259 were due to the more severe tornadoes.

$$\frac{259}{305} \approx 0.8492$$

Roughly 85% of the deaths during the outbreak were due to the more severe tornadoes. This may be a little high, but it is consistent since it is greater than 70%.

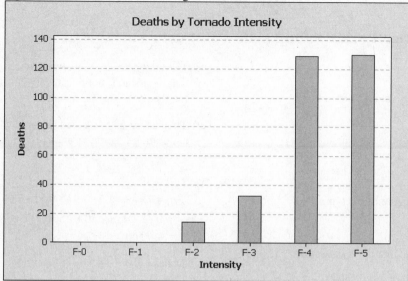

5. The provided data is not sufficient to determine whether or not tornadoes are more likely to strike rural areas. Some research at Texas A&M University indicates that tornadoes are more likely to occur in urban or suburban areas, possibly due to greater temperature differences. The data does indicate that the number of deaths decreases as the population of the community increases. The higher the population density, the greater the chance that a tornado is detected and reported early, thereby providing more time for residents to take shelter.

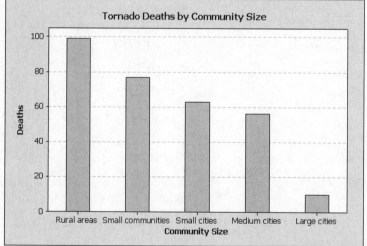

6. Answers will vary. The outbreak of April 3-4, 1974 seemed to be more severe in intensity than usual with 20% of the tornadoes being classified as F-4 or F-5. While the shape of the duration distribution was roughly the same for each intensity level, the duration of a tornado increased with its intensity. The number of deaths decreased as the community size increased.

Chapter 3

Numerically Describing Data from One Variable

Section 3.1

1. A statistic is resistant if its value is not sensitive to extreme data values.

3. HUD uses the median because the data are skewed. Explanations will vary. One possibility is that the price of homes has a distribution that is skewed to the right, so the median is more representative of the typical price of a home.

5. $\dfrac{10,000+1}{2} = 5000.5$. The median is between the 5000th and the 5001st ordered values.

7. $\bar{x} = \dfrac{20+13+4+8+10}{5} = \dfrac{55}{5} = 11$

9. $\mu = \dfrac{3+6+10+12+14}{5} = \dfrac{45}{5} = 9$

11. $\dfrac{165.5}{55} = 3.01$
    The mean price per ad slot was $3.01 million.

13. $\sum x_i = 2529+1889+2610+1073 = 8101$
    Mean = $\bar{x} = \dfrac{\sum x_i}{n} = \dfrac{8101}{4} = \$2025.25$
    Data in order: 1073, 1889, 2529, 2610
    Median = $\dfrac{1889+2529}{2} = \dfrac{4418}{2} = \$2209$
    No data value occurs more than once, so there is no mode.

15. $\sum x_i = 3960+4090+3200+3100+2940$
    $\qquad +3830+4090+4040+3780$
    $\qquad = 33,030 \text{ psi}$
    Mean = $\bar{x} = \dfrac{\sum x_i}{n} = \dfrac{33,030}{9} = 3670 \text{ psi}$
    Data in order: 2940, 3100, 3200, 3780, 3830, 3960, 4040, 4090, 4090
    Median = the 5th ordered data value = 3830 psi
    Mode = 4090 psi (because it is the only data value to occur twice)

17. (a) The histogram is skewed to the right, suggesting that the mean is greater than the median. That is, $\bar{x} > M$.

    (b) The histogram is symmetric, suggesting that the mean is approximately equal to the median. That is, $\bar{x} = M$.

    (c) The histogram is skewed to the left, suggesting that the mean is less than the median. That is, $\bar{x} < M$.

19. (a) Tap Water:
    $\sum x_i = 7.64+7.45+7.47+7.50+7.68+7.69$
    $\qquad +7.45+7.10+7.56+7.47+7.52+7.47$
    $\qquad = 90.00$
    Mean = $\bar{x} = \dfrac{\sum x_i}{n} = \dfrac{90.00}{12} = 7.50$
    Data in order: 7.10, 7.45, 7.45, 7.47, 7.47, 7.47, 7.50, 7.52, 7.56, 7.64, 7.68, 7.69
    Median = $\dfrac{7.47+7.50}{2} = \dfrac{14.97}{2} = 7.485$
    Mode = 7.47 (because it occurs three times, which is the most)

    Bottled Water:
    $\sum x_i = 5.15+5.09+5.26+5.20+5.02+5.23$
    $\qquad +5.28+5.26+5.13+5.26+5.21+5.24$
    $\qquad = 62.33$
    Mean = $\dfrac{\sum x_i}{n} = \dfrac{62.33}{12} \approx 5.194$
    Data in order: 5.02, 5.09, 5.13, 5.15, 5.20, 5.21, 5.23, 5.24, 5.26, 5.26, 5.26, 5.28
    Median = $\dfrac{5.21+5.23}{2} = \dfrac{10.44}{2} = 5.22$
    Mode = 5.26 (because it occurs three times, which is the most)

    The pH of the sample of tap water is somewhat higher than the pH of the sample of bottled water.

    (b) $\sum x_i = 7.64+7.45+7.47+7.50+7.68+7.69$
    $\qquad +7.45+1.70+7.56+7.47+7.52+7.47$
    $\qquad = 84.60$
    Mean = $\bar{x} = \dfrac{\sum x_i}{n} = \dfrac{84.60}{12} = 7.05$
    Data in order: 1.70, 7.45, 7.45, 7.47, 7.47, 7.47, 7.50, 7.52, 7.56, 7.64, 7.68, 7.69
    Median = $\dfrac{7.47+7.50}{2} = \dfrac{14.97}{2} = 7.485$

The incorrect entry causes the mean to decrease substantially while the median does not change. This illustrates that the median is resistant while the mean is not resistant.

21. (a) $\sum x_i = 76 + 60 + 60 + 81 + 72 + 80 + 80$
$+ 68 + 73$
$= 650$
$\mu = \dfrac{\sum x_i}{N} = \dfrac{650}{9} \approx 72.2$ beats per minute

   (b) Samples and sample means will vary.

   (c) Answers will vary.

23. (a) $\sum x_i = 1,785,029 + 1,591,756 + 439,695$
$+ 419,241 + 342,117 + 214,872$
$+ 151,988 + 147,155 + 137,257$
$+ 135,257$
$= 5,362,367$
$\mu = \dfrac{\sum x_i}{N} = \dfrac{5,362,367}{10} = 536,236.7$
thousand metric tons.

   (b) Per capita is a better gauge because it adjusts for CO_2 emissions for population. After all, countries with more people will, in general, have higher CO_2 emissions.

   (c) $\sum x_i = 1.35 + 5.20 + 0.391 + 2.95 + 2.71$
$+ 2.61 + 4.61 + 2.41 + 2.82 + 1.88$
$= 26.93$
$\mu = \dfrac{26.93}{10} = 2.693$ thousand metric tons
Data in order: 0.391, 1.35, 1.88, 2.41, 2.61, 2.71, 2.82, 2.95, 4.61, 5.20
$\text{Median} = \dfrac{2.61 + 2.71}{2} = \dfrac{5.32}{2} = 2.660$
thousand metric tons.
The environmentalist will likely use the mean to support the position that CO_2 emissions
are too high because the mean is higher.

25. The distribution is relatively symmetric as is evidenced by both the histogram and the fact that the mean and median are approximately equal. Therefore, the mean is the better measure of central tendency.

27. To create the histogram, we choose the lower class limit of the first class to be 0.78 and the class width to be 0.02. The resulting classes and frequencies follow:

Class	Freq.	Class	Freq.
0.78 – 0.79	1	0.88 – 0.89	10
0.80 – 0.81	1	0.90 – 0.91	9
0.82 – 0.83	4	0.92 – 0.93	4
0.84 – 0.85	8	0.94 – 0.95	2
0.86 – 0.87	11		

Weight of Plain M&Ms

To find the mean, we add all of the data values and divide by the sample size: $\sum x_i = 43.73$;
$\bar{x} = \dfrac{\sum x_i}{n} = \dfrac{43.73}{50} \approx 0.875$ grams

To find the median, we arrange the data in order. The median is the mean of the 25th and 26th data values:
$M = \dfrac{0.87 + 0.88}{2} = 0.875$ grams

The mean is approximately equal to the median suggesting that the distribution is symmetric. This is confirmed by the histogram (though is does appear to be slightly skewed left). The mean is the better measure of central tendency.

29. To create the histogram, we choose the lower class limit of the first class to be 0 and the class width to be 5. The resulting classes and frequencies follow:

Class	Freq.	Class	Freq.
0 – 4	3	25 – 29	4
5 – 9	1	30 – 34	5
10 – 14	0	35 – 39	3
15 – 19	4	40 – 44	1
20 – 24	4		

Hours Worked per Week

To find the mean, we add all of the data values and divide by the sample size: $\sum x_i = 550$;

$$\overline{x} = \frac{\sum x_i}{n} = \frac{550}{25} = 22 \text{ hours}$$

To find the median, we arrange the data in order. The median is the 13th data value:
$M = 25$ hours

The mean is smaller than the median suggesting that the distribution is skewed left. This is confirmed by the histogram. The median is the better measure of central tendency.

31. **(a)** The frequencies are:
   Liberal = 10
   Moderate = 12
   Conservative = 8
   The mode political view is Moderate.

   (b) Yes. Rotating the choices will help to avoid response bias that might be caused by the wording of the question.

33. Sample size of 5:
   All data recorded correctly: $\overline{x} = 99.8$; $M = 100$.
   106 recorded at 160: $\overline{x} = 110.6$; $M = 100$.
   Sample size of 12:
   All data recorded correctly: $\overline{x} \approx 100.4$; $M = 101$.
   106 recorded at 160: $\overline{x} \approx 104.9$; $M = 101$.
   Sample size of 30:
   All data recorded correctly: $\overline{x} = 100.6$; $M = 99$.
   106 recorded at 160: $\overline{x} = 102.4$; $M = 99$.
   For each sample size, the mean becomes larger while the median remains the same. As the sample size increases, the impact of the incorrectly recorded data value on the mean decreases.

35. The sum of the nineteen readable scores is $19 \cdot 84 = 1596$. The sum of all twenty scores is $20 \cdot 82 = 1640$. Therefore, the unreadable score is $1640 - 1596 = 44$.

37. Answers will vary.

39. **(a)** Mean:
   $$\sum x_i = 30 + 30 + 45 + 50 + 50 + 50 + 55 + 55$$
   $$+ 60 + 75$$
   $$= 500$$
   $$\mu = \frac{\sum x_i}{N} = \frac{500}{10} = 50$$
   The mean is $50,000.

   Median: The ten data values are in order. The median is the mean of the 5th and 6th data values: $M = \frac{50 + 50}{2} = \frac{100}{2} = 50$.
   The median is $50,000.

   Mode: The mode is $50,000 (the most frequent salary).

   (b) Add $2500 ($2.5 thousand) to each salary to form the 2nd: New data set: 32.5, 32.5, 47.5, 52.5, 52.5, 52.5, 57.5, 57.5, 62.5, 77.5

   2nd Mean:
   $$\sum x_i = 32.5 + 32.5 + 47.5 + 52.5 + 52.5$$
   $$+ 52.5 + 57.5 + 57.5 + 62.5 + 77.5$$
   $$= 525$$

   $\mu_{2nd} = \frac{\sum x_i}{N} = \frac{525}{10} = 52.5$ The mean for the 2nd data set is $52,500.

   2nd Median: The ten data values are in order. The median is the mean of the 5th and 6th data values:
   $$M_{2nd} = \frac{52.5 + 52.5}{2} = \frac{105}{2} = 52.5 .$$
   The median for the 2nd data set is $52,500.

   2nd Mode: The mode for the 2nd data set is $52,500 (the most frequent new salary).

   All three measures of central tendency increased by $2500, which was the amount of the raises.

   (c) Multiply each original data value by 1.05 to generate the 3rd data set: 31.5, 31.5, 47.25, 52.5, 52.5, 52.5, 57.75, 57.75, 63, 78.75

3rd Mean:
$$\sum x_i = 31.5 + 31.5 + 47.25 + 52.5 + 52.5$$
$$+ 52.5 + 57.75 + 57.75 + 63 + 78.75$$
$$= 525$$
$$\mu_{3rd} = \frac{\sum x_i}{N} = \frac{525}{10} = 52.5$$

The mean of the 3rd data set is $52,500.

3rd Median: The ten data values are in order. The median is the mean of the 5th and 6th data values:
$$M_{3rd} = \frac{52.5 + 52.5}{2} = \frac{105}{2} = 52.5 .$$

The median of the 3rd data set is $52,500.

3rd Mode: The mode of the 3rd data set is $52,500 (the most frequent new salary).

All three measures of central tendency increased by 5%, which was the amount of the raises.

(d) Add $25 thousand to the largest data value to form the new data set: 30, 30, 45, 50, 50, 50, 55, 55, 60, 100

4th Mean:
$$\sum x_i = 30 + 30 + 45 + 50 + 50 + 50 + 55 + 55$$
$$+ 60 + 100$$
$$= 525$$
$$\mu_{4th} = \frac{\sum x_i}{N} = \frac{525}{10} = 52.5$$

The mean of the 4th data set is $52,500.

4th Median: The ten data values are in order. The median is the mean of the 5th and 6th data values:
$$M_{4th} = \frac{50 + 50}{2} = \frac{100}{2} = 50 .$$

The median of the 4th data set is $50,000.

4th Mode: The mode of the 4th data set is $50,000 (the most frequent salary).

The mean was increased by $2500, but the median and mode remained unchanged.

41. The largest value is 0.95 and the smallest is 0.79. After deleting these two values, we have:
$$\sum x_i = 41.99; \quad \bar{x} = \frac{\sum x_i}{n} = \frac{41.99}{48} \approx 0.875 \text{ grams} .$$
The mean after deleting these two data values

is 0.875 grams. The trimmed mean is more resistant than the regular mean because the most extreme values are omitted before the mean is computed.

43. (a) The data are discrete.

(b) To construct the histogram, we first organize the data into a frequency table:

Number of Drinks	Frequency
0	23
1	17
2	4
3	3
4	2
5	1

Number of Days High School Students Consume Alcohol Each Week

The distribution is skewed right.

(c) Since the distribution is skewed right, we would expect the mean to be greater than the median.

(d) To find the mean, we add all of the data values and divide by the sample size.
$$\sum x_i = 47; \quad \bar{x} = \frac{\sum x_i}{n} = \frac{47}{50} = 0.94$$

To find the median, we arrange the data in order. The median is the mean of the 25th and 26th data values. $M = \frac{1+1}{2} = 1 .$

This tells us that the mean can be less than the median in skewed-right data. Therefore, the rule *mean greater than median implies the data are skewed right* is not always true.

(e) The mode is 0 (the most frequent value).

(f) Yes, Carlos' survey likely suffers from sampling bias. It is difficult to get truthful responses to this type of question. Carlos would need to ensure that the identity of the respondents is anonymous.

45. The median is resistant because it is the "middle observation" and increasing the largest value or decreasing the smallest value does not affect the median. The mean is not resistant because it is a function of the sum of the data values. Changing the magnitude of one value changes the sum of the values, and thus affects the mean.

47. (a) The median describes the typical U.S. household's net worth better. Household net worth is expected to be skewed right, and the mean will be affected by the outliers.

(b) Household net worth is expected to be skewed right.

(c) There are a few households with very high net worth.

49. The distribution is skewed right, so the median amount of money lost is less than the mean amount lost.

Section 3.2

1. zero

3. True

5. From Section 3.1, Exercise 7, we know $\bar{x} = 11$.

x_i	$\bar{x}$	$x_i - \bar{x}$	$(x_i - \bar{x})^2$
20	11	$20 - 11 = 9$	$9^2 = 81$
13	11	$13 - 11 = 2$	$2^2 = 4$
4	11	$4 - 11 = -7$	$(-7)^2 = 49$
8	11	$8 - 11 = -3$	$(-3)^2 = 9$
10	11	$10 - 11 = -1$	$(-1)^2 = 1$
		$\sum(x_i - \bar{x}) = 0$	$\sum(x_i - \bar{x})^2 = 144$

$$s^2 = \frac{\sum(x_i - \bar{x})^2}{n-1} = \frac{144}{5-1} = \frac{144}{4} = 36$$

$$s = \sqrt{s^2} = \sqrt{\frac{\sum(x_i - \bar{x})^2}{n-1}} = \sqrt{\frac{144}{5-1}} = \sqrt{36} = 6$$

7. From Section 3.1, Exercise 9, we know $\mu = 9$.

x_i	μ	$x_i - \mu$	$(x_i - \mu)^2$
3	9	$3 - 9 = -6$	$(-6)^2 = 36$
6	9	$6 - 9 = -3$	$(-3)^2 = 9$
10	9	$10 - 9 = 1$	$1^2 = 1$
12	9	$12 - 9 = 3$	$3^2 = 9$
14	9	$14 - 9 = 5$	$5^2 = 25$
		$\sum(x_i - \mu) = 0$	$\sum(x_i - \mu)^2 = 80$

$$\sigma^2 = \frac{\sum(x_i - \mu)^2}{N} = \frac{80}{5} = 16$$

$$\sigma = \sqrt{\sigma^2} = \sqrt{\frac{\sum(x_i - \mu)^2}{N}} = \sqrt{\frac{80}{5}} = \sqrt{16} = 4$$

9. $\bar{x} = \dfrac{6 + 52 + 13 + 49 + 35 + 25 + 31 + 29 + 31 + 29}{10}$

$= \dfrac{300}{10} = 30$

x_i	$\bar{x}$	$x_i - \bar{x}$	$(x_i - \bar{x})^2$
6	30	$6 - 30 = -24$	$(-24)^2 = 576$
52	30	$52 - 30 = 22$	$22^2 = 484$
13	30	$13 - 30 = -17$	$(-17)^2 = 289$
49	30	$49 - 30 = 19$	$19^2 = 361$
35	30	$35 - 30 = 5$	$5^2 = 25$
25	30	$25 - 30 = -5$	$(-5)^2 = 25$
31	30	$31 - 30 = 1$	$1^2 = 1$
29	30	$29 - 30 = -1$	$(-1)^2 = 1$
31	30	$31 - 30 = 1$	$1^2 = 1$
29	30	$29 - 30 = -1$	$(-1)^2 = 1$
		$\sum(x_i - \bar{x}) = 0$	$\sum(x_i - \bar{x})^2 = 1764$

$$s^2 = \frac{\sum(x_i - \bar{x})^2}{n-1} = \frac{1764}{10-1} = \frac{1764}{9} = 196$$

$$s = \sqrt{s^2} = \sqrt{\frac{\sum(x_i - \bar{x})^2}{N}} = \sqrt{\frac{1764}{10-1}} = \sqrt{196} = 14$$

11. Range = Largest Value – Smallest Value
= $2610 - 1073 = \$1537$.

To calculate the sample variance and the sample standard deviation, we use the computational formulas:

x_i	x_i^2
2529	6,395,841
1889	3,568,321
2610	6,812,100
1073	1,151,329
$\sum x_i = 8101$	$\sum x_i^2 = 17,927,591$

$$s^2 = \frac{\sum x_i^2 - \frac{(\sum x_i)^2}{n}}{n-1} = \frac{17,927,591 - \frac{(8101)^2}{4}}{4-1}$$

$$\approx 507,013.6 \ (\$)^2$$

$$s = \sqrt{s^2} = \sqrt{507,013.6} \approx \$712.0$$

13. Range = Largest Value – Smallest Value
$$= 4090 - 2940 = 1150 \text{ psi}$$

To calculate the sample variance and the sample standard deviation, we use the computational formulas:

x_i	x_i^2
3960	15,681,600
4090	16,728,100
3200	10,240,000
3100	9,610,000
2940	8,643,600
3830	14,668,900
4090	16,728,100
4040	16,321,600
3780	14,288,400
$\sum x_i = 33,020$	$\sum x_i^2 = 122,910,300$

$$s^2 = \frac{\sum x_i^2 - \frac{(\sum x_i)^2}{n}}{n-1} = \frac{122,910,300 - \frac{(33,030)^2}{9}}{9-1}$$

$$\approx 211,275 \ (\text{psi})^2$$

$$s = \sqrt{\frac{122,910,300 - \frac{(33,030)^2}{9}}{9-1}} \approx 459.6 \text{ psi}$$

15. Histogram (b) depicts a higher standard deviation because the data is more dispersed, with data values ranging from 30 to 75. In Histogram (a), the data values only range from 40 to 60.

17. **(a)** pH of tap water:
Range = Largest Value – Smallest Value
$$= 7.69 - 7.10 = 0.59$$

pH of bottled water:
Range = Largest Value – Smallest Value
$$= 5.28 - 5.02 = 0.26$$

Using range as the measure, the pH of tap water has more dispersion.

(b) To calculate the sample standard deviation, we use the computational formula:

pH of tap water:

x_i	x_i^2
7.64	58.3696
7.45	55.5025
7.47	55.8009
7.50	56.25
7.68	58.9824
7.69	59.1361
7.45	55.5025
7.10	50.41
7.56	57.1536
7.47	55.8009
7.52	56.5504
7.47	55.8009
$\sum x_i = 90$	$\sum x_i^2 = 675.2598$

$$s = \sqrt{\frac{\sum x_i^2 - \frac{(\sum x_i)^2}{n}}{n-1}} = \sqrt{\frac{675.2598 - \frac{(90)^2}{12}}{12-1}}$$

$$\approx 0.154$$

pH of bottled water:

x_i	x_i^2
5.15	26.5225
5.09	25.9081
5.26	27.6676
5.20	27.04
5.02	25.2004
5.23	27.3529
5.28	27.8784
5.26	27.6676
5.13	26.3169
5.26	27.6676
5.21	27.1441
5.24	27.4576
$\sum x_i = 62.33$	$\sum x_i^2 = 323.8237$

$$s = \sqrt{\frac{\sum x_i^2 - \frac{(\sum x_i)^2}{n}}{n-1}} = \sqrt{\frac{323.8237 - \frac{(62.33)^2}{12}}{12-1}}$$

$$\approx 0.081$$

Using standard deviation as the measure, the pH of tap water has more dispersion.

19. (a) We use the computational formula:

x_i	x_i^2
76	5776
60	3600
60	3600
81	6561
72	5184
80	6400
80	6400
68	4624
73	5329
$\sum x_i = 650$	$\sum x_i^2 = 47{,}474$

$$\sigma = \sqrt{\frac{\sum x_i^2 - \frac{(\sum x_i)^2}{N}}{N}} = \sqrt{\frac{47{,}474 - \frac{(650)^2}{9}}{9}}$$

$$\approx 7.7 \text{ beats/minute}$$

(b) Samples and sample standard deviations will vary.

(c) Answers will vary.

21. (a) Ethan:

$$\sum x_i = 9+24+8+9+5+8+9+10+8+10$$
$$= 100$$
$$\mu = \frac{\sum x_i}{N} = \frac{100}{10} = 10 \text{ fish}$$

Range = Largest Value – Smallest Value
$$= 24 - 5 = 19 \text{ fish}$$

Drew:
$$\sum x_i = 15+2+3+18+20+1+17+2+19+3$$
$$= 100$$
$$\mu = \frac{\sum x_i}{N} = \frac{100}{10} = 10 \text{ fish}$$

Range = Largest Value – Smallest Value
$$= 20 - 1 = 19 \text{ fish}$$

Both fishermen have the same mean and range, so these values do not indicate any differences between their catches per day.

(b) Ethan:
$$N = 10 ; \quad \sum x_i = 100$$
$$\sum x_i^2 = 9^2 + 24^2 + 8^2 + 9^2 + 5^2 + 8^2 + 9^2$$
$$+ 10^2 + 8^2 + 10^2$$
$$= 1236$$
$$\sigma = \sqrt{\frac{\sum x_i^2 - \frac{(\sum x_i)^2}{N}}{N}} = \sqrt{\frac{1236 - \frac{(100)^2}{10}}{10}}$$
$$\approx 4.9 \text{ fish}$$

Drew:
$$N = 10 ; \quad \sum x_i = 100$$
$$\sum x_i^2 = 15^2 + 2^2 + 3^2 + 18^2 + 20^2 + 1^2 + 17^2$$
$$+ 2^2 + 19^2 + 3^2$$
$$= 1626$$
$$\sigma = \sqrt{\frac{\sum x_i^2 - \frac{(\sum x_i)^2}{N}}{N}} = \sqrt{\frac{1626 - \frac{(100)^2}{10}}{10}}$$
$$\approx 7.9 \text{ fish}$$

Yes, now there appears to be a difference in the two fishermen's records. Ethan had a more consistent fishing record, which is indicated by the smaller standard deviation.

(c) Answers will vary. One possibility follows: The range is limited as a measure of dispersion because it does not take all of the data values into account. It is obtained by using only the two most extreme data values. Since the standard deviation utilizes all of the data values, it provides a better overall representation of dispersion.

23. (a) We use the computational formula:

$\sum x_i = 43.73$; $\sum x_i^2 = 38.3083$; $n = 50$;

$$s = \sqrt{\frac{\sum x_i^2 - \frac{(\sum x_i)^2}{n}}{n-1}} = \sqrt{\frac{38.3083 - \frac{(43.73)^2}{50}}{50-1}}$$

≈ 0.036 g

(b) The histogram is approximately symmetric, so the Empirical Rule is applicable.

(c) Since 0.803 is two standard deviations below the mean [0.803 = 0.875 − 2(0.036)] and 0.943 is two standard deviations above the mean [0.947 = 0.875 + 2(0.036)], the Empirical Rule predicts that approximately 95% of the M&Ms will weigh between 0.803 and 0.947 grams.

(d) All except 2 of the M&Ms weigh between 0.803 and 0.947 grams. Thus, the actual percentage is 48/50 = 96%.

(e) Since 0.911 is one standard deviation above the mean [0.911 = 0.875 + 0.036], the Empirical Rule predicts that 13.5% + 2.35% + 0.15% = 16% of the M&Ms will weigh more than 0.911 grams.

(f) Six of the M&Ms weigh more than 0.911 grams. Thus, the actual percentage is 6/50 = 12%.

25. Car 1:

$\sum x_i = 3352$; $\sum x_i^2 = 755,712$; $n = 15$

Measures of Center:

$\bar{x} = \dfrac{\sum x_i}{n} = \dfrac{3352}{15} \approx 223.5$ miles

$M = 223$ miles (8th value in the ordered data)
Mode: 223 and 233

Measures of Dispersion:
Range = Largest Value − Smallest Value
= 271 − 178 = 93 miles

$$s^2 = \frac{\sum x_i^2 - \frac{(\sum x_i)^2}{n}}{n-1} = \frac{755,712 - \frac{(3352)^2}{15}}{15-1}$$

≈ 475.1 (miles)^2

$$s = \sqrt{s^2} = \sqrt{\frac{755,712 - \frac{(3352)^2}{15}}{15-1}} \approx 21.8 \text{ miles}$$

Car 2:
$\sum x_i = 3558$; $\sum x_i^2 = 877,654$; $n = 15$

Measures of Center:

$\bar{x} = \dfrac{\sum x_i}{n} = \dfrac{3558}{15} = 237.2$ miles

$M = 230$ miles (8th value in the ordered data)
Mode: 217 and 230

Measures of Dispersion:
Range = Largest Value − Smallest Value
= 326 − 160 = 166 miles

$$s^2 = \frac{\sum x_i^2 - \frac{(\sum x_i)^2}{n}}{n-1} = \frac{877,654 - \frac{(3558)^2}{15}}{15-1}$$

≈ 2406.9 (miles)^2

$$s = \sqrt{s^2} = \sqrt{\frac{877,654 - \frac{(3558)^2}{15}}{15-1}} \approx 49.1 \text{ miles}$$

We expect that the distribution for Car 1 is symmetric since the mean and median are approximately equal. We expect that the distribution for Car 2 is skewed right slightly since the mean is larger than the median. Both distributions have similar measures of center, but Car 2 has more dispersion which can be seen by its larger range, variance, and standard deviation. This means that the distance Car 1 can be driven on 10 gallons of gas is more consistent. Thus, Car 1 is probably the better car to buy.

27. (a) Consumer Goods Stocks:

$\sum x_i = 245.12$; $\sum x_i^2 = 4344.23$; $n = 25$

$\bar{x} = \dfrac{\sum x_i}{n} = \dfrac{245.12}{25} \approx 9.805\%$

$M = 10.11\%$ (13th value in the ordered data)

Energy Stocks:

$\sum x_i = 291.97$; $\sum x_i^2 = 5503.732$; $n = 25$

$$\bar{x} = \frac{\sum x_i}{n} = \frac{291.97}{25} \approx 11.679\%$$

$M = 10.76\%$ (13th value in the ordered data)
Energy Stocks have higher mean and
median rates of return.

(b) Consumer Goods Stocks:

$$s = \sqrt{\frac{\sum x_i^2 - \frac{\left(\sum x_i\right)^2}{n}}{n-1}}$$

$$= \sqrt{\frac{4344.23 - \frac{(245.12)^2}{25}}{25-1}} \approx 8.993\%$$

Energy Stocks:

$$s = \sqrt{\frac{\sum x_i^2 - \frac{\left(\sum x_i\right)^2}{n}}{n-1}}$$

$$= \sqrt{\frac{5503.732 - \frac{(291.97)^2}{25}}{25-1}} \approx 9.340\%$$

Energy Stocks are riskier since they have a
larger standard deviation. The investor is
paying for the higher return. The higher
returns are probably worth the cost.

29. (a) Since 70 is two standard deviations below
the mean [70 = 100 – 2(15)] and 130 is
two standard deviations above the mean
[130 = 100 + 2(15)], the Empirical Rule
predicts that approximately 95% of
people has an IQ between 70 and 130.

(b) Since about 95% of people has an IQ score
between 70 and 30, then approximately 5%
of people has an IQ score either less than
70 or greater than 130.

(c) Approximately $5\%/2 = 2.5\%$ of people
has an IQ score greater than 130.

31. (a) Approximately 95% of the data will be
within two standard deviations of the mean.
Now, 325 – 2(30) = 265 and 325 + 2(30) =
385. Thus, about 95% of pairs of kidneys
will be between 265 and 385 grams.

(b) Since 235 is three standard deviations
below the mean [235 = 325 – 3(30)] and
415 is three standard deviations above the
mean [415 = 325 + 3(30)], the Empirical
Rule predicts that about 99.7% of pairs of
kidneys weighs between 235 and 415
grams.

(c) Since about 99.7% of pairs of kidneys
weighs between 235 and 415 grams, then
about 0.3% of pairs of kidneys weighs
either less than 235 or more than 415
grams.

(d) Since 295 is one standard deviation below
the mean [295 = 325 – 30] and 385 is two
standard deviations above the mean [385
= 325 + 2(30)], the Empirical Rule
predicts that approximately 34% + 34% +
13.5% = 81.5% of pairs of kidneys weighs
between 295 and 385 grams.

33. In Professor Alpha's class, only about 2.5% of
the students will get an A, since 90% is 2
standard deviations above the mean. In
Professor Omega's class, approximately 16%
will get an A, since 90% in that class is only 1
standard deviation above the mean. Assuming
you intend to earn an A, you will likely choose
Professor Omega's class.
On the other hand, if a student only wants to get
a passing grade, then Professor Alpha would be
a better choice. Approximately 95% of
Professor Alpha's class scores between 70% and
90%.

35. (a) By Chebyshev's inequality, at least

$$\left(1 - \frac{1}{k^2}\right) \cdot 100\% = \left(1 - \frac{1}{3^2}\right) \cdot 100\% \approx 88.9\%$$

of gasoline prices have prices within three
standard deviations of the mean.

(b) By Chebyshev's inequality, at least

$$\left(1 - \frac{1}{k^2}\right) \cdot 100\% = \left(1 - \frac{1}{2.5^2}\right) \cdot 100\% = 84\%$$

of gasoline prices has prices within
$k = 2.5$ standard deviations of the mean.
Now, $3.06 - 2.5(0.06) = 2.91$ and
$3.06 + 2.5(0.06) = 3.21$. Thus, the
gasoline prices that are within 2.5
standard deviations of the mean are from
$2.91 to $3.21.

(c) Since 2.94 is $k = 2$ standard deviations
below the mean [2.94 = 3.06 – 2(0.06)]
and 3.18 is $k = 2$ standard deviations
above the mean [3.18 = 3.06 + 2(0.06)],
Chebyshev's theorem predicts that at least

$$\left(1 - \frac{1}{k^2}\right) \cdot 100\% = \left(1 - \frac{1}{2^2}\right) \cdot 100\% = 75\%$$

of gas stations has prices between $2.94
and $3.18 per gallon.

37. When calculating the variability in team batting averages, we are finding the variability of means. When calculating the variability of all players, we are finding the variability of individuals. Since there is more variability among individuals than among means, the teams will have less variability.

39. Sample size of 5:
All data recorded correctly: $s \approx 5.3$.
106 recorded incorrectly as 160: $s \approx 27.9$.

Sample size of 12:
All data recorded correctly: $s \approx 14.7$.
106 recorded incorrectly as 160: $s \approx 22.7$.

Sample size of 30:
All data recorded correctly: $s \approx 15.9$.
106 recorded incorrectly as 160: $s \approx 19.2$.

As the sample size increases, the impact of the misrecorded observation on the standard deviation decreases.

41. **(a)** We use the computational formula:
$\sum x_i = 1991.6$; $\sum x_i^2 = 199,033.1$; $n = 20$;

$$s = \sqrt{\dfrac{\sum x_i^2 - \dfrac{(\sum x_i)^2}{n}}{n-1}}$$

$$= \sqrt{\dfrac{199,033.1 - \dfrac{(1991.6)^2}{20}}{20-1}} \approx 6.11 \text{ cm}$$

(b) $\sum x_i = 969.4$; $\sum x_i^2 = 94,262.8$; $n = 10$;

$$s = \sqrt{\dfrac{\sum x_i^2 - \dfrac{(\sum x_i)^2}{n}}{n-1}}$$

$$= \sqrt{\dfrac{94,262.8 - \dfrac{(969.4)^2}{10}}{10-1}} \approx 5.67 \text{ cm}$$

(c) $\sum x_i = 1022.2$; $\sum x_i^2 = 104,770.3$; $n = 10$;

$$s = \sqrt{\dfrac{\sum x_i^2 - \dfrac{(\sum x_i)^2}{n}}{n-1}}$$

$$= \sqrt{\dfrac{104,770.3 - \dfrac{(1022.2)^2}{10}}{10-1}} \approx 5.59 \text{ cm}$$

(d) The standard deviation is lower for each block than it is for the combined group.

43. **(a)** Skewness $= \dfrac{3(50-40)}{10} = 3$.
The distribution is skewed to the right.

(b) Skewness $= \dfrac{3(100-100)}{15} = 0$.
The distribution is symmetric.

(c) Skewness $= \dfrac{3(400-500)}{120} = -2.5$.
The distribution is skewed to the left.

(d) From Section 3.1, Problem 27, and Section 3.2, Problem 23, we know $\bar{x} = 0.875$ gram, $M = 0.875$ gram, and $s \approx 0.036$ gram.

Skewness $= \dfrac{3(0.875 - 0.875)}{0.036} = 0$.
The distribution is symmetric.

(e) From Section 3.1, Problem 28, and Section 3.2, Problem 24, we know $\bar{x} = 104.1$ sec. $M = 104$ sec., and $s \approx 6$ hours.

Skewness $= \dfrac{3(104.1 - 104)}{6} = 0.05$.
The distribution is symmetric.

45. **(a)** The mean for the bond mutual funds is
$\bar{x} = \dfrac{\sum x_i}{n} = \dfrac{19}{8} = 2.38$
The standard deviation for the bond mutual funds is found by:
$\sum x_i = 19$; $\sum x_i^2 = 48.26$; $n = 8$;

$$s = \sqrt{\dfrac{\sum x_i^2 - \dfrac{(\sum x_i)^2}{n}}{n-1}}$$

$$= \sqrt{\dfrac{48.26 - \dfrac{(19)^2}{8}}{8-1}} \approx 0.669$$

The mean for the stock mutual funds is
$\bar{x} = \dfrac{\sum x_i}{n} = \dfrac{64.1}{8} = 8.01$
The standard deviation for the stock mutual funds is found by:

$\sum x_i = 64.1$; $\sum x_i^2 = 519.31$; $n = 8$;

$$s = \sqrt{\frac{\sum x_i^2 - \frac{\left(\sum x_i\right)^2}{n}}{n-1}}$$

$$= \sqrt{\frac{519.31 - \frac{(64.1)^2}{8}}{8-1}} \approx 0.903$$

(b) Based on the standard deviation, the stock mutual funds have more spread.

(c) We want to find the proportion of bond mutual funds between
$\overline{x} - s = 2.375 - 0.669 = 1.706$ and
$\overline{x} + s = 2.375 + 0.669 = 3.044$.
A total of 5 of the 8 observations, are between these values, so the proportion of bond mutual funds that are within one standard deviation of the mean is
$5/8 = 0.625$.

We want to find the proportion of stock mutual funds between
$\overline{x} - s = 8.0125 - 0.903 = 7.109$ and
$\overline{x} + s = 8.0125 + 0.903 = 8.916$
A total of 5 of the 8 observations, are between these values, so the proportion of stock mutual funds that are within one standard deviation of the mean is
$5/8 = 0.625$.

(d) The coefficient of variation for the bond mutual funds is $\dfrac{s}{\overline{x}} = \dfrac{0.669}{2.375} = 0.28$. The coefficient of variation for the stock mutual funds is $\dfrac{s}{\overline{x}} = \dfrac{0.903}{8.0125} = 0.11$.

(e) The mean for the heights in inches is
$$\overline{x} = \frac{\sum x_i}{n} = \frac{560}{8} = 70$$
and the standard deviation is given by:

$\sum x_i = 560$; $\sum x_i^2 = 39244$; $n = 8$;

$$s = \sqrt{\frac{\sum x_i^2 - \frac{\left(\sum x_i\right)^2}{n}}{n-1}}$$

$$= \sqrt{\frac{39244 - \frac{(560)^2}{8}}{8-1}} \approx 2.5$$

The mean for the heights in centimeters is

$$\overline{x} = \frac{\sum x_i}{n} = \frac{1422.4}{8} = 177.8$$

and the standard deviation is given by:

$\sum x_i = 1422.4$; $\sum x_i^2 = 253186.6$; $n = 8$;

$$s = \sqrt{\frac{\sum x_i^2 - \frac{\left(\sum x_i\right)^2}{n}}{n-1}}$$

$$= \sqrt{\frac{253186.6 - \frac{(1422.4)^2}{8}}{8-1}} \approx 6.368$$

So, the heights of the males appear more dispersed based on the standard deviation when measured in centimeters than when measured in inches, because the standard deviation is larger when measured in centimeters.

The coefficient of variation of the heights when measured in inches is:

$$\frac{s}{\overline{x}} = \frac{2.5}{70} = 0.036$$

The coefficient of variation of the heights when measured in centimeters is:

$$\frac{s}{\overline{x}} = \frac{6.368}{177.8} = 0.036$$

The coefficient of variation is the same, whether the heights are measured in inches or centimeters.

47. Answers will vary.

49. Degrees of freedom refers to the number of data values that are free to be any value while requiring the entire data set to have a specified mean. For example, if a data set is composed of 10 observations, 9 of the observations are free to be any value, while the tenth must be a specific value to obtain a specific mean.

51. A statistic is said to be biased if it consistently underestimates or overestimates the value of the corresponding parameter.

53. There is more spread among heights when gender is not accounted for. Think about the range of heights from the shortest female to the tallest male (in general) versus the range of heights from the shortest female to the tallest female and the shortest male to the tallest male.

55. The IQ of residents in your home town would have a higher standard deviation because the college campus likely has certain admission requirements which make the individuals more homogenous as far as intellect goes.

57. Answers will vary. The histogram with the larger spread will have the larger standard deviation. In the example below, Histogram I has the larger standard deviation.

Consumer Reports®: Basement Waterproofing Coatings

(a) $\bar{x}_A = \dfrac{\sum x_i}{n} = \dfrac{546.2}{6} \approx 91.03$ g

$M_A = \dfrac{90.9 + 91.2}{2} = \dfrac{182.1}{2} = 91.05$ g ;

There are 2 modes: 90.8 g and 91.2 g (each value occurs twice).

(b) $\sum x_i = 546.2$; $\sum x_i^2 = 49,722.66$; $n = 6$;

$s_A = \sqrt{\dfrac{\sum x_i^2 - \dfrac{\left(\sum x_i\right)^2}{n}}{n-1}}$

$= \sqrt{\dfrac{49,722.66 - \dfrac{(546.2)^2}{6}}{6-1}}$

≈ 0.23 g

(c) $\bar{x}_B = \dfrac{\sum x_i}{n} = \dfrac{522.3}{6} = 87.05$ g

$M_B = \dfrac{87.0 + 87.1}{2} = \dfrac{174.1}{2} = 87.05$ g

There are 2 modes: 87.0 g and 87.2 g (each value occurs twice).

(d) $\sum x_i = 522.2$; $\sum x_i^2 = 45,466.33$; $n = 6$;

$s_B = \sqrt{\dfrac{\sum x_i^2 - \dfrac{\left(\sum x_i\right)^2}{n}}{n-1}}$

$= \sqrt{\dfrac{45,466.33 - \dfrac{(522.3)^2}{6}}{6-1}}$

≈ 0.15 g

(e)

	A		B				
		86	8				
		87	0	0	1	2	2
		88					
		89					
9	8 8	90					
3	2 2	91					

Yes, there appears to be a difference in these two products' abilities to mitigate water seepage. All 6 of the measurements for product B are less than the measurements for product A. Although it is not clear whether there is any practical difference in these two products' abilities to mitigate water seepage, product B appears to do a better job.

Section 3.3

1. To find the mean, we use the formula $\mu = \dfrac{\sum x_i f_i}{\sum f_i}$. To find the standard deviation, we choose to use the

computational formula $\sigma = \sqrt{\sigma^2} = \sqrt{\dfrac{\sum x_i^2 f_i - \dfrac{\left(\sum x_i f_i\right)^2}{\sum f_i}}{\sum f_i}}$. We organize our computations of x_i, $\sum f_i$, $\sum x_i f_i$,

and $\sum x_i^2 f_i$ in the table that follows:

Class	Midpoint, x_i	Frequency, f_i	$x_i f_i$	x_i^2	$x_i^2 f_i$
0 – 999	$\dfrac{0+1,000}{2}=500$	30	15,000	250,000	7,500,000
1,000 – 1,999	$\dfrac{1,000+2,000}{2}=1,500$	97	145,500	2,250,000	218,250,000
2,000 – 2,999	2,500	935	2,337,500	6,250,000	5,843,750,000
3,000 – 3,999	3,500	2,698	9,443,000	12,250,000	33,050,500,000
4,000 – 4,999	4,500	344	1,548,000	20,250,000	6,966,000,000
5,000 – 5,999	5,500	5	27,500	30,250,000	151,250,000
		$\sum f_i = 4109$	$\sum x_i f_i = 13,516,500$		$\sum x_i^2 f_i = 46,237,250,000$

With the table complete, we compute the population mean and population standard deviation:

$$\mu = \frac{\sum x_i f_i}{\sum f_i} = \frac{13,516,500}{4109} \approx 3,289.5 \text{ grams}$$

$$\sigma = \sqrt{\sigma^2} = \sqrt{\frac{\sum x_i^2 f_i - \frac{\left(\sum x_i f_i\right)^2}{\sum f_i}}{\sum f_i}} = \sqrt{\frac{46,237,250,000 - \frac{(13,516,500)^2}{4109}}{4109}} \approx 657.2 \text{ grams}$$

3. To find the mean, we use the formula $\bar{x} = \dfrac{\sum x_i f_i}{\sum f_i}$. To find the standard deviation, we choose to use the

computational formula $s = \sqrt{s^2} = \sqrt{\dfrac{\sum x_i^2 f_i - \dfrac{\left(\sum x_i f_i\right)^2}{\sum f_i}}{\left(\sum f_i\right)-1}}$. We organize our computations of x_i, $\sum f_i$, $\sum x_i f_i$,

and $\sum x_i^2 f_i$ in the table that follows:

Class	Midpoint, x_i	Frequency, f_i	$x_i f_i$	x_i^2	$x_i^2 f_i$
61 – 64	$\dfrac{61+65}{2}=63$	31	1,953	3,969	123,039
65 – 67	$\dfrac{65+68}{2}=66.5$	67	4,455.5	4,422.25	296,290.75
68 – 69	69	198	13,662	4,761	942,678
70	70	195	13,650	4,900	955,500
71 – 72	72	120	8,640	5,184	622,080
73 – 76	75	89	6,675	5,625	500,625
77 – 80	79	50	3,950	6,241	312,050
		$\sum f_i = 750$	$\sum x_i f_i = 52,985.5$		$\sum x_i^2 f_i = 3,752,262.75$

With the table complete, we compute the sample mean and sample standard deviation:

$$\bar{x} = \frac{\sum x_i f_i}{\sum f_i} = \frac{52{,}985.5}{750} \approx 70.6°F; \quad s = \sqrt{s^2} = \sqrt{\frac{\sum x_i^2 f_i - \frac{(\sum x_i f_i)^2}{\sum f_i}}{(\sum f_i) - 1}} = \sqrt{\frac{3{,}752{,}262.75 - \frac{(52{,}985.5)^2}{750}}{750 - 1}} \approx 3.5°F$$

5. **(a)** To find the mean, we use the formula $\mu = \dfrac{\sum x_i f_i}{\sum f_i}$. To find the standard deviation, we choose to use the

computational formula $\sigma = \sqrt{\sigma^2} = \sqrt{\dfrac{\sum x_i^2 f_i - \dfrac{(\sum x_i f_i)^2}{\sum f_i}}{\sum f_i}}$. We organize our computations of x_i, $\sum f_i$,

$\sum x_i f_i$, and $\sum x_i^2 f_i$ in the table that follows:

Class	Midpoint, x_i	Frequency, f_i	$x_i f_i$	x_i^2	$x_i^2 f_i$
15 – 19	$\frac{15+20}{2} = 17.5$	100	1,750	306.25	30,625
20 – 24	$\frac{20+25}{2} = 22.5$	467	10,507.5	506.25	236,418.75
25 – 29	27.5	1,620	44,550	756.25	1,225,125
30 – 34	32.5	2,262	73,515	1,056.25	2,389,237.5
35 – 39	37.5	1,545	57,937.5	1,406.25	2,172,656.25
40 – 44	42.5	328	13,940	1,806.25	592,450
45 – 49	47.5	85	4,037.5	2,256.25	191,781.25
50 – 54	52.5	20	1,050	2,756.25	55,125
		$\sum f_i = 6{,}427$	$\sum x_i f_i = 207{,}287.5$		$\sum x_i^2 f_i = 6{,}893{,}418.75$

With the table complete, we compute the population mean and population standard deviation:

$$\mu = \frac{\sum x_i f_i}{\sum f_i} = \frac{207{,}287.5}{6{,}427} \approx 32.3 \text{ years}$$

$$\sigma = \sqrt{\sigma^2} = \sqrt{\frac{\sum x_i^2 f_i - \frac{(\sum x_i f_i)^2}{\sum f_i}}{\sum f_i}} = \sqrt{\frac{6{,}893{,}418.75 - \frac{(207{,}287.5)^2}{6{,}427}}{6{,}427}} \approx 5.7 \text{ years}$$

(b)

Number of Multiple Births

(Age of Mother vs. Frequency)

(c) By the Empirical Rule, 95% of the observations will be within 2 standard deviations of the mean. Now, $\mu - 2\sigma = 32.3 - 2(5.7) = 20.9$ and $\mu + 2\sigma = 32.3 + 2(5.7) = 43.7$, so 95% of the mothers of multiple births will be between 20.9 and 43.7 years of age.

7. We organize our computations of x_i, $\sum f_i$, $\sum x_i f_i$, and $\sum x_i^2 f_i$ in the table that follows:

Class	Midpoint, x_i	Frequency, f_i	$x_i f_i$	x_i^2	$x_i^2 f_i$
0 – 0.499	$\dfrac{0+0.5}{2}=0.25$	7	1.75	0.0625	0.4375
0.5 – 0.999	0.75	14	10.5	0.5625	7.875
1 – 1.499	1.25	7	8.75	1.5625	10.9375
1.5 – 1.999	1.75	8	14	3.0625	24.5
2 – 2.499	2.25	6	13.5	5.0625	30.375
2.5 – 2.999	2.75	4	11	7.5625	30.25
3 – 3.499	3.25	4	13	10.5625	42.25
3.5 – 3.999	3.75	0	0	14.0625	0
4 – 4.499	4.25	1	4.25	18.0625	18.0625
		$\sum f_i = 51$	$\sum x_i f_i = 76.75$		$\sum x_i^2 f_i = 164.6875$

With the table complete, we compute the estimated population mean and standard deviation:

$$\mu = \frac{\sum x_i f_i}{\sum f_i} = \frac{76.75}{51} \approx \$1.5049 \; ; \; \sigma = \sqrt{\sigma^2} = \sqrt{\frac{\sum x_i^2 f_i - \dfrac{\left(\sum x_i f_i\right)^2}{\sum f_i}}{\left(\sum f_i\right)}} = \sqrt{\frac{164.6875 - \dfrac{(76.75)^2}{51}}{51}} \approx \$0.982$$

From the raw data, we find: $\sum x_i = 73.774$; $\sum x_i^2 = 150.388$; $n = 51$

$$\mu = \frac{\sum x_i}{n} = \frac{73.774}{51} \approx \$1.447 \; ; \; \sigma = \sqrt{\sigma^2} = \sqrt{\frac{\sum x_i^2 - \dfrac{\left(\sum x_i\right)^2}{n}}{n-1}} = \sqrt{\frac{150.388 - \dfrac{(73.774)^2}{51}}{51}} \approx 0.925$$

The approximations from the grouped data are good estimates of the actual results from the raw data.

9. GPA $= \bar{x}_w = \dfrac{\sum w_i x_i}{\sum w_i} = \dfrac{5(3)+3(4)+4(4)+3(2)}{5+3+4+3} = \dfrac{49}{15} \approx 3.27$

11. Cost per pound $= \bar{x}_w = \dfrac{\sum w_i x_i}{\sum w_i} = \dfrac{4(\$3.50)+3(\$2.75)+2(\$2.25)}{4+3+2} = \dfrac{\$26.75}{9} \approx \$2.97 / \text{pound}$

13. (a) Male: We organize our computations of x_i, $\sum f_i$, $\sum x_i f_i$, and $\sum x_i^2 f_i$ in the following table.

Class	Midpoint, x_i	Frequency, f_i	$x_i f_i$	x_i^2	$x_i^2 f_i$
0–9	5	20,929	104,645	25	523,225
10–19	15	21,074	316,110	225	4,741,650
20–29	25	21,105	527,625	625	13,190,625
30–39	35	19,780	692,300	1,225	24,230,500
40–49	45	21,754	978,930	2,025	44,051,850
50–59	55	19,303	1,061,665	3,025	58,391,575
60–69	65	12,388	805,220	4,225	52,339,300
70–79	75	6,940	520,500	5,625	39,037,500
80–89	85	3,106	264,010	7,225	22,440,850
≥ 90	95	479	45,505	9,025	4,322,975
		$\sum f_i = 146,858$	$\sum x_i f_i = 5,316,510$		$\sum x_i^2 f_i = 263,270,050$

With the table complete, we compute the population mean and population standard deviation:

$$\mu = \frac{\sum x_i f_i}{\sum f_i} = \frac{5,316,510}{146,858} \approx 36.2 \text{ years}$$

$$\sigma = \sqrt{\sigma^2} = \sqrt{\frac{\sum x_i^2 f_i - \frac{\left(\sum x_i f_i\right)^2}{\sum f_i}}{\sum f_i}} = \sqrt{\frac{263,270,050 - \frac{(5,316,510)^2}{146,858}}{146,858}} \approx 22.0 \text{ years}$$

(b) Female: We organize our computations of x_i, $\sum f_i$, $\sum x_i f_i$, and $\sum x_i^2 f_i$ in the following table.

Class	Midpoint, x_i	Frequency, f_i	$x_i f_i$	x_i^2	$x_i^2 f_i$
0–9	5	19,992	99,960	25	499,800
10–19	15	20,278	304,170	225	4,562,550
20–29	25	20,482	512,050	625	12,801,250
30–39	35	20,042	701,470	1,225	24,551,450
40–49	45	22,346	1,005,570	2,025	45,250,650
50–59	55	20,302	1,116,610	3,025	61,413,550
60–69	65	13,709	891,085	4,225	57,920,525
70–79	75	8,837	662,775	5,625	49,708,125
80–89	85	9,154	778,090	7,225	66,137,650
≥ 90	95	1,263	119,985	9,025	11,398,575
		$\sum f_i = 156,405$	$\sum x_i f_i = 6,191,765$		$\sum x_i^2 f_i = 334,244,125$

With the table complete, we compute the population mean and population standard deviation:

$$\mu = \frac{\sum x_i f_i}{\sum f_i} = \frac{6,191,765}{156,405} \approx 39.6 \text{ years}$$

$$\sigma = \sqrt{\sigma^2} = \sqrt{\frac{\sum x_i^2 f_i - \frac{\left(\sum x_i f_i\right)^2}{\sum f_i}}{\sum f_i}} = \sqrt{\frac{334,244,125 - \frac{(6,191,765)^2}{156,405}}{156,405}} \approx 23.9 \text{ years}$$

(c) Females have a higher mean age.

(d) Females also have more dispersion in age, which is indicated by the larger standard deviation.

15.

Class	f_i	CF
0 – 999	30	30
1,000 – 1,999	97	127
2,000 – 2,999	935	1,062
3,000 – 3,999	2,698	3,760
4,000 – 4,999	344	4,104
5,000 – 5,999	5	4,109

The distribution contains $n = 4109$ data values. The position of the median is
$$\frac{n+1}{2} = \frac{4109+1}{2} = 2055,$$ which is in the fourth class, $3,000 - 3,999$. Then,

$$M = L + \frac{\frac{n}{2} - CF}{f} \cdot i$$

$$= 3,000 + \frac{\frac{4109}{2} - 1,062}{2,698}(4,000 - 3,000)$$

$$\approx 3,367.9 \text{ grams}$$

17.

Class	f_i	CF
61 – 64	31	31
65 – 67	67	98
68 – 69	198	296
70	195	491
71 – 72	120	611
73 – 76	89	700
77 – 80	50	750

The distribution contains $n = 750$ data values. The position of the median is
$$\frac{n+1}{2} = \frac{750+1}{2} = 375.5,$$ which is in the fourth class, 70. Then,

$$M = L + \frac{\frac{n}{2} - CF}{f} \cdot i$$

$$= 70 + \frac{\frac{750}{2} - 296}{195}(71 - 70)$$

$$= 70.4°F$$

19. From the table in Problem 1, the highest frequency is 2,698. So, the modal class is $3,000 - 3,999$ grams.

Section 3.4

1. z-score

3. quartiles

5. 34-week gestation:
$$z = \frac{x - \mu}{\sigma} = \frac{2,400 - 2,600}{660} \approx -0.30$$

40-week gestation:
$$z = \frac{x - \mu}{\sigma} = \frac{3,300 - 3,500}{470} \approx -0.43$$

The weight of the 34-week gestation baby is 0.30 standard deviations below the mean, while the weight of the 40-week gestation baby is 0.43 standard deviations below the mean. Thus, the 40-week gestation baby weighs less relative to the gestation period.

7. 75-inch man: $z = \dfrac{x - \mu}{\sigma} = \dfrac{75 - 69.6}{3.0} = 1.8$

70-inch woman: $z = \dfrac{x - \mu}{\sigma} = \dfrac{70 - 64.1}{3.8} \approx 1.55$

The height of the 75-inch man is 1.8 standard deviations above the mean, while the height of a 70-inch woman is 1.55 standard deviations above the mean. Thus, the 75-inch man is relatively taller than the 70-inch woman.

9. Josh Johnson: $z = \dfrac{x - \mu}{\sigma} = \dfrac{2.30 - 3.622}{0.743} \approx -1.78$

Felix Hernandez:
$$z = \frac{x - \mu}{\sigma} = \frac{2.27 - 3.929}{0.775} \approx -2.14$$

Felix Hernandez's ERA was 2.14 standard deviations below the mean for his league, while Josh Johnson's was 1.78 standard deviations below the mean for his league. Since lower ERA's are better, Hernandez's had the better year relative to his peers.

11. Indianapolis 500 (Dario Franchitti):
$$z = \frac{x - \mu}{\sigma} = \frac{185.62 - 186.15}{0.359} \approx -1.48$$
Indy Grand Prix of Sonoma (Will Power):
$$z = \frac{x - \mu}{\sigma} = \frac{112.57 - 112.8}{0.131} \approx -1.76$$
Will power had the more convincing victory, because his finishing time was 1.76 standard deviations below the mean, while Franchitti's victory was 1.48 standard deviations below the mean.

13. $z = \dfrac{x - \mu}{\sigma}$

$\dfrac{x - 200}{26} = 1.5$

$x - 200 = 1.5(26)$

$x - 200 = 39$

$x = 239$

An applicant must make a minimum score of 239 to be accepted into the school.

15. (a) 15% of 3- to 5-month-old males have a head circumference that is 41.0 cm or less, and $(100 - 15)\% = 85\%$ of 3- to 5-month-old males have a head circumference that is greater than 41.0 cm.

(b) 90% of 2-year-old females have a waist circumference that is 52.7 cm or less, and $(100 - 90)\% = 10\%$ of 2-year-old females have a waist circumference that is more than 52.7 cm.

(c) The heights at each percentile decrease (except for the 40-49 age group) as the age increases. This implies that adults males are getting taller.

17. (a) 25% of the states have a violent crime rate that is 255.3 crimes per 100,000 population or less, and $(100 - 25)\% = 75\%$ of the states have a violent crime rate more than 255.3. 50% of the states have a violent crime rate that is 335.5 crimes per 100,000 population or less, while $(100 - 50)\% = 50\%$ of the states have a violent crime rate more than 335.5. 75% of the states have a violent crime rate that is 497.2 crimes per 100,000 population or less, and $(100 - 75)\% = 25\%$ of the states have a violent crime rate more than 497.2.

(b) $IQR = Q_3 - Q_1 = 497.2 - 255.3 = 241.9$ crimes per 100,000 population. This means that the middle 50% of all observations have a range of 241.9 crimes per 100,000 population.

(c) $LF = Q_1 - 1.5(IQR)$
$= 255.3 - 1.5(241.9) = -107.55$
$UF = Q_3 + 1.5(IQR)$
$= 497.2 + 1.5(241.9) = 860.05$
Since 1,459 is above the upper fence, the Washington, D.C. crime rate is an outlier.

(d) Skewed right. The difference between Q_1 and Q_2 (80.2) is quite a bit less than the difference between Q_2 and Q_3 (141.7), and the outlier is in the right tail of the distribution, which implies that the distribution is skewed right.

19. (a) An IQ of 100 corresponds to the 50th percentile. A person with an IQ of 100 has an IQ that is as high or higher than 50 percent of the population.

(b) An IQ of 120 corresponds to roughly to the 90th percentile. A person with an IQ of 120 has an IQ that is as high or higher than 90 percent of the population. (Answers will vary slightly, but they should be near the 90th percentile.)

(c) If an individual has an IQ in the 60th percentile, their score would be 105. A person with an IQ of 105 has an IQ that is as high or higher than 60 percent of the population. (Answers will vary slightly, but they should be near 105.)

21. (a) $z = \dfrac{x - \mu}{\sigma} \approx \dfrac{36.3 - 38.775}{3.416} \approx -0.72$

(b) By hand/TI-83 or 84/StatCrunch:

$Q_1 = \dfrac{36.3 + 37.4}{2} = 36.85$ mpg

$Q_2 = \dfrac{38.3 + 38.4}{2} = 38.35$ mpg

$Q_3 = \dfrac{40.6 + 41.4}{2} = 41.0$ mpg

Note: Results from MINITAB differ:
$Q_1 = 36.575$ mpg, $Q_2 = 38.35$ mpg,
$Q_3 = 41.2$ mpg

(c) By hand/TI-83 or 84/StatCrunch:
$IQR = Q_3 - Q_1 = 41.0 - 36.85 = 4.15$ mpg
Note: Results from MINITAB differ:
$IQR = 41.2 - 36.575 = 4.625$ mpg

(d) By hand/TI-83 or 84/StatCrunch:

$$LF = Q_1 - 1.5(IQR)$$

$$= 36.85 - 1.5(4.15) = 30.625 \text{ mpg}$$

$$UF = Q_3 + 1.5(IQR)$$

$$= 41.0 + 1.5(4.15) = 47.225 \text{ mpg}$$

Yes, 47.5 mpg is an outlier.
Note: Results from MINITAB differ:

$$LF = Q_1 - 1.5(IQR)$$

$$= 36.575 - 1.5(4.625) = 29.6375 \text{ mpg}$$

$$UF = Q_3 + 1.5(IQR)$$

$$= 41.2 + 1.5(4.625) = 48.1375 \text{ mpg}$$

There are no outliers using MINITAB's quartiles.

23. (a) There are $n = 45$ data values, and we put them in ascending order:

-0.18	-0.08	0.00	0.05	0.09
-0.18	-0.08	0.00	0.05	0.09
-0.17	-0.07	0.01	0.05	0.10
-0.15	-0.07	0.01	0.05	0.10
-0.14	-0.07	0.02	0.06	0.14
-0.10	-0.02	0.02	0.06	0.17
-0.10	-0.02	0.03	0.07	0.17
-0.10	-0.02	0.03	0.08	0.25
-0.10	-0.01	0.04	0.08	0.30

The second quartile (median) is the value that lies the 23rd position, which is 0.02. So, $Q_2 = M = 0.02$.

The first quartile is the median of the bottom 23 data values, which is the value that lies in the 12th position.
This value is -0.07, so $Q_1 = -0.07$.

The third quartile is the median of the top 23 data values, which is the value that lies in the 34th position.
This value is 0.07, so $Q_3 = 0.07$.

Note: Results from MINITAB differ:
$Q_1 = -0.075$, $Q_2 = 0.02$, and $Q_3 = 0.075$.

Interpretation: Using the by-hand results, 25% of the monthly returns are less than or equal to the first quartile, –0.07, and about 75% of the monthly returns are greater than –0.07; 50% of the monthly returns are less than or equal to the second quartile, 0.02, and about 50% of the monthly returns are greater than 0.02; about 75% of the monthly returns are less

than or equal to the third quartile, 0.07, and about 25% of the monthly returns are greater than 0.07.

(b) $IQR = Q_3 - Q_1 = 0.07 - (-0.07) = 0.14$

$$LF = Q_1 - 1.5(IQR)$$

$$= -0.07 - 1.5(0.14) = -0.28$$

$$UF = Q_3 + 1.5(IQR)$$

$$= 0.07 + 1.5(0.14) = 0.28$$

The return 0.3 is an outlier because it is greater than the upper fence.

Note: If using MINITAB, the result will be:
$IQR = Q_3 - Q_1 = 0.075 - (-0.075) = 0.15$

$$LF = Q_1 - 1.5(IQR)$$

$$= -0.075 - 1.5(0.15) = -0.3$$

$$UF = Q_3 + 1.5(IQR)$$

$$= 0.075 + 1.5(0.15) = 0.3$$

There are no outliers using MINITAB's quartiles.

25. To find the upper fence, we must find the third quartile and the interquartile range. There are $n = 20$ data values, and we put them in ascending order:

345	429	461	471	505
346	437	466	480	515
358	442	466	489	516
372	442	470	490	549

The first quartile is the median of the bottom 10 data values, which is the mean of the data values that lie in the 5th and 6th positions. These values are 429 and 437, so $Q_1 = \dfrac{429 + 437}{2} = 433$ min.

The third quartile is the median of the top 10 data values, which is the mean of the data values that lie in the 15th and 16st positions. These values are 489 and 490, so

$$Q_3 = \frac{489 + 490}{2} = 489.5 \text{ min.}$$

$IQR = 489.5 - 433 = 56.5$ min.

$$UF = Q_3 + 1.5(IQR)$$

$$= 489.5 + 1.5(56.5) = 574.25 \text{ min.}$$

The customer is contacted if more than 574 minutes are used.

Note: Results from MINITAB differ:
$Q_1 = 431$ minutes and $Q_3 = 489.8$ minutes
$IQR = 489.8 - 431 = 58.8$ minutes
$UF = 489.8 + 1.5(58.8) = 578$ min.

Using MINITAB, the customer is contacted if more than 578 minutes are used.

27. (a) To find outliers, we must find the first and third quartiles and the interquartile range. There are $n = 50$ data values, and we put them in ascending order:

0	0	188	347	547
0	0	203	367	567
0	67	244	375	579
0	82	262	389	628
0	83	281	403	635
0	95	289	454	650
0	100	300	476	671
0	149	310	479	719
0	159	316	521	736
0	181	331	527	12,777

The first quartile is the median of the bottom 25 data values, which is the value that lies in the 13th position. So, $Q_1 = \$67$.

The third quartile is the median of the top 25 data values, which is value that lies in the 38th position. So, $Q_3 = \$479$.

$IQR = 479 - 67 = \$412$
$LF = Q_1 - 1.5(IQR)$
$\quad = 67 - 1.5(412) = -\$551$
$UF = Q_3 + 1.5(IQR)$
$\quad = 479 + 1.5(412) = \$1,097$

Note: Results from MINITAB differ:
$Q_1 = \$50$ and $Q_3 = \$490$
$IQR = 490 - 50 = \$440$
$LF = 50 - 1.5(440) = -\$610$
$UF = 490 + 1.5(440) = \$1,150$

So, the only outlier is \$12,777 because it is greater than the upper fence.

(b) To create the histogram, we choose the lower class limit of the first class to be 0 and the class width to be 100. The resulting classes and frequencies follow:

Class	Freq.	Class	Freq.
0 – 99	16	500 – 599	5
100 – 199	5	600 – 699	4
200 – 299	5	700 – 799	2
300 – 3999	8	⋮	
400 – 499	4	12,700 – 12,799	1

Students Weekly Income

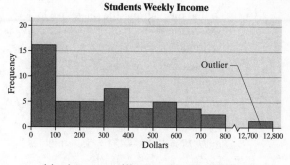

(c) Answers will vary. One possibility is that a student may have provided his or her annual income instead of his or her weekly income.

29. From Problem 23 in Section 3.1 and Problem 25 in Section 3.2, we have $\mu \approx 72.2$ beats per minute and $\sigma \approx 7.7$ beats per minute.

Student	Pulse, x_i	z-score, z_i
P. Bernpah	76	$\dfrac{76 - 72.2}{7.7} = 0.49$
M. Brooks	60	$\dfrac{60 - 72.2}{7.7} = -1.58$
J. Honeycutt	60	$\dfrac{60 - 72.2}{7.7} = -1.58$
C. Jefferson	81	$\dfrac{81 - 72.2}{7.7} = 1.14$
C. Kurtenbach	72	$\dfrac{72 - 72.2}{7.7} = -0.03$
J.Laotka	80	$\dfrac{80 - 72.2}{7.7} = 1.01$
K. McCarthy	80	$\dfrac{80 - 72.2}{7.7} = 1.01$
T. Ohm	68	$\dfrac{68 - 72.2}{7.7} = -0.55$
K. Wojdyla	73	$\dfrac{73 - 72.2}{7.7} = 0.10$

The mean of the z-scores is 0.001 and the population standard deviation is 0.994. These are off slightly from the true mean of 0 and the true standard deviation of 1 because of rounding.

31. (a) To find the standard deviation, we use the computational formula:

$$\sum x_i = 9{,}049 \; ; \; \sum x_i^2 = 4{,}158{,}129 \; ; \; n = 20$$

$$s = \sqrt{\dfrac{\sum x_i^2 - \dfrac{\left(\sum x_i\right)^2}{n}}{n-1}} = \sqrt{\dfrac{4{,}158{,}129 - \dfrac{(9{,}049)^2}{20}}{20-1}}$$

$$\approx 58.0 \text{ minutes}$$

To find the interquartile range, we look back to the solution of Problem 25. There we found $Q_1 = 433$ minutes, $Q_3 = 489.5$ minutes, and $IQR = 489.5 - 433 = 56.5$ minutes.

Note: Results from MINITAB differ: $Q_1 = 431$ minutes, $Q_3 = 489.8$ minutes, and $IQR = 489.8 - 431 = 58.8$ minutes.

(b) Again, to find the standard deviation, we use the computational formula:

$$\sum x_i = 8{,}703 \; ; \; \sum x_i^2 = 4{,}038{,}413 \; ; \; n = 20$$

$$s = \sqrt{\dfrac{\sum x_i^2 - \dfrac{\left(\sum x_i\right)^2}{n}}{n-1}} = \sqrt{\dfrac{4{,}038{,}413 - \dfrac{(8{,}703)^2}{20}}{20-1}}$$

$$\approx 115.0 \text{ minutes}$$

To find the interquartile range, we first put the $n = 20$ data values in ascending order:

0	429	461	471	505
345	437	466	480	515
358	442	466	489	516
372	442	470	490	549

The first quartile is the median of the bottom 10 data values, which is the mean of the data values that lie in the 5th and 6th positions. These values are 429 and 437, so

$$Q_1 = \frac{429 + 437}{2} = 433 \text{ minutes.}$$

The third quartile is the median of the top 10 data values, which is the mean of the data values that lie in the 15th and 16st positions. These values are 489 and 490,

so $Q_3 = \dfrac{489 + 490}{2} = 489.5$ minutes.

So, $IQR = 489.5 - 433 = 56.5$ minutes.

Note: Results from MINITAB differ: $Q_1 = 431$ minutes and $Q_3 = 489.8$ minutes $IQR = 489.8 - 431 = 58.8$ minutes.

Changing the value 346 to 0 causes the standard deviation to nearly double in size, while the interquartile range does not change at all. This illustrates the property of resistance. The standard deviation is not resistant, but the interquartile range is resistant.

Section 3.5

1. The five-number summary consists of the minimum value in the data set, the first quartile, the median, the third quartile, and the maximum value in the data set.

3. (a) The median is to the left of the center of the box and the right line is substantially longer than the left line, so the distribution is skewed right.

(b) Reading the boxplot, the five-number summary is: 0, 1, 3, 6, 16.

5. (a) For the variable x: $M = 40$

(b) For the variable y: $Q_3 = 52$

(c) The variable y has more dispersion. This can be seen by the much broader range (span of the lines) and the much broader interquartile range (span of the box).

(d) The distribution of the variable x is symmetric. This can be seen because the median is near the center of the box and the horizontal lines are approximately the same in length.

(e) The distribution of the variable y is skewed right. This can be seen because the median is to the left of the center of the box and the right line is substantially longer than the left line.

7.

Statistics Exams Scores

Percent

9. (a) Notice that the $n = 44$ data values are already arranged in order (moving down the columns). The smallest value (youngest president) in the data set is 42. The largest value (oldest president) in the data set is 69.

The second quartile (median) is data value that lies in between the 22nd and 23nd positions. These values are 54 and 55, so

$$Q_2 = M = \frac{54+55}{2} = 54.5 .$$

The first quartile is the median of the bottom 22 data values, which is the data value that lies between the 11th and 12th positions. So, $Q_1 = \frac{50+51}{2} = 50.5$.

The third quartile is the median of the top 22 data values, which is the data value that lies in the 33rd and 34th positions. So,

$$Q_3 = \frac{57+58}{2} = 57.5 .$$

So, the five-number summary is:
42, 50.5, 54.5, 57.5, 69

(b) $IQR = 57.5 - 50.5 = 7$
$LF = Q_1 - 1.5(IQR) = 50.5 - 1.5(7) = 40$
$UF = Q_3 + 1.5(IQR) = 57.5 + 1.5(7) = 68$
Thus, 69 is an outlier.

Age of Presidents at Inauguration

(c) The median is near the center of the box and the horizontal lines are approximately the same in length, so the distribution is symmetric, with an outlier.

11. (a) We arrange the *n* = 30 data values into ascending order.

16	20	22	25	29
17	20	23	25	30
18	21	23	25	32
19	21	24	25	33
19	21	24	26	33
20	22	25	29	35

The smallest data value is 16. The largest data value is 35.

The second quartile (median) is mean of the data values that lie in the 15th and 16th positions. So, $Q_2 = M = 23.5$.

The first quartile is the median of the bottom 15 data values, which is the mean of the data values that lie in the $\frac{15+1}{2} = 8$th position, which is 20. So, $Q_1 = 20$.

The third quartile is the median of the top 15 data values, which is the data value that lies in the 23rd position, which is 26. So, $Q_3 = 26$.

So, the five-number summary is:
16, 20, 23.5, 26, 35

$IQR = 26 - 20 = 6$
$LF = Q_1 - 1.5(IQR)$
$ = 20 - 1.5(6) = 11$
$UF = Q_3 + 1.5(IQR)$
$ = 26 + 1.5(6) = 35$
Thus, there are no outliers.

(b) The right line is a little longer than the left line. The distribution is slightly skewed right.

13. To find the five-number summary, we arrange the *n* = 50 data values in ascending order:

0.79	0.84	0.87	0.88	0.91
0.81	0.84	0.87	0.88	0.91
0.82	0.85	0.87	0.89	0.91
0.82	0.85	0.87	0.89	0.91
0.83	0.86	0.87	0.89	0.92
0.83	0.86	0.88	0.90	0.93
0.84	0.86	0.88	0.90	0.93
0.84	0.86	0.88	0.90	0.93
0.84	0.86	0.88	0.90	0.94
0.84	0.86	0.88	0.91	0.95

The smallest data value is 0.79 grams, and the largest data value is 0.95 grams.

The second quartile (median) is the mean of the data values that lie in the 25th and 26th positions, which are 0.87 and 0.88. So,

$$Q_2 = M = \frac{0.87 + 0.88}{2} = 0.875 \text{ grams}.$$

The first quartile is the median of the bottom 25 data values, which is the value that lies in the 13th position. So, $Q_1 = 0.85$ grams.

The third quartile is the median of the top 25 data values, which is value that lies in the 38th position. So, $Q_3 = 0.90$ grams.

So, the five-number summary is:
0.79, 0.85, 0.875, 0.90, 0.95
$IQR = 0.90 - 0.85 = 0.05$ grams
$LF = Q_1 - 1.5(IQR)$
  $= 0.85 - 1.5(0.05) = 0.775$ grams
$UF = Q_3 + 1.5(IQR)$
  $= 0.90 + 1.5(0.05) = 0.975$ grams

So, there are no outliers.

Note: Results from MINITAB differ:
0.79, 0.8475, 0.875, 0.90, 0.95
$IQR = 0.0525$, $LF = 0.76875$, $UF = 0.97875$

Using the by-hand computations for the five-number summary, the boxplot follows:

Weight of M&Ms

Since the range of the data between the minimum value and the median is roughly the same as the range between the median and the maximum value, and because the range of the data between the first quartile and median is the same as the range of the data between the median and third quartile, the distribution is symmetric.

15. (a) To find the five-number summary for each vitamin type, we arrange each data set in ascending order:

Centrum

2.15	2.57	2.80	3.12	3.85
2.15	2.60	2.95	3.25	3.92
2.23	2.63	3.02	3.30	4.00
2.25	2.67	3.02	3.35	4.02
2.30	2.73	3.03	3.53	4.17
2.38	2.73	3.07	3.63	4.33

Generic Brand

4.97	5.55	6.17	6.50	7.17
5.03	5.57	6.23	6.50	7.18
5.25	5.77	6.30	6.57	7.25
5.35	5.78	6.33	6.60	7.42
5.38	5.92	6.35	6.73	7.42
5.50	5.98	6.47	7.13	7.58

For Centrum, the smallest value is 2.15, and the largest value is 4.33. For the generic brand, the smallest value is 4.97, and the largest value is 7.58.

Since both sets of data contain $n = 30$ data points, the quartiles are in the same positions for both sets.

The second quartile (median) is the mean of the values that lie in the 15th and 16th positions. For Centrum, these values are both 3.02. So, $Q_2 = M = \frac{3.02 + 3.02}{2} = 3.02$.

For the generic brand, these values are 6.30 and 6.33, so $Q_2 = M = \frac{6.30 + 6.33}{2} = 6.315$.

The first quartile is the median of the bottom 15 data values. This is the value that lies in the 8th position. So, for Centrum, $Q_1 = 2.60$. For the generic brand, $Q_1 = 5.57$.

The third quartile is the median of the top 15 data values. This is the value that lies in the 23rd position. So, for Centrum, $Q_3 = 3.53$. For the generic brand, $Q_3 = 6.73$.

So, the five-number summaries are:
Centrum: 2.15, 2.60, 3.02, 3.53, 4.33
Generic: 4.97, 5.57, 6.315, 6.73, 7.58

The fences for Centrum are:

$$LF = Q_1 - 1.5(IQR)$$
$$= 2.60 - 1.5(3.53 - 2.60) = 1.205$$

$$UF = Q_3 + 1.5(IQR)$$
$$= 3.53 + 1.5(3.53 - 2.60) = 4.925$$

The fences for the generic brand are:

$$LF = Q_1 - 1.5(IQR)$$
$$= 5.57 - 1.5(6.73 - 5.57) = 3.83$$

$$UF = Q_3 + 1.5(IQR)$$
$$= 6.73 + 1.5(6.73 - 5.57) = 8.47$$

So, neither data set has any outliers.

Note: Results from MINITAB differ:
Centrum: 2.15, 2.593, 3.02, 3.555, 4.33
  $LF = 1.15$ and $UF = 4.998$
Generic: 4.97, 5.565, 6.315, 6.83, 7.58
  $LF = 3.6675$ and $UF = 8.7275$

Using the by-hand computations for the five-number summaries, the side-by-side boxplots follow:

Dissolving Time of Vitamins

(b) From the boxplots, we can see that the generic brand has both a larger range and a larger interquartile range. Therefore, the generic brand has more dispersion.

(c) From the boxplots, we can see that the Centrum vitamins dissolve in less time than the generic vitamins. That is, Centrum vitamins dissolve faster.

17. (a) This is an observational study. The researchers did not attempt to manipulate the outcomes, but merely observed them.

(b) The explanatory variable is whether or not the father smoked. The response variable is the birth weight.

(c) Answers will vary. Some possible lurking variables are eating habits, exercise habits, and whether the mother received prenatal care.

(d) This means that the researchers attempted to adjust their results for any variable that may also be related to birth weight.

(e) Nonsmokers:
Adding up the $n = 30$ data values, we obtain: $\sum x_i = 109,965$. So,

$$\bar{x} = \frac{\sum x_i}{n} = \frac{109,965}{30} = 3,665.5 \text{ grams}$$

To find the standard deviation, we use the computational formula. To do so, we square each data value and add them to obtain:

$$\sum x_i^2 = 406,751,613.$$

So,

$$s = \sqrt{s^2} = \sqrt{\frac{\sum x_i^2 - \frac{(\sum x_i)^2}{n}}{n-1}}$$

$$= \sqrt{\frac{406,751,613 - \frac{(109,965)^2}{30}}{30-1}}$$

$$\approx 356.0 \text{ grams}$$

To find the median and quartiles, we arrange the data in order:

Nonsmokers

2976	3423	3544	3771	4019
3062	3436	3544	3783	4054
3128	3454	3668	3823	4067
3263	3471	3719	3884	4194
3290	3518	3732	3976	4248
3302	3522	3746	3994	4354

The median (second quartile) is the mean of the data values that lie in the 15th and 16th positions, which are 3668 and 3719. So,

$$M = Q_2 = \frac{3668 + 3719}{2} = 3693.5 \text{ grams.}$$

The first quartile is the median of the bottom 15 data values, which is the value in the 8th position. So, $Q_1 = 3436$ grams.

The third quartile is the median of the top 15 data values, which is the value in the 23rd position. So, $Q_3 = 3976$ grams.

Note: Results from MINITAB for the quartiles differ: $Q_1 = 3432.8$ grams and $Q_3 = 3980.5$ grams.

Smokers:
Adding up the $n = 30$ data values, we obtain: $\sum x_i = 103,812$. So,

$$\bar{x} = \frac{\sum x_i}{n} = \frac{103,812}{30} = 3,460.4 \text{ grams}$$

To find the standard deviation, we use the computational formula. To do so, we square each data value and add them to obtain: $\sum x_i^2 = 365,172,886$.

So,

$$s = \sqrt{s^2} = \sqrt{\frac{\sum x_i^2 - \frac{(\sum x_i)^2}{n}}{n-1}}$$

$$= \sqrt{\frac{365,172,886 - \frac{(103,812)^2}{30}}{30-1}}$$

$$\approx 452.6 \text{ grams}$$

To find the median and quartiles, we arrange the data in order:

Smokers

2746	3066	3282	3548	3963
2768	3129	3455	3629	3998
2851	3145	3457	3686	4104
2860	3150	3493	3769	4131
2918	3234	3502	3807	4216
2986	3255	3509	3892	4263

The median (second quartile) is the mean of the data values that lie in the 15th and 16th positions, which are 3457 and 3493. So,

$$M = Q_2 = \frac{3457 + 3129}{2} = 3475 \text{ grams.}$$

The first quartile is the median of the bottom 15 data values, which is the value in the 8th position. So, $Q_1 = 3129$ grams.

The third quartile is the median of the top 15 data values, which is the value in the 23rd position. So, $Q_3 = 3807$ grams.

Note: Results from MINITAB for the quartiles differ: $Q_1 = 3113.3$ grams and $Q_3 = 3828.3$ grams.

(f) For nonsmoking fathers, 25% of infants have a birth weight that is 3,436 grams or less (or 3,432.8 grams, if using MINITAB). Furthermore, 75% of infants have a birth weight that is more than 3,436 grams.

For smoking fathers, 25% of infants have a birth weight that is 3,129 grams or less. Furthermore, 75% of infants have a birth weight that is more than 3,129 grams(or 3,113.3 grams, if using MINITAB).

(g) Using the by-hand results from part (e), the five number summaries are:
Nonsmokers: 2976, 3436, 3693.5, 3976, 4354
Smokers: 2746, 3129, 3475, 3807, 4263

The side-by-side boxplots follow:

The side-by-side boxplots confirm the results of the study.

19. **Using the boxplot:** If the median is left of center in the box, and the right whisker is longer than the left whisker, the distribution is skewed right. If the median is in the center of the box, and the left and right whiskers are roughly the same length, the distribution is symmetric. If the median is right of center in the box, and the left whisker is longer than the right whisker, the distribution is skewed left.
Using the quartiles: If the distance from the median to the first quartile is less than the distance from the median to the third quartile, or the distance from the median to the minimum value in the data set is less than the distance from the median to the maximum value in the data set, then the distribution is skewed right. If the distance from the median to the first quartile is the same as the distance from the median to the third quartile, or the distance from the median to the minimum is the same as the distance from the median to the maximum, the distribution is symmetric. If the distance from the median to the first quartile is more than the distance from the median to the third quartile, or the distance from the median to the minimum value in the data set is more than the distance from the median to the maximum value in the data set, the distribution is skewed left.

Chapter 3 Review Exercises

1. (a) $\sum x_i = 793.8 + 793.1 + 792.4 + 794.0 + 791.4$
$\qquad + 792.4 + 791.7 + 792.3 + 789.6 + 794.4$
$\qquad = 7,925.1$

Mean $= \bar{x} = \dfrac{\sum x_i}{n} = \dfrac{7,925.1}{10} = 792.51$ m/sec.

Data in order:
789.6, 791.4, 791.7, 792.3, 792.4,
792.4, 793.1, 793.8, 794.0, 794.4

Median $= \dfrac{792.4 + 792.4}{2} = 792.4$ m/sec.

(b) Range = Largest Value − Smallest Value
$\qquad = 794.4 - 789.6 = 4.8$ m/sec

To calculate the sample variance and the sample standard deviation, we use the computational formulas:

x_i	x_i^2
793.8	630,118.44
793.1	629,007.61
792.4	627,897.76
794.0	630,436
791.4	626,313.96
792.4	627,897.76
791.7	626,788.89
792.3	627,739.29
789.6	623,468.16
794.4	631,071.36
$\sum x_i = 7925.1$	$\sum x_i^2 = 6,280,739.23$

$$s^2 = \frac{\sum x_i^2 - \dfrac{(\sum x_i)^2}{n}}{n-1}$$

$$= \frac{6,280,739.23 - \dfrac{(7,925.1)^2}{10}}{10-1}$$

$$\approx 2.03 \text{ (m/sec)}^2$$

$$s = \sqrt{\frac{6,280,739.23 - \dfrac{(7,925.1)^2}{10}}{10-1}}$$

$$\approx 1.42 \text{ m/sec}$$

2. (a) Add up the 9 data values: $\sum x = 91,610$

$$\bar{x} = \frac{\sum x}{n} = \frac{91,610}{9} \approx \$10,178.9$$

Data in order:
5500, 7200, 7889, 8998, 9980, 10995,
12999, 13999, 14050
The median is in the 5th position, so
$M = \$9,980$.

(b) Range = Largest Value − Smallest Value
$\qquad = 14,050 - 5,500 = \$8,550$

To find the interquartile range, we must find the first and third quartiles. The first quartile is the median of the bottom four data values, which is the mean of the values in the 2nd and 3rd positions. So,

$$Q_1 = \frac{7,200 + 7,889}{2} = 7,544.5.$$

The third quartile is the median of the top four data values, which is the mean of the values in the 7th and 8th positions. So,

$$Q_3 = \frac{12,999 + 13,999}{2} = 13,499.$$

Finally, the interquartile range is:
IQR = $13,499 - 7,544.5 = \$5,954.5$.

Note: Results from MINITAB differ:
$Q_1 = \$7,545$ and $Q_3 = \$13,499$, so
$IQR = \$5,954$.

To calculate the sample standard deviation, we use the computational formulas:

Data, x_i	x_i^2
14,050	197,402,500
13,999	195,972,001
12,999	168,974,001
10,995	120,890,025
9,980	99,600,400
8,998	80,964,004
7,889	62,236,321
7,200	51,840,000
5,550	30,250,000
$\sum x_i = 91,610$	$\sum x_i^2 = 1,008,129,252$

$$s = \sqrt{\frac{\sum x_i^2 - \dfrac{(\sum x_i)^2}{n}}{n-1}}$$

$$= \sqrt{\frac{1,008,129,252 - \dfrac{(91,610)^2}{9}}{9-1}} \approx \$3,074.9$$

(c) Add up the 9 data values: $\sum x = 118,610$

$$\bar{x} = \frac{\sum x}{n} = \frac{118,610}{9} \approx \$13,178.9$$

Data in order:
5500, 7200, 7889, 8998, 9980, 10995, 12999, 13999, 41050

The median is in the 5th position, so
$M = \$9,980$.

Range $= 41,050 - 5,500 = \$35,550$.

The first quartile is the mean of the values in the 2nd and 3rd positions. So,

$$Q_1 = \frac{7,200 + 7,889}{2} = 7,544.5.$$

The third quartile is the mean of the values in the 7th and 8th positions. So,

$$Q_3 = \frac{12,999 + 13,999}{2} = 13,499.$$

Finally, the interquartile range is:
IQR $= 13,499 - 7,544.5 = \$5,954.5$
(or $\$5,954$ if using MINITAB)

To calculate the sample standard deviation, we again use the computational formulas:

Data, x_i	x_i^2
41,050	1,685,102,500
13,999	195,972,001
12,999	168,974,001
10,995	120,890,025
9,980	99,600,400
8,998	80,964,004
7,889	62,236,321
7,200	51,840,000
5,550	30,250,000
$\sum x_i = 118,610$	$\sum x_i^2 = 2,495,829,252$

$$s = \sqrt{\frac{\sum x_i^2 - \frac{\left(\sum x_i\right)^2}{n}}{n-1}}$$

$$= \sqrt{\frac{2,495,829,252 - \frac{(118,610)^2}{9}}{9-1}}$$

$$\approx \$10,797.5$$

The mean, range, and standard deviation are all changed considerably by the incorrectly entered data value. The median and interquartile range did not change. The median and interquartile range are resistant, while the mean, range and standard deviation are not resistant.

3. (a) $\mu = \dfrac{\sum x}{N} = \dfrac{983}{17} \approx 57.8$ years

Data in order:
44, 46, 50, 51, 55, 56, 56, 56, 58, 59, 62, 62, 62, 64, 65, 68, 69

The median is the data value in the

$\dfrac{17+1}{2} = 9$th position. So, $M = 58$ years.

The data is bimodal: 56 years and 62 years. Both have frequencies of 3.

(b) Range $= 69 - 44 = 25$ years
To calculate the population standard deviation, we use the computational formula:

Data, x_i	x_i^2
44	1936
56	3136
51	2601
46	2116
59	3481
56	3136
58	3364
55	3025
65	4225
64	4096
68	4624
69	4761
56	3136
62	3844
62	3844
62	3844
50	2500
$\sum x_i = 983$	$\sum x_i^2 = 57,669$

$$\sigma = \sqrt{\frac{\sum x_i^2 - \frac{\left(\sum x_i\right)^2}{N}}{N}} = \sqrt{\frac{57,669 - \frac{(983)^2}{17}}{17}}$$

$$\approx 6.98 \text{ years}$$

(c) Answers will vary.

4. (a) To construct the histogram, we first organize the data into a frequency table:

Tickets Issued	Frequency
0	18
1	11
2	1

Number of Tickets Issued in a Month

The distribution is skewed right.

(b) Since the distribution is skewed right, we would expect the mean to be greater than the median.

(c) To find the mean, we add all of the data values and divide by the sample size.

$$\sum x_i = 13 \, ; \; \overline{x} = \frac{\sum x_i}{n} = \frac{13}{30} = 0.4$$

To find the median, we arrange the data in order. The median is the mean of the mean of the 15th and 16th data values.

$$M = \frac{0+0}{2} = 0 \, .$$

(d) The mode is 0 (the most frequent value).

5. (a) By the Empirical Rule, approximately 99.7% of the data will be within 3 standard deviations of the mean. Now, $600 - 3(53) = 441$ and $600 + 3(53) = 759$. Thus, about 99.7% of light bulbs have lifetimes between 441 and 759 hours.

(b) Since 494 is exactly 2 standard deviations below the mean [$494 = 600 - 2(53)$] and 706 is exactly 2 standard deviations above the mean [$706 = 600 + 2(53)$], the Empirical Rule predicts that about 95% of the light bulbs will have lifetimes between 494 and 706 hours.

(c) Since 547 is exactly 1 standard deviations below the mean [$547 = 600 - 1(53)$] and 706 is exactly 2 standard deviations above the mean [$706 = 600 + 2(53)$], the Empirical Rule predicts that about $34 + 47.5 = 81.5\%$ of the light bulbs will have lifetimes between 547 and 706 hours.

(d) Since 441 hours is 3 standard deviations below the mean [$441 = 600 - 3(53)$], the Empirical Rule predicts that 0.15% of light bulbs will last less than 441 hours. Thus, the company should expect to replace about 0.15% of the light bulbs.

(e) By Chebyshev's theorem, at least

$$\left(1 - \frac{1}{k^2}\right) \cdot 100\% = \left(1 - \frac{1}{2.5^2}\right) \cdot 100\% = 84\%$$

of all the light bulbs are within $k = 2.5$ standard deviations of the mean.

(f) Since 494 is exactly $k = 2$ standard deviations below the mean [$494 = 600 - 2(53)$] and 706 is exactly 2 standard deviations above the mean [$706 = 600 + 2(53)$], Chebyshev's inequality indicates that at least

$$\left(1 - \frac{1}{k^2}\right) \cdot 100\% = \left(1 - \frac{1}{2^2}\right) \cdot 100\% = 75\%$$

of the light bulbs will have lifetimes between 494 and 706 hours.

6. (a) To find the mean, we use the formula $\bar{x} = \dfrac{\sum x_i f_i}{\sum f_i}$. To find the standard deviation in part (b), we will

choose to use the computational formula $s = \sqrt{s^2} = \sqrt{\dfrac{\sum x_i^2 f_i - \dfrac{\left(\sum x_i f_i\right)^2}{\sum f_i}}{\left(\sum f_i\right)-1}}$. We organize our

computations of x_i, $\sum f_i$, $\sum x_i f_i$, and $\sum x_i^2 f_i$ in the table that follows:

Class	Midpoint, x_i	Frequency, f_i	$x_i f_i$	x_i^2	$x_i^2 f_i$
$0-9$	$\dfrac{0+10}{2}=5$	125	625	25	3,125
$10-19$	$\dfrac{10+20}{2}=15$	271	4,065	225	60,975
$20-29$	25	186	4,650	625	116.250
$30-39$	35	121	4,235	1,225	148,225
$40-49$	45	54	2,430	2,025	109,350
$50-59$	55	62	3,410	3,025	187,550
$60-69$	65	43	2,795	4,225	181,675
$70-79$	75	20	1,500	5,625	112,500
$80-89$	85	13	1,105	7,225	93,925
		$\sum f_i = 895$	$\sum x_i f_i = 24{,}815$		$\sum x_i^2 f_i = 1{,}013{,}575$

With the table complete, we compute the sample mean:

$$\bar{x} = \frac{\sum x_i f_i}{\sum f_i} = \frac{24{,}815}{895} = 27.7 \text{ minutes}$$

(b) $s = \sqrt{s^2} = \sqrt{\dfrac{\sum x_i^2 f_i - \dfrac{\left(\sum x_i f_i\right)^2}{\sum f_i}}{\left(\sum f_i\right)-1}} = \sqrt{\dfrac{1{,}013{,}575 - \dfrac{(24{,}815)^2}{895}}{895-1}} \approx 19.1 \text{ minutes}$

---

7. GPA $= \bar{x}_w = \dfrac{\sum w_i x_i}{\sum w_i}$

$= \dfrac{5(4)+4(3)+3(4)+3(2)}{5+4+3+3} = \dfrac{50}{15} \approx 3.33$

8. Female: $z = \dfrac{x-\mu}{\sigma} = \dfrac{160-156.5}{51.2} \approx 0.07$

Male: $z = \dfrac{x-\mu}{\sigma} = \dfrac{185-183.4}{40.0} = 0.04$

The weight of the 160-pound female is 0.07 standard deviations above the mean, while the weight of the 185-pound male is 0.04 standard deviations above the mean. Thus, the 160-pound female is relatively heavier.

9. (a) The two-seam fastball.

(b) The two-seam fastball.

(c) The four-seam fastball.

(d) There is an outlier at approximately 88 mph.

(e) Symmetric

(f) Skewed right

10. (a) Add up the 54 data values: $\sum x_i = 132,090$

$$\mu = \frac{\sum x_i}{n} = \frac{127,611}{54} \approx 2,358.8 \text{ words}$$

To find the median and quartiles, we must arrange the data in order:

135	1,340	1,883	2,449	3,838
559	1,355	2,015	2,463	3,967
698	1,425	2,073	2,480	4,059
985	1,437	2,130	2,546	4,388
996	1,507	2,158	2,821	4,467
1,087	1,526	2,170	2,906	4,776
1,125	1,571	2,217	2,978	5,433
1,128	1,668	2,242	3,217	8,445
1,172	1,681	2,283	3,318	
1,175	1,729	2,308	3,319	
1,209	1,802	2,406	3,634	
1,337	1,807	2,446	3,801	

The median is the mean of the data values that lie in the 28th and 29th positions, which are 2,130 and 2,158. So,

$$M = \frac{2,130 + 2,158}{2} = 2,144 \text{ words.}$$

(b) The second quartile is the median, $Q_2 = M = 2,144$ words.

The first quartile is the median of the bottom 28 data values, which is the mean of the data values that lie in the 14th and 15th positions, which are 2,130 and 2,158. So,

$$Q_1 = \frac{1,355 + 1,425}{2} = 1,390 \text{ words.}$$

The third quartile is the median of the top 28 data values, which is the mean of the data values that lie in the 42nd and 43rd positions, which are 2,906 and 2,978. So,

$$Q_3 = \frac{2,906 + 2,978}{2} = 2,942 \text{ words.}$$

Note: Results from MINITAB differ: $Q_1 = 1,373$, $Q_2 = 2,144$, and $Q_3 = 2,960$.

We use the by-hand quartiles for the interpretations: 25% of the inaugural addresses had 1,390 words or less, 75% of the inaugural addresses had more than 1,390 words; 50% of the inaugural addresses had 2,144 words or less, 50% of the inaugural addresses had more than 2,144 words; 75% of the inaugural addresses had 2,942 words or less, 25% of the inaugural addresses had more than 2,942 words.

(c) The smallest value is 135 and the largest is 8,445. Combining these with the quartiles, we obtain the five-number summary:
135, 1,390, 2,144, 2,942, 8,445

Note: results from MINITAB differ:
135, 1,373, 2,144, 2,960, 8,445

(d) To calculate the sample standard deviation, we use the computational formulas:

We add up the 56 data values:
$\sum x_i = 132,090$.

We square each of the 56 data values and add up the results: $\sum x_i^2 = 420,258,932$

$$\sigma = \sqrt{\frac{\sum x_i^2 - \frac{(\sum x_i)^2}{N}}{N}}$$

$$= \sqrt{\frac{420,258,932 - \frac{(132,090)^2}{56}}{56}}$$

$$\approx 1,393.2 \text{ words}$$

The interquartile range is:
$$IQR = Q_3 - Q_1$$
$$= 2,942 - 1,390 = 1,552 \text{ words}$$

Note: results from MINITAB differ:
$IQR = 2,960 - 1,373 = 1,587$ words

(e) $LF = Q_1 - 1.5(IQR)$
$= 1,390 - 1.5(1,552) = -938$ words
$UF = Q_3 + 1.5(IQR)$
$= 2,942 + 1.5(1,552) = 5270$ words

So, there are two outliers: 5,433 and 8,445.

Note: results from MINITAB differ:
$LF = 1,373 - 1.5(1,587) = -1007.5$ words
$UF = 2,960 + 1.5(1,587) = 5,340.5$ words
Using MINITAB, there are two outliers: 5,433 and 8,445.

(f) We use the by-hand results to construct the boxplot.

Words in Inaugural Address

(g) The distribution is skewed right. The distance between the first and second quartile is 754. The distance between the second and third quartile is slightly larger at 798. We can tell because the median is slightly left of center in the box and the right whisker is longer than the left whisker (even without considering the outlier.)

(h) The median is the better measure because the distribution is skewed right and the outliers inflate the value of the mean.

(i) The interquartile range is the better measure because the outliers inflate the value of the standard deviation.

11. This means that 85% of 19-year-old females have a height that is 67.1 inches or less, and $(100 - 85)\% = 15\%$ of 19-year-old females have a height that is more than 67.1 inches.

12. The median is used for three measures since it is likely the case that one of the three measures is extreme relative to the other two, thus substantially affecting the value of the mean. Since the median is resistant to extreme values, it is the better measure of central tendency.

Chapter 3 Test

1. (a) $\sum x_i = 48 + 88 + 57 + 109 + 111 + 93 + 71 + 63$
$= 640$

Mean = $\bar{x} = \dfrac{\sum x_i}{n} = \dfrac{640}{8} = 80 \,\text{min}$

(b) Data in order:
48, 57, 63, 71, 88, 93, 109, 111
The median is the mean of the values in the 4th and 5th positions:

Median = $\dfrac{71 + 88}{2} = 79.5 \,\text{min}$

(c) $\sum x_i = 48 + 88 + 57 + 1009 + 111 + 93 + 71 + 63$
$= 1540$

Mean = $\bar{x} = \dfrac{\sum x_i}{n} = \dfrac{1540}{8} = 192.5 \,\text{min}$

Data in order:
48, 57, 63, 71, 88, 93, 111, 1009
The median is the mean of the values in the 4th and 5th positions:

Median = $\dfrac{71 + 88}{2} = 79.5 \,\text{min}$

The mean was changed substantially by the incorrectly entered data value. The median did not change. The median is resistant, while the mean is not resistant.

2. The mode type of larceny is "From motor vehicles." It has the highest frequency.

3. Range = Largest Value – Smallest Value
$= 111 - 48 = 63 \,\text{minutes}$

4. (a) To calculate the sample standard deviation, we use the computational formula:

x_i	x_i^2
48	2,304
88	7,744
57	3,249
109	11,881
111	12,321
93	8,649
71	5,041
63	3,969
$\sum x_i = 640$	$\sum x_i^2 = 55,158$

$$s = \sqrt{\dfrac{\sum x_i^2 - \dfrac{\left(\sum x_i\right)^2}{n}}{n-1}}$$

$$= \sqrt{\dfrac{55{,}158 - \dfrac{(640)^2}{8}}{8-1}} \approx 23.8 \,\text{min}$$

(b) To find the interquartile range, we must find the first and third quartiles. The first quartile is the median of the bottom four data values, which is the mean of the values in the 2nd and 3rd positions. So,

$$Q_1 = \frac{57+63}{2} = 60 \text{ min.}$$

The third quartile is the median of the top four data values, which is the mean of the values in the 6th and 7th positions. So,

$$Q_3 = \frac{93+109}{2} = 101 \text{ min.}$$

Finally, the interquartile range is:
IQR = 101 − 60 = 41 min.
Interpretation: The middle 50% of all the times students spent on the assignment have a range of 41 inches.

(c) The interquartile range is resistant; the standard deviation is not resistant.

5. (a) By the Empirical Rule, approximately 99.7% of the data will be within 3 standard deviations of the mean. Now, 4302 − 3(340) = 3282 and 4302 + 3(340) = 5322. So, about 99.7% of toner cartridges will print between 3282 and 5322 pages.

(b) Since 3622 is =2 standard deviations below the mean [3622 = 4302 − 2(340)] and 4982 is 2 standard deviations above the mean [4982 = 4302 + 2(340)], the Empirical Rule predicts that about 95% of the toner cartridges will print between 3622 and 4982 pages.

(c) Since 3622 is 2 standard deviations below the mean [3622 = 4302 − 2(340)], the Empirical Rule predicts that 0.15 + 2.35 = 2.5% of the toner cartridges will last less than 3622 pages. So, the company can expect to replace about 2.5% of the toner cartridges.

(d) By Chebyshev's theorem, at least

$$\left(1-\frac{1}{k^2}\right)\cdot 100\% = \left(1-\frac{1}{1.5^2}\right)\cdot 100\% \approx 55.6\%$$

of all the toner cartridges are within $k = 1.5$ standard deviations of the mean.

(e) Since 3282 is $k = 3$ standard deviations below the mean [3282 = 4302 − 3(340)] and 5322 is 3 standard deviations above the mean [5322 = 4302 + 3(340)], so by Chebyshev's inequality at least

$$\left(1-\frac{1}{k^2}\right)\cdot 100\% = \left(1-\frac{1}{3^2}\right)\cdot 100\% \approx 88.9\%$$

of the toner cartridges will print between 3282 and 5322 pages.

---

6. (a) To find the mean, we use the formula $\bar{x} = \frac{\sum x_i f_i}{\sum f_i}$. To find the standard deviation in part (b), we will use

the computational formula $s = \sqrt{s^2} = \sqrt{\dfrac{\sum x_i^2 f_i - \dfrac{\left(\sum x_i f_i\right)^2}{\sum f_i}}{\left(\sum f_i\right)-1}}$. We organize our computations of x_i, $\sum f_i$,

$\sum x_i f_i$, and $\sum x_i^2 f_i$ in the table that follows:

Class	Midpoint, x_i	Frequency, f_i	$x_i f_i$	x_i^2	$x_i^2 f_i$
40 − 49	$\frac{40+50}{2} = 45$	8	360	2,025	16,200
50 − 59	$\frac{50+60}{2} = 55$	44	2,420	3,025	133,100
60 − 69	65	23	1,495	4,225	97,175
70 − 79	75	6	450	5,625	33,750
80 − 89	85	107	9,095	7,225	773,075
90 − 99	95	11	1,045	9,025	99,275
100 − 109	105	1	105	11,025	11,025
		$\sum f_i = 200$	$\sum x_i f_i = 14,970$		$\sum x_i^2 f_i = 1,163,600$

With the table complete, we compute the sample mean: $\bar{x} = \dfrac{\sum x_i f_i}{\sum f_i} = \dfrac{14{,}970}{200} \approx 74.9$ minutes

(b) $s = \sqrt{s^2} = \sqrt{\dfrac{\sum x_i^2 f_i - \dfrac{\left(\sum x_i f_i\right)^2}{\sum f_i}}{\left(\sum f_i\right) - 1}} = \sqrt{\dfrac{1{,}163{,}600 - \dfrac{(14{,}970)^2}{200}}{200 - 1}} \approx 14.7$ minutes

---

7. Cost per pound $= \bar{x}_w = \dfrac{\sum w_i x_i}{\sum w_i x_i}$

$= \dfrac{2(\$2.70) + 1(\$1.30) + \frac{1}{2}(\$1.80)}{2 + 1 + \frac{1}{2}}$

$\approx \$2.17/\text{lb}$

8. (a) Material A: $\sum x_i = 64.04$ and $n = 10$, so

$\bar{x}_A = \dfrac{64.04}{10} = 6.404$ million cycles .

Material B: $\sum x_i = 113.32$ and $n = 10$,

so $\bar{x}_B = \dfrac{113.32}{10} = 11.332$ million cycles .

(b) Notice that each set of $n = 10$ data values is already arranged in order. For each set, the median is the mean of the values in the 5th and 6th positions.

$M_A = \dfrac{5.69 + 5.88}{2} = 5.785$ million cycles

$M_B = \dfrac{8.20 + 9.65}{2} = 8.925$ million cycles

(c) To find the sample standard deviation, we choose to use the computational formula:
Material A:

$\sum x_i = 64.04$ and $\sum x_i^2 = 472.177$, so

$s_A = \sqrt{\dfrac{\sum x_i^2 \sum x_i^2 - \dfrac{(\sum x_i)^2}{n}}{n - 1}}$

$= \sqrt{\dfrac{472.177 - \dfrac{(64.04)^2}{10}}{10 - 1}}$

≈ 2.626 million cycles

Material B:

$\sum x_i = 113.32$ and $\sum x_i^2 = 1{,}597.4002$, so

$s_B = \sqrt{\dfrac{1597.4002 - \dfrac{(113.32)^2}{10}}{10 - 1}}$

≈ 5.900 million cycles

Material B has more dispersed failure times because it has a much larger standard deviation.

(d) For each set, the first quartile is the data value in the 3rd position and the third quartile is the data value in the 8th position.

Material A:
3.17; 4.52; 5.785; 8.01; 11.92

Material B:
5.78; 6.84; 8.925; 14.71; 24.37

(e) Before drawing the side-by-side boxplots, we check each data set for outliers.

Fences for Material A:
$LF = Q_1 - 1.5(IQR)$
$= 4.52 - 1.5(8.01 - 4.52)$
$= -0.715$ million cycles
$UF = Q_3 + 1.5(IQR)$
$= 8.01 + 1.5(8.01 - 4.52)$
$= 13.245$ million cycles
Material A has no outliers.

Fences for Material B:
$LF = 6.84 - 1.5(14.71 - 6.84)$
$= -4.965$ million cycles
$UF = 14.71 + 1.5(14.71 - 6.84)$
$= 26.515$ million cycles
Material B has no outliers

Bearing Failures

Material A

Material B

Failure Time (million cycles)

Annotated remarks will vary. Possible remarks. Material A generally has lower failure times. Material B has more dispersed failure times.

(f) In both boxplots, the median is to the left of the center of the box and the right line is substantially longer than the left line, so both distributions are skewed right.

9. Notice that the data set is already arranged in order. The second quartile (median) is the mean of the values in the 25th and 26th positions, which are both 5.60. So,

$$Q_2 = \frac{5.60 + 5.60}{2} = 5.60 \text{ grams.}$$

The first quartile is the median of the bottom 25 data values, which is the value in the 13th position. So, $Q_1 = 5.58$ grams.

The third quartile is the median of the top 25 data values, which is the value in the 38th position. So, $Q_3 = 5.66$ grams.

$$LF = Q_1 - 1.5(IQR)$$
$$= 5.58 - 1.5(5.66 - 5.58) = 5.46 \text{ grams}$$
$$UF = Q_3 + 1.5(IQR)$$
$$= 5.66 + 1.5(5.66 - 5.58) = 5.78 \text{ grams}$$

So, the quarter that weighs 5.84 grams is an outlier.

Note: Results from MINITAB differ:
$Q_1 = 5.58$, $Q_2 = 5.60$, $Q_3 = 5.6625$,
$LF = 5.45625$, and $UF = 5.78625$.

For the interpretations, we use the by-hand quartiles: 25% of quarters have a weight that is 5.58 grams or less, 75% of the quarters have a weight more than 5.58 grams; 50% of the quarters have a weight that is 5.60 grams or less, 50% of the quarters have a weight more than 5.60 grams; 75% of the quarters have a weight that is 5.66 grams or less, 25% of the quarters have a weight that is more than 5.66 grams.

10. SAT: $z = \dfrac{x - \mu}{\sigma} = \dfrac{610 - 515}{114} \approx 0.83$

ACT: $z = \dfrac{x - \mu}{\sigma} = \dfrac{27 - 21.0}{5.1} \approx 1.18$

Armando's SAT score is 0.83 standard deviation above the mean, his ACT score is 1.18 standard deviations above the mean. So, Armando should report his ACT score since it is more standard deviations above the mean.

11. This means that 15% of 10-year-old males have a height that is 53.5 inches or less, and $(100 - 15)\% = 85\%$ of 10-year-old males have a height that is more than 53.5 inches.

12. You should report the median. Income data will be skewed right which means the median will be less than the mean.

13. **(a)** Report the mean since the distribution is symmetric.

(b) Histogram I has more dispersion. The range of classes is larger.

Case Study: Who Was "A Mourner"?

1. The table that follows gives the length of each word, line by line in the passage. A listing is also provided of the proper names, numbers, abbreviation, and titles that have been omitted from the data set.

Word length:
Line 1:   3, 7, 8, 3, 7, 3, 3, 6, 2, 3, 3, 2, 3, 4
Line 2:   3, 8, 2, 3, 7, 4, 2, 10, 6, 3
Line 3:   4, 9, 3, 7, 4, 2, 4, 2, 6, 4, 3, 7, 5, 2
Line 4:   8, 2, 4, 4, 3, 3, 7, 2
Line 5:   7, 3, 4, 2, 11, 2, 6, 5, 4, 8, 2, 3, 7
Line 6:   2, 4, 6, 4, 3, 5, 6, 2, 3, 5, 5, 5, 5
Line 7:   6, 5, 4, 8, 8, 2, 3, 8, 7, 2, 3
Line 8:   6, 3, 6, 2, 3, 9, 3, 6, 4, 3, 3, 7
Line 9:   3, 5, 2, 9, 3, 8, 8, 2, 6, 4, 3, 4, 5
Line 10: 2, 3, 3, 4, 2, 7, 5, 6, 8, 4, 3, 7, 6
Line 11: 6, 5, 2, 3, 6, 12
Line 12: 5, 6, 2, 5, 2, 3, 1, 7, 6, 3, 5, 4, 4, 1
Line 13: 6, 3

Omission list:
Line 2:   Richardson, 22d
Line 4:   Frogg Lane, Liberty-Tree, Monday
Line 7:   appear'd
Line 11: *Wolfe's Summit of human Glory*

2. Mean = 4.5; Median = 4; Mode = 3; Standard deviation ≈ 2.21; Sample variance ≈ 4.90; Range = 11; Minimum = 1; Maximum = 12; Sum = 649; Count = 143

   Answers will vary. None of the provided authors match both the measures of central tendency and the measures of dispersion well. In other words, there is no clear cut choice for the author based on the information provided. Based on measures of central tendency, James Otis or Samuel Adams would appear to be the more likely candidates for A MOURNER. Based on measures of dispersion, Tom Sturdy seems the more likely choice. Still, the unknown author's mean word length differs considerably from that of Sturdy, and the unknown author's standard deviation differs considerably from those of Otis and Adams.

3. Comparing the two Adams summaries, both the measures of center and the measures of variability differ considerably for the two documents. For example, the means differ by 0.08 and the standard deviations differ by 0.19, not to mention the differences in word counts and the maximum length. This calls into question the viability of word-length analysis as a tool for resolving disputed documents. Word-length may be a part of the analysis needed to determine unknown authors, but other variables should also be taken into consideration.

4. Other information that would be useful to identify A MOURNER would be the style of the rhetoric, vocabulary choices, use of particular phrases, and the overall flow of the writing. In other words, identifying an unknown author requires qualitative analysis in addition to quantitative analysis.

Chapter 4

Describing the Relation between Two Variables

Section 4.1

1. Univariate data measures the value of a single variable for each individual in the study. Bivariate data measures values of two variables for each individual.

3. Scatter diagram

5. -1

7. Lurking

9. Nonlinear

11. Linear; positive

13. (a) III (b) IV (c) II (d) I

15. (a) Looking at the scatter diagram, the points tend to increase at a relatively consistent rate from left to right. So, there appears to be a positive, linear association between level of education and median income.

   (b) The point with roughly the coordinates (48, 56000) appears to stick out. Reasons may vary. One possibility:
   A high concentration of government jobs that require a bachelor's degree but pay less than the private sector in a region with a high cost of living.

   (c) This illustrates that the correlation coefficient is not resistant.

17. (a)

   (b) We compute the mean and standard deviation for both variables: $\bar{x} = 5$, $s_x = 2$, $\bar{y} = 11$, and $s_y = 6$. We determine $\dfrac{x_i - \bar{x}}{s_x}$ and $\dfrac{y_i - \bar{y}}{s_y}$ in columns 3 and 4. We multiply these entries to determine to obtain the entries in column 5.

x_i	y_i	$\dfrac{x_i - \bar{x}}{s_x}$	$\dfrac{y_i - \bar{y}}{s_y}$	$\left(\dfrac{x_i - \bar{x}}{s_x}\right)\left(\dfrac{y_i - \bar{y}}{s_y}\right)$
2	4	−1.5	−1.16667	1.75
4	8	−0.5	−0.5	0.25
6	10	0.5	−0.16667	−0.08334
6	13	0.5	0.33333	0.16667
7	20	1	1.5	1.5

We add the entries in column 5 to obtain

$$\sum\left(\frac{x_i - \bar{x}}{s_x}\right)\left(\frac{y_i - \bar{y}}{s_y}\right) = 3.58333 .$$

Finally, we use this result to compute r:

$$r = \frac{\sum\left(\dfrac{x_i - \bar{x}}{s_x}\right)\left(\dfrac{y_i - \bar{y}}{s_y}\right)}{n-1} = \frac{3.58333}{5-1} \approx 0.896$$

(c) A linear relation exists between x and y.

19. (a)

   (b) We compute the mean and standard deviation for both variables: $\bar{x} = 6$, $s_x = 2.54951$, $\bar{y} = 7$, and $s_y = 1.58114$.

   We determine $\dfrac{x_i - \bar{x}}{s_x}$ and $\dfrac{y_i - \bar{y}}{s_y}$ in columns 3 and 4. We multiply these entries to determine to obtain the entries in column 5.

x_i	y_i	$\dfrac{x_i - \bar{x}}{s_x}$	$\dfrac{y_i - \bar{y}}{s_y}$	$\left(\dfrac{x_i - \bar{x}}{s_x}\right)\left(\dfrac{y_i - \bar{y}}{s_y}\right)$
2	8	−1.56893	0.63246	−0.99228
6	7	0	0	0
6	6	0	−0.63246	0
7	9	0.39223	1.26491	0.49614
9	5	1.17670	−1.26491	−1.48842

We add the entries in column 5 to obtain

$$\sum\left(\frac{x_i - \bar{x}}{s_x}\right)\left(\frac{y_i - \bar{y}}{s_y}\right) = -1.98456 .$$

Finally, we use this result to compute r:

$$r = \frac{\sum \left(\frac{x_i - \bar{x}}{s_x} \right) \left(\frac{y_i - \bar{y}}{s_y} \right)}{n-1} = \frac{-1.98456}{5-1}$$

$$\approx -0.496$$

(c) No linear relation exists between x and y.

21. (a) Positive correlation. The more infants the more diapers will be needed.

(b) Negative correlation. The lower the interest rates the more people can afford to buy a car.

(c) Negative correlation. More exercise is associated with lower cholesterol.

(d) Negative correlation. The higher the price of a Big Mac, the fewer Big Macs and French fries will be sold.

(e) No correlation. There is no correlation between shoe size and intelligence.

23. The correlation coefficient is high, suggesting that there is a linear relation between student task persistence and achievement score. This means that if a student has high task persistence, they will tend to have a high achievement score. Students with low task persistence will tend to have a low achievement score.

25. (a) Explanatory variable: commute time; Response: well-being score

(b) The relationship between the commute time and the well-being score is illustrated below:

(c) We compute the mean and standard deviation for both variables:
$\bar{x} = 43.85714$, $s_x = 34.98775$,
$\bar{y} = 66.92857$, and $s_y = 1.70950$. We

determine $\dfrac{x_i - \bar{x}}{s_x}$ and $\dfrac{y_i - \bar{y}}{s_y}$ in columns

3 and 4. We multiply these entries to determine to obtain the entries in column 5.

x_i	y_i	$\dfrac{x_i - \bar{x}}{s_x}$	$\dfrac{y_i - \bar{y}}{s_y}$	$\left(\dfrac{x_i - \bar{x}}{s_x}\right)\left(\dfrac{y_i - \bar{y}}{s_y}\right)$
5	69.2	-1.11059	1.32871	-1.47566
15	68.3	-0.82477	0.80224	-0.66167
25	67.5	-0.53896	0.33427	-0.18016
35	67.1	-0.25315	0.10028	-0.02539
50	66.4	0.17557	-0.30920	-0.05429
72	66.1	0.80436	-0.48469	-0.38986
105	63.9	1.74755	-1.77162	-3.09599

We add the entries in column 5 to obtain

$$\sum \left(\frac{x_i - \bar{x}}{s_x} \right) \left(\frac{y_i - \bar{y}}{s_y} \right) = -5.88302 .$$ Finally,

we use this result to compute r:

$$r = \frac{\sum \left(\frac{x_i - \bar{x}}{s_x} \right) \left(\frac{y_i - \bar{y}}{s_y} \right)}{n-1} = \frac{-5.88302}{7-1}$$

$$\approx -0.981$$

(d) Yes, because $|-0.981| = 0.981 > 0.754$ (0.754 is the critical value from Table II), so a negative association exists between the commute time and the well-being index score.

27. (a) Explanatory variable: height; Response variable: head circumference

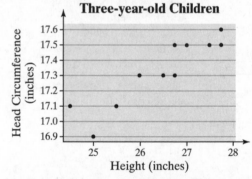

(c) To calculate the correlation coefficient, we use the computational formula:

x_i	y_i	x_i^2	y_i^2	$x_i y_i$
27.75	17.5	770.0625	306.25	485.625
24.5	17.1	600.25	292.41	418.95
25.5	17.1	650.25	292.41	436.05
26	17.3	676	299.29	449.8
25	16.9	625	285.61	422.5
27.75	17.6	770.0625	309.76	488.4
26.5	17.3	702.25	299.29	458.45
27	17.5	729	306.25	472.5
26.75	17.3	715.5625	299.29	462.775
26.75	17.5	715.5625	306.25	468.125
27.5	17.5	756.25	306.25	481.25
291	190.6	7710.25	3303.06	5044.425

From the table, we have $n = 11$, $\sum x_i = 291$,

$\sum y_i = 190.6$, $\sum x_i^2 = 7710.25$,

$\sum y_i^2 = 3303.06$, and $\sum x_i y_i = 5044.425$.

So, the correlation coefficient is

$$r = \frac{\sum x_i y_i - \dfrac{\sum x_i \sum y_i}{n}}{\sqrt{\left(\sum x_i^2 - \dfrac{(\sum x_i)^2}{n}\right)\left(\sum y_i^2 - \dfrac{(\sum y_i)^2}{n}\right)}}$$

$$= \frac{5044.425 - \dfrac{(291)(190.6)}{11}}{\sqrt{\left(7710.25 - \dfrac{(291)^2}{11}\right)\left(3303.06 - \dfrac{(190.6)^2}{11}\right)}}$$

$$\approx 0.911$$

(d) There is a strong positive linear association between the height and head circumference of a child.

29. (a) Explanatory variable: weight;
Response variable: gas mileage

(b)

(c) To calculate the correlation coefficient, we use the computational formula:

x_i	y_i	x_i^2	y_i^2	$x_i y_i$
3,735	17	13,950,225	289	63,495
3,860	16	14,899,600	256	61,760
2,721	25	7,403,841	625	68,025
3,555	19	12,638,025	361	67,545
3,319	21	11,015,761	441	69,699
2,966	23	8,797,156	529	68,218
3,727	17	13,890,529	289	63,359
2,605	24	6,786,025	576	62,520
3,473	19	12,061,729	361	65,987
3,796	17	14,409,616	289	64,532
3,310	18	10,956,100	324	59,580
37,067	216	126,808,607	4,340	714,720

From the table, we have $n = 11$,

$\sum x_i = 37,067$, $\sum y_i = 216$,

$\sum x_i^2 = 126,808,607$, $\sum y_i^2 = 4,340$, and

$\sum x_i y_i = 714,720$. So, the correlation coefficient is

$$r = \frac{\sum x_i y_i - \dfrac{\sum x_i \sum y_i}{n}}{\sqrt{\left(\sum x_i^2 - \dfrac{(\sum x_i)^2}{n}\right)\left(\sum y_i^2 - \dfrac{(\sum y_i)^2}{n}\right)}}$$

$$= \frac{714,720 - \dfrac{(37,067)(216)}{11}}{\sqrt{\left(126,808,607 - \dfrac{(37,067)^2}{11}\right)\left(4,340 - \dfrac{(216)^2}{11}\right)}}$$

$$\approx -0.960$$

(d) Yes, there is a strong negative linear association between the weight of a car and its miles per gallon in the city.

31. (a) The explanatory variable is stock return.

(b)

(c) To calculate the correlation coefficient, we use the computational formula.

x_i	y_i	x_i^2	y_i^2	$x_i y_i$
26.35	5.91	694.3225	34.9281	155.7285
12.48	30.39	155.7504	923.5521	379.2672
19.44	31.72	377.9136	1006.1584	616.6368
13.37	79.76	178.7569	6361.6576	1066.3912
12.21	−8.4	149.0841	70.5600	−102.5640
11.89	2.69	141.3721	7.2361	31.9841
26.21	4.53	686.9641	20.5209	118.7313
14.95	10.8	223.5025	116.6400	161.4600
17.57	4.01	308.7049	16.0801	70.4557
14.36	11.76	206.2096	138.2976	168.8736
168.83	173.17	3122.5807	8695.6309	2666.9644

From the table, we have $n = 10$,

$\sum x_i = 168.83$, $\sum y_i = 173.17$,

$\sum x_i^2 = 3,122.5807$, $\sum y_i^2 = 8,695.631$,

and $\sum x_i y_i = 2,666.964$. So, the correlation coefficient is

$$r = \frac{\sum x_i y_i - \frac{\sum x_i \sum y_i}{n}}{\sqrt{\left(\sum x_i^2 - \frac{(\sum x_i)^2}{n}\right)\left(\sum y_i^2 - \frac{(\sum y_i)^2}{n}\right)}}$$

$$= \frac{2,666.964 - \frac{(168.83)(173.17)}{10}}{\sqrt{\left(3,122.5807 - \frac{(168.83)^2}{10}\right)\left(8,695.631 - \frac{(173.17)^2}{10}\right)}}$$

$$\approx -0.206$$

(d) No, there is not a linear relation between compensation and stock return; Stock performance does not appear to play a role in determining the compensation of a CEO.

33. (a)

Brain Size vs. IQ

There is a moderate positive linear relation between the MRI count and the IQ.

(b) To calculate the correlation coefficient, we use the computational formula:

x_i	y_i	x_i^2	y_i^2	$x_i y_i$
816,932	133	667,377,892,624	17,689	108,651,956
951,545	137	905,437,887,025	18,769	130,361,665
991,305	138	982,685,603,025	19,044	136,800,090
833,868	132	695,335,841,424	17,424	110,070,576
856,472	140	733,544,286,784	19,600	119,906,080
852,244	132	726,319,835,536	17,424	112,496,208
790,619	135	625,078,403,161	18,225	106,733,565
866,662	130	751,103,022,244	16,900	112,666,060
857,782	133	735,789,959,524	17,689	114,085,006
948,066	133	898,829,140,356	17,689	126,092,778
949,395	140	901,350,866,025	19,600	132,915,300
1,001,121	140	1,002,243,256,641	19,600	140,156,940
1,038,437	139	1,078,351,402,969	19,321	144,342,743
965,353	133	931,906,414,609	17,689	128,391,949
955,466	133	912,915,277,156	17,689	127,076,978
1,079,549	141	1,165,426,043,401	19,881	152,216,409
924,059	135	853,885,035,481	18,225	124,747,965
955,003	139	912,030,730,009	19,321	132,745,417
935,494	141	875,149,024,036	19,881	131,904,654
949,589	144	901,719,268,921	20,736	136,740,816
18,518,961	2,728	17,256,479,190,951	372,396	2,529,103,155

From the table, we have $n = 20$,

$\sum x_i = 18,518,961$, $\sum y_i = 2,728$, $\sum x_i^2 = 17,256,479,190,951$, $\sum y_i^2 = 373,396$,

and $\sum x_i y_i = 2,529,103,155$. So, the correlation coefficient is

$$r = \frac{\sum x_i y_i - \frac{\sum x_i \sum y_i}{n}}{\sqrt{\left(\sum x_i^2 - \frac{(\sum x_i)^2}{n}\right)\left(\sum y_i^2 - \frac{(\sum y_i)^2}{n}\right)}}$$

$$= \frac{2,529,103,155 - \frac{(18,518,961)(2,728)}{20}}{\sqrt{\left(17,256,479,190,951 - \frac{(18,518,961)^2}{20}\right)\left(372,396 - \frac{(2,728)^2}{20}\right)}}$$

$$\approx 0.548$$

A weak positive linear relation exists between MRI count and IQ.

(c)

Brain Size vs. IQ

Looking at the scatter diagram, we can see that females tend to have lower MRI counts. When separating the two groups, even the weak linear relation seems to disappear. Neither group presents any clear relation between IQ and MRI counts.

(d) We first use the computational formula on the data from the female participants:

x_i	y_i	x_i^2	y_i^2	$x_i y_i$
816,932	133	667,377,892,624	17,689	108,651,956
951,545	137	905,437,887,025	18,769	130,361,665
991,305	138	982,685,603,025	19,044	136,800,090
833,868	132	695,335,841,424	17,424	110,070,576
856,472	140	733,544,286,784	19,600	119,906,080
852,244	132	726,319,835,536	17,424	112,496,208
790,619	135	625,078,403,161	18,225	106,733,565
866,662	130	751,103,022,244	16,900	112,666,060
857,782	133	735,789,959,524	17,689	114,085,006
948,066	133	898,829,140,356	17,689	126,092,778
8,765,495	1,343	7,721,501,871,703	180,453	1,177,863,984

From the table, we have $n = 10$,

$\sum x_i = 8,765,495$, $\sum y_i = 1,343$,

$\sum x_i^2 = 7,721,501,871,703$,

$\sum y_i^2 = 180,453$, and

$\sum x_i y_i = 1,177,863,984$.

So, the correlation coefficient for the female data is

$$r = \frac{\sum x_i y_i - \dfrac{\sum x_i \sum y_i}{n}}{\sqrt{\left(\sum x_i^2 - \dfrac{(\sum x_i)^2}{n}\right)\left(\sum y_i^2 - \dfrac{(\sum y_i)^2}{n}\right)}}$$

$$= \frac{1,177,863,984 - \dfrac{(8,765,495)(1,343)}{10}}{\sqrt{\left(7,721,501,871,703 - \dfrac{(8,765,495)^2}{10}\right)\left(180,453 - \dfrac{(1,343)^2}{10}\right)}}$$

≈ 0.359

We next use the computational formula on the data from the male participants:

x_i	y_i	x_i^2	y_i^2	$x_i y_i$
949,395	140	901,350,866,025	19,600	132,915,300
1,001,121	140	1,002,243,256,641	19,600	140,156,940
1,038,437	139	1,078,351,402,969	19,321	144,342,743
965,353	133	931,906,414,609	17,689	128,391,949
955,466	133	912,915,277,156	17,689	127,076,978
1,079,549	141	1,165,426,043,401	19,881	152,216,409
924,059	135	853,885,035,481	18,225	124,747,965
955,003	139	912,030,730,009	19,321	132,745,417
935,494	141	875,149,024,036	19,881	131,904,654
949,589	144	901,719,268,921	20,736	136,740,816
9,753,466	1,385	9,534,977,319,248	191,943	1,351,239,171

From the table, we have $n = 10$,

$\sum x_i = 9,753,466$, $\sum y_i = 1,385$,

$\sum x_i^2 = 9,534,977,319,248$,

$\sum y_i^2 = 191,943$, and

$\sum x_i y_i = 1,351,239,171$. So, the correlation coefficient for the male data is

$$r = \frac{\sum x_i y_i - \dfrac{\sum x_i \sum y_i}{n}}{\sqrt{\left(\sum x_i^2 - \dfrac{(\sum x_i)^2}{n}\right)\left(\sum y_i^2 - \dfrac{(\sum y_i)^2}{n}\right)}}$$

$$= \frac{1,351,239,171 - \dfrac{(9,753,466)(1,385)}{10}}{\sqrt{\left(9,534,977,319,248 - \dfrac{(9,753,466)^2}{10}\right)\left(191,943 - \dfrac{(1,385)^2}{10}\right)}}$$

≈ 0.236

There is no linear relation between MRI count and IQ. It is important to beware of lurking variables. Mixing distinct populations can produce misleading results that result in incorrect conclusions.

35. (a)

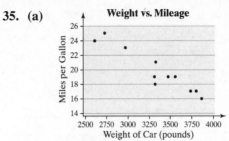

(b) To calculate the correlation coefficient, we use the computational formula:

x_i	y_i	x_i^2	y_i^2	$x_i y_i$
3,735	17	13,950,225	289	63,495
3,860	16	14,899,600	256	61,760
2,721	25	7,403,841	625	68,025
3,555	19	12,638,025	361	67,545
3,319	21	11,015,761	441	69,699
2,966	23	8,797,156	529	68,218
3,727	17	13,890,529	289	63,359
2,605	24	6,786,025	576	62,520
3,473	19	12,061,729	361	65,987
3,796	17	14,409,616	289	64,532
3,310	18	10,956,100	324	59,580
3,305	19	10,923,025	361	62,795
40,372	235	137,731,632	4,701	777,515

From the table, we have $n = 12$, $\sum x_i = 40,372$, $\sum y_i = 235$, $\sum x_i^2 = 137,731,632$, $\sum y_i^2 = 4701$, and $\sum x_i y_i = 777,515$. So, the correlation coefficient is

$$r = \frac{\sum x_i y_i - \dfrac{\sum x_i \sum y_i}{n}}{\sqrt{\left(\sum x_i^2 - \dfrac{(\sum x_i)^2}{n}\right)\left(\sum y_i^2 - \dfrac{(\sum y_i)^2}{n}\right)}}$$

$$= \frac{777,515 - \dfrac{(40,372)(235)}{12}}{\sqrt{\left(137,731,632 - \dfrac{(40,372)^2}{12}\right)\left(4701 - \dfrac{(235)^2}{12}\right)}}$$

≈ -0.954

(c) The scatter diagram in part (a) looks very similar to the one from problem 29, and the correlation coefficient from part (b) is similar to the one computed in problem 29 (-0.960). These results are reasonable because the Taurus follows the overall pattern of the data.

(d)

Weight vs. Mileage

(e) To recalculate the correlation coefficient, we use the computational formula:

x_i	y_i	x_i^2	y_i^2	$x_i y_i$
3,735	17	13,950,225	289	63,495
3,860	16	14,899,600	256	61,760
2,721	25	7,403,841	625	68,025
3,555	19	12,638,025	361	67,545
3,319	21	11,015,761	441	69,699
2,966	23	8,797,156	529	68,218
3,727	17	13,890,529	289	63,359
2,605	24	6,786,025	576	62,520
3,473	19	12,061,729	361	65,987
3,796	17	14,409,616	289	64,532
3,310	18	10,956,100	324	59,580
3,720	41	13,838,400	1,681	152,520
40,787	257	140,647,007	6,021	867,240

From the table, we have $n = 12$, $\sum x_i = 40,787$, $\sum y_i = 257$, $\sum x_i^2 = 140,647,007$, $\sum y_i^2 = 6,021$, and $\sum x_i y_i = 867,240$. So, the correlation coefficient is

$$r = \frac{\sum x_i y_i - \frac{\sum x_i \sum y_i}{n}}{\sqrt{\left(\sum x_i^2 - \frac{(\sum x_i)^2}{n}\right)\left(\sum y_i^2 - \frac{(\sum y_i)^2}{n}\right)}}$$

$$= \frac{867,240 - \frac{(40,787)(257)}{12}}{\sqrt{\left(140,647,007 - \frac{(40,787)^2}{12}\right)\left(6,021 - \frac{(257)^2}{12}\right)}}$$

$$\approx -0.195$$

(f) The Ford Fusion is a hybrid car, so it gets much better gas mileage than the other cars which are not hybrids.

37. (a) Data Set 1:

x_i	y_i	x_i^2	y_i^2	$x_i y_i$
10	8.04	100	64.6416	80.40
8	6.95	64	48.3025	55.60
13	7.58	169	57.4564	98.54
9	8.81	81	77.6161	79.29
11	8.33	121	69.3889	91.63
14	9.96	196	99.2016	139.44
6	7.24	36	52.4176	43.44
4	4.26	16	18.1476	17.04
12	10.84	144	117.5056	130.08
7	4.82	49	23.2324	33.74
5	5.68	25	32.2624	28.40
99	82.51	1001	660.1727	797.60

From the table, we have $n = 11$, $\sum x_i = 99$, $\sum y_i = 82.51$, $\sum x_i^2 = 1001$, $\sum y_i^2 = 660.1727$, and $\sum x_i y_i = 797.60$. So, the correlation coefficient is

$$r = \frac{\sum x_i y_i - \frac{\sum x_i \sum y_i}{n}}{\sqrt{\left(\sum x_i^2 - \frac{(\sum x_i)^2}{n}\right)\left(\sum y_i^2 - \frac{(\sum y_i)^2}{n}\right)}}$$

$$= \frac{797.60 - \frac{(99)(82.51)}{11}}{\sqrt{\left(1001 - \frac{(99)^2}{11}\right)\left(660.1727 - \frac{(82.51)^2}{11}\right)}}$$

$$\approx 0.816$$

Data Set 2:

x_i	y_i	x_i^2	y_i^2	$x_i y_i$
10	9.14	100	83.5396	91.40
8	8.14	64	66.2596	65.12
13	8.74	169	76.3876	113.62
9	8.77	81	76.9129	78.93
11	9.26	121	85.7476	101.86
14	8.10	196	65.6100	113.40
6	6.13	36	37.5769	36.78
4	3.10	16	9.6100	12.40
12	9.13	144	83.3569	109.56
7	7.26	49	52.7076	50.82
5	4.47	25	19.9809	22.35
99	82.24	1001	657.6896	796.24

From the table, we have $n = 11$, $\sum x_i = 99$, $\sum y_i = 82.24$, $\sum x_i^2 = 1001$, $\sum y_i^2 = 657.6896$, and $\sum x_i y_i = 796.24$. So, the correlation coefficient is

$$r = \frac{\sum x_i y_i - \dfrac{\sum x_i \sum y_i}{n}}{\sqrt{\left(\sum x_i^2 - \dfrac{(\sum x_i)^2}{n}\right)\left(\sum y_i^2 - \dfrac{(\sum y_i)^2}{n}\right)}}$$

$$= \frac{796.24 - \dfrac{(99)(82.24)}{11}}{\sqrt{\left(1001 - \dfrac{(99)^2}{11}\right)\left(657.6896 - \dfrac{(82.24)^2}{11}\right)}}$$

$$\approx 0.817$$

Data Set 3:

x_i	y_i	x_i^2	y_i^2	$x_i y_i$
10	7.46	100	55.6516	74.60
8	6.77	64	45.8329	54.16
13	12.74	169	162.3076	165.62
9	7.11	81	50.5521	63.99
11	7.81	121	60.9961	85.91
14	8.84	196	78.1456	123.76
6	6.08	36	36.9664	36.48
4	5.39	16	29.0521	21.56
12	8.15	144	66.4225	97.80
7	6.42	49	41.2164	44.94
5	5.73	25	32.8329	28.65
99	82.50	1001	659.9762	797.47

From the table, we have $n = 11$,
$\sum x_i = 99$, $\sum y_i = 82.50$, $\sum x_i^2 = 1001$,
$\sum y_i^2 = 659.9762$, and $\sum x_i y_i = 797.47$.
So, the correlation coefficient is

$$r = \frac{\sum x_i y_i - \dfrac{\sum x_i \sum y_i}{n}}{\sqrt{\left(\sum x_i^2 - \dfrac{(\sum x_i)^2}{n}\right)\left(\sum y_i^2 - \dfrac{(\sum y_i)^2}{n}\right)}}$$

$$= \frac{797.47 - \dfrac{(99)(82.50)}{11}}{\sqrt{\left(1001 - \dfrac{(99)^2}{11}\right)\left(659.9762 - \dfrac{(82.50)^2}{11}\right)}}$$

$$\approx 0.816$$

Data Set 4:

x_i	y_i	x_i^2	y_i^2	$x_i y_i$
8	6.58	64	43.2964	52.64
8	5.76	64	33.1776	46.08
8	7.71	64	59.4441	61.68
8	8.84	64	78.1456	70.72
8	8.47	64	71.7409	67.76
8	7.04	64	49.5616	56.32
8	5.25	64	27.5625	42.00
8	5.56	64	30.9136	44.48
8	7.91	64	62.5681	63.28
8	6.89	64	47.4721	55.12
19	12.5	361	156.2500	237.50
99	82.51	1001	660.1325	797.58

From the table, we have $n = 11$,
$\sum x_i = 99$, $\sum y_i = 82.51$, $\sum x_i^2 = 1001$,
$\sum y_i^2 = 660.1325$, and $\sum x_i y_i = 797.58$.
So, the correlation coefficient is

$$r = \frac{\sum x_i y_i - \dfrac{\sum x_i \sum y_i}{n}}{\sqrt{\left(\sum x_i^2 - \dfrac{(\sum x_i)^2}{n}\right)\left(\sum y_i^2 - \dfrac{(\sum y_i)^2}{n}\right)}}$$

$$= \frac{797.58 - \dfrac{(99)(82.51)}{11}}{\sqrt{\left(1001 - \dfrac{(99)^2}{11}\right)\left(660.1325 - \dfrac{(82.51)^2}{11}\right)}}$$

$$\approx 0.817$$

(b)

Data Set I:

Data Set II:

Data Set III:

Data Set IV:

Even though the correlation coefficients for the four data sets are roughly equal, the scatter plots are clearly different. Thus, linear correlation coefficients and scatter diagrams must be used together in a statistical analysis of bivariate data to determine whether a linear relations exists.

39. Begin by finding the linear correlation coefficient between each pair of variables. The following correlation matrix summarizes the correlation coefficients.

	Cisco	Disney	GE	Exxon	TECO
Disney	0.551				
GE	0.746	0.511			
Exxon	0.362	0.470	0.702		
TECO	−0.121	0.343	0.148	0.283	
Dell	0.063	0.291	0.145	−0.044	−0.177

These data are illustrated in a matrix of scatterplots. The scatterplots show the relations between each of these variables.

If the goal is to have the lowest correlation between two stocks (i.e., a correlation close to zero), then invest in Dell and Exxon Mobil, since their correlation coefficient is only −0.044.

If the goal is to have one stock go up when the other goes down (i.e., a negative relation), then invest in Dell and TECO Energy since they have the strongest negative relation, −0.271.

41. $r = 0.599$ implies that a positive linear relation exists between the number of television stations and life expectancy. However, this is correlation, not causation. The more television stations a country has, the more affluent it is. The more affluent, the better the healthcare is, which in turn helps increase the life expectancy. So, wealth is a likely lurking variable.

43. No, increasing the percentage of the population that has a cell phone will not decrease the violent crime rate. A likely lurking variable is the economy. In a strong economy, crime rates tend to decrease, and consumers are better able to afford cell phones.

45. **(a)**

To calculate the correlation coefficient, we use the computational formula:

x_i	y_i	x_i^2	y_i^2	$x_i y_i$
2.2	3.9	4.84	15.21	8.58
3.7	4.0	13.69	16.00	14.80
3.9	1.4	15.21	1.96	5.46
4.1	2.8	16.81	7.84	11.48
2.6	1.5	6.76	2.25	3.90
4.1	3.3	16.81	10.89	13.53
2.9	3.6	8.41	12.96	10.44
4.7	4.9	22.09	24.01	23.03
28.2	25.4	104.62	91.12	91.22

From the table, we have $n = 8$, $\sum x_i = 28.2$, $\sum y_i = 25.4$, $\sum x_i^2 = 104.62$, $\sum y_i^2 = 91.12$, and $\sum x_i y_i = 91.22$. So, the correlation coefficient is

$$r = \frac{\sum x_i y_i - \dfrac{\sum x_i \sum y_i}{n}}{\sqrt{\left(\sum x_i^2 - \dfrac{(\sum x_i)^2}{n}\right)\left(\sum y_i^2 - \dfrac{(\sum y_i)^2}{n}\right)}}$$

$$= \frac{91.22 - \dfrac{(28.2)(25.4)}{8}}{\sqrt{\left(104.62 - \dfrac{(28.2)^2}{8}\right)\left(91.12 - \dfrac{(25.4)^2}{8}\right)}}$$

$$\approx 0.228$$

(b)

To calculate the correlation coefficient, we use the computational formula:

x_i	y_i	x_i^2	y_i^2	$x_i y_i$
10.4	9.3	108.16	86.49	96.72
2.2	3.9	4.84	15.21	8.58
3.7	4.0	13.69	16.00	14.80
3.9	1.4	15.21	1.96	5.46
4.1	2.8	16.81	7.84	11.48
2.6	1.5	6.76	2.25	3.90
4.1	3.3	16.81	10.89	13.53
2.9	3.6	8.41	12.96	10.44
4.7	4.9	22.09	24.01	23.03
38.6	34.7	212.78	177.61	187.94

From the table, we have $n=9$, $\sum x_i = 38.6$, $\sum y_i = 34.7$, $\sum x_i^2 = 212.78$, $\sum y_i^2 = 177.61$, and $\sum x_i y_i = 187.94$. So, the correlation coefficient is

$$r = \frac{\sum x_i y_i - \dfrac{\sum x_i \sum y_i}{n}}{\sqrt{\left(\sum x_i^2 - \dfrac{(\sum x_i)^2}{n}\right)\left(\sum y_i^2 - \dfrac{(\sum y_i)^2}{n}\right)}}$$

$$= \frac{187.94 - \dfrac{(38.6)(34.7)}{9}}{\sqrt{\left(212.78 - \dfrac{(38.6)^2}{9}\right)\left(177.61 - \dfrac{(34.7)^2}{9}\right)}}$$

$$\approx 0.860$$

The additional data point increases r from a value that suggests no linear correlation to one that suggests a fairly strong linear correlation. However, the second scatter diagram shows that the new data point is located very far away from the rest of the data, so this new data point has a strong influence on the value of r, even though there is no apparent correlation between the variables. Therefore, correlations should always be reported along with scatter diagrams in order to check for potentially influential observations.

47. Answers may vary. For those who feel "in-class" evaluations are meaningful, a correlation of 0.68 would lend some validity, because it indicates that students respond in the same general way on RateMyProfessors.com (high with high, low with low). The correlations between quality and easiness or hotness tend to indicate that evaluations at RateMyProfessors.com are based more on likability rather than actual quality of teaching.

49. If $r = 0$, then no *linear* relation exists between the variables.

51. Answers will vary. Answers should include that correlation measures the strength of the linear relation between two quantitative variables.

53. Answers will vary. The plot of calories burned versus time spent exercising should show a strong positive relation.

Answers will vary. We might expect that the plot of GPA versus time on Facebook would show a moderate negative relation.

55. No, the negative relation between the variables is a general trend in the data. There may be specific data points that do not fit in that trend.

Section 4.2

1. Residual

3. True; this is a property of the least-squares regression line.

5. (a)

There appears to be a negative linear relation between the *x* and *y*.

(b) $b_1 = r \cdot \dfrac{s_y}{s_x} = -0.9477\left(\dfrac{1.8239}{2.4221}\right) \approx -0.7136$

$b_0 = \bar{y} - b_1\bar{x}$

$= 3.9333 - (-0.7136)(3.6667)$

≈ 6.5496

So, the least-squares regression line is

$\hat{y} = -0.7136x + 6.5496$

(c)

7. (a)

(b) Using the points $(3,4)$ and $(8,14)$:

$m = \dfrac{14-4}{8-3} = 2$

$y - y_1 = m(x - x_1)$

$y - 4 = 2(x - 3)$

$y - 4 = 2x - 6$

$y = 2x - 2$

(c)

(d) We compute the mean and standard deviation for each variable and the correlation coefficient to be $\bar{x} = 5.4$, $\bar{y} = 8.6$, $s_x \approx 2.07364$, $s_y \approx 4.21901$, and $r \approx 0.99443$. Then the slope and intercept for the least-squares regression line are:

$b_1 = r \cdot \dfrac{s_y}{s_x} = 0.99443 \left(\dfrac{4.21901}{2.07364} \right)$

≈ 2.02326

$b_0 = \bar{y} - b_1 \bar{x} = 8.6 - (2.02326)(5.4)$

≈ -2.3256

Rounding to four decimal places, the least-squares regression line is

$\hat{y} = 2.0233x - 2.3256$

(e)

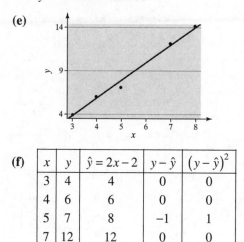

(f)

x	y	$\hat{y} = 2x - 2$	$y - \hat{y}$	$(y - \hat{y})^2$
3	4	4	0	0
4	6	6	0	0
5	7	8	−1	1
7	12	12	0	0
8	14	14	0	0

Total = 1

Sum of squared residuals (computed line): 1.0

(g)

x	y	$\hat{y}$	$y - \hat{y}$	$(y - \hat{y})^2$
3	4	3.7443	0.2557	0.0654
4	6	5.7676	0.2324	0.0540
5	7	7.7909	−0.7909	0.6255
7	12	11.8375	0.1625	0.0264
8	14	13.8608	0.1392	0.0194

Total = 0.7907

Sum of squared residuals (regression line): 0.7907

(h) Answers will vary. The regression line gives a smaller sum of squared residuals, so it has a better fit.

9. (a)

(b) Using the points $(-2, -4)$ and $(2, 5)$:

$m = \dfrac{5 - (-4)}{2 - (-2)} = \dfrac{9}{4}$

$y - y_1 = m(x - x_1)$

$y - 5 = \dfrac{9}{4}(x - 2)$

$y - 5 = \dfrac{9}{4}x - \dfrac{9}{2}$

$y = \dfrac{9}{4}x + \dfrac{1}{2}$ or $y = 2.25x + 0.5$

(c)

(d) We compute the mean and standard deviation for each variable and the correlation coefficient to be $\bar{x} = 0$, $\bar{y} = 1.2$, $s_x \approx 1.58114$, $s_y \approx 3.56371$, and $r \approx 0.97609$. Then the slope and intercept for the least-squares regression line are:

$b_1 = r \cdot \dfrac{s_y}{s_x} = 0.97609 \left(\dfrac{3.56371}{1.58114} \right) \approx 2.20000$

$b_0 = \bar{y} - b_1 \bar{x} = 1.2 + (2.20000)(0) = 1.2$

So, the least-squares regression line is

$\hat{y} = 2.2x + 1.2$

(e)

(f)

x	y	$y = \dfrac{9}{4}x + \dfrac{1}{2}$	$y - \hat{y}$	$(y - \hat{y})^2$
−2	−4	−4.00	0.00	0.0000
−1	0	−1.75	1.75	3.0625
0	1	0.50	0.50	0.2500
1	4	2.75	1.25	1.5625
2	5	5.00	0.00	0.0000

Total = 4.8750

Sum of squared residuals (calculated line):
4.8750

(g)

x	y	$\hat{y} = 2.2x + 1.2$	$y - \hat{y}$	$(y - \hat{y})^2$
−2	−4	−3.2	−0.8	0.64
−1	0	−1.0	1.0	1.00
0	1	1.2	−0.2	0.04
1	4	3.4	0.6	0.36
2	5	5.6	−0.6	0.36

Total = 2.40

Sum of squared residuals (regression line):
2.4

(h) Answers will vary. The regression line gives a smaller sum of squared residuals, so it has a better fit.

11. (a)

(b) Using the points $(30, 95)$ and $(60, 70)$:

$$m = \frac{70 - 95}{60 - 30} = \frac{-25}{30} = -\frac{5}{6}$$

$$y - y_1 = m(x - x_1)$$

$$y - 95 = -\frac{5}{6}(x - 30)$$

$$y - 95 = -\frac{5}{6}x + 25$$

$$y = -\frac{5}{6}x + 120$$

(c)

(d) We compute the mean and standard deviation for each variable and the correlation coefficient to be $\bar{x} = 40$, $\bar{y} = 87.8$, $s_x \approx 15.81139$, $s_y \approx 11.73456$, and $r \approx -0.97014$. Then the slope and intercept for the least-squares regression line are:

$$b_1 = r \cdot \frac{s_y}{s_x} = -0.97014\left(\frac{11.73456}{15.81139}\right)$$

$$\approx -0.72000$$

$$b_0 = \bar{y} - b_1\bar{x} = 87.8 - (-0.72000)(40)$$

$$\approx 116.6$$

So, the least-squares regression line is
$$\hat{y} = -0.72x + 116.6$$

(e)

(f)

x	y	$\hat{y} = -\frac{5}{6}x + 120$	$y - \hat{y}$	$(y - \hat{y})^2$
20	100	103.333	−3.333	11.1111
30	95	95.000	0.000	0.0000
40	91	86.667	4.333	18.7778
50	83	78.333	4.667	21.7778
60	70	70.000	0.000	0.0000

Total = 51.6667

Sum of squared residuals (calculated line):
51.6667.

(g)

x	y	$\hat{y}$	$y - \hat{y}$	$(y - \hat{y})^2$
20	100	102.2	−2.2	4.84
30	95	95.0	0.0	0.00
40	91	87.8	3.2	10.24
50	83	80.6	2.4	5.76
60	70	73.4	−3.4	11.56

Total = 32.40

Sum of squared residuals (regression line):
32.4

(h) Answers will vary. The regression line gives a smaller sum of squared residuals, so it has a better fit.

13. (a) Let $x = 8$ in the regression equation.
$$\hat{y} = -0.0526(8) + 2.9342 \approx 2.51$$
So, we predict that the GPA of a student who plays video games 8 hours per week will be 2.51.

(b) The slope is −0.0526. For each additional hour a student spends playing video games in a week, the GPA will decrease by 0.0526 points, on average.

(c) The *y*-intercept is 2.9342. The mean grade-point average for a student who does not play video games is 2.9342.

(d) Let $x = 7$ in the regression equation.
$$\hat{y} = -0.0526(7) + 2.9342 \approx 2.57$$
So, the average GPA among all students who play video games 7 hours per week is 2.57. Therefore, a student with a grade-point average of 2.68 is above average for those who play video games 7 hours per week.

15. (a) The slope is 3.4722. For each percentage point increase in on-base percentage, the winning percentage will increase by 3.4722 percentage points, on average.

(b) The *y*-intercept is -0.6294. This is outside the scope of the model. An on-base percentage of 0 is not reasonable for this model, since a team cannot win if players do not reach base.

(c) No, it would not be a good idea to use this model to predict the winning percentage of a team whose on-base percentage was 0.250 because 0.250 is outside the scope of the model.

(d) Let $x = 0.322$ in the regression equation.
$$\hat{y} = 3.4722(0.321) - 0.6294 \approx 0.4852$$
So, the residual is
$$y - \hat{y} = 0.568 - 0.4852 = 0.0828$$
This residual indicates that San Diego's winning percentage is above average for teams with an on-base percentage of 0.321.

17. (a) $\hat{y} = -0.0479x + 69.0296$

(b) Slope: For each one-minute increase in commute time, the index score decreases by 0.0479, on average; Y-intercept: The mean index score for a person with a 0-minute commute is 69.0296. A person who works from home would have a 0-minute commute so the y-intercept has a meaningful interpretation.

(c) If the commute time is 30 minutes, the predicted well-being index score would be
$$\hat{y} = -0.0479(30) + 69.0296 \approx 67.6$$

(d) The mean well-being index for a person with a 20 minute commute is
$\hat{y} = -0.0479(20) + 69.0296 \approx 68.1$ So, she is less well off than would be expected.

19. (a) In Problem 27, Section 4.1, we computed the correlation coefficient. Rounded to six decimal places, it is $r = 0.911073$. We compute the mean and standard deviation for each variable to be $\bar{x} \approx 26.454545$, $s_x \approx 1.094407$, $\bar{y} \approx 17.327273$, and $s_y \approx 0.219504$.

$$b_1 = r \cdot \frac{s_y}{s_x} = 0.911073 \left(\frac{0.219504}{1.094407} \right)$$
$$\approx 0.182733$$
$$b_0 = \bar{y} - b_1 \bar{x}$$
$$= 17.327273 - (0.182733)(26.454545)$$
$$\approx 12.4932$$
Rounding to four decimal places, the least-squares regression line is
$$\hat{y} = 0.1827x + 12.4932$$

(b) If height increases by 1 inch, head circumference increases by about 0.1827 inches, on average. It is not appropriate to interpret the y-intercept since a height of 0 is outside the scope of the model. In addition, it makes no sense to consider the head circumference of a child with a height of 0 inches.

(c) Let $x = 25$ in the regression equation.
$$\hat{y} = 0.1827(25) + 12.4932 \approx 17.06$$
We predict the head circumference of a child who is 25 inches tall is 17.06 inches.

(d) $y - \hat{y} = 16.9 - 17.06 = -0.16$ inches
This indicates that the head circumference for this child is below average.

(e)

(f) The head circumferences of children who are 26.75 inches tall naturally vary.

(g) No, a height of 32 inches is well outside the scope of the model.

21. (a) In Problem 29, Section 4.1, we computed the correlation coefficient. Rounded to seven decimal places, it is $r = -0.9596263$. We compute the mean and standard deviation for each variable to be $\bar{x} \approx 3369.727$, $s_x \approx 436.2254$, $\bar{y} \approx 19.63636$, and $s_y \approx 3.139195$.

$$b_1 = r \cdot \frac{s_y}{s_x} = -0.95963 \left(\frac{3.139195}{4362254} \right) \approx -0.00691$$

$$b_0 = \bar{y} - b_1 \bar{x}$$
$$= 19.63636 - (-0.00691)(3369.727)$$
$$\approx 42.90678$$

Rounding to four decimal places, the least-squares regression line is
$$\hat{y} = -0.0069x + 42.9068$$

(b) For every pound added to the weight of the car, gas mileage in the city will decrease by about 0.0069 miles per gallon, on average. It is not appropriate to interpret the y-intercept because it is well beyond the scope of the model.

(c) Let $x = 3600$ in the regression equation.
$$\hat{y} = -0.0069(3600) + 42.9068 \approx 18.0668$$
mpg (or 18.0462 mpg when software is used.) The mileage for the Regal is only slightly below average.

(d) No, it is not reasonable to use this least-squares regression line to predict the miles per gallon of a Toyota Prius. The data given are for domestic cars with traditional internal combustion engines. The Toyota Prius is a foreign-made hybrid car.

23. (a) We compute the correlation coefficient, rounded to seven decimal places, to be $r = -0.8056468$. We compute the mean and standard deviation for each variable to be $\bar{x} = 3.6$, $s_x \approx 2.6403463$, $\bar{y} \approx 0.8756667$, and $s_y \approx 0.0094693$.

$$b_1 = r \cdot \frac{s_y}{s_x} = -0.8056468 \left(\frac{0.0094693}{2.6403463} \right)$$
$$\approx -0.0028893$$

$$b_0 = \bar{y} - b_1 \bar{x}$$
$$= 0.8756667 - (-0.0028893)(3.6)$$
$$\approx 0.8861$$

Rounding to four decimal places, the least-squares regression line is
$$\hat{y} = -0.0029x + 0.8861$$

(b) For each additional cola consumed per week, bone mineral density will decrease by 0.0029 g/cm^2, on average.

(c) For a woman who does not drink cola, the mean bone mineral density will be 0.8861 g/cm^2.

(d) Let $x = 4$ in the regression equation.
$$\hat{y} = -0.0029(4) + 0.8861 = 0.8745$$
We predict the bone mineral density of the femoral neck of a woman who consumes four colas per week is 0.8745 g/cm^2.

(e) Since 0.873 is smaller than the result in part (d), this woman's bone mineral density is below average among women who consume four colas per week.

(f) No. Two cans of soda per day equates to 14 cans of soda per week, which is outside the scope of the model.

25. In Problem 31, Section 4.1, we computed the correlation coefficient: $r \approx -0.206$. Since this value is so close to zero, we conclude that there is not a linear relation between the CEO's compensation and the stock return. For both \$15 million and \$25 million, $\hat{y} = \bar{y} = 17.3\%$.

27. (a) <u>Males:</u> In Problem 30(c), Section 4.1, we computed the correlation coefficient for males. Unrounded, it is $r \approx 0.8833643940$. We compute the mean and standard deviation for each variable for males to be $\bar{x} = 11,009.555...$, $s_x \approx 7358.553596175$, $\bar{y} = 4772.111...$, and $s_y \approx 2855.260997021$.

$$b_1 = r \cdot \frac{s_y}{s_x} \approx 0.342762456$$

$$b_0 = \bar{y} - b_1 \bar{x} \approx 998.4488$$
Rounding to four decimal places, the least-squares regression line for males is
$$\hat{y} = 0.3428x + 998.4488$$

<u>Females:</u> In Problem 30(c), Section 4.1, we computed the correlation coefficient for males. Unrounded, it is $r \approx 0.8361242533$. We compute the mean and standard deviation for each variable for males to be $\bar{x} \approx 10,988.222...$, $s_x \approx 7240.2546532870$, $\bar{y} = 1662.222...$, and $s_y \approx 904.7405398480$.

$b_1 = r \cdot \dfrac{s_y}{s_x} \approx 0.1044818925$

$b_0 = \bar{y} - b_1 \bar{x} \approx 514.1520$

Rounding to four decimal places, the least-squares regression line is

$\hat{y} = 0.1045x + 514.1520$

(b) <u>Males</u>: If the number of licensed drivers increases by 1 (thousand), then the number of fatal crashes increases by 0.3428 (thousand), on average.

<u>Females</u>: If the number of licensed drivers increases by 1 (thousand), then the number of fatal crashes increases by 0.1045 (thousand), on average.

Since females tend to be involved in fewer fatal crashes, an insurance company may use this information to argue for higher rates for male customers.

(c) <u>16 to 20 year old males</u>: Let $x = 6424$ in the regression equation for males:
$\hat{y} = 0.3428(6424) + 998.4488 \approx 3200.6$
The actual number of fatal crashes (5180) is above this result, so it is above average.

<u>21 to 24 year old males</u>: Let $x = 6941$ in the regression equation for males:
$\hat{y} = 0.3428(6941) + 998.4488 \approx 3377.8$
The actual number of fatal crashes (5016) is above this result, so it is above average.

<u>> 74 year old males</u>: Let $x = 4803$ in the regression equation for males:
$\hat{y} = 0.3428(4803) + 998.4488 \approx 2644.9$
The actual number of fatal crashes (2022) is below this result, so it is below average.

An insurance company might use these results to argue for higher rates for younger drivers and lower rates for older drivers.

<u>16 to 20 year old females</u>: Let $x = 6139$ in the regression equation for females:
$\hat{y} = 0.1045(6139) + 514.1520 \approx 1155.7$
The actual number of fatal crashes (2113) is above this result, so it is above average.

<u>21 to 24 year old females</u>: Let $x = 6816$ in the regression equation for females:
$\hat{y} = 0.1045(6816) + 514.1520 \approx 1226.4$
The actual number of fatal crashes (1531) is above this result, so it is above average.

<u>> 74 year old females</u>: Let $x = 5375$ in the regression equation for females:
$\hat{y} = 0.1045(5375) + 514.1520 \approx 1075.8$
The actual number of fatal crashes (980) is below this result, so it is below average.

The same relationship holds for female drivers as for male drivers.

29. (a) The distribution is bell shaped. The class width is 200.

(b) This is an observational study, since the researchers did not impose a treatment but only observed the results.

(c) A prospective study is one in which the subjects are followed forward in time.

(d) Since cotinine leaves the body through fluids, a urinalysis is a less intrusive way to detect cotinine than a blood test would be. In addition, a urinalysis provides a means to verify the statements of the subjects.

(e) The explanatory variable in the study is the number of cigarettes smoked during the third trimester of pregnancy. The response variable is the birth weight.

(f) A negative relation appears to exist between cigarette consumption in the third trimester of pregnancy and birth weight.

(g) From the output, we have
$\hat{y} = -31.014x + 3456.04$

(h) For each additional cigarette smoked, birth weight decreases by 31.014 grams, on average.

(i) The mean birth weight of a baby whose mother smokes but does not smoke in the third trimester of pregnancy is 3456.04 grams, on average.

(j) No, this model should not be used to predict the birth weight of a baby whose mother smoked 10 cigarettes per day during the third trimester, because 10 cigarettes per day equates to roughly 900 cigarettes in the third trimester, which is outside of the scope of the model.

(k) No, an observational study cannot establish causation.

(l) Answers may vary. Some lurking variables could be genetic factors or the general health of the mothers.

31. Values of the explanatory variable that are much larger or much smaller than those observed are considered *outside the scope of the model*. It is dangerous to make such predictions because we do not know the behavior of the data for which we have no observations.

33. Each point's *y*-coordinate represents the mean value of the response variable for a given value of the explanatory variable.

Section 4.3

1. Coefficient of determination

3. Residual plot

5. The variance of the residuals is not constant. The variance of the residuals increases as the value of *x* increases.

7. There is an outlier in the residual plot (a residual of 4).

9. (a) III   **(b)** II

  (c) IV   **(d)** I

11. Yes, the point is influential because the two lines have substantially different slopes and *y*-intercepts.

13. No, the point is not influential because there is very little change in either the slope or the *y*-intercept. The lines are almost identical.

15. (a) From the scatter diagram, a positive linear relation appears to exist between time between eruptions and length of eruption.

  (b) Yes, the residual plot does not show any obvious departures from the requirements of the model.

  (c) 83.0% of the variation in the length of eruption is explained by the least-squares regression equation.

17. (a)

No relation appears to exist between calories and sugar content.

(b) To calculate the correlation coefficient, we use the computational formula.

x_i	y_i	x_i^2	y_i^2	$x_i y_i$
200	18	40,000	324	3600
210	23	44,100	529	4830
170	17	28,900	289	2890
190	20	36,100	400	3800
200	18	40,000	324	3600
180	19	32,400	361	3420
210	23	44,100	529	4830
210	16	44,100	256	3360
210	17	44,100	289	3570
190	12	36,100	144	2280
190	11	36,100	121	2090
200	11	40,000	121	2200
2360	205	466,000	3687	40,470

From the table, we have $n = 12$,

$\sum x_i = 2306$, $\sum y_i = 205$, $\sum x_i^2 = 466,000$,

$\sum y_i^2 = 3687$, and $\sum x_i y_i = 40,470$. So, the correlation coefficient is

$$r = \frac{\sum x_i y_i - \dfrac{\sum x_i \sum y_i}{n}}{\sqrt{\left(\sum x_i^2 - \dfrac{(\sum x_i)^2}{n}\right)\left(\sum y_i^2 - \dfrac{(\sum y_i)^2}{n}\right)}}$$

$$= \frac{40,470 - \dfrac{(2360)(205)}{12}}{\sqrt{\left(466,000 - \dfrac{(2360)^2}{12}\right)\left(3687 - \dfrac{(205)^2}{12}\right)}}$$

$$\approx 0.261$$

No, there is no significant linear relation between calories and sugar content.

(c)

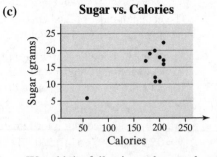

We add the following values to the table in part (a): $x_{13} = 80$, $y_{13} = 6$, $x_{13}^2 = 6400$, $y_{13}^2 = 36$, and $x_{13} y_{13} = 480$. We now have $n = 13$, $\sum x_i = 2440$, $\sum y_i = 211$, $\sum x_i^2 = 472,400$, $\sum y_i^2 = 3723$, and $\sum x_i y_i = 40,950$.

So, the new correlation coefficient is

$$r = \frac{40{,}950 - \dfrac{(2440)(211)}{13}}{\sqrt{\left(472{,}400 - \dfrac{(2440)^2}{13}\right)\left(3723 - \dfrac{(211)^2}{13}\right)}}$$

$$\approx 0.649$$

Yes, according to the new correlation coefficient, there is a linear relation between calories and sugar content. All-Bran is an influential observation. It substantially changed the correlation coefficient.

(d) It is necessary to show a scatter diagram with the correlation coefficient. Influential observations can cause the correlation coefficient to increase substantially, thereby increasing the apparent strength of the linear relation between the two variables.

19. (a) From Problem 25(c) in Section 4.1, we have $r = -0.980501$ (unrounded). So, $R^2 = (-0.980501)^2 \approx 0.961 = 96.1\%$.

(b) From Problem 17(a) in Section 4.2, we have $\hat{y} = -0.0479x + 69.0296$. We compute the residuals and construct the residual plot.

x_i	y_i	$\hat{y} = -0.0479x + 69.0296$	$y - \hat{y}$
5	69.2	68.79	0.41
15	68.3	68.31	−0.01
25	67.5	67.83	−0.33
35	67.1	67.35	−0.25
50	66.4	66.63	−0.23
72	66.1	65.58	0.52
105	63.9	64.00	−0.10

(c) 96.1% of the variation in well-being score is explained by the least-squares regression line. The residual plot does not reveal any problems, so the linear model appears to be appropriate.

21. (a) From Problem 27(c) in Section 4.1, we have $r = 0.911073$ (unrounded). So, $R^2 = (0.911073)^2 \approx 0.830 = 83.0\%$.

(b) From Problem 19(a) in Section 4.2, we have $\hat{y} = 0.1827x + 12.4932$. We compute the residuals and construct the residual plot.

x_i	y_i	$\hat{y} = 0.1827x + 12.4932$	$y - \hat{y}$
27.75	17.5	17.56	−0.06
24.50	17.1	16.97	0.13
25.50	17.1	17.15	−0.05
26.00	17.3	17.24	0.06
25.00	16.9	17.06	−0.16
27.75	17.6	17.56	0.04
26.50	17.3	17.33	−0.03
27.00	17.5	17.43	0.07
26.75	17.3	17.38	−0.08
26.75	17.5	17.38	0.12
27.50	17.5	17.52	−0.02

(c) 83.0% of the variation in head circumference is explained by the least-squares regression line. The residual plot does not reveal any problems, so the linear model appears to be appropriate.

23. (a) From Problem 29(c) in Section 4.1, we have $r = -0.959626$ (unrounded). So, $R^2 = (-0.959626)^2 \approx 0.921 = 92.1\%$.

(b) From Problem 21(a) in Section 4.2, we have $\hat{y} = -0.0069x + 42.9068$. We compute the residuals and construct the residual plot.

x_i	y_i	$\hat{y} = -0.0069x + 42.9068$	$y - \hat{y}$
3735	17	17.11	−0.11
3860	16	16.25	−0.25
2721	25	24.12	0.88
3555	19	18.36	0.64
3319	21	19.99	1.01
2966	23	22.42	0.58
3727	17	17.17	−0.17
2605	24	24.92	−0.92
3473	19	18.92	0.08
3796	17	16.69	0.31
3310	18	20.05	−2.05

Residuals vs. Weight

(c) 92.1% of the variation in gas mileage is explained by the least-squares regression line. The residual plot does not reveal any problems, so the linear model appears to be appropriate.

25. (a)

Planetary Motion

(b) To calculate the correlation coefficient, we use the computational formula.

x_i	y_i	x_i^2	y_i^2	$x_i y_i$
36	0.24	1,296	0.0576	8.64
67	0.62	4,489	0.3844	41.54
93	1.00	8,649	1	93
142	1.88	20,164	3.5344	266.96
483	11.9	233,289	141.61	5,747.7
887	29.5	786,769	870.25	26,166.5
1,785	84	3,186,225	7,056	14,9940
2,797	165	7,823,209	27,225	46,1505
3,675	248	13,505,625	61,504	91,1400
9,965	542.14	25,569,715	96,801.8364	1,555,169.34

From the table, we have $n = 9$,
$\sum x_i = 9,965$, $\sum y_i = 542.14$,
$\sum x_i^2 = 25,569,715$, $\sum y_i^2 = 96,801.8364$,
and $\sum x_i y_i = 1,555,169.34$. So, the correlation coefficient is

$$r = \frac{\sum x_i y_i - \frac{\sum x_i \sum y_i}{n}}{\sqrt{\left(\sum x_i^2 - \frac{(\sum x_i)^2}{n}\right)\left(\sum y_i^2 - \frac{(\sum y_i)^2}{n}\right)}}$$

$$= \frac{1,555,169.346 - \frac{(9965)(542.14)}{9}}{\sqrt{\left(25,569,715 - \frac{(9965)^2}{9}\right)\left(96,801.8364 - \frac{(542.14)^2}{9}\right)}}$$

$$= 0.9888989 \approx 0.989$$

Yes, there appears to be a linear relation between distance and sidereal year.

(c) We compute the mean and standard deviation for each variable to be
$\bar{x} \approx 1107.2222222$, $s_x \approx 1347.972809$,
$\bar{y} \approx 60.237778$, and $s_y \approx 89.543654$.

$$b_1 = r \cdot \frac{s_y}{s_x} = 0.9888989\left(\frac{89.543654}{1347.972809}\right)$$
$$\approx 0.06569096$$
$$b_0 = \bar{y} - b_1\bar{x}$$
$$= 60.237778 - (0.06569095)(1107.222222)$$
$$\approx -12.4967$$

Rounding to four decimal places, the least-squares regression line is
$$\hat{y} = 0.0657x - 12.4967$$

(d) We compute the residuals and construct the residual plot.

x_i	y_i	$\hat{y} = 0.0657x - 12.4967$	$y - \hat{y}$
36	0.24	−10.132	10.372
67	0.62	−8.095	8.715
93	1.00	−6.387	7.387
142	1.88	−3.167	5.047
483	11.9	19.236	−7.336
887	29.5	45.779	−16.279
1,785	84	104.778	−20.778
2,797	165	171.266	−6.266
3,675	248	228.951	19.049

Residuals vs. Distance

(e) No; the U-shaped pattern in the residual plot indicates that a linear model is not appropriate.

27. (a) From Problem 29(c) in Section 4.1, we found $r = -0.959626$ (unrounded). In Problem 23 of Section 4.3, we found $R^2 = 92.1\%$.

Including the Dodge Viper data ($x_{12} = 3425$, $y_{12} = 11$), we obtain $n = 12$, $\sum x_i = 40{,}492$, $\sum y_i = 227$, $\sum x_i^2 = 138{,}539{,}232$, $\sum y_i^2 = 4{,}461$, and $\sum x_i y_i = 752{,}395$. So, the new correlation coefficient is

$$r = \frac{\sum x_i y_i - \dfrac{\sum x_i \sum y_i}{n}}{\sqrt{\left(\sum x_i^2 - \dfrac{(\sum x_i)^2}{n}\right)\left(\sum y_i^2 - \dfrac{(\sum y_i)^2}{n}\right)}}$$

$$= \frac{752{,}395 - \dfrac{(40{,}492)(227)}{12}}{\sqrt{\left(138{,}539{,}232 - \dfrac{(40{,}492)^2}{12}\right)\left(4461 - \dfrac{(227)^2}{12}\right)}}$$

$$\approx -0.76134$$

Therefore, the coefficient of determination with the Dodge Viper included is $R^2 = (-0.76134)^2 \approx 0.580 = 58.0\%$.

Adding the Viper reduces the amount of variability explained by the model by approximately 35%.

(b) With the Viper included, we recomputed the means and standard deviations for each variable to be $\bar{x} = 3374.333$, $s_x \approx 416.2306$, $\bar{y} = 18.91667$, and $s_y \approx 3.895413$.

$$b_1 = r \cdot \frac{s_y}{s_x} = -0.76133706\left(\frac{3.8954130}{416.2305597}\right)$$

$$\approx -0.00712519$$

$$b_0 = \bar{y} - b_1\bar{x}$$

$$= 18.9166667 - (-0.00712519)(3374.333333)$$

$$\approx 42.95943647$$

Rounding to four decimal places, the new least-squares regression line with the Viper included is $\hat{y} = -0.0071x + 42.9594$

Since neither the slope or y-intercept are altered significantly from the original equation found in Problem 23(a) of Section 4.2, the Viper is not influential. We illustrate this further with the following scatter plot showing the original regression line (solid) and the new regression line that includes the Viper (dashed).

Gas Mileage (MPG) vs Weight (in lbs)

Looking at the scatter plot above, since the point (3425, 11) that corresponds to the Dodge Viper, lies far below the rest of the data, it is an outlier.

29. Using all 14 data points, we compute the regression line for carbon dioxide emissions, y, versus energy generated, x, to be $\hat{y} = 0.9245x - 64.075$. The correlation coefficient is $r = 0.964$ and the coefficient of determination is $R^2 = 92.9\%$.

Excluding the Boardman Power Plant, the regression line for the remaining 13 data points becomes $\hat{y} = 0.5029x - 22.938$. The correlation coefficient is $r = 0.989$ and the coefficient of determination is $R^2 = 97.7\%$.

By removing the Boardman Power Plant, the slope and y-intercept both change substantially. Likewise, removing this plant increases the amount of variability in carbon emissions that is explained by the model from 92.9% to 97.7%.

The Boardman Power Plant is unusual in that it emits more carbon that expected from the overall trend. The observation is an outlier and it is influential. We illustrate this further with the following scatter plot showing the original regression line (solid) and the new regression line with Boardman Power Plant removed (dashed).

Excluding the Hermiston power plant (and putting back in the Boardman Power Plant), the regression line for the remaining 13 data points becomes $\hat{y} = 1.0455x - 39.208$. The correlation coefficient is $r = 0.999$ and the coefficient of determination is $R^2 = 99.8\%$.

By removing the Hermiston power plant, the slope and the y-intercept change slightly. Removing this plant substantially increases the amount of variability in carbon missions that is explained by the model.

The carbon emissions for the Hermiston power plant are not unusual, in the sense that they follow the overall trend of the data. Although it has a large residual when all the data are included, this effect goes away if the Boardman observation is removed from the data. The following scatter plot showing the original regression line (solid) and the new regression line with Hermiston Power Plant removed (dashed).

We also show a comparison of the regression line with only the Boardman plant removed (solid) and the regression line with both the Boardman plant and the Hermiston plant removed (dashed).

31. (a)

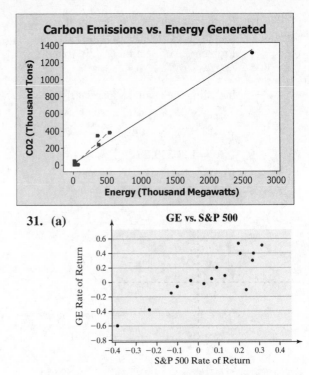

(b) To calculate the correlation coefficient, we use the computational formula.

x_i	y_i	x_i^2	y_i^2	x_iy_i
0.203	0.402	0.041209	0.161604	0.081606
0.310	0.510	0.096100	0.260100	0.158100
0.267	0.410	0.071289	0.168100	0.109470
0.195	0.536	0.038025	0.287296	0.104520
−0.101	−0.060	0.010201	0.003600	0.006060
−0.130	−0.151	0.016900	0.022801	0.019630
−0.234	−0.377	0.054756	0.142129	0.088218
0.264	0.308	0.069696	0.094864	0.081312
0.090	0.207	0.008100	0.042849	0.018630
0.030	−0.014	0.000900	0.000196	−0.000420
0.128	0.093	0.016384	0.008649	0.011904
−0.035	0.027	0.001225	0.000729	−0.000945
−0.385	−0.593	0.148225	0.351649	0.228305
0.235	−0.102	0.055225	0.010404	−0.02397
0.067	0.053	0.004489	0.002809	0.003551
0.904	1.249	0.632724	1.557779	0.885971

From the table, we have $n = 15$,
$\sum x_i = 0.904$, $\sum y_i = 1.249$,
$\sum x_i^2 = 0.632724$, $\sum y_i^2 = 1.557779$, and
$\sum x_iy_i = 0.885971$. So, the correlation coefficient is

$$r = \dfrac{\sum x_i y_i - \dfrac{\sum x_i \sum y_i}{n}}{\sqrt{\left(\sum x_i^2 - \dfrac{(\sum x_i)^2}{n}\right)\left(\sum y_i^2 - \dfrac{(\sum y_i)^2}{n}\right)}}$$

$$= \dfrac{0.885971 - \dfrac{(0.904)(1.249)}{12}}{\sqrt{\left(0.632724 - \dfrac{(0.904)^2}{12}\right)\left(1.557779 - \dfrac{(1.249)^2}{12}\right)}}$$

$$= 0.884208557$$

$$\approx 0.884$$

(c) Yes, since $0.884 > 0.514$, (from Table II.)
A linear relation exists between the rates of return of the S&P 500 and GE.

(d) We compute the mean and standard deviation for each variable to be
$\bar{x} = 0.0602667$, $s_x \approx 0.2032316$,
$\bar{y} \approx 0.0832667$, and $s_y \approx 0.3222442$.

$$b_1 = r \cdot \dfrac{s_y}{s_x} = 0.8842086 \left(\dfrac{0.3222442}{0.2032316}\right)$$

$$\approx 1.4020023$$

$$b_0 = \bar{y} - b_1 \bar{x}$$

$$= 0.0832667 - (1.4020023)(0.08225)$$

$$\approx -0.001227$$

Rounding to four decimal places, the least-squares regression line is
$\hat{y} = 1.4020x - 0.0012$

(e) Let $x = 0.10$ in the regression equation:
$\hat{y} = 1.4020(0.10) - 0.0012 \approx 0.1390$
We predict the rate of return of GE stock will be 0.1390 (or 13.90%) if the rate of return of the S&P 500 is 0.10 (or 10%).

(f) Since the actual rate of return of 13.2% (or 0.132) is below the predicted rate of return from part (e), GE's performance was below average among all years for which the S&P 500's returns were 10%.

(g) The slope indicates that, for each percentage point increase in the rate of return for the S&P 500, the rate of return of GE stock will increase by about 1.40 percentage points, on average.

(h) The y-intercept indicates that the rate of return for GE stock will be -0.0012 (or -0.12%) when there is no change to the S&P 500 (i.e., when there is a 0% rate of return for the S&P 500).

(i) $R^2 = (0.8842086)^2 \approx 0.782 = 78.2\%$

(j) We compute the residuals and construct the residual plot.

x_i	y_i	$\hat{y} = 1.4020x + -0.0012$	$y - \hat{y}$
0.203	0.402	0.28338	0.11862
0.310	0.510	0.43339	0.07661
0.267	0.410	0.37311	0.03689
0.195	0.536	0.27216	0.26384
−0.101	−0.060	−0.14283	0.08283
−0.130	−0.151	−0.18349	0.03249
−0.234	−0.377	−0.32930	−0.04770
0.264	0.308	0.36890	−0.06090
0.090	0.207	0.12495	0.08205
0.030	−0.014	0.04083	−0.05483
0.128	0.093	0.17823	−0.08523
−0.035	0.027	−0.05030	0.07730
−0.385	−0.593	−0.54100	−0.05200
0.235	−0.102	0.32824	−0.43024
0.067	0.053	0.09271	−0.03971

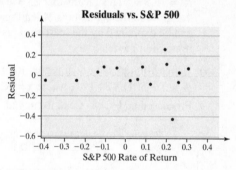

Residuals vs. S&P 500

Overall, the residual plot supports the linear relation. However, there may be a slight concern of a nonconstant variance.

(k) The five-number summary for the residuals is -0.4302, -0.0548, 0.0325, 0.0821, and 0.2638. We compute the fences to look for outliers:

Lower Fence
$= -0.0548 - 1.5(0.0821 - (-0.0548))$
$= -0.2602$

Upper Fence
$= 0.0821 + 1.5(0.0821 - (-0.0548))$
$= 0.2874$

One of the residuals fall outside the fences: -0.4302, the residual for the observations from 2009. So, the data from 2009 represents an outlier.

Consumer Reports®: Fit to Drink

(a)

(b) No, the relation appears to be nonlinear.

(c) $r = 0.868$. This measure is not very useful since the relation between No. Gallons Processed and % Lead Removed is nonlinear. $R^2 = 75.3\%$. This means that 75.3% of the variation in the percentage of lead removed can be explained by the least-squares regression model.

(d) Using Minitab, the linear regression model is: $\hat{y} = 0.04501x + 84.124$, where $x =$ No. Gallons Processed and $y =$ % Lead Removed.

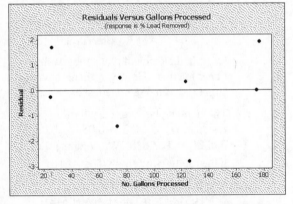

Notice the bowed pattern in the residual plot. This suggests that the linear model does not adequately describe the relation between No. Gallons Processed and % Lead Removed.

(e) Using Minitab, the quadratic regression model is: $\hat{y} = 86.46 - 0.02220x + 0.000333x^2$, where $x =$ No. Gallons Processed and $y =$ % Lead Removed.

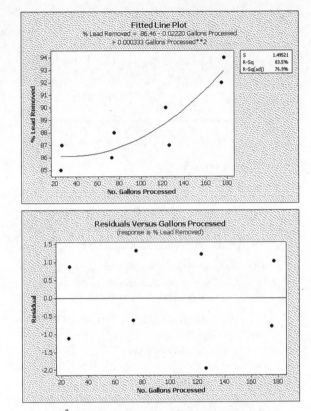

$R^2 = 83.5\%$. Based on the regression plots, the residual plots, and the values of R^2, the quadratic model clearly fits the data better. However, one has to be careful using this model. For example, it isn't possible to remove more than 100% of the lead, regardless of how many gallons are processed.

(f) The shape of the true relationship is most likely s-shaped since the response, i.e., the percent of lead removed, must be between 0% and 100%. Since you cannot remove more than 100% of the lead, this relation must eventually "level off," creating an s-shape.

Section 4.4

1. A *marginal distribution* is a frequency or relative frequency distribution of either the row or column variable in a contingency table. A *conditional distribution* is the relative frequency of each category of one variable, given a specific value of the other variable in a contingency table.

3. The term correlation is used with quantitative variables. In this section, we are considering the relation between qualitative variables (variables which take non-numeric values).

5. (a) We find the frequency marginal distribution for the row variable, y, by finding the total of each row. We find the frequency marginal distribution for the column variable, x, by finding the total for each column.

	x_1 x_2 x_3	Frequency Marg. Dist.
y_1	20 25 30	75
y_2	30 25 50	105
Frequency Marg. Dist.	50 50 80	180

(b) We find the relative frequency marginal distribution for the row variable, y, by dividing the row total for each y_i by the total table, 180. For example, for the relative frequency for y_1 is $\frac{75}{180} \approx 0.417$.

We find the relative frequency marginal distribution for the column variable, x, by dividing the column total for each x_i by the table total. For example, for the relative frequency for x_1 is $\frac{50}{180} \approx 0.278$.

	x_1	x_2	x_3	Rel. Freq. Marg. Dist.
y_1	20	25	30	0.417
y_2	30	25	50	0.583
Rel. Freq. Marg. Dist.	0.278	0.278	0.444	1

(c) Beginning with x_1, we determine the relative frequency of each y_i given x_1 by dividing each frequency by the column total for x_1. For example, the relative frequency of y_1 given x_1 is $\frac{20}{50} = 0.4$.

Next, we compute each relative frequency of each y_i given x_2. For example, the relative frequency of y_1 given x_2 is $\frac{25}{50} = 0.5$. Finally, we compute the relative frequency of each y_i given x_3.

	x_1	x_2	x_3
y_1	$\frac{20}{50} = 0.4$	$\frac{25}{50} = 0.5$	$\frac{30}{80} = 0.375$
y_2	$\frac{30}{50} = 0.6$	$\frac{25}{50} = 0.5$	$\frac{50}{80} = 0.625$
Total	1	1	1

(d) We draw two bars, side by side, for each y_i. The horizontal axis represents x_i and the vertical axis represents the relative frequency.

7. (a) 2160 Americans were surveyed. 536 were 55 and older.

(b) The relative frequency marginal distribution for the row variable, response to immigration question, is found by dividing the row total for each answer choice by the table total, 2160. For example, the relative frequency for the response "More Likely" is $\frac{1329}{2160} \approx 0.615$. The relative frequency marginal distribution for the column variable, age, is found by dividing the column total for each age group by the table total. For example, the relative frequency for 18-34 year-olds is $\frac{542}{2160} \approx 0.251$.

Age					
Likely to Buy	18-34	35 – 44	45 – 54	55+	Rel. Freq. Marg. Dist.
More Likely	238	329	360	402	0.615
Less Likely	22	6	22	16	0.031
Neither	282	201	164	118	0.354
Rel. Freq. Marg. Dist.	0.251	0.248	0.253	0.248	1

(c) The proportion of Americans who are more likely to buy a product when it says 'Made in America' is 0.615.

(d) Beginning with respondents between the ages of 18 and 34, we determine the relative frequency for each response to the likely to buy question by dividing each frequency by the column total for respondents who are between the ages of 18 and 34. For example, the relative frequency for the response "More Likely" given that the respondent is age 18-34 is $\frac{238}{542} \approx 0.439$.
We continue with this calculation for each of the answers to the Likely to Buy question. Next, we compute the each relative frequency for each response given the respondent is between the ages of 35 and 44. We continue in this pattern until we have completed this for all ages.

Age				
Likely to Buy	18-34	35 – 44	45 – 54	55+
More Likely	0.439	0.614	0.659	0.750
Less Likely	0.041	0.011	0.040	0.030
Neither	0.520	0.375	0.300	0.220
Total	1	1	1	1

(e) We draw four bars, side by side, for each response to the age question. The horizontal axis represents the likelihood to buy question and the vertical axis represents the relative frequency.

Likelihood to Buy "Made in America"

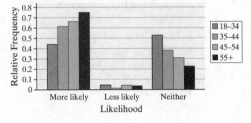

(f) The number of people more likely to buy a product because it is made in America increases with age. On the other hand, age does not seem to be a significant factor in whether a person is less likely to buy a product because it is made in America.

9. (a) We find the frequency marginal distribution for the row variable, party affiliation, by finding the total of each row. We find the frequency marginal distribution for the column variable, gender, by finding the total for each column.

	Female	Male	Frequency Marg. Dist.
Republican	105	115	220
Democrat	150	103	253
Independent	150	179	329
Frequency Marg. Dist.	405	397	802

(b) The relative frequency marginal distribution for the row variable, party affiliation, is found by dividing the row total for each party by the table total, 802. For example, the relative frequency for Republican is $\frac{220}{802} \approx 0.274$. The relative frequency marginal distribution for the column variable, gender, is found by dividing the column total for each gender by the table total. For example, the relative frequency for female is $\frac{405}{802} \approx 0.505$.

	Female	Male	Rel. Freq. Marg. Dist.
Republican	105	115	0.274
Democrat	150	103	0.315
Independent	150	179	0.410
Rel. Freq. Marg. Dist.	0.505	0.495	1

(c) The proportion of registered voters who consider themselves Independent is 0.410.

(d) Beginning with females, we determine the relative frequency for each political party by dividing each frequency by the column total for females. For example, the relative frequency for Republicans given that the voter is female is $\frac{105}{405} \approx 0.259$.

We then compute each relative frequency for each political party given that the voter is male.

	Female	Male
Republican	$\frac{105}{405} \approx 0.259$	$\frac{115}{397} \approx 0.290$
Democrat	$\frac{150}{405} \approx 0.370$	$\frac{103}{397} \approx 0.259$
Independent	$\frac{150}{405} \approx 0.370$	$\frac{179}{397} \approx 0.451$
Total	1	1

(e) We draw three bars, side by side, for each political affiliation. The horizontal axis represents gender and the vertical axis represents the relative frequency.

Party Affiliation

(f) Yes, gender is associated with party affiliation. Male voters are more likely to be Independents and less likely to be Democrats.

11. To determine whether healthier people tend to also be happier, we construct a conditional distribution of people's happiness by health and draw a bar graph of the conditional distribution.

	Poor Health	Fair Health	Good Health	Excellent Health
Not too happy	$\frac{696}{1,996} \approx 0.349$	$\frac{1,386}{6,585} \approx 0.210$	$\frac{1,629}{15,791} \approx 0.103$	$\frac{732}{11,022} \approx 0.066$
Pretty happy	$\frac{950}{1,996} \approx 0.476$	$\frac{3,817}{6,585} \approx 0.580$	$\frac{9,642}{15,791} \approx 0.611$	$\frac{5,195}{11,022} \approx 0.471$
Very happy	$\frac{350}{1,996} \approx 0.175$	$\frac{1,382}{6,585} \approx 0.210$	$\frac{4,520}{15,791} \approx 0.286$	$\frac{5,095}{11,022} \approx 0.462$
Total	1	1	1	1

We draw three bars, side by side, for each level of happiness. The horizontal axis represents level of health and the vertical axis represents the relative frequency.

Happiness

There is a relation between health and happiness. Based on the conditional distribution by health, we can see that healthier people tend to be happier. As health increases, the proportion who are very happy increases, while the proportion who are not happy decreases.

13. (a) Of the 582 smokers in the study, 139 were dead after 20 years, so the proportion of smokers who were dead after 20 years was $\frac{139}{582} \approx 0.239$. Of the 732 nonsmokers in the study, 230 were dead after 20 years, so the proportion of nonsmokers who were dead after 20 years was $\frac{230}{732} \approx 0.314$. This result implies that it is healthier to smoke.

(b) Of the $2 + 53 = 55$ smokers in the 18- to 24-year-old category, 2 were dead after 20 years, so the proportion who were dead was $\frac{2}{55} \approx 0.036$. Of the $1 + 61 = 62$ nonsmokers in the 18- to 24-year old category, 1 was dead after 20 years, so the proportion was $\frac{1}{62} \approx 0.016$.

(c) We repeat the procedure from part (b) for each age category and organize the results in the table that follows:

Age	Smoker	Nonsmoker
18-24	$\frac{2}{2+53} \approx 0.036$	$\frac{1}{1+61} \approx 0.016$
25-34	$\frac{3}{3+121} \approx 0.024$	$\frac{5}{5+152} \approx 0.032$
35-44	$\frac{14}{14+95} \approx 0.128$	$\frac{7}{7+114} \approx 0.058$
45-54	$\frac{27}{27+103} \approx 0.208$	$\frac{12}{12+66} \approx 0.154$
55-64	$\frac{51}{51+64} \approx 0.443$	$\frac{40}{40+81} \approx 0.331$
65-74	$\frac{29}{29+7} \approx 0.806$	$\frac{101}{101+28} \approx 0.783$
75+	$\frac{13}{13+0} \approx 1$	$\frac{64}{64+0} \approx 1$

(d) We draw two bars, side by side, for smokers and nonsmokers. The horizontal axis represents age and the vertical axis represents the relative frequency of deaths.

(e) Reports will vary. When taking age into account, the direction of relation changed. In almost all age groups, smokers had a higher death rate than nonsmokers. The most notable exception is for the 25 to 34 age group, the largest age group for the nonsmokers. A possible explanation could be rigorous physical activity (e.g., rock climbing) that nonsmokers are more likely to participate in than smokers.

Section 4.5 (on CD)

1. The logarithm to the base a, where $a > 0$, $a \neq 1$, is $y = \log_a x$ if and only if $x = a^y$. The logarithm is only defined for $x > 0$.

3. $\log y = \log(ab^x) = \log a + \log b^x$
$$= \log a + x \log b$$

The linear form of $y = ab^x$

is: $\log y = \log a + x \log b$.

If we let $Y = \log y$, $B = \log b$ and $A = \log a$,

then the linear form is $Y = A + Bx$.

5. $4^2 = 16 \iff 2 = \log_4 16$

7. $x^5 = 18 \iff 5 = \log_x 18$

9. $7^x = 15 \iff x = \log_7 15$

11. $5^3 = y \iff 3 = \log_5 y$

13. $\log_3 81 = 4 \iff 3^4 = 81$

15. $\log_a 14 = 3 \iff a^3 = 14$

17. $\log_4 c = 3 \iff 4^3 = c$

19. $\log_2 5 = k \iff 2^k = 5$

21. $\log_3(uv) = \log_3 u + \log_3 v$ (Property 1)

23. $\log_5 x^4 = 4\log_5 x$ (Property 2)

25. $\log_a(uv^2) = \log_a u + \log_a v^2$ (Property 1)
$$= \log_a u + 2\log_a v \quad \text{(Property 2)}$$

27. $\log(x^4 y^3) = \log x^4 + \log y^3$ (Property 1)
$$= 4\log x + 3\log y \quad \text{(Property 2)}$$

29. $\log 34 \approx 1.531$

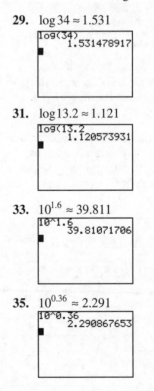

31. $\log 13.2 \approx 1.121$

33. $10^{1.6} \approx 39.811$

35. $10^{0.36} \approx 2.291$

37. (a)

E-coli Population vs. Time

(b)

x	y	$Y = \log y$
0	0.09	$\log 0.09 \approx -1.0458$
2.5	0.18	$\log 0.18 \approx -0.7447$
3.5	0.26	$\log 0.26 \approx -0.5850$
4.5	0.35	$\log 0.35 \approx -0.4559$
6	0.50	$\log 0.50 \approx -0.3010$

Logarithm of E-coli
Population vs. Time

(c) Using technology we find the least-squares regression line of the transformed data to be: $\hat{Y} = 0.1266x - 1.0443$

(d) $\hat{Y} = 0.1266x - 1.0443$, where
$A = -1.0443$ and $B = 0.1266$
$a = 10^A = 10^{-1.0443} \approx 0.0903$
$b = 10^B = 10^{0.1266} \approx 1.3384$
$\hat{y} = a \cdot (b)^x$

$\hat{y} = 0.0903(1.3384)^x$

(e) $y = 0.0903(1.3384)^7 = 0.69$

The predicted population after 7 hours is 0.69.

39. (a)

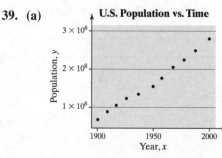

U.S. Population vs. Time

(b)

x	y	$Y = \log y$
1900	$76,212,168$	7.8820
1910	$92,228,496$	7.9649
1920	$106,021,537$	8.0254
1930	$123,202,624$	8.0906
1940	$132,164,569$	8.1211
1950	$151,325,798$	8.1799
1960	$179,323,175$	8.2536
1970	$203,302,031$	8.3081
1980	$226,542,203$	8.3551
1990	$248,709,873$	8.3957
2000	$281,421,906$	8.4494

Logarithm of
U.S. Population vs. Time

(c) Using technology we find the least-squares regression line of the transformed data to be: $\hat{Y} = 0.0056x - 2.6593$

(d) $\hat{Y} = 0.0056x - 2.6593$, where
$A = -2.6593$ and $B = 0.0056$
$a = 10^A = 10^{-2.65933} \approx 0.0022$
$b = 10^B = 10^{0.0056} \approx 1.013$
$\hat{y} = a \cdot (b)^x$

$\hat{y} = 0.0022(1.013)^x$

(e) $\hat{y} = 0.0022(1.013)^{2010} = 414,388,653.$
The predicted population for the U.S. in 2010 is 414,388,653, or about 414 million.

(f) Answers may vary.

(g) Using the U.S. Census Bureau website, the predicted U.S. population in 2010 (as of August 14, 2008) was 310,233,000. This is substantially lower than the prediction from our regression equation.

41. (a)

Planetary Motion

(b)

x	y	X = log x	Y = log y
36	0.24	1.5563	−0.6198
67	0.62	1.8261	−0.2076
93	1.00	1.9685	0
142	1.88	2.1523	0.2742
483	11.9	2.6839	1.0755
887	29.5	2.9479	1.4698
1785	84.0	3.2516	1.9243
2797	165.0	3.4467	2.2175
3675	248.0	3.5653	2.3945

Planetary Motion

(c) Using technology we find the least-squares regression line of the transformed data to be: $\hat{Y} = 1.4994X - 2.9506$

(d) $\hat{Y} = 1.4994X - 2.9506$, where
$A = -2.9506$ and $B = 1.4994$
$a = 10^A = 10^{-2.9506} \approx 0.0011$
$\hat{y} = a \cdot (x)^B$
$\quad = 0.0011x^{1.4994}$

43. (a)

Intensity vs. Distance

(b)

x	y	X = log x	Y = log y
1.0	0.29645	0	−0.5280
1.1	0.25215	0.0414	−0.5983
1.2	0.20547	0.0792	−0.6873
1.3	0.17462	0.1139	−0.7579
1.4	0.15342	0.1461	−0.8141
1.5	0.13521	0.1761	−0.8690
1.6	0.11450	0.2041	−0.9412
1.7	0.10243	0.2304	−0.9896
1.8	0.09231	0.2553	−1.0348
1.9	0.08321	0.2788	−1.0798
2.0	0.07342	0.3010	−1.1342

Logarithm of Intensity vs. Logarithm of Distance

(c) Using technology we find the least-squares regression line of the transformed data to be: $\hat{Y} = -2.012X - 0.5236$

(d) $\hat{Y} = -2.012X - 0.5236$, where
$A = -0.5239$ and $B = -2.012$
$a = 10^A = 10^{-0.5236} \approx 0.2995$
$\hat{y} = a \cdot (x)^B$
$\hat{y} = 0.2995x^{-2.012}$

(e) $\hat{y} = 0.2995(2.3)^{-2.012} = 0.05605$
The predicted intensity of a 100-watt light bulb that is 2.3 meters away is 0.05605.

45. (a) Using technology we find the least-squares regression line to be:
$\hat{y} = -3711.8x + 30984.3$

We compute the residuals and construct the residual plot.

x	y	$\hat{y} = -3.711.8x + 30,984.3$	$y - \hat{y}$
1	27,417	27272.5	144.5
1	28,199	27272.5	927
2	22,995	23560.7	−565.7
2	23,195	23560.7	−365.7
3	18,499	19848.9	−1,349.9
4	16,995	16137.1	857.9
4	16,490	16137.1	352.9

The residual plot shows a clear U-shaped pattern, so the relation between asking price and age is not linear.

(b) To transform the data, we determine the logarithm of the y-values.

x	y	$Y = \log y$
1	27,417	4.43802
1	28,199	4.45023
2	22,995	4.36163
2	23,195	4.36539
3	18,499	4.26715
4	16,995	4.23032
4	16,490	4.21722

Using technology we find the least-squares regression line of the transformed data to be:

$\hat{Y} = -0.074589216x + 4.513998311$.

So, the exponential equation of best fit is:

$\hat{Y} = -0.074589216x + 4.513998311$,

where $A = 4.513998311$ and $B = -0.074589216$

$a = 10^A = 10^{4.513998311} \approx 32658.6562$

$b = 10^B = 10^{-0.074589216} \approx 0.8422$

$\hat{y} = a \cdot (b)^x$

$y = 32658.6562(0.8422)^x$

We compute the residuals and construct the residual plot.

x	$Y = \log y$	$\hat{Y} = -0.074589x + 4.513998$	$Y - \hat{Y}$
1	4.43802	4.43941	−0.00139
1	4.45023	4.43941	0.01082
2	4.36163	4.36482	−0.00319
2	4.36539	4.36482	0.00057
3	4.26715	4.29023	−0.02308
4	4.23032	4.21564	0.01468
4	4.21722	4.21564	0.00158

The relation between asking price and age could be exponential since the residual plot does not show a discernable pattern.

(c) The coefficient of determination for the linear model found in part (a) is $R^2 \approx 0.971$. The coefficient of determination for the exponential model found in part (b) is $R^2 \approx 0.984$. Thus, the exponential model does a better job of explaining the variation in asking price. In fact, 98.4% of the variation in asking price is explained by the age of the car in the exponential model.

Chapter 4 Review Exercises

1. (a) The predicted winning margin is
$\hat{y}=1.007(3) - 0.012 = 3.009$ points

(b) The predicted winning margin is
$\hat{y}=1.007(-7) - 0.012 = -7.061$ points, suggesting that the visiting team is predicted to win by 7.061 points.

(c) For each 1-point increase in the spread, the winning margin increases by 1.007 points, on average.

(d) If the spread is 0, the home team is expected to lose by 0.012 points, on average.

(e) 39% of the variation in winning margins can be explained by the least-squares regression equation.

2. (a) The explanatory variable is fat content.

(b)

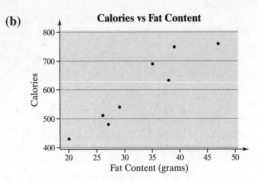

Calories vs Fat Content

3. (a)

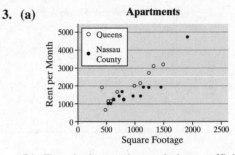

Apartments

(b) To calculate each correlation coefficient, we use the computational formula.

(c) To calculate the correlation coefficient, we use the computational formula.

x_i	y_i	x_i^2	y_i^2	$x_i y_i$
20	430	400	184,900	8,600
39	750	1521	562,500	29,250
27	480	729	230,400	12,960
29	540	841	291,600	15,660
26	510	676	260,100	13,260
47	760	2209	577,600	35,720
35	690	1225	476,100	24,150
38	632	1444	399,424	24,016
261	4792	9045	2,982,624	163,616

From the table, we have $n = 8$, $\sum x_i = 261$, $\sum y_i = 4792$, $\sum x_i^2 = 9045$, $\sum y_i^2 = 2,982,624$, and $\sum x_i y_i = 163,616$. So, the correlation coefficient is

$$r = \frac{\sum x_i y_i - \frac{\sum x_i \sum y_i}{n}}{\sqrt{\left(\sum x_i^2 - \frac{(\sum x_i)^2}{n}\right)\left(\sum y_i^2 - \frac{(\sum y_i)^2}{n}\right)}}$$

$$= \frac{163,616 - \frac{(261)(4792)}{8}}{\sqrt{\left(9045 - \frac{(261)^2}{8}\right)\left(2,982,624 - \frac{(4792)^2}{8}\right)}}$$

$$\approx 0.944$$

(d) Yes, a strong positive linear relation exists between fat content and calories in fast-food restaurant sandwiches.

Queens (New York City)

x_i	y_i	x_i^2	y_i^2	$x_i y_i$
500	650	250,000	422,500	325,000
588	1215	345,744	1,476,225	714,420
1000	2000	1,000,000	4,000,000	2,000,000
688	1655	473,344	2,739,025	1,138,640
825	1250	680,625	1,562,500	1,031,250
460	1805	211,600	3,258,025	830,300
1259	2700	1,585,081	7,290,000	3,399,300
650	1200	422,500	1,440,000	780,000
560	1250	313,600	1,562,500	700,000
1073	2350	1,151,329	5,522,500	2,521,550
1452	3300	2,108,304	10,890,000	4,791,600
1305	3100	1,703,025	9,610,000	4,045,500
10,360	22,475	10,245,152	49,773,275	22,277,560

From the table for the Queens (New York City) apartments, we have $n = 12$, $\sum x_i = 10,360$, $\sum y_i = 22,475$, $\sum x_i^2 = 10,245,152$, $\sum y_i^2 = 49,773,275$, and $\sum x_i y_i = 22,277,560$.

So, the correlation coefficient for Queens is:

$$r = \frac{\sum x_i y_i - \frac{\sum x_i \sum y_i}{n}}{\sqrt{\left(\sum x_i^2 - \frac{(\sum x_i)^2}{n}\right)\left(\sum y_i^2 - \frac{(\sum y_i)^2}{n}\right)}}$$

$$= \frac{22,277,560 - \frac{(10,360)(22,475)}{12}}{\sqrt{\left(10,245,152 - \frac{(10,360)^2}{12}\right)\left(49,773,275 - \frac{(22,475)^2}{12}\right)}}$$

$$\approx 0.909$$

Nassau County (Long Island)

x_i	y_i	x_i^2	y_i^2	$x_i y_i$
1100	1875	1,210,000	3,515,625	2,062,500
588	1075	345,744	1,155,625	632,100
1250	1775	1,562,500	3,150,625	2,218,750
556	1050	309,136	1,102,500	583,800
825	1300	680,625	1,690,000	1,072,500
743	1475	552,049	2,175,625	1,095,925
660	1315	435,600	1,729,225	867,900
975	1400	950,625	1,960,000	1,365,000
1429	1900	2,042,041	3,610,000	2,715,100
800	1650	640,000	2,722,500	1,320,000
1906	4625	3,632,836	21,390,625	8,815,250
1077	1395	1,159,929	1,946,025	1,502,415
11,909	20,835	13,521,085	46,148,375	24,251,240

From the table for the Nassau County (Long Island) apartments, we have $n = 12$, $\sum x_i = 11{,}909$, $\sum y_i = 20{,}835$, $\sum x_i^2 = 13{,}521{,}085$, $\sum y_i^2 = 46{,}148{,}375$, and $\sum x_i y_i = 24{,}251{,}240$. So, the correlation coefficient for Nassau County is:

$$r = \frac{\sum x_i y_i - \dfrac{\sum x_i \sum y_i}{n}}{\sqrt{\left(\sum x_i^2 - \dfrac{(\sum x_i)^2}{n}\right)\left(\sum y_i^2 - \dfrac{(\sum y_i)^2}{n}\right)}}$$

$$= \frac{24{,}251{,}240 - \dfrac{(11{,}909)(20{,}835)}{12}}{\sqrt{\left(13{,}521{,}085 - \dfrac{(11{,}909)^2}{12}\right)\left(46{,}148{,}375 - \dfrac{(20{,}835)^2}{12}\right)}}$$

$$\approx 0.867$$

(c) Yes, both locations have a positive linear relation between square footage and monthly rent.

(d) For small apartments (those less than 1000 square feet in area), there seems to be no difference in rent between Queens and Nassau County. In larger apartments, however, Queens seems to have higher rents than Nassau County.

4. (a) In Problem 1, we computed the correlation coefficient. Rounded to eight decimal places, it is $r = 0.94370937$. We compute the mean and standard deviation for each variable to be $\bar{x} = 32.625$, $s_x \approx 8.7003695$, $\bar{y} = 599$, and $s_y \approx 126.6130212$.

$$b_1 = r \cdot \frac{s_y}{s_x} = 0.94370937 \left(\frac{126.6130212}{8.7003695}\right)$$

$$\approx 13.7334276$$

$$b_0 = \bar{y} - b_1 \bar{x}$$

$$= 599 - (13.7334276)(32.625)$$

$$\approx 150.9469$$

Rounding to four decimal places, the least-squares regression line is

$$\hat{y} = 13.7334x + 150.9469$$

(b)

Calories vs Fat Content

(c) The slope indicates that each additional gram of fat in a sandwich adds 13.73 calories, on average. The y-intercept indicates that a sandwich with no fat will contain about 151 calories.

(d) Let $x = 30$ in the regression equation.

$$\hat{y} = 13.7334(30) + 150.9469 \approx 562.9$$

We predict that a sandwich with 30 grams of fat will have 562.9 calories.

(e) Let $x = 42$ in the regression equation.

$$\hat{y} = 13.7334(42) + 150.9469 \approx 727.7$$

Sandwiches with 42 grams of fat have an average of 727.7 calories. So, the number of calories for this cheeseburger from Sonic is below average.

5. (a) In Problem 2, we computed the correlation coefficient. Rounded to eight decimal places, it is $r = 0.90928694$. We compute the mean and standard deviation for each variable to be $\bar{x} \approx 863.333333$, $s_x \approx 343.9104887$, $\bar{y} \approx 1872.916667$, and $s_y \approx 835.5440752$.

$$b_1 = r \cdot \frac{s_y}{s_x} = 0.90928694 \left(\frac{835.5440751}{343.9104887}\right)$$

$$\approx 2.20914843$$

$$b_0 = \bar{y} - b_1 \bar{x}$$

$$= 1872.916667 - (2.2091484)(863.333333)$$

$$\approx -34.3148$$

Rounding to four decimal places, the least-squares regression line is

$$\hat{y} = 2.2091x - 34.3148.$$

(b) The slope of the least-squares regression equation indicates that, for each additional square foot of floor area, the rent increases by $2.21, on average. It is not appropriate to interpret the y-intercept since it is not possible to have an apartment with an area of 0 square feet. The rent value of $ -34.3148$ is outside the scope of the model.

(c) Let $x = 825$ in the regression equation.
$$\hat{y} = 2.2091(825) - 34.3148 \approx 1788$$

Apartments in Queens with 825 square feet have an average rent of $1788. Since the rent of the apartment in the data set with 825 square feet is $1250, it is below average.

6. (a)

(b) Using the two points (10,105) and (18, 76) gives:
$$m = \frac{76 - 105}{18 - 10} = \frac{-29}{8} = -\frac{29}{8}$$
$$y - 105 = -\frac{29}{8}(x - 10)$$
$$y - 105 = -\frac{29}{8}x + \frac{145}{4}$$
$$y = -\frac{29}{8}x + \frac{565}{4}$$

(c)

(d) We compute the mean and standard deviation for each variable and the correlation coefficient to be $\bar{x} = 16$, $\bar{y} = 84$, $s_x \approx 4.18330$, $s_y \approx 16.20185$, and $r \approx -0.99222$. Then the slope and intercept for the least-squares regression line are:

$$b_1 = r \cdot \frac{s_y}{s_x} = -0.992221\left(\frac{16.20185}{4.18330}\right)$$
$$\approx -3.842855$$
$$b_0 = \bar{y} - b_1\bar{x} = 84 - (-3.842855)(16)$$
$$\approx 145.4857$$

So, rounding to four decimal places, the least-squares regression line is
$$\hat{y} = -3.8429x + 145.4857.$$

(e)

(f)

x	y	$\hat{y} = -\dfrac{29}{8}x + \dfrac{565}{4}$	$y - \hat{y}$	$(y - \hat{y})^2$
10	105	105	0	0
14	94	90.5	3.5	12.25
17	82	79.625	2.375	5.6406
18	76	76	0	0
21	63	65.125	-2.125	4.5156

Total = 22.4062

(g)

x	y	$\hat{y}$	$y - \hat{y}$	$(y - \hat{y})^2$
10	105	107.0571	-2.0571	4.2318
14	94	91.6857	2.3143	5.3559
17	82	80.1571	1.8429	3.3961
18	76	76.3143	-0.3143	0.0988
21	63	64.7857	-1.7857	3.1888

Total = 16.2714

(h) The regression line gives a smaller sum of squared residuals, so it is a better fit.

7. A linear model is appropriate. There is no pattern to the residuals and the variability stays the same.

8. A linear model is not appropriate because a discernible pattern is present in the plot.

9. A linear model is not appropriate because the amount of scatter in the residuals is clearly decreasing, indicating that the error variance is not constant.

10. (a) In Problem 1(c), we found $r \approx 0.944$.
So, $R^2 = (0.944)^2 \approx 0.891 = 89.1\%$.
This means that 89.1% of the variation in calories is explained by the least-squares regression line.

(b) From Problem 3(a), we have
$\hat{y} = 13.7334x + 150.9469$. We compute
the residuals and construct the residual plot.

x_i	y_i	$\hat{y} = 13.7334x + 150.9469$	$y - \hat{y}$
20	430	425.6	4.4
39	750	686.5	63.5
27	480	521.7	−41.7
29	540	549.2	−9.2
26	510	508.0	2.0
47	760	796.4	−36.4
35	690	631.6	58.4
38	632	672.8	−40.8

Residuals versus Fat Content

(c) Yes, based on the residual plot, a linear model is appropriate.

(d) There are no outliers or influential observations present.

(e) The observation (9, 80) is an outlier since it is far away from the rest of the data. Adding it to the data set, the regression equation becomes $\hat{y} = 17.7154x + 9.8713$. Since there is a significant change in both the slope and y-intercept, the observation is influential.

11. (a) From Problem 2(b), the correlation coefficient for the Queens apartments is $r = 0.90928694$ (unrounded). Therefore, $R^2 = (0.90928694)^2 \approx 0.827 = 82.7\%$. This means that 82.7% of the variation in monthly rent is explained by the least-squares regression equation.

(b) From Problem 4(a), we have $\hat{y} = 2.2091x - 34.3148$. We compute the residuals and construct the residual plot.

x_i	y_i	$\hat{y} = 2.2091x - 34.3148$	$y - \hat{y}$
500	650	1070.2	−420.2
588	1215	1264.6	−49.6
1000	2000	2174.8	−174.8
688	1655	1485.5	169.5
825	1250	1788.2	−538.2
460	1805	981.9	823.1
1259	2700	2746.9	−46.9
650	1200	1401.6	−201.6
560	1250	1202.8	47.2
1073	2350	2336.0	14.0
1452	3300	3173.3	126.7
1305	3100	2848.6	251.4

Apartments

Square Footage

(c) A linear model is appropriate. There is no pattern to the residuals and the variability stays the same.

(d) The observation (460, 1805) is an outlier because it lies far away from the rest of the data. It is also an influential point because, if it is removed from the data set, the linear regression equation becomes $\hat{y} = 2.5315x - 399.2496$ in which the y-intercept is significantly different.

12. (a) In Problem 2(b), we computed the correlation coefficient. Rounded to eight decimal places, it is $r = 0.86741573$. We compute the mean and standard deviation for each variable to be $\bar{x} \approx 992.416667$, $s_x \approx 393.399510$, $\bar{y} = 1736.25$, and $s_y \approx 952.203485$.

$$b_1 = r \cdot \frac{s_y}{s_x} = 0.86741573 \left(\frac{952.203485}{393.399510} \right)$$

$$\approx 2.0995356$$

$$b_0 = \bar{y} - b_1 \bar{x}$$

$$= 1736.25 - (2.0995356)(992.416667)$$

$$\approx -347.3641$$

Rounding to four decimal places, the least-squares regression line is
$\hat{y} = 2.0995x - 347.3641$.

(b) The observation (1906, 4625) is an outlier because it lies far away from the rest of the data. It is also influential point because, if it is removed from the data set, the linear regression equation becomes $\hat{y} = 0.8779x + 675.2945$ in which both the slope and y-intercept are significantly different.

13. (a)

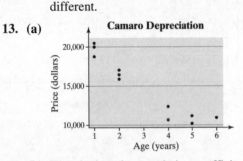

Camaro Depreciation

(b) To calculate the correlation coefficient, we use the computational formula.

x_i	y_i	x_i^2	y_i^2	$x_i y_i$
2	15,900	4	252,810,000	31,800
5	10,988	25	120,736,144	54,940
2	16,980	4	288,320,400	33,960
5	9,995	25	99,900,025	49,975
4	11,995	16	143,880,025	47,980
5	10,995	25	120,890,025	54,975
1	20,365	1	414,733,225	20,365
2	16,463	4	271,030,369	32,926
6	10,824	36	117,158,976	64,944
1	19,995	1	399,800,025	19,995
1	18,650	1	347,822,500	18,650
4	10,488	16	109,998,144	41,952
38	173,638	158	2,687,079,858	472,462

From the table, we have $n = 12$, $\sum x_i = 38$, $\sum y_i = 173{,}638$, $\sum x_i^2 = 158$, $\sum y_i^2 = 2{,}687{,}079{,}858$, and $\sum x_i y_i = 472{,}462$. So, the correlation coefficient is

$$r = \frac{472{,}462 - \dfrac{(38)(173{,}638)}{12}}{\sqrt{\left(158 - \dfrac{(38)^2}{12}\right)\left(2{,}687{,}079{,}858 - \dfrac{(173{,}638)^2}{12}\right)}}$$

$\approx -0.95440918 \approx -0.954$

Yes, there is strong negative linear relation between age and price.

(c) We compute the mean and standard deviation for each variable: $\bar{x} \approx 3.1666667$, $\bar{y} \approx 14{,}469.8333333$, $s_x \approx 1.85047087$, and $s_y \approx 3983.68191577$. Then the slope and intercept for the least-squares regression line are:

$$b_1 = r \cdot \frac{s_y}{s_x} = -0.95440918\left(\frac{3983.68191577}{1.85047087}\right)$$

≈ -2054.646011

$$b_0 = \bar{y} - b_1\bar{x}$$
$$= 14{,}469.8333333 - (-2054.646011)(3.1666667)$$
$\approx 20{,}976.2124$

So, rounding to four decimal places, the least-squares regression line is $\hat{y} = -2054.6460x + 20{,}976.2124$.

(d) We compute the residuals and construct the residual plot.

x_i	y_i	$\hat{y} = -2054.6460x + 20{,}976.2124$	$y - \hat{y}$
2	15,900	16,866.9	−966.9
5	10,988	10,703.0	285.0
2	16,980	16,866.9	113.1
5	9,995	10,703.0	−708.0
4	11,995	12,757.6	−762.6
5	10,995	10,703.0	292.0
1	20,365	18,921.6	1,443.4
2	16,463	16,866.9	−403.9
6	10,824	8,648.3	2,175.7
1	19,995	18,921.6	1,073.4
1	18,650	18,921.6	−271.6
4	10,488	12,757.6	−2,269.6

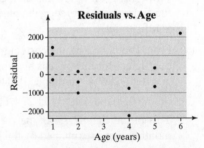

Residuals vs. Age

(e) No, the residuals follow a U-shaped pattern indicating that there is a nonlinear relation.

14. No. Correlation does not imply causation. Florida has a large number of tourists in warmer months, times when more people will be in the water to cool off. The larger number of people in the water splashing around lends to a larger number of shark attacks.

15. (a) 203 buyers were extremely satisfied with their automobile purchase.

(b) The relative frequency marginal distribution for the row variable, level of satisfaction, is found by dividing the row total for each level of satisfaction by the table total, 396. For example, the relative frequency for "Not too satisfied" is $\frac{36}{396} \approx 0.091$. The relative frequency marginal distribution for the column variable, purchase type (new versus used), is found by dividing the column total for each purchase type by the table total. For example, the relative frequency for "New" is $\frac{207}{396} \approx 0.523$.

	New	Used	Rel. Freq. Marg. Dist.
Not Too Satisfied	11	25	$\frac{36}{396} \approx 0.091$
Pretty Satisfied	78	79	$\frac{157}{396} \approx 0.396$
Extremely Satisfied	118	85	$\frac{203}{396} \approx 0.513$
Rel. Freq. Marg. Dist.	$\frac{207}{396} \approx 0.523$	$\frac{189}{396} \approx 0.477$	1

(c) The proportion of consumers who were extremely satisfied with their automobile purchase was 0.513.

(d) Beginning with new automobiles, we determine the relative frequency for level of satisfaction by dividing each frequency by the column total for new automobiles. For example, the relative frequency for "Not Too Satisfied" given that the consumer purchased a new automobile is $\frac{11}{207} \approx 0.053$. We compute each relative frequency for the used automobiles.

	New	Used
Not Too Satisfied	$\frac{11}{207} \approx 0.053$	$\frac{25}{189} \approx 0.132$
Pretty Satisfied	$\frac{78}{207} \approx 0.377$	$\frac{79}{189} \approx 0.418$
Extremely Satisfied	$\frac{118}{207} \approx 0.570$	$\frac{85}{189} \approx 0.450$
Total	1	1

(e) We draw three bars, side by side, for each level of satisfaction. The horizontal axis represents purchase type (new versus used) and the vertical axis represents the relative frequency.

Customer Satisfaction

(f) Yes, there appears to be some association between purchase type (new versus used) and level of satisfaction. Buyers of used cars are more likely to be dissatisfied and are less likely to be extremely satisfied than buyers of new cars.

16. (a) In 1982, the unemployment rate can be found by dividing the number of people who are unemployed by the total number of people who are either employed or unemployed

$$\frac{11.3 \text{ thousand}}{(99.1 + 11.3) \text{ thousand}} \approx 0.102 = 10.2\%$$

In 2009, the unemployment rate was:

$$\frac{14.5 \text{ thousand}}{(130.1 + 14.5) \text{ thousand}} \approx 0.100 = 10.0\%$$

(b) To find the unemployment rate for the recession of 1982 for those who do not have a high school diploma, for example, we divide the number in the group who are unemployed by the total number of people in the group.

$$\frac{3.9 \text{ thousand}}{(20.3 + 3.9) \text{ thousand}} \approx 0.161 = 16.1\%$$

	High School or Less	High School	Bachelor's or Higher
Recession of 1982	$\frac{3.9}{24.2} \approx 16.1\%$	$\frac{6.6}{64.8} \approx 10.2\%$	$\frac{0.8}{21.2} \approx 3.8\%$
Recession of 2009	$\frac{2.0}{12.0} \approx 16.7\%$	$\frac{10.3}{87.0} \approx 11.8\%$	$\frac{2.2}{45.6} \approx 4.8\%$

(c) We draw two bars, side by side, for each of the two years. The horizontal axis represents the level of education and the vertical axis represents the relative frequency.

Unemployment by Level of Education

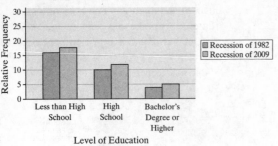

(d) Answers will vary. The discussion should include the observation that although the overall unemployment rate was higher in 1982, the unemployment rate within each level of education was higher in 2009.

17. (a) A positive linear relation appears to exist between number of marriages and number unemployed.

(b) Population is highly correlated with both the number of marriages and the number unemployed. The size of the population affects both variables.

(c) It appears that no relation exists between the two variables.

(d) Answers may vary. A strong correlation between two variables may be due to a third variable that is highly correlated with the two original variables.

18. The eight properties of a linear correlation coefficient are:

1. The linear correlation coefficient is always between -1 and 1, inclusive. That is, $-1 \le r \le 1$.

2. If $r = +1$, there is a perfect positive linear relation between the two variables.

3. If $r = -1$, there is a perfect negative linear relation between the two variables.

4. The closer r is to $+1$, the stronger is the evidence of positive association between the two variables.

5. The closer r is to -1, the stronger is the evidence of negative association between the two variables.

6. If r is close to 0, there is no evidence of a *linear* relation between the two variables. Because the linear correlation coefficient is a measure of the strength of the linear relation; an r close to 0 does not imply no relation, just no linear relation.

7. The linear correlation coefficient is a unitless measure of association. So, the unit of measure for x and y plays no role in the interpretation of r.

8. The correlation coefficient is not resistant.

19. (a) Answers will vary.

(b) The slope can be interpreted as "the school day decreases by about 0.01 hours for every 1 percent increase in the percentage of the district with low income", on average. The y-intercept can be interpreted as the length of the school day for a district with no low income families.

(c) $\hat{y} = -0.0102(20) + 7.11 \approx 6.91$

The school day is predicted to be about 6.91 hours long if 20% of the district is low income.

(d) Answers will vary. Because so many of the data values are clumped together, it is difficult to see any patterns. As shown, it does not appear that any of the model requirements are badly violated.

(e) Answers will vary. The likely candidates will be the three points with the highest percentage of low income population.

(f) Answers will vary.

(g) Answers will vary. There is some indication that there is a positive association between length of school day and PSAE score.

(h) Answers will vary. The scatter diagram indicates that there is some negative association between PSAE score and percentage of population as low income. However, it is likely not as strong as the correlation coefficient indicates. A few influential observations could be inflating the correlation coefficient.

(i) Answers will vary.

Chapter 4 Test

1. **(a)** The likely explanatory variable is temperature because crickets would likely chirp more frequently in warmer temperatures and less frequently in colder temperatures, so the temperature could be seen as explaining the number of chirps.

   (b)

   (c) To calculate the correlation coefficient, we use the computational formula.

x_i	y_i	x_i^2	y_i^2	$x_i y_i$
88.6	20.0	7849.96	400.00	1772.00
93.3	19.8	8704.89	392.04	1847.34
80.6	17.1	6496.36	292.41	1378.26
69.7	14.7	4858.09	216.09	1024.59
69.4	15.4	4816.36	237.16	1068.76
79.6	15.0	6336.16	225.00	1194.00
80.6	16.0	6496.36	256.00	1289.60
76.3	14.4	5821.69	207.36	1098.72
71.6	16.0	5126.56	256.00	1145.60
84.3	18.4	7106.49	338.56	1551.12
75.2	15.5	5655.04	240.25	1165.60
82.0	17.1	6724.00	292.41	1402.20
83.3	16.2	6938.89	262.44	1349.46
82.6	17.2	6822.76	295.84	1420.72
83.5	17.0	6972.25	289.00	1419.50
1200.6	249.8	96,725.86	4200.56	20,127.47

From the table, we have $n = 15$, $\sum x_i = 1200.6$, $\sum y_i = 249.8$, $\sum x_i^2 = 96,725.86$, $\sum y_i^2 = 4200.56$, and $\sum x_i y_i = 20,127.47$. So, the correlation coefficient is

$$r = \frac{\sum x_i y_i - \frac{\sum x_i \sum y_i}{n}}{\sqrt{\left(\sum x_i^2 - \frac{(\sum x_i)^2}{n}\right)\left(\sum y_i^2 - \frac{(\sum y_i)^2}{n}\right)}}$$

$$= \frac{20,127.47 - \frac{(1200.6)(249.8)}{15}}{\sqrt{\left(96,725.86 - \frac{(1200.6)^2}{15}\right)\left(4200.56 - \frac{(249.8)^2}{15}\right)}}$$

$$\approx 0.8351437868$$

$$\approx 0.835$$

(d) Based on the scatter diagram and the linear correlation coefficient, a positive linear relation between temperature and chirps per second.

(e) We compute the mean and standard deviation for each variable: $\bar{x} = 80.04$, $\bar{y} \approx 16.6533333$, $s_x \approx 6.70733074$, and $s_y \approx 1.70204359$. The slope and intercept for the least-squares regression line are:

$$b_1 = r \cdot \frac{s_y}{s_x} = 0.8351437868\left(\frac{1.70204359}{6.70733074}\right)$$

$$\approx 0.21192501$$

$$b_0 = \bar{y} - b_1\bar{x}$$

$$= 16.6533333 - (0.21192501)(80.04)$$

$$\approx -0.3091$$

So, rounding to four decimal places, the least-squares regression line is $\hat{y} = 0.2119x - 0.3091$.

(f) If the temperature increases $1°F$, the number of chirps per second increases by 0.2119, on average. Since there are no observations near $0°F$, it is outside the scope of the model. So, it does not make sense to interpret the y-intercept.

(g) Let $x = 83.3$ in the regression equation.
$\hat{y} = 0.2119(83.3) - 0.3091 \approx 17.3$
We predict that, if the temperature is $83.3°F$, then there will be 17.3 chirps per second.

(h) Let $x = 82$ in the regression equation.
$\hat{y} = 0.2119(82) - 0.3091 \approx 17.1$
There will be an average of 17.1 chirps per second when the temperature is $82°F$. Therefore, 15 chirps per second is below average.

(i) No, we should not use this model to predict the number of chirps when the temperature is 55°F, because 55°F is outside the scope of the model.

(j) $R^2 = (0.8351437868)^2 \approx 0.697 = 69.7\%$

69.7% of the variation in number of chirps per second is explained by the least-squares regression line.

(k) Yes, a linear model is appropriate. There is no pattern to the residuals and the variability stays the same.

2. If we add the point (92, 14.5) to the data set, then the linear regression equation becomes $\hat{y} = 0.1431x + 4.9565$. Since both the slope and y-intercept change significantly, the observation is influential.

3. We compute the correlation coefficient to be $r \approx 0.992$, so from this it appears that there is a strong positive linear relation between speed and braking distance. However, we must analyze this further by computing the least-squares regression equation and construct a residual plot. The least-squares regression line is $\hat{y} = 5.3943x - 131.3524$. We compute the residuals and construct the residual plot.

x_i	y_i	$\hat{y} = 5.3943x - 131.3524$	$y - \hat{y}$
30	48	30.5	17.5
40	79	84.4	−5.4
50	123	138.4	−15.4
60	184	192.3	−8.3
70	243	246.2	−3.2
80	315	300.2	14.8

The residual plot has a clear U-shaped pattern, so a linear model is not appropriate despite the correlation coefficient.

4. There is a problem with the politician's reasoning. Correlation does not imply causation. It is possible that a lurking variable, such as income level or educational level, is affecting both the explanatory and response variables.

5. (a) The relative frequency marginal distribution for the row variable, level of education, is found by dividing the row total for each level of satisfaction by the table total, 2375. For example, the relative frequency for "Less than high school" is

$\dfrac{412}{2375} \approx 0.173$. The relative frequency marginal distribution for the column variable, response concerning belief in Heaven, is found by dividing the column total for each response by the table total. For example, the relative frequency for "Yes, Definitely" is $\dfrac{1539}{2375} = 0.648$.

	Yes, Definitely	Yes, Probably	No, Probably Not	No, Definitely Not	Rel. Freq. Marg. Dist.
Less than High School	316	66	21	9	0.173
High School	956	296	122	65	0.606
Bachelor's	267	131	62	64	0.221
Rel. Freq. Marg. Dist.	0.648	0.208	0.086	0.058	1

(b) The proportion of adult Americans in the survey who definitely believe in Heaven is 0.648.

(c) Beginning with "Less than high school", we determine the relative frequency for responses concerning belief in Heaven by dividing each frequency by the row total for "Less than high school." For example, the relative frequency for "Yes, Definitely" given that the respondent has less than a high school education is $\dfrac{316}{412} \approx 0.767$.

We then compute each relative frequency for the other levels of education.

	Yes, Definitely	Yes, Probably	No, Probably Not	No, Definitely Not	Total
< HS	$\dfrac{316}{412} \approx 0.767$	$\dfrac{66}{412} \approx 0.160$	$\dfrac{21}{412} \approx 0.051$	$\dfrac{9}{412} \approx 0.022$	1
HS	$\dfrac{956}{1439} \approx 0.664$	$\dfrac{296}{1439} \approx 0.206$	$\dfrac{122}{1439} \approx 0.085$	$\dfrac{65}{1439} \approx 0.045$	1
BS	$\dfrac{267}{524} \approx 0.510$	$\dfrac{131}{524} \approx 0.250$	$\dfrac{62}{524} \approx 0.118$	$\dfrac{64}{524} \approx 0.122$	1

(d) We draw four bars, side by side, for each response concerning belief in Heaven. The horizontal axis represents level of education and the vertical axis represents the relative frequency.

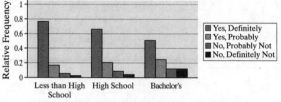

(e) Yes, as education level increases, the proportion who definitely believe in Heaven decreases (i.e., doubt in Heaven increases).

6. (a) For each gender, we determine the proportions for acceptance status by dividing the number of applicants in each acceptance status by the total number of applicants of that gender:

	Accepted	Denied
Male	$\dfrac{98}{98+522} \approx 0.158$	$\dfrac{522}{98+522} \approx 0.842$
Female	$\dfrac{90}{90+200} \approx 0.310$	$\dfrac{200}{90+200} \approx 0.690$

(b) The proportion of males that was accepted is 0.158. The proportion of females that was accepted is 0.310.

(c) The college accepted a higher proportion of females than males.

(d) The proportion of male applicants accepted into the business school was:

$$\frac{90}{90+510} = 0.15$$

The proportion of female applicants accepted into the business school was:

$$\frac{10}{10+60} \approx 0.143$$

(e) The proportion of male applicants accepted into the social work school was:

$$\frac{8}{8+12} = 0.4$$

The proportion of male applicants accepted into the social work school was:

$$\frac{80}{80+140} \approx 0.364$$

(f) Answers will vary. A larger number of males applied to the business school, which has an overall lower acceptance rate than the social work school, so more male applicants were declined.

7. A set of quantitative bivariate data whose linear coefficient is −1 would have a perfect negative linear relation. A scatter diagram would show all the observations being collinear (falling on the same line) with a negative slope.

8. If the slope of the least-squares regression line is negative, then the correlation between the explanatory and response variables is also negative.

9. If a linear correlation is close to zero, then this means that there is no linear relation between the explanatory and response variables. This does not necessarily mean that there is no relation at all, however, just no *linear* relation.

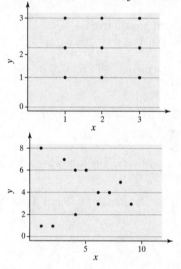

Case Study: Thomas Malthus, Population, and Subsistence

(b)

Log (Population) vs Time

$$\log P = 0.00877t + 6.78385$$

$$a = 10^{6.78385} \approx 6,079,249.95$$

$$\log b = 0.00877$$
$$b = 10^{0.00877} \approx 1.0204$$

The annual growth rate is
$b - 1 = 1.0204 - 1 = 0.0204 = 2.04\%$.

Thus, $P = 6,079,249.95(1.0204)^t$

(c) $A = 1.2P$
$$= 1.2\left(6,079,249.95(1.0204)^t\right)$$
$$= 7,295,099.94(1.0204)^t$$

(d) Answers will vary. The following is for the years 1995–2003.

	t (years after 1790)	Farm Acres (millions)
1995	205	963
1996	206	959
1997	207	956
1998	208	952
1999	209	948
2000	210	945
2001	211	942
2002	212	940
2003	213	939

$$\hat{y} = -3.1333t + 1604.2$$

(e) Since the number of acres is in millions, we need to divide our model for A by one million to keep the same units.

$$\hat{y}_1 = \frac{A}{1,000,000} = 7.295099(1.0204)^t$$

$$\hat{y}_2 = -3.1333t + 1604.2$$

Based on this information, the United States will no longer be able to provide a diverse diet for its entire population $t = 237$ years after 1790. That is, in the year 2027.

(f) Answers will vary. As technology improves, particularly in regards to agriculture, it becomes possible to generate a higher yield from a smaller piece of land. Therefore, food production may actually increase even if the number of acres of farmland is decreasing.

Chapter 5

Probability

Section 5.1

1. The probability of an impossible event is zero. No. An event with probability approximately zero is a very unlikely event, but it is not necessarily an impossible event.

3. True

5. experiment

7. Rule 1 is satisfied since all of the probabilities in the model are greater than or equal to zero and less than or equal to one. Rule 2 is satisfied since the sum of the probabilities in the model is one: $0.3 + 0.15 + 0 + 0.15 + 0.2 + 0.2 = 1$. In this model, the outcome "blue" is an impossible event.

9. This cannot be a probability model because $P(\text{green}) < 0$.

11. Probabilities must be between 0 and 1, inclusive, so the only values which could be probabilities are: 0, 0.01, 0.35, and 1.

13. The probability of 0.42 means that approximately 42 out of every 100 hands will contain two cards of the same value and three cards of different value. No, probability deals with long-term behavior. While we would expect to see about 42 such hands out of every 100, on average, this does not mean we will always get a pair exactly 42 out of every 100 hands.

15. The empirical probability that the next flip would result in a head is $\dfrac{95}{100} \approx 0.95$.

17. $P(2) \neq \dfrac{1}{11}$ because the 11 possible outcomes are not equally likely.

19. The sample space is $S = \{1H, 2H, 3H, 4H, 5H, 6H, 1T, 2T, 3T, 4T, 5T, 6T\}$

21. The probability that a randomly selected three-year-old is enrolled in daycare is $P = 0.428$.

23. Event E contains 3 of the 10 equally likely outcomes, so $P(E) = \dfrac{3}{10} = 0.3$.

25. There are 10 equally likely outcomes and 4 are even numbers less than 9, so
$$P(E) = \frac{4}{10} = \frac{2}{5} = 0.4.$$

27. (a) $P(\text{plays organized sports}) = \dfrac{288}{500} = 0.576$

   (b) If we sampled 100 high school students, we would expect that about 576 of the students play organized sports.

29. (a) Since 40 of the 100 equally likely tulip bulbs are red, $P(\text{red}) = \dfrac{40}{100} = 0.4$.

   (b) Since 25 of the 100 equally likely tulip bulbs are purple,
   $$P(\text{purple}) = \frac{25}{100} = 0.25.$$

   (c) If we sampled 100 tulip bulbs, we would expect about 40 of the bulbs to be red and about 25 of the bulbs to be purple.

31. (a) The sample space is $S = \{0, 00, 1, 2, 3, 4, \ldots, 36\}$.

   (b) Since the slot marked 8 is one of the 38 equally likely outcomes,
   $P(8) = \dfrac{1}{38} \approx 0.0263$. This means that, in many spins of such a roulette wheel, the long-run relative frequency of the ball landing on "8" will be close to
   $\dfrac{1}{38} \approx 0.0263 = 2.63\%$. That is, if we spun the wheel 100 times, we would expect about 3 of those times to result in the ball landing in slot 8.

   (c) Since there are 18 odd slots (1, 3, 5, 7, 9, 11, 13, 15, 17, 19, 21, 23, 25, 27, 29, 31, 33, 35) in the 38 equally likely outcomes,
   $P(\text{odd}) = \dfrac{18}{38} = \dfrac{9}{19} \approx 0.4737$. This means that, in many spins of such a roulette wheel, the long-run relative frequency of the ball landing in an odd slot will be

close to $\dfrac{9}{19} \approx 0.4737 = 47.37\%$. That is, if we spun the wheel 100 times, we would expect about 47 of those times to result in an odd number.

33. **(a)** The sample space of possible genotypes is $\{SS, Ss, sS, ss\}$.

    (b) Only one of the four equally likely genotypes gives rise to sickle-cell anemia, namely ss. Thus, the probability is $P(ss) = \dfrac{1}{4} = 0.25$. This means that of the many children who are offspring of two Ss parents, approximately 25% will have sickle-cell anemia.

    (c) Two of the four equally likely genotypes result in a carrier, namely Ss and sS. Thus, the probability of this is $P(Ss \text{ or } sS) = \dfrac{2}{4} = \dfrac{1}{2} = 0.5$. This means that of the many children who are offspring of two Ss parents, approximately 50% will not themselves have sickle-cell anemia, but will be carriers of sickle-cell anemia.

35. **(a)** There are $125 + 324 + 552 + 1257 + 2518 = 4776$ college students in the survey. The individuals can be thought of as the trials of the probability experiment. The relative frequency of "Never" is $\dfrac{125}{4776} \approx 0.026$. We compute the relative frequencies of the other outcomes similarly and obtain the probability model below.

Response	Probability
Never	0.026
Rarely	0.068
Sometimes	0.116
Most of the time	0.263
Always	0.527

    (b) Yes, it is unusual to find a college student who never wears a seatbelt when riding in a car driven by someone else. The approximate probability of this is only 0.026, which is less than 0.05.

37. **(a)** There are $4 + 6 + 133 + 219 + 90 + 42 + 143 + 5 = 642$ police records included in the survey. The individuals can be thought of as the trials of the probability experiment. The relative frequency of "Pocket picking" is $\dfrac{4}{642} \approx 0.006$. We compute the relative frequencies of the other outcomes similarly and obtain the probability model below.

Type of Larceny Theft	Probability
Pocket picking	0.006
Purse snatching	0.009
Shoplifting	0.207
From motor vehicles	0.341
Motor vehicle accessories	0.140
Bicycles	0.065
From buildings	0.223
From coin-operated machines	0.008

    (b) Yes, purse-snatching larcenies are unusual since the probability is only $0.009 < 0.05$.

    (c) No, bicycle larcenies are not unusual since the probability is $0.065 > 0.05$.

39. Assignments A, B, C, and F are consistent with the definition of a probability model. Assignment D cannot be a probability model because it contains a negative probability, and Assignment E cannot be a probability model because is does not add up to 1.

41. Assignment B should be used if the coin is known to always come up tails.

43. **(a)** The sample space is $S = \{$John-Roberto; John-Clarice; John-Dominique; John-Marco; Roberto-Clarice; Roberto-Dominique; Roberto-Marco; Clarice-Dominique; Clarice-Marco; Dominique-Marco$\}$.

    (b) Clarice-Dominique is one of the ten possible samples from part (a). Thus, $P(\text{Clarice and Dominique}) = \dfrac{1}{10} = 0.1$.

    (c) Four of the ten samples from part (a) include Clarice. Thus, $P(\text{Clarice attends}) = \dfrac{4}{10} = \dfrac{2}{5} = 0.4$.

(d) Six of the ten samples from part (a) do not include John. Thus,

$$P(\text{John stays home}) = \frac{6}{10} = \frac{3}{5} = 0.6 \ .$$

45. (a) Since 24 of the 73 homeruns went to right field, $P(\text{right field}) = \frac{24}{73} \approx 0.329$.

(b) Since 2 of the 73 homeruns went to left field, $P(\text{right field}) = \frac{2}{73} \approx 0.027$.

(c) Yes, it was unusual for Barry Bonds to his a homerun to left field. The probability is below 0.05.

47. (a)-(d) Answers will vary depending on the results from the simulation.

49. If the dice were fair, then each outcome should occur approximately $\frac{400}{6} \approx 67$ times. Since 1 and 6 occurred with considerably higher frequency, the dice appear to be loaded.

51. Half of all families are above the median and half are below, so

$$P(\text{Income greater than } \$58{,}500) = \frac{1}{2} = 0.5 \ .$$

53. Answers will vary.

55. The Law of Large Numbers states that as the number of repetitions of a probability experiment increases (in the long term), the proportion with which a certain outcome is observed (i.e. the relative frequency) gets closer to the probability of the outcome. The games at a gambling casino are designed to benefit the casino in the long run; the risk to the casino is minimal because of the large number of gamblers.

57. An event is unusual if it has a low probability of occurring. The same "cut-off" should not always be used to identify unusual events. Selecting a "cut-off" is subjective and should take into account the consequence s of incorrectly identifying an event as unusual.

59. Empirical probability is based on the outcomes of a probability experiment and is the relative frequency of the event. Classical probability is based on counting techniques and is equal to the ratio of the number of ways an event can occur to the number of possible outcomes of the experiment.

Section 5.2

1. Two events are disjoint (mutually exclusive) if they have no outcomes in common.

3. $P(E) + P(F) - P(E \text{ and } F)$

5. E and $F = \{5, 6, 7\}$. No, E and F are not mutually exclusive because they have simple events in common.

7. F or $G = \{5, 6, 7, 8, 9, 10, 11, 12\}$.
$P(F \text{ or } G) = P(F) + P(G) - P(F \text{ and } G)$
$$= \frac{5}{12} + \frac{4}{12} - \frac{1}{12} = \frac{8}{12} = \frac{2}{3} \ .$$

9. E and $G = \{ \ \}$. Yes, E and G are mutually exclusive because they have no simple events in common.

11. $E^c = \{1, 8, 9, 10, 11, 12\}$.
$$P(E^c) = 1 - P(E) = 1 - \frac{6}{12} = \frac{1}{2}$$

13. $P(E \text{ or } F) = P(E) + P(F) - P(E \text{ and } F)$
$$= 0.25 + 0.45 - 0.15 = 0.55$$

15. $P(E \text{ or } F) = P(E) + P(F) = 0.25 + 0.45 = 0.7$

17. $P(E^c) = 1 - P(E) = 1 - 0.25 = 0.75$

19. $P(E \text{ or } F) = P(E) + P(F) - P(E \text{ and } F)$
$$0.85 = 0.60 + P(F) - 0.05$$
$$P(F) = 0.85 - 0.60 + 0.05 = 0.30$$

21. $P(\text{Titleist or Maxfli}) = \frac{9+8}{20} = \frac{17}{20} = 0.85$

23. $P(\text{not Titleist}) = 1 - P(\text{Titleist})$
$$= 1 - \frac{9}{20} = \frac{11}{20} = 0.55$$

25. (a) Rule 1 is satisfied since all of the probabilities in the model are between 0 and 1.
Rule 2 is satisfied since the sum of the probabilities in the model is one: $0.473 + 0.026 + 0.031 + 0.141 + 0.134 + 0.059 + 0.136 = 1$.

(b) $P(\text{rifle or shotgun}) = 0.026 + 0.031$
$= 0.057$
If 100 murders in 2009 were randomly selected, we would expect 57 of them to be committed with a rifle or a shotgun.

(c) $P(\text{handgun, rifle, or shotgun})$.

$= 0.473 + 0.026 + 0.031$

$= 0.530$

If 100 murders in 2009 were randomly selected, we would expect 53 of them to be committed with a handgun, a rifle, or a shotgun.

(d) To find the probability that a randomly selected murder was committed with a weapon other than a gun, we subtract the probability that a gun was used from 1.

$P(\text{not a gun}) = 1 - P(\text{a gun was used})$

$= 1 - (0.473 + 0.026 + 0.031 + 0.141)$

$= 1 - 0.671 = 0.329$

This means that there is a 32.9% probability of randomly selecting a murder that was not committed with a gun.

(e) Yes, murders with a shotgun are unusual since the probability is $0.031 < 0.05$.

27. No; for example, on one draw of a card from a standard deck, let $E = \text{diamond}$, $F = \text{club}$, and $G = \text{red card}$. Here, E and F are disjoint, as are F and G. However, E and G are *not* disjoint since diamond cards are red.

29. (a) $P(30 - 39) = P(30 - 34) + P(35 - 39)$

$= \dfrac{2262 + 1545}{6427} \approx 0.592$

If we randomly select 1000 mothers involved in a multiple birth, we would expect about 592 to be between 30 and 39 years old.

(b) $P(\text{not } 30 - 39) = 1 - P(30 - 39)$

$\approx 1 - 0.592 = 0.408$

This means that there is a 40.8% chance that a mother involved in a multiple birth was not between 30 and 39. That is, the mother was younger than 30 or older than 39.

(c) $P(\text{younger than } 45) = 1 - P(45 \text{ or older})$

$= 1 - \dfrac{105}{6427}$

≈ 0.984

This means that there is a 98.4% chance that a randomly selected mother involved in a multiple birth was younger than 45.

(d) $P(\text{at least } 20) = 1 - \dfrac{100}{6694} \approx 0.984$

This means that there is a 98.4% chance that a randomly selected mother involved in a multiple birth was at least 20 years old.

31. (a) $P(\text{Heart or Club}) = P(\text{Heart}) + P(\text{Club})$

$= \dfrac{13}{52} + \dfrac{13}{52}$

$= \dfrac{1}{2} = 0.5$

(b) $P(\text{Heart or Club or Diamond})$
$= P(\text{Heart}) + P(\text{Club}) + P(\text{Diamond})$

$= \dfrac{13}{52} + \dfrac{13}{52} + \dfrac{13}{52}$

$= \dfrac{3}{4} = 0.75$

(c) $P(\text{Ace or Heart})$
$= P(\text{Ace}) + P(\text{Heart}) - P(\text{Ace of Hearts})$

$= \dfrac{4}{52} + \dfrac{13}{52} - \dfrac{1}{52}$

$= \dfrac{4}{13} \approx 0.308$

33. (a) $P(\text{not on Nov. 8}) = 1 - P(\text{on Nov. 8})$

$= 1 - \dfrac{1}{365}$

$= \dfrac{364}{365} \approx 0.997$

(b) $P\left(\text{not on the } 1^{\text{st}}\right) = 1 - P\left(\text{on the } 1^{\text{st}}\right)$

$= 1 - \dfrac{12}{365}$

$= \dfrac{353}{365} \approx 0.967$

(c) $P\left(\text{not on the } 31^{\text{st}}\right) = 1 - \left(\text{on the } 31^{\text{st}}\right)$

$= 1 - \dfrac{7}{365}$

$= \dfrac{358}{365} \approx 0.981$

(d) $P(\text{not in Dec.}) = 1 - P(\text{in Dec.})$

$= 1 - \dfrac{31}{365}$

$= \dfrac{334}{365} \approx 0.915$

35. No, we cannot compute the probability of randomly selecting a citizen of the U.S. who has hearing problems or vision problems by adding the given probabilities because the events "hearing problems" and "vision problems" are not disjoint. That is, some people have both vision and hearing problems, but we do not know the proportion.

37. (a) $P(\text{between 6 and 10 yrs old}) = \dfrac{5}{24} \approx 0.208$

This is not unusual because $0.208 > 0.05$.

(b) $P(\text{more than 5 yrs old}) = \dfrac{5+7+5}{24}$

$= \dfrac{17}{24} \approx 0.708$

(c) $P(\text{less than 1 yr old}) = \dfrac{1}{24} \approx 0.042$

This is unusual because $0.042 < 0.05$.

39. (a) $P(\text{drives or takes public transportation})$
$= P(\text{drives}) + P(\text{takes public transportation})$
$= 0.867 + 0.048 = 0.915$

(b) $P(\text{neither drives nor takes pub. trans.})$
$= 1 - P(\text{drives or takes pub. trans.})$
$= 1 - 0.915 = 0.085$

(c) $P(\text{does not drive}) = 1 - P(\text{drives})$
$= 1 - 0.867 = 0.133$

(d) No, the probability that a randomly selected worker walks to work cannot equal 0.15 because the sum of the probabilities would be more than 1 and there would be no probability model.

41. (a) Of the 137,243 men included in the study, $782 + 91 + 141 = 1014$ died from cancer. Thus,

$P(\text{died from cancer}) = \dfrac{1014}{137,243} \approx 0.007$.

(b) Of the 137,243 men included in the study, $141 + 7725 = 7866$ were current cigar smokers. Thus,

$P(\text{current cigar smoker}) = \dfrac{7866}{137,243}$

≈ 0.057.

(c) Of the 137,243 men included in the study, 141 were current cigar smokers who died from cancer. Thus,

$P(\text{died from cancer and current smoker})$

$= \dfrac{141}{137,243} \approx 0.001$.

(d) Of the 137,243 men included in the study, 1014 died from cancer, 7866 were current cigar smokers, and 141 were current cigar smokers who died from cancer. Thus,

$P(\text{died from cancer or current smoker})$

$= P(\text{died from cancer}) + P(\text{current smoker})$

$\quad - P(\text{died from cancer and current smoker})$

$= \dfrac{1014}{137,243} + \dfrac{7866}{137,243} - \dfrac{141}{137,243}$

$= \dfrac{8739}{137,243} \approx 0.064$.

43. (a) Of the 197 people surveyed, 115 were female. Thus, $P(\text{female}) = \dfrac{115}{197} = 0.584$.

(b) Of the 197 people surveyed, 94 were issued one ticket in the last year. Thus,

$P(1 \text{ ticket}) = \dfrac{21}{197} \approx 0.107$.

(c) Of the 197 people surveyed, 14 were females who were issued one ticket in the last year. Thus,

$P(\text{female and 1 ticket}) = \dfrac{14}{197} \approx 0.071$.

(d) Of the 197 people surveyed, 115 were female, 21 were issued one ticket, and 14 were females who were issued one ticket in the last year. Thus,

$P(\text{female or 1 ticket})$

$= P(\text{female}) + P(1 \text{ ticket})$

$\quad - P(\text{female and 1 ticket})$

$= \dfrac{115}{197} + \dfrac{21}{197} - \dfrac{14}{197}$

$= \dfrac{122}{197}$

≈ 0.619

45. (a) Of the 530 adults surveyed, 338 use social media. Thus,

$P(\text{use social media}) = \dfrac{338}{530} \approx 0.638$.

(b) Of the 530 adults surveyed, 140 are 45-54. Thus, $P(\text{age 45-54}) = \dfrac{140}{530} \approx 0.264$.

(c) Of the 530 adults surveyed, 89 are 35-44 years old and use social media. Thus,

$P(\text{age 35-44 and social media user})$

$= \dfrac{89}{530} \approx 0.168$

(d) Of the 530 adults surveyed, 125 are 35 – 44 years old, 338 use social media, and 89 are 35 – 44 and use social media. Thus,

$P(\text{35-44 years old or uses social media})$

$= P(\text{35-44 years old}) + P(\text{uses social media})$

$\quad - P(\text{35-44 years old and uses social media})$

$= \dfrac{338}{530} + \dfrac{125}{530} - \dfrac{89}{530}$

$= \dfrac{374}{530} \approx 0.706.$

47. (a) The variables presented in the table are crash type, crashes with current system, and projected crashes with red-light cameras.

(b) Crash type is qualitative because it describes an attribute or characteristic.

Crashes with current system is discrete quantitative variable because it describes the number of crashes with the current system.

Projected crashes with red-light cameras is quantitative because it is a numerical measure. It is discrete because the numerical measurement is the result of a count.

(c) Of the 1,121 projected crashes under the current system, 289 were projected to have reported injuries. Of the 922 projected crashes under the new system, 221 were projected to have reported injuries. The relative frequency for reported injury crashes would be

$\dfrac{289}{1,121} \approx 0.26$ under the current system

and $\dfrac{221}{922} \approx 0.24$ under the new system. Similar computations can be made for the remaining crash types for both systems.

Crash Type	Current System	w/Red-Light Cameras
Reported Injury	0.26	0.24
Reported Property Damage Only	0.35	0.36
Unreported Injury	0.07	0.07
Unreported Property Damage Only	0.32	0.33
Total	1	1

(d)

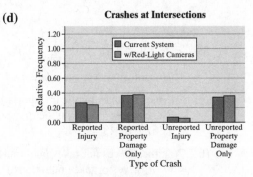

(e) Mean for Current System:

$\dfrac{1121}{13} \approx 86.2$ crashes per intersection

Mean for Red-light Camera System:

$\dfrac{922}{13} \approx 70.9$ crashes per intersection

(f) It is not possible to compute the standard deviation because we do not know the number of crashes at each intersection.

(g) Since the mean number of crashes is less with the cameras, it appears that the program will be beneficial.

(h) There are 1,121 crashes under the current system and 289 with reported injuries. Thus,

$P(\text{reported injuries}) = \dfrac{289}{1121} \approx 0.258.$

(i) There are 922 crashes under the camera system. Of these, 333 had reported property damage only and 308 had unreported property damage only. Thus,

$$P\binom{\text{property damage}}{\text{only}} = \frac{333 + 308}{922}$$

$$= \frac{641}{922} \approx 0.695$$

(j) Simpson's Paradox refers to situations when conclusions reverse or change direction when accounting for specific variable. When accounting for the cause of the crash (rear-end vs. red-light running), the camera system does not reduce all types of accidents. Under the camera system, red-light running crashes decreased, but rear-end crashes increased.

(k) Recommendations may vary. The benefits of the decrease in red-light running crashes must be weighed against the negative of increased rear-end crashes. Seriousness of injuries and amount of property damage may need to be considered.

Section 5.3

1. independent

3. Addition

5. $P(E) \cdot P(F)$

7. (a) Dependent. Speeding on the interstate increases the probability of being pulled over by a police officer.

(b) Dependent: Eating fast food affects the probability of gaining weight.

(c) Independent: Your score on a statistics exam does not affect the probability that the Boston Red Sox win a baseball game.

9. Since E and F are independent.,
$$P(E \text{ and } F) = P(E) \cdot P(F)$$
$$= (0.3)(0.6)$$
$$= 0.18$$

11. $P(5 \text{ heads in a row}) = \left(\frac{1}{2}\right)\left(\frac{1}{2}\right)\left(\frac{1}{2}\right)\left(\frac{1}{2}\right)\left(\frac{1}{2}\right)$

$$= \left(\frac{1}{2}\right)^5 = \frac{1}{32} = 0.03125$$

If we flipped a coin five times, 100 different times, we would expect to observe 5 heads in a row about 3 times.

13. $P(2 \text{ left-handed people}) = (0.13)(0.13)$
$$= 0.0169$$
$$P\binom{\text{At least 1 is}}{\text{right-handed}} = 1 - P\binom{2 \text{ left-handed}}{\text{people}}$$
$$= 1 - 0.0169 = 0.9831$$

15. (a) $P(\text{all 5 negative})$
$$= (0.995)(0.995)(0.995)(0.995)(0.995)$$
$$= (0.995)^5 \approx 0.9752$$

(b) $P(\text{at least one positive})$
$$= 1 - P(\text{all 5 negative})$$
$$= 1 - 0.9752$$
$$= 0.0248$$

17. (a) $P(\text{Both live to be 41}) = (0.99757)(0.99757)$
$$\approx 0.99515$$

(b) $P(\text{All 5 live to be 41}) = (0.99757)^5$
$$\approx 0.98791$$

(c) This is the complement of the event in (b), so the probability is $1 - 0.98791 = 0.01209$ which is unusual since $0.01209 < 0.05$.

19. (a) $P(\text{Both have mental illness}) = (0.3)(0.3)$
$$= 0.09$$

(b) $P(\text{All 6 have mental illness}) = (0.3)^6$
$$= 0.000729$$

(c) $P(\text{At least 1 has mental illness})$
$$= 1 - P(\text{None has mental illness})$$
$$= 1 - (1 - 0.3)^6$$
$$= 1 - (0.7)^6$$
$$\approx 0.882$$

(d) The probability that none have some form of mental illness is:
$P(\text{None of 4 have mental illness})$
$$= (1 - 0.3)^4 = 0.2401$$
This is not unusual, since 0.2401 is greater than 0.05.

21. (a) $P(\text{one failure}) = 0.15$; this is not unusual because $0.15 > 0.05$.
Since components fail independent of each other, we get
$P(\text{two failures}) = (0.15)(0.15) = 0.0225$;
this is unusual because $0.0225 < 0.05$.

(b) This is the complement of both components failing, so
$P(\text{system succeeds}) = 1 - P(\text{both fail})$

$$= 1 - (0.15)^2$$
$$= 1 - 0.0225$$
$$= 0.9775$$

(c) From part (b) we know that two components are not enough, so we increase the number.
3 components:

$P(\text{system succeeds}) = 1 - (0.15)^3$
$$\approx 0.99663$$
4 components:

$P(\text{system succeeds}) = 1 - (0.15)^4$
$$\approx 0.99949$$
5 components:

$P(\text{system succeeds}) = 1 - (0.15)^5$
$$\approx 0.99992$$
Therefore, 5 components would be needed to make the probability of the system succeeding greater than 0.9999.

23. (a) $P\left(\begin{array}{c}\text{batter makes 10}\\\text{consecutive outs}\end{array}\right) = (0.70)^{10} \approx 0.02825$

If we randomly selected 100 different at bats of 10, we would expect about 3 to result in a streak of 10 consecutive runs.

(b) Yes, cold streaks are unusual since the probability is $0.02825 < 0.05$.

(c) The probability that the hitter makes five consecutive outs and then reaches base safely is the same as the probability that the hitter gets five outs in a row and then gets to base safely on the sixth attempt. Assuming these events are independent, it is the product of these two probabilities:

$P(\text{5 consecutive outs then has a base hit})$

$$= P\left(\begin{array}{c}\text{player makes}\\\text{5 consecutive}\\\text{outs}\end{array}\right) \cdot P\left(\begin{array}{c}\text{player}\\\text{reaches}\\\text{base safely}\end{array}\right)$$
$$= (0.7)^5 (0.3) \approx 0.050421$$

25. (a) $P\left(\begin{array}{c}\text{two strikes}\\\text{in a row}\end{array}\right) = (0.3)(0.3) = 0.09$

(b) $P(\text{turkey}) = (0.3)^3 = 0.027$

(c) $P\left(\begin{array}{c}\text{gets a turkey}\\\text{but fails to get}\\\text{a clover}\end{array}\right) = P\left(\begin{array}{c}\text{three strikes}\\\text{followed by}\\\text{a non strike}\end{array}\right)$

$$= P\left(\begin{array}{c}\text{three strikes}\\\text{in a row}\end{array}\right) \cdot P(\text{non-strike})$$
$$= (0.3)^3 (0.7) = 0.0189$$

27. (a) $P\left(\begin{array}{c}\text{all 3 have}\\\text{driven under}\\\text{the influence}\\\text{of alcohol}\end{array}\right) = (0.29)^3 \approx 0.0244$

(b) $P\left(\begin{array}{c}\text{at least one has not driven}\\\text{under the influence of alcohol}\end{array}\right)$

$$= 1 - P\left(\begin{array}{c}\text{all 3 have driven}\\\text{under the influence}\\\text{of alcohol}\end{array}\right)$$
$$= 1 - (0.29)^3$$
$$\approx 1 - 0.0244 = 0.9756$$

(c) The probability that an individual 21- to 25-year-old has not driven while under the influence of alcohol is $1 - 0.29 = 0.71$, so

$$P\left(\begin{array}{c}\text{none of the}\\\text{3 have driven}\\\text{under the}\\\text{influence of}\\\text{alcohol}\end{array}\right) = (0.71)^3 \approx 0.3579$$

(d) $P\left(\begin{array}{c}\text{at least one has driven under}\\\text{the influence of alcohol}\end{array}\right)$

$$= 1 - P\left(\begin{array}{c}\text{none has driven under}\\\text{the influence of alcohol}\end{array}\right)$$
$$= 1 - (0.71)^3$$
$$\approx 1 - 0.3579 = 0.6421$$

29. Since, the events are independent, the probability that a randomly selected household is audited and owns a dog is:

P(audited and owns a dog)

$= P$(audited)$\cdot P$(owns a dog)

$= (0.0642)(0.39) \approx 0.025$

31. The probability that all 3 stocks increase by 10% is:

P(all 3 stocks increase by 10%)

$= P$(#1 up 10%)$\cdot P$(#2 up 10%)$\cdot P$(#3 up 10%)

$= (0.70)(0.55)(0.20) = 0.077$

This is not unusual.

33. (a) $P\left(\begin{matrix} \text{all 24 squares} \\ \text{filled correctly} \end{matrix}\right) = \left(\dfrac{1}{2}\right)^{24} \approx 5.96 \times 10^{-8}$

(b) $P\left(\begin{matrix} \text{determine complete} \\ \text{configuration} \end{matrix}\right)$

$= \left(\dfrac{1}{2}\right)^{24} \cdot \left(\dfrac{1}{2}\right)^{4} \cdot \left(\dfrac{1}{2}\right)^{8}$

$= \left(\dfrac{1}{2}\right)^{36}$

$\approx 1.46 \times 10^{-11}$

Section 5.4

1. $F; E$

3. $P(F \mid E) = \dfrac{P(E \text{ and } F)}{P(E)} = \dfrac{0.6}{0.8} = 0.75$

5. $P(F \mid E) = \dfrac{N(E \text{ and } F)}{N(E)} = \dfrac{420}{740} = 0.568$

7. $P(E \text{ and } F) = P(E) \cdot P(F \mid E)$

$= (0.8)(0.4)$

$= 0.32$

9. No, the events "earn more than $100,000 per year" and "earned a bachelor's degree" are not independent because P("earn more than $100,000 per year" | "earned a bachelor's degree") $\ne$ P("earn more than $100,000 per year").

11. P(club) $= \dfrac{13}{52} = \dfrac{1}{4}$;

P(club | black card) $= \dfrac{13}{26} = \dfrac{1}{2}$

13. P(rainy | cloudy) $= \dfrac{P(\text{rainy and cloudy})}{P(\text{cloudy})}$

$= \dfrac{0.21}{0.37} \approx 0.568$

15. P(unemployed | high school dropout)

$= \dfrac{P(\text{umployed and high school dropout})}{P(\text{high school dropout})}$

$= \dfrac{0.021}{0.080} \approx 0.263$

17. (a) P(age 35–44 | more likely)

$= \dfrac{N(\text{age 35} - 44 \text{ and more likely})}{N(\text{more likely})}$

$= \dfrac{329}{1329} \approx 0.248$

(b) P(more likely | age 35–44)

$= \dfrac{N(\text{more likely and age 35} - 44)}{N(\text{age 35} - 44)}$

$= \dfrac{329}{536} \approx 0.614$

(c) For 18 – 34 year olds, the probability that they are more likely to buy a product that is 'Made in America' is:

$P\left(\text{More likely} \mid \text{18-34 years old}\right)$

$= \dfrac{238}{542} \approx 0.439$

For individuals in general, the probability is:

$P\left(\text{More likely}\right) = \dfrac{1329}{2160} \approx 0.615$

18 – 34 year olds are less likely to buy a product that is labeled 'Made in America' than individuals in general.

19. (a) $P\left(\text{driver} \mid \text{female}\right)$

$= \dfrac{N\left(\text{driver and female}\right)}{N\left(\text{female}\right)}$

$= \dfrac{11,856}{18,219} \approx 0.651$

Copyright © 2013 Pearson Education, Inc.

(b) $P(\text{female} \mid \text{passenger})$

$$= \frac{N(\text{female and passenger})}{N(\text{passengers})}$$

$$= \frac{6,363}{12,816} \approx 0.496$$

(c) $P(\text{male} \mid \text{driving})$

$$= \frac{N(\text{male and driving})}{N(\text{driving})}$$

$$= \frac{32,873}{44,729} \approx 0.735$$

The probability of the driver being a male is 0.735 and therefore the probability of the driver being a female is

$$P(\text{female}) = 1 - P(\text{male})$$

$$= 1 - 0.735$$

$$= 0.265$$

The victim is more likely to be male.

21. $P(\text{both televisions work})$

$= P(\text{1st works}) \cdot P(\text{2nd works} \mid \text{1st works})$

$$= \frac{4}{6} \cdot \frac{3}{5} = 0.4$$

$P(\text{at least one television does not work})$

$= 1 - P(\text{both televisions work})$

$= 1 - 0.4 = 0.6$

23. (a) $P(\text{both kings})$

$= P(\text{first king}) \cdot P(\text{2nd king} \mid \text{first king})$

$$= \frac{4}{52} \cdot \frac{3}{51} = \frac{1}{221} \approx 0.005$$

(b) $P(\text{both kings})$

$= P(\text{first king}) \cdot P(\text{2nd king} \mid \text{first king})$

$$= \frac{4}{52} \cdot \frac{4}{52} = \frac{1}{169} \approx 0.006$$

25. $P(\text{Dave 1st and Neta 2nd})$

$= P(\text{Dave 1st}) \cdot P(\text{Neta 2nd} \mid \text{Dave 1st})$

$$= \frac{1}{5} \cdot \frac{1}{4} = \frac{1}{20} = 0.05$$

27. (a) $P(\text{like both songs})$

$= P(\text{like 1st}) \cdot P(\text{like 2nd} \mid \text{like 1st})$

$$= \frac{5}{13} \cdot \frac{4}{12} = \frac{5}{39} \approx 0.128$$

The probability is greater than 0.05. This is not a small enough probability to be considered unusual.

(b) $P(\text{dislike both songs})$

$= P(\text{dislike 1st}) \cdot P(\text{dislike 2nd} \mid \text{dislike 1st})$

$$= \frac{8}{13} \cdot \frac{7}{12} = \frac{14}{39} \approx 0.359$$

(c) Since you either like both or neither or exactly one (and these are disjoint) then the probability that you like exactly one is given by

$P(\text{like exactly one song})$

$$= 1 - \left(P(\text{like both}) + P(\text{dislike both}) \right)$$

$$= 1 - \left(\frac{5}{39} + \frac{14}{39} \right) = \frac{20}{39} \approx 0.513$$

(d) $P(\text{like both songs})$

$= P(\text{like 1st}) \cdot P(\text{like 2nd} \mid \text{like 1st})$

$$= \frac{5}{13} \cdot \frac{5}{13} = \frac{25}{169} \approx 0.148$$

The probability is greater than 0.05. This is not a small enough probability to be considered unusual.

$P(\text{dislike both songs})$

$= P(\text{dislike 1st}) \cdot P(\text{dislike 2nd} \mid \text{dislike 1st})$

$$= \frac{8}{13} \cdot \frac{8}{13} = \frac{64}{169} \approx 0.379$$

$P(\text{like exactly one song})$

$$= 1 - \left(P(\text{like both}) + P(\text{dislike both}) \right)$$

$$= 1 - \left(\frac{25}{169} + \frac{64}{169} \right) = \frac{80}{169} \approx 0.473$$

29.

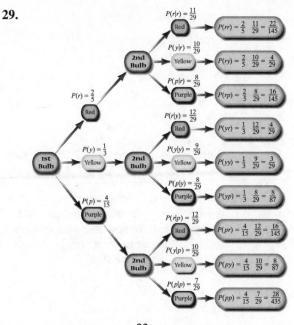

(a) $P(\text{both red}) = \dfrac{22}{145} \approx 0.152$

(b) $P(\text{1st red and 2nd yellow}) = \dfrac{4}{29} \approx 0.138$

(c) $P(\text{1st yellow and 2nd red}) = \dfrac{4}{29} \approx 0.138$

(d) Since one each of red and yellow must be either 1st red, 2nd yellow or vice versa, by the addition rule this probability is

$$P(\text{one red and one yellow}) = \dfrac{4}{29} + \dfrac{4}{29}$$
$$= \dfrac{8}{29} \approx 0.276$$

31. $P(\text{female and smoker})$
$= P(\text{female} \mid \text{smoker}) \cdot P(\text{smoker})$
$= (0.445)(0.203) \approx 0.090$
The probability is greater than 0.05. It is not unusual to select a female who smokes.

33. (a) $P(\text{10 different birthdays})$

$$= \dfrac{365}{365} \cdot \dfrac{364}{365} \cdot \dfrac{363}{365} \cdots \dfrac{358}{365} \cdot \dfrac{357}{365} \cdot \dfrac{356}{365}$$
$$\approx 0.883$$

(b) $P(\text{at least 2})$
$= 1 - P(\text{none})$
$\approx 1 - 0.883$
$= 0.117$

35. (a) $P(\text{male}) = \dfrac{200}{400} = \dfrac{1}{2}$

$P(\text{male} \mid 0 \text{ activ.}) = \dfrac{21}{42} = \dfrac{1}{2}$

$P(\text{male}) = P(\text{male} \mid 0 \text{ activities})$, so the events "male" and "0 activities" are independent.

(b) $P(\text{female}) = \dfrac{200}{400} = \dfrac{1}{2} = 0.5$

$P(\text{female} \mid 5+ \text{ activ.}) = \dfrac{71}{109} \approx 0.651$

$P(\text{female}) \neq P(\text{female} \mid 5+ \text{ activ.})$, so the events "female" and "5+ activities" are not independent.

(c) Yes, the events "1-2 activities" and "3-4 activities" are mutually exclusive because the two events cannot occur at the same time. $P(\text{1-2 activ. and 3-4 activ.}) = 0$

(d) No, the events "male" and "1-2 activities" are not mutually exclusive because the two events can happen at the same time.

$P(\text{male and 1-2 activ.}) = \dfrac{81}{400}$
$= 0.2025 \neq 0$

37. (a) $P(\text{being dealt 5 clubs})$

$= \dfrac{13}{52} \cdot \dfrac{12}{51} \cdot \dfrac{11}{50} \cdot \dfrac{10}{49} \cdot \dfrac{9}{48} = \dfrac{33}{66,640}$
≈ 0.000495

(b) $P(\text{bing dealt a flush})$

$= 4\left(\dfrac{33}{66,640}\right) = \dfrac{33}{16,660} \approx 0.002$

39. (a) $P(\text{selecting two defective chips})$

$= \dfrac{50}{10,000} \cdot \dfrac{49}{9,999} \approx 0.0000245$

(b) Assuming independence,
$P(\text{selecting two defective chips})$
$\approx (0.005)^2$
$= 0.000025$
The difference in the results of parts (a) and (b) is only 0.0000005, so the assumption of independence did not significantly affect the probability.

41. $P(\text{45-54 yrs old}) = \dfrac{546}{2,160} \approx 0.253$

$P(\text{45-54 years old} \mid \text{more likely}) = \dfrac{360}{1,329}$

≈ 0.271

No, the events "45-54 years old" and "more likely" are not independent since the preceding probabilities are not equal.

43. $P(\text{female}) = P(\text{female}) = \dfrac{18,219}{57,545} \approx 0.317$

$P(\text{female} \mid \text{driver}) = \dfrac{11,856}{44,729} \approx 0.265$

No, the events "female" and "driver" are not independent since the preceding probabilities are not equal.

45. (a)-(d) Answers will vary.
For part (d), of the 18 boxes in the outermost ring, 12 indicate you win if you switch while 6 indicate you lose if you switch. Assuming random selection so each box is equally likely, we get $P(\text{win if switch}) = \dfrac{12}{18} = \dfrac{2}{3} \approx 0.667$.

Consumer Reports®: His N' Hers Razor?

(a) $P(\text{Excellent}) = \dfrac{21+19+13}{90} = \dfrac{53}{90} \approx 0.589$

(b) $P(\text{Poor}) = \dfrac{1+0+6}{90} = \dfrac{7}{90} \approx 0.078$

(c) $P(\text{Razor B} \mid \text{Fair}) = \dfrac{N(\text{Razor B and Fair})}{N(\text{Fair})}$

$= \dfrac{11}{30} \approx 0.367$

(d) $P(\text{Excellent} \mid \text{Razor C})$

$= \dfrac{N(\text{Excellent and Razor C})}{N(\text{Razor C})}$

$= \dfrac{13}{30} \approx 0.433$

(e) $P(\text{Poor} \mid \text{Razor A})$

$= \dfrac{N(\text{Poor and Razor A})}{N(\text{Razor A})}$

$= \dfrac{1}{30} \approx 0.033$

$P(\text{Poor}) = \dfrac{7}{90} \approx 0.078$

Since $P(\text{Poor} \mid \text{Razor A}) \neq P(\text{Poor})$, the razor type and rating are not independent.

(f) Razor A has the highest percentage of Very Good to Excellent ratings. However, the difference between razors A and B is small and could be due to the particular sample of panelists used for the study.

Section 5.5

1. permutation

3. True

5. $5! = 5 \cdot 4 \cdot 3 \cdot 2 \cdot 1 = 120$

7. $10! = 10 \cdot 9 \cdot 8 \cdot 7 \cdot 6 \cdot 5 \cdot 4 \cdot 3 \cdot 2 \cdot 1 = 3,628,800$

9. $0! = 1$

11. $_6P_2 = \dfrac{6!}{(6-2)!} = \dfrac{6!}{4!} = 6 \cdot 5 = 30$

13. $_4P_4 = \dfrac{4!}{(4-4)!} = \dfrac{4!}{0!} = \dfrac{24}{1} = 24$

15. $_5P_0 = \dfrac{5!}{(5-0)!} = \dfrac{5!}{5!} = 1$

17. $_8P_3 = \dfrac{8!}{(8-3)!} = \dfrac{8!}{5!} = 8 \cdot 7 \cdot 6 = 336$

19. $_8C_3 = \dfrac{8!}{3!(8-3)!} = \dfrac{8!}{3!5!} = \dfrac{8 \cdot 7 \cdot 6}{3 \cdot 2 \cdot 1} = 56$

21. $_{10}C_2 = \dfrac{10!}{2!(10-2)!} = \dfrac{10!}{2!8!} = \dfrac{10 \cdot 9}{2 \cdot 1} = 45$

23. $_{52}C_1 = \dfrac{52!}{1!(52-1)!} = \dfrac{52!}{1!51!} = \dfrac{52}{1} = 52$

25. $_{48}C_3 = \dfrac{48!}{3!(48-3)!} = \dfrac{48!}{3!45!} = \dfrac{48 \cdot 47 \cdot 46}{3 \cdot 2 \cdot 1}$

$= 17,296$

27. *ab, ac, ad, ae, ba, bc, bd, be, ca, cb, cd,*
ce, da, db, dc, de, ea, eb, ec, ed
Since there are 20 permutations, $_5P_2 = 20$.

29. *ab, ac, ad, ae, bc, bd, be, cd, ce, de*
Since there are 10 combinations, $_5C_2 = 10$.

31. Here we use the Multiplication Rule of
Counting. There are six shirts and four ties, so
there are $6 \cdot 4 = 24$ different shirt-and-tie
combinations the man can wear.

33. There are 12 ways Dan can select the first
song, 11 ways to select the second song, etc.
From the Multiplication Rule of Counting,
there are $12 \cdot 11 \cdot \ldots \cdot 2 \cdot 1 = 12! = 479,001,600$
ways that Dan can arrange the 12 songs.

35. There are 8 ways to pick the first city, 7 ways
to pick the second, etc. From the
Multiplication Rule of Counting, there are
$8 \cdot 7 \cdot \ldots \cdot 2 \cdot 1 = 8! = 40,320$ different routes
possible for the salesperson.

37. Since the company name can be represented
by 1, 2, or 3 letters, we find the total number
of 1 letter abbreviations, 2 letter abbreviations,
and 3 letter abbreviations, then we sum the
results to obtain the total number of
abbreviations possible. There are 26 letters
that can be used for abbreviations, so there are
26 one-letter abbreviations. Since repetitions
are allowed, there are $26 \cdot 26 = 26^2$ different
two-letter abbreviations, and $26 \cdot 26 \cdot 26 = 26^3$
different three-letter abbreviations. Therefore,
the maximum number of companies that can
be listed on the New York Stock Exchange is
26 (one letter) $+ 26^2$ (two letters)
$+ 26^3$ (three letters) $= 18,278$ companies.

39. **(a)** $10 \cdot 10 \cdot 10 \cdot 10 = 10^4 = 10,000$ different
codes are possible.

(b) $P(\text{guessing the correct code}) = \dfrac{1}{10,000}$
$= 0.0001$

41. Since lower and uppercase letters are
considered the same, each of the 8 letters to be
selected has 26 possibilities. Therefore, there
are $26^8 = 208,827,064,576$ different
usernames possible for the local area network.

43. **(a)** $50 \cdot 50 \cdot 50 = 50^3 = 125,000$ different
combinations are possible.

(b) $P(\text{guessing combination}) = \dfrac{1}{50^3} = \dfrac{1}{125,000}$
$= 0.000008$

45. Since order matters, we use the permutation
formula $_nP_r$.
$_{40}P_3 = \dfrac{40!}{(40-3)!} = \dfrac{40!}{37!} = 40 \cdot 39 \cdot 38 = 59,280$
There are 59,280 ways in which the top three
cars can result.

47. Since the order of selection determines the
office of the member, order matters.
Therefore, we use the permutation formula
$_nP_r$.
$_{20}P_4 = \dfrac{20!}{(20-4)!} = \dfrac{20!}{16!} = 20 \cdot 19 \cdot 18 \cdot 17$
$= 116,280$
There are 116,280 different leadership
structures possible.

49. Since the problem states the numbers must be
matched in order, this is a permutation
problem.
$_{25}P_4 = \dfrac{25!}{(25-4)!} = \dfrac{25!}{21!} = 25 \cdot 24 \cdot 23 \cdot 22$
$= 303,600$
There are 303,600 different outcomes possible
for this game.

51. Since order of selection does not matter, this is
a combination problem.
$_{50}C_5 = \dfrac{50 \cdot 49 \cdot 48 \cdot 47 \cdot 46}{5 \cdot 4 \cdot 3 \cdot 2 \cdot 1} = 2,118,760$
There are 2,118,760 different simple random
samples of size 5 possible.

53. There are 6 children and we need to determine
the number of ways two can be boys. The
order of the two boys does not matter, so this
is a combination problem. There are
$_6C_2 = \dfrac{6!}{2!(6-2)!} = \dfrac{6!}{2! \, 4!} = \dfrac{6 \cdot 5}{2 \cdot 1} = 15$ ways to
select 2 of the 6 children to be boys (the rest
are girls), so there are 15 different birth and
gender orders possible.

55. Since there are three A's, two C's, two G's,
and three T's from which to form a DNA
sequence, we can make $\dfrac{10!}{3! \, 2! \, 2! \, 3!} = 25,200$
distinguishable DNA sequences.

57. Arranging the trees involves permutations with repetitions. Using Formula (3), we find that there are $\dfrac{11!}{4!5!2!} = 6930$ different ways the landscaper can arrange the trees.

59. Since the order of the balls does not matter, this is a combination problem. Using Formula (2), we find there are $_{39}C_5 = 575,757$ possible choices (without regard to order), so

$$P(\text{winning}) = \frac{1}{575,757} \approx 0.00000174 \,.$$

61. (a) $P(\text{all students}) = \dfrac{_8C_5}{_{18}C_5}$

$$= \frac{8\cdot 7\cdot 6}{3\cdot 2\cdot 1}\cdot\frac{5\cdot 4\cdot 3\cdot 2\cdot 1}{18\cdot 17\cdot 16\cdot 15\cdot 14}$$

$$= \frac{1}{153} \approx 0.0065$$

(b) $P(\text{all faculty}) = \dfrac{_{10}C_5}{_{18}C_5}$

$$= \frac{10\cdot 9\cdot 8\cdot 7\cdot 6}{5\cdot 4\cdot 3\cdot 2\cdot 1}\cdot\frac{5\cdot 4\cdot 3\cdot 2\cdot 1}{18\cdot 17\cdot 16\cdot 15\cdot 14}$$

$$= \frac{1}{34} \approx 0.0294$$

(c) $P(\text{2 students and 3 faculty}) = \dfrac{_8C_2 \cdot {_{10}C_3}}{_{18}C_5}$

$$= \frac{8\cdot 7}{2\cdot 1}\cdot\frac{10\cdot 9\cdot 8}{3\cdot 2\cdot 1}\cdot\frac{5\cdot 4\cdot 3\cdot 2\cdot 1}{18\cdot 17\cdot 16\cdot 15\cdot 14}$$

$$= \frac{20}{51} \approx 0.3922$$

63. $P(\text{one or more defective})$

$= 1 - P(\text{none defective})$

$$= 1 - \frac{_{116}C_4}{_{120}C_4}$$

$$= 1 - \frac{116\cdot 115\cdot 114\cdot 113}{4\cdot 3\cdot 2\cdot 1}\cdot\frac{4\cdot 3\cdot 2\cdot 1}{120\cdot 119\cdot 118\cdot 117}$$

≈ 0.1283

There is a probability of 0.1283 that the shipment is rejected.

65. (a) $P(\text{you like 2 of the 4 songs}) = \dfrac{_5C_2 \cdot {_8C_2}}{_{13}C_4}$

$$\approx 0.3916$$

There is a probability of 0.3916 that you will like 2 of the 4 songs played.

(b) $P(\text{you like 3 of the 4 songs}) = \dfrac{_5C_3 \cdot {_8C_1}}{_{13}C_4}$

$$\approx 0.1119$$

There is a probability of 0.1119 that you will like 3 of the 4 songs played.

(c) $P(\text{you like all 4 songs}) = \dfrac{_5C_4 \cdot {_8C_0}}{_{13}C_4}$

$$\approx 0.0070$$

There is a probability of 0.007 that you will like all 4 songs played.

67. (a) Five cards can be selected from a deck in $_{52}C_5 = 2,598,960$ ways.

(b) There are $_4C_3 = 4$ ways of choosing 3 two's, and so on for each denomination. Hence, there are $13\cdot 4 = 52$ ways of choosing three of a kind.

(c) There are $_{12}C_2 = 66$ choices of two additional denominations (different from that of the three of a kind) and 4 choices of suit for the first remaining card and then, for each choice of suit for the first remaining card, there are 4 choices of suit for the last card. This gives a total of $66\cdot 4\cdot 4 = 1056$ ways of choosing the last two cards.

(d) $P(\text{three of a kind}) = \dfrac{52\cdot 1056}{2,598,960} \approx 0.0211$

69. $P(\text{all 4 modems work}) = \dfrac{17}{20}\cdot\dfrac{16}{19}\cdot\dfrac{15}{18}\cdot\dfrac{14}{17}$

$$\approx 0.4912$$

There is a probability of 0.4912 that the shipment will be accepted.

71. (a) Using the Multiplication Rule of Counting and the given password format, there are $21\cdot 5\cdot 21\cdot 21\cdot 5\cdot 21\cdot 10\cdot 10 = 486,202,500$ different passwords that are possible.

(b) If letters are case sensitive, there are $42\cdot 10\cdot 42\cdot 42\cdot 10\cdot 42\cdot 10\cdot 10 = 420^4 = 31,116,960,000$ different passwords that are possible.

Section 5.6

1. In a permutation, the order in which the objects are chosen matters; in a combination, the order in which the objects are chosen is unimportant.

3. 'AND' is generally associated with multiplication, while 'OR' is generally associated with addition.

5. $P(E) = \dfrac{4}{10} = \dfrac{2}{5} = 0.4$

7. *abc, acb, abd, adb, abe, aeb, acd, adc, ace, aec, ade, aed, bac, bca, bad, bda, bae, bea, bcd, bdc, bce, bec, bde, bed, cab, cba, cad, cda, cae, cea, cbd, cdb, cbe, ceb, cde, ced, dab, dba, dac, dca, dae, dea, dbc, dcb, dbe, deb, dce, dec, eab, eba, eac, eca, ead, eda, ebc, ecb, ebd, edb, ecd, edc*

9. $P(E \text{ or } F) = P(E) + P(F) - P(E \text{ and } F)$
$= 0.7 + 0.2 - 0.15$
$= 0.75$

11. $_7P_3 = \dfrac{7!}{(7-3)!} = \dfrac{7!}{4!} = 7 \cdot 6 \cdot 5 = 210$

13. $P(E \text{ and } F) = P(E) \cdot P(F)$
$= (0.8)(0.5)$
$= 0.4$

15. $P(E \text{ and } F) = P(E) \cdot P(F \mid E)$
$= (0.9)(0.3)$
$= 0.27$

17. $P(\text{soccer}) \approx \dfrac{22}{500} = \dfrac{11}{250} = 0.044$

19. Since the actual order of the men in their three seats matters, there are $_3P_3 = 6$ ways to arrange the men. Similarly, there are $_3P_3 = 6$ ways to arrange the women among their seats. Therefore, there are $6 \cdot 6 = 36$ ways to arrange the men and women.

Alternatively, using the Multiplication Rule of Counting, and starting with a female, we get:
$$\underset{F}{\dfrac{3}{}} \cdot \underset{M}{\dfrac{3}{}} \cdot \underset{F}{\dfrac{2}{}} \cdot \underset{M}{\dfrac{2}{}} \cdot \underset{F}{\dfrac{1}{}} \cdot \underset{M}{\dfrac{1}{}} = 3 \cdot 3 \cdot 2 \cdot 2 \cdot 1 \cdot 1 = 36$$
Again, there are 36 ways to arrange the men and women.

21. **(a)** $P(\text{survived}) = \dfrac{711}{2224} \approx 0.320$

(b) $P(\text{female}) = \dfrac{425}{2224} \approx 0.191$

(c) $P(\text{female or child}) = \dfrac{425 + 109}{2224} = \dfrac{534}{2224}$
$= \dfrac{267}{1112} \approx 0.240$

(d) $P(\text{female and survived}) = \dfrac{316}{2224}$
$= \dfrac{79}{556} \approx 0.142$

(e) $P(\text{female or survived})$
$= P(\text{female}) + P(\text{survived})$
$\quad - P(\text{female and survived})$
$= \dfrac{425}{2224} + \dfrac{711}{2224} - \dfrac{316}{2224}$
$= \dfrac{820}{2224} = \dfrac{205}{556} \approx 0.369$

(f) $P(\text{survived} \mid \text{female}) = \dfrac{316}{425} \approx 0.744$

(g) $P(\text{survived} \mid \text{child}) = \dfrac{57}{109} \approx 0.523$

(h) $P(\text{survived} \mid \text{male}) = \dfrac{338}{1690} = \dfrac{1}{5} = 0.2$

(i) Yes; the survival rate was much higher for women and children than for men.

(j) $P(\text{both females survived})$
$= P(\text{1st surv.}) \cdot P(\text{2nd surv.} \mid \text{1st surv.})$
$= \dfrac{316}{425} \cdot \dfrac{315}{424} \approx 0.552$

23. Since the order in which the colleges are selected does not matter, there are $_{12}C_3 = 220$ different ways to select 3 colleges from the 12 he is interested in.

25. **(a)** Since the games are independent of each other, we get
$$P(\text{win both}) = \dfrac{1}{5,200,000} \cdot \dfrac{1}{705,600}$$
$$\approx 0.00000000000027$$

(b) Since the games are independent of each other, we get

$$P\left(\begin{array}{c}\text{win Jubilee}\\\text{twice}\end{array}\right)=\left(\frac{1}{705,600}\right)^2$$

$$\approx 0.000000000002$$

27. $P(\text{parent owns} \mid \text{teenager owns})$

$$=\frac{P(\text{both own})}{P(\text{teenager owns})}=\frac{0.43}{0.79}\approx 0.544$$

29. The order in which the questions are answered does not matter so there are $_{12}C_8 = 495$ different sets of questions that could be answered.

31. Using the Multiplication Rule of Counting, there are $2\cdot 2\cdot 3\cdot 8\cdot 2 = 192$ different Hyundai Genesis cars that are possible.

Section 5.7 (on CD)

1.

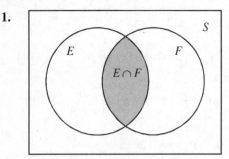

3. To partition a sample space S into three subsets A_1, A_2, and A_3 means that S is separated into three nonempty subsets where the subsets are all pairwise-disjoint. That is, $A_1 \cup A_2 \cup A_3 = S$.

5. Answers will vary. The Rule of Total Probability expresses a probability as a weighted average of conditional probabilities.

7. From the figure, $P(E\mid A)=0.6$.

9. From the figure, $P(E^c \mid A)=0.4$.

11. From the figure, $P(E\mid C)=0.4$.

13. $P(E)=P(A)\cdot P(E\mid A)$
$\qquad +P(B)\cdot P(E\mid B)$
$\qquad +P(B)\cdot P(E\mid B)$
$\qquad =0.2(0.6)+0.55(0.7)+0.25(0.4)$
$\qquad =0.12+0.385+0.1$
$\qquad =0.605$

15. $P(A\mid E)=\dfrac{P(A)\cdot P(E\mid A)}{P(E)}$

$$=\frac{0.2(0.6)}{0.2(0.6)+0.55(0.7)+0.25(0.4)}$$

$$=\frac{0.12}{0.605}\approx 0.198$$

17. $P(C\mid E)=\dfrac{P(C)\cdot P(E\mid C)}{P(E)}$

$$=\frac{0.25(0.4)}{0.2(0.6)+0.55(0.7)+0.25(0.4)}$$

$$=\frac{0.1}{0.605}\approx 0.165$$

19. $P(B\mid E)=\dfrac{P(B)\cdot P(E\mid B)}{P(E)}$

$$=\frac{0.55(0.7)}{0.2(0.6)+0.55(0.7)+0.25(0.4)}$$

$$=\frac{0.385}{0.605}\approx 0.636$$

21. $P(E)=P(A_1)\cdot P(E\mid A_1)+P(A_2)\cdot P(E\mid A_2)$
$\qquad =0.55(0.06)+0.45(0.08)$
$\qquad =0.069$

23. $P(E)=P(A_1)\cdot P(E\mid A_1)+P(A_2)\cdot P(E\mid A_2)$
$\qquad +P(A_3)\cdot P(E\mid A_3)$
$\qquad =0.35(0.25)+0.45(0.18)+0.2(0.14)$
$\qquad =0.1965$

25. (a) $P(A_1\mid E)=\dfrac{P(A_1)\cdot P(E\mid A_1)}{P(E)}$

$$=\frac{0.55(0.06)}{0.069}\approx 0.4783$$

(b) $P(A_2 \mid E) = \dfrac{P(A_2) \cdot P(E \mid A_2)}{P(E)}$

$= \dfrac{0.45(0.08)}{0.069} \approx 0.5217$

27. (a) $P(A_1 \mid E) = \dfrac{P(A_1) \cdot P(E \mid A_1)}{P(E)}$

$= \dfrac{0.35(0.25)}{0.1965} \approx 0.4453$

(b) $P(A_2 \mid E) = \dfrac{P(A_2) \cdot P(E \mid A_2)}{P(E)}$

$= \dfrac{0.45(0.18)}{0.1965} \approx 0.4122$

(c) $P(A_3 \mid E) = \dfrac{P(A_3) \cdot P(E \mid A_3)}{P(E)}$

$= \dfrac{0.2(0.14)}{0.1965} \approx 0.1425$

29. (a)

$\tfrac{1}{2}$ — U_1 — $\tfrac{5}{12}$ Black

$\tfrac{7}{12}$ White

$\tfrac{1}{2}$ — U_2 — $\tfrac{6}{14}$ Black

$\tfrac{5}{14}$ — White

$\tfrac{3}{14}$ Red

$P(B) = B \cap U_1 + B \cap U_2$

$= \dfrac{1}{2}\left(\dfrac{5}{12}\right) + \dfrac{1}{2}\left(\dfrac{6}{14}\right) = \dfrac{71}{168} \approx 0.4226$

(b) $P(B) = P(U_1) \cdot P(B \mid U_1)$

$\qquad + P(U_2) \cdot P(B \mid U_2)$

$= \dfrac{1}{2}\left(\dfrac{5}{12}\right) + \dfrac{1}{2}\left(\dfrac{6}{14}\right)$

$= \dfrac{71}{168} \approx 0.4226$

(c) $P(U_1 \mid B) = \dfrac{P(U_1) \cdot P(B \mid U_1)}{P(B)}$

$= \dfrac{(1/2)(5/12)}{71/168} = \dfrac{35}{71} \approx 0.4930$

(d) $P(U_1 \mid R) = 0$ since there are no red balls in urn I.

$P(U_2 \mid R) = 1$ since the only red balls are in urn II.

31. (a)

$P(B) = B \cap U_1 + B \cap U_2 + B \cap U_3$

$= \dfrac{3}{6}\left(\dfrac{6}{15}\right) + \dfrac{2}{6}\left(\dfrac{5}{26}\right) + \dfrac{1}{6}\left(\dfrac{11}{29}\right)$

$= \dfrac{617}{1885} \approx 0.3273$

(b) $P(B) = P(U_1) \cdot P(B \mid U_1)$

$\qquad + P(U_2) \cdot P(B \mid U_2)$

$\qquad + P(U_3) \cdot P(B \mid U_3)$

$= \dfrac{3}{6}\left(\dfrac{6}{15}\right) + \dfrac{2}{6}\left(\dfrac{5}{26}\right) + \dfrac{1}{6}\left(\dfrac{11}{29}\right)$

$= \dfrac{617}{1885} \approx 0.3273$

(c) $P(U_1 \mid B) = \dfrac{P(U_1) \cdot P(B \mid U_1)}{P(B)}$

$= \dfrac{(1/2)(6/15)}{617/1885} = \dfrac{377}{617} \approx 0.6110$

(d)
$$P(R) = P(U_1) \cdot P(R \mid U_1)$$
$$+ P(U_2) \cdot P(R \mid U_2)$$
$$+ P(U_3) \cdot P(R \mid U_3)$$
$$= \frac{3}{6}(0) + \frac{2}{6}\left(\frac{9}{26}\right) + \frac{1}{6}\left(\frac{12}{29}\right)$$
$$= \frac{139}{754} \approx 0.1844;$$

$$P(U_1 \mid R) = \frac{P(U_1) \cdot P(R \mid U_1)}{P(R)}$$
$$= \frac{(1/2)(0)}{139/754} = 0$$

(note that no red balls are in urn I);

$$P(U_2 \mid R) = \frac{P(U_2) \cdot P(R \mid U_2)}{P(R)} \quad ;$$
$$= \frac{(1/3)(9/26)}{139/754} = \frac{87}{139} \approx 0.6259$$

$$P(U_3 \mid R) = \frac{P(U_3) \cdot P(R \mid U_3)}{P(R)}$$
$$= \frac{(1/6)(12/29)}{139/754} = \frac{52}{139} \approx 0.3741$$

33. (a) $P(\text{colorblind})$
$$= P(\text{male}) \cdot P(\text{colorblind} \mid \text{male})$$
$$+ P(\text{female}) \cdot P(\text{colorblind} \mid \text{female})$$
$$= 0.491(0.08) + 0.509(0.0064)$$
$$= 0.0425376 \approx 0.0425$$

(b) $P(\text{female} \mid \text{colorblind})$
$$= \frac{P(\text{female}) \cdot P(\text{colorblind} \mid \text{female})}{P(\text{colorblind})}$$
$$= \frac{0.509(0.0064)}{0.0425376} \approx 0.0766$$

35. (a)
$$P(M) = P(A_1) \cdot P(M \mid A_1)$$
$$+ P(A_2) \cdot P(M \mid A_2)$$
$$+ P(A_3) \cdot P(M \mid A_3)$$
$$+ P(A_4) \cdot P(M \mid A_4)$$
$$+ P(A_5) \cdot P(M \mid A_5)$$
$$+ P(A_6) \cdot P(M \mid A_6)$$
$$= 0.134(0.505) + 0.312(0.482)$$
$$+ 0.172(0.467) + 0.088(0.433)$$
$$+ 0.191(0.480) + 0.103(0.513)$$
$$= 0.481001 \approx 0.4810$$

(b) $P(A_6 \mid M) = \dfrac{P(A_6) \cdot P(M \mid A_6)}{P(M)}$
$$= \frac{0.513(0.103)}{0.481001} \approx 0.1099$$

(c) $P(A_1 \mid M) = \dfrac{P(A_1) \cdot P(M \mid A_1)}{P(M)}$
$$= \frac{0.134(0.505)}{0.481001} \approx 0.1407$$

37. (a)
$$P(D) = P(A_1) \cdot P(D \mid A_1)$$
$$+ P(A_2) \cdot P(D \mid A_2)$$
$$+ P(A_3) \cdot P(D \mid A_3)$$
$$= 0.039(0.720) + 0.590(0.570)$$
$$+ 0.341(0.510)$$
$$= 0.55989 \approx 0.560$$

(b) $P(A_3 \mid D) = \dfrac{P(A_3) \cdot P(D \mid A_3)}{P(D)}$
$$= \frac{0.341(0.510)}{0.55989} \approx 0.311$$

(c) $P(A_1 \mid D) = \dfrac{P(A_1) \cdot P(D \mid A_1)}{P(D)}$
$$= \frac{0.069(0.720)}{0.55989} \approx 0.089$$

39.
$$\begin{aligned}
P(\text{found } G) &= P(G) \cdot P(\text{found } G \mid G) \\
&\quad + P(I) \cdot P(\text{found } G \mid I) \\
&= 0.95(0.90) + 0.05(0.01) \\
&= 0.8555
\end{aligned}$$

$$\begin{aligned}
P(I \mid \text{found } G) &= \frac{P(I) \cdot P(\text{found } G \mid I)}{P(\text{found } G)} \\
&= \frac{0.05(0.01)}{0.8555} \approx 0.00058
\end{aligned}$$

Chapter 5 Review Exercises

1. (a) Probabilities must be between 0 and 1, so the possible probabilities are: 0, 0.75, 0.41.

(b) Probabilities must be between 0 and 1, so the possible probabilities are: $\frac{2}{5}, \frac{1}{3}, \frac{6}{7}$.

2. Event E contains 1 of the 5 equally likely outcomes, so $P(E) = \frac{1}{5} = 0.2$.

3. Event F contains 2 of the 5 equally likely outcomes, so $P(F) = \frac{2}{5} = 0.4$.

4. Event E contains 3 of the 5 equally likely outcomes, so $P(E) = \frac{3}{5} = 0.6$.

5. Since $P(E) = \frac{1}{5} = 0.2$, we have
$$P(E^c) = 1 - \frac{1}{5} = \frac{4}{5} = 0.8.$$

6. $P(E \text{ or } F) = P(E) + P(F) - P(E \text{ and } F)$
$$= 0.76 + 0.45 - 0.32 = 0.89$$

7. Since events E and F are mutually exclusive,
$P(E \text{ or } F) = P(E) + P(F)$
$$= 0.36 + 0.12 = 0.48$$

8. Since events E and F are independent,
$P(E \text{ and } F) = P(E) \cdot P(F) = 0.45 \cdot 0.2 = 0.09$.

9. No, events E and F are not independent because
$P(E) \cdot P(F) = 0.8 \cdot 0.5 = 0.40 \neq P(E \text{ and } F)$.

10. $P(E \text{ and } F) = P(E) \cdot P(F \mid E)$
$$= 0.59 \cdot 0.45 = 0.2655$$

11. $P(E \mid F) = \dfrac{P(E \text{ and } F)}{P(F)} = \dfrac{0.35}{0.7} = 0.5$

12. (a) $7! = 7 \cdot 6 \cdot 5 \cdot 4 \cdot 3 \cdot 2 \cdot 1 = 5040$

(b) $0! = 1$

(c) $_9C_4 = \dfrac{9!}{4!(9-4)!} = \dfrac{9!}{4!5!} = \dfrac{9 \cdot 8 \cdot 7 \cdot 6}{4 \cdot 3 \cdot 2 \cdot 1} = 126$

(d) $_{10}C_3 = \dfrac{10!}{3!(10-3)!} = \dfrac{10!}{3!7!} = \dfrac{10 \cdot 9 \cdot 8}{3 \cdot 2 \cdot 1} = 120.$

(e) $_9P_2 = \dfrac{9!}{(9-2)!} = \dfrac{9!}{7!} = 9 \cdot 8 = 72$

(f) $_{12}P_4 = \dfrac{12!}{(12-4)!} = \dfrac{12!}{8!}$
$$= 12 \cdot 11 \cdot 10 \cdot 9 = 11{,}880.$$

13. (a) $P(\text{green}) = \dfrac{2}{38} = \dfrac{1}{19} \approx 0.0526$. If the wheel is spun 100 times, we would expect about 5 spins to end with the ball in a green slot.

(b) $P(\text{green or red}) = \dfrac{2+18}{38} = \dfrac{20}{38}$
$$= \dfrac{10}{19} \approx 0.5263$$
If the wheel is spun 100 times, we would expect about 53 spins to end with the ball in a green or a red slot.

(c) $P(00 \text{ or red}) = \dfrac{1+18}{38} = \dfrac{19}{38} = \dfrac{1}{2} = 0.5$
If the wheel is spun 100 times, we would expect about 50 spins to end with the ball in either a 00 or a red slot.

(d) Since 31 is an odd number and the odd slots are colored red,
$P(31 \text{ and black}) = 0$. This is called an impossible event.

14. (a) Of the 33,722 accidents, 9,817 were alcohol related, so
$$P(\text{alcohol related}) = \frac{9{,}817}{33{,}722} \approx 0.291.$$

(b) Of the 33,722 accidents, $33{,}722 - 9{,}817 = 23{,}905$ were not alcohol related, so
$$P(\text{not alcohol related}) = \frac{23{,}905}{33{,}722} \approx 0.709.$$

(c) $P\left(\text{both alcohol related}\right) = \dfrac{9,817}{33,722} \cdot \dfrac{9,817}{33,722}$
≈ 0.085

(d) $P(\text{neither was alcohol related})$
$= \dfrac{23,905}{33,722} \cdot \dfrac{23,905}{33,722} \approx 0.503$

(e) $P\left(\text{at least one of the two was alcohol related}\right)$
$= 1 - P\left(\text{neither was alcohol related}\right)$
$= 1 - \dfrac{23,905}{33,722} \cdot \dfrac{23,905}{33,722} \approx 0.497$

15. (a) There are $126 + 262 + 263 + 388 = 1039$ people who were surveyed. The individuals can be thought of as the trials of the probability experiment. The relative frequency of "Age 18-79" is
$\dfrac{126}{1,039} \approx 0.121$.
We compute the relative frequencies of the other outcomes similarly and obtain the probability model:

Age	Probability
18 − 79	0.121
80 − 89	0.252
90 − 99	0.253
100 +	0.373

(b) No, since the probability that an individual will want to live between 18 and 79 years is 0.121 (which is greater than 0.05), this is not unusual.

16. (a) Of the 4,305,340 births included in the table, 242,139 are postterm. Thus,
$P(\text{postterm}) = \dfrac{242,139}{4,305,340} \approx 0.056$.

(b) Of the 4,305,340 births included in the table, 2,825,549 weighed 3000 to 3999 grams. Thus, $P(3000 \text{ to } 3999 \text{ grams})$
$= \dfrac{2,825,549}{4,305,340} \approx 0.656$.

(c) Of the 4,305,340 births included in the table, 175,221 both weighed 3000 to 3999 grams and were postterm. Thus,

$P(3000 \text{ to } 3999 \text{ grams and postterm}) =$
$\dfrac{175,221}{4,305,340} \approx 0.041$.

(d) $P(3000 \text{ to } 3999 \text{ grams or postterm})$
$= P(3000 \text{ to } 3999 \text{ grams}) + P(\text{postterm})$
$- P(3000 \text{ to } 3999 \text{ grams and postterm})$
$= \dfrac{2,825,549}{4,305,340} + \dfrac{242,139}{4,305,340} - \dfrac{175,221}{4,305,340}$
$= \dfrac{2,892,467}{4,305,340} \approx 0.672$

(e) Of the 4,305,340 births included in the table, 31 both weighed less than 1000 grams and were postterm. Thus,
$P(\text{less than 1000 grams and postterm})$
$= \dfrac{31}{4,305,340} \approx 0.000007$.
This event is highly unlikely, but not impossible.

(f) $P(3000 \text{ to } 3999 \text{ grams} \mid \text{postterm})$
$= \dfrac{N(3000 \text{ to } 3999 \text{ grams and postterm})}{N(\text{postterm})}$
$= \dfrac{175,221}{242,139} \approx 0.724$

(g) No, the events "postterm baby" and "weighs 3000 to 3999 grams" are not independent since
$P(3000 \text{ to } 3999 \text{ grams}) \cdot P(\text{postterm})$
$\approx 0.656(0.055) = 0.036$
$\neq P(3000 \text{ to } 3999 \text{ grams and postterm})$
≈ 0.041.

17. (a) $P(\text{trust}) = 0.18$

(b) $P(\text{not trust}) = 1 - 0.18 = 0.82$

(c) $P(\text{all 3 trust}) = (0.18)^3 \approx 0.006$. This is surprising, since the probability is less than 0.05.

(d) $P(\text{at least one of the three does not trust})$
$= 1 - P(\text{all 3 trust})$
$= 1 - 0.006 = 0.994$

(e) $P(\text{none of the 5 trust}) = (0.82)^5 \approx 0.371$.
It is not surprising.

(f) $P(\text{at least one of the five trust})$
$= 1 - P(\text{none of 5 trust})$
$= 1 - 0.371 = 0.629$

18. $P\left(\text{matching the winning PICK 3 numbers}\right)$

$= \dfrac{1}{10} \cdot \dfrac{1}{10} \cdot \dfrac{1}{10} = \dfrac{1}{1000} = 0.001$

19. $P\left(\text{matching the winning PICK 4 numbers}\right)$

$= \dfrac{1}{10} \cdot \dfrac{1}{10} \cdot \dfrac{1}{10} \cdot \dfrac{1}{10} = \dfrac{1}{10,000} = 0.0001$

20. $P\left(\text{three aces}\right) = \dfrac{4}{52} \cdot \dfrac{3}{51} \cdot \dfrac{2}{50} \approx 0.00018$

21. $26 \cdot 26 \cdot 10^4 = 6,760,000$ license plates are possible.

22. $_{10}P_4 = 10 \cdot 9 \cdot 8 \cdot 7 = 5040$ seating arrangements are possible.

23. $\dfrac{10!}{4!3!2!} = 12,600$ different vertical arrangements of flags are possible.

24. $_{55}C_8 = 1,217,566,350$ simple random samples are possible.

25. $P\left(\begin{array}{c}\text{winning Arizona's}\\ \text{Pick 5}\end{array}\right) = \dfrac{1}{_{35}C_5}$

$= \dfrac{1}{324,632} \approx 0.000003$

26. (a) $P\left(\text{all three are Merlot}\right) = \dfrac{5}{12} \cdot \dfrac{4}{11} \cdot \dfrac{3}{10}$

$= \dfrac{1}{22} \approx 0.0455$

(b) $P\left(\begin{array}{c}\text{exactly two}\\ \text{are Merlot}\end{array}\right) = \dfrac{_5C_2 \cdot {_7}C_1}{_{12}C_3}$

$= \dfrac{5 \cdot 4}{2 \cdot 1} \cdot \dfrac{7}{1} \cdot \dfrac{3 \cdot 2 \cdot 1}{12 \cdot 11 \cdot 10}$

$= \dfrac{7}{22} \approx 0.3182$

(c) $P\left(\text{none are Merlot}\right) = \dfrac{7}{12} \cdot \dfrac{6}{11} \cdot \dfrac{5}{10}$

$= \dfrac{7}{44} \approx 0.1591$

27. (a), (b) Answers will vary depending on the results of the simulation. Results should be reasonably close to $\frac{1}{38}$ for part (a) and $\frac{1}{19}$ for part (b).

28. Subjective probabilities are probabilities based on personal experience or intuition.

Examples will vary. Some examples are the likelihood of life on other planets or the chance that the Packers will make it to the 2012 NFL playoffs.

29. (a) There are 13 clubs in the deck.

(b) There are 37 cards remaining in the deck. There are also 4 cards unknown by you. So, there are $37 + 4 = 41$ cards not known to you. Of the unknown cards, $13 - (3 + 2) = 8$ are clubs.

(c) $P\left(\text{next card dealt is a club}\right) = \dfrac{8}{41}$

≈ 0.1951

(d) $P\left(\text{two clubs in a row}\right) = \dfrac{8}{41} \cdot \dfrac{7}{40} \approx 0.0341$

(e) Answers will vary. One possibility follows: No, you should not stay in the game because the probability of completing the flush is low (0.0341 is less than 0.05).

30. (a) Since 20 of the 70 homeruns went to left field, $P\left(\text{left field}\right) = \dfrac{34}{70} = \dfrac{17}{35} \approx 0.486$. Mark McGwire hit 48.6% of his home runs to left field that year.

(b) Since none of the 70 homeruns went to right field, $P\left(\text{right field}\right) = \dfrac{0}{70} = 0$.

(c) No. While Mark McGwire did not hit any home runs to right field in 1998, this does not imply that it is impossible for him to hit a right-field homerun. He just never did it.

31. Someone winning a lottery twice is not that unlikely considering millions of people play lotteries who have already won a lottery (sometimes more than once) each week, and many lotteries have multiple drawings each week.

32. (a) $P\left(\text{Bryce}\right) = \dfrac{119}{1009} \approx 0.118$

This is not unusual since the probability is greater than 0.05.

(b) $P\left(\text{Gourmet}\right) = \dfrac{264}{1009} \approx 0.262$

(c) $P(\text{Mallory} \mid \text{Single Cup}) = \dfrac{75}{625} = 0.12$

(d) $P(\text{Bryce} \mid \text{Gourmet}) = \dfrac{9}{264} = \dfrac{3}{88} \approx 0.034$
This is unusual since the probability is less than 0.05.

(e) While it is not unusual for Bryce to sell a case, it is unlikely that he will sell a Gourmet case.

(f) $P(\text{Mallory}) = \dfrac{186}{1009} \approx 0.184$

$P(\text{Mallory} \mid \text{Filters}) = \dfrac{40}{120} = \dfrac{1}{3} \approx 0.333$

No, the events 'Mallory' and 'Filters' are not independent since
$P(\text{Mallory}) \neq P(\text{Mallory} \mid \text{Filters})$.

(g) No, the events 'Paige' and 'Gourmet' are not mutually exclusive because the two events can happen at the same time.

$P(\text{Paige and Gourmet}) = \dfrac{42}{1009} \approx 0.042 \neq 0$

Chapter 5 Test

1. Probabilities must be between 0 and 1, so the possible probabilities are: 0.23, 0, and $\dfrac{3}{4}$.

2. Event E contains 1 of the 5 equally likely outcomes, so $P(E) = \dfrac{1}{5} = 0.2$.

3. Event F contains 2 of the 5 equally likely outcomes, so $P(F) = \dfrac{2}{5} = 0.4$.

4. Since $P(E) = \dfrac{1}{5} = 0.2$, we have

$P(E^c) = 1 - \dfrac{1}{5} = \dfrac{4}{5} = 0.8$.

5. (a) Since events E and F are mutually exclusive, $P(E \text{ or } F) = P(E) + P(F)$
$= 0.37 + 0.22$
$= 0.59$

(b) Since events E and F are independent,
$P(E \text{ and } F) = P(E) \cdot P(F)$
$= (0.37)(0.22)$
$= 0.0814$.

6. (a) $P(E \text{ and } F) = P(E) \cdot P(F \mid E)$
$= 0.15 \cdot 0.70$
$= 0.105$

(b) $P(E \text{ or } F) = P(E) + P(F) - P(E \text{ and } F)$
$= 0.15 + 0.45 - 0.105$
$= 0.495$

(c) $P(E \mid F) = \dfrac{P(E \text{ and } F)}{P(F)} = \dfrac{0.105}{0.45}$

$= \dfrac{7}{30} \approx 0.233$

(d) $P(E \mid F) = \dfrac{7}{30} \approx 0.233$ and $P(E) = 0.15$.
Since $P(E \mid F) \neq F(E)$, the events E and F are not independent.

7. (a) $8! = 8 \cdot 7 \cdot 6 \cdot 5 \cdot 4 \cdot 3 \cdot 2 \cdot 1 = 40,320$

(b) $_{12}C_6 = \dfrac{12!}{6!(12-6)!} = \dfrac{12!}{6!6!}$

$= \dfrac{12 \cdot 11 \cdot 10 \cdot 9 \cdot 8 \cdot 7}{6 \cdot 5 \cdot 4 \cdot 3 \cdot 2 \cdot 1}$

$= 924$

(c) $_{14}P_8 = \dfrac{14!}{(14-8)!} = \dfrac{14!}{6!}$

$= 14 \cdot 13 \cdot 12 \cdot 11 \cdot 10 \cdot 9 \cdot 8 \cdot 7$

$= 121,080,960$

8. (a) A "7" can occur in six ways: $(1, 6)$, $(2, 5)$, $(3, 4)$, $(4, 3)$, $(5, 2)$, or $(6, 1)$. An "11" can occur two ways: $(6, 5)$ or $(5, 6)$. Thus, 8 of the 36 possible outcomes of the first roll results in a win, so

$P(\text{wins on first roll}) = \dfrac{8}{36} = \dfrac{2}{9} \approx 0.2222$.

If the dice are thrown 100 times, we would expect that the player will win on the first roll about 22 times.

(b) A "2" can occur in only one way: $(1, 1)$. A "3" can occur in two ways: $(1, 2)$ or $(2, 1)$. A "12" can occur in only one way: $(6, 6)$. Thus, 4 of the 36 possible outcomes of the first roll results in a loss, so

$P(\text{loses on first roll}) = \dfrac{4}{36} = \dfrac{1}{9} \approx 0.1111$.

If the dice are thrown 100 times, we would expect that the player will lose on the first roll about 11 times.

9. $P(\text{structure}) = 0.32$;

$P(\text{not in a structure}) = 1 - P(\text{structure})$
$= 1 - 0.32$
$= 0.68$

10. (a) Rule 1 is satisfied since all of the probabilities in the model are between 0 and 1.
Rule 2 is satisfied since the sum of the probabilities in the model is one:
$0.25 + 0.19 + 0.13 + 0.11 + 0.09 + 0.23 = 1$.

(b) $P\left(\begin{array}{l}\text{PB Patties/Tagalongs or}\\\text{PB Sandwich/Do-si-dos}\end{array}\right) = 0.13 + 0.11$
$= 0.24$

(c) $P\left(\begin{array}{l}\text{Thin Mints, Samoas,}\\\text{Shortbread}\end{array}\right) = 0.25 + 0.19 + 0.09$
$= 0.53$

(d) $P(\text{not Thin Mints}) = 1 - 0.25 = 0.75$

11. (a) Of the 297 people surveyed, 155 thought the ideal number of children is 2. Thus,
$P(\text{ideal is 2}) = \dfrac{155}{297} \approx 0.522$.

(b) Of the 297 people surveyed, 87 were females and thought the ideal number of children is 2. Thus,
$P(\text{female and ideal is 2}) = \dfrac{87}{297} \approx 0.293$.

(c) Of the 297 people surveyed, 155 thought the ideal number of children is 2, 188 were females, and 87 were females and thought the ideal number of children is 2. Thus,
$P(\text{female or ideal is 2})$
$= P(\text{female}) + P(\text{ideal is 2})$
$\quad - P(\text{female and ideal is 2})$
$= \dfrac{188}{297} + \dfrac{155}{297} - \dfrac{87}{297} = \dfrac{256}{297} \approx 0.862$

(d) Of the 188 females, 87 thought the ideal number of children is 2. Thus,
$P(\text{ideal is 2} \mid \text{female}) = \dfrac{87}{188} \approx 0.463$.

(e) Of the 36 people who said the ideal number of children was 4, 8 were males. Thus,
$P(\text{male} \mid \text{ideal is 4}) = \dfrac{8}{36} \approx 0.222$.

12. (a) Assuming the outcomes are independent, we get $P(\text{win two games in a row})$
$= (0.58)^2 \approx 0.336$.

(b) Assuming the outcomes are independent, we get $P(\text{win seven games in a row})$
$= (0.58)^7 \approx 0.022$.

(c) $P(\text{lose at least one of next seven games})$
$P(\text{lose at least one of next seven games})$
$= 1 - P(\text{win seven games in a row})$
$= 1 - (0.58)^7 \approx 1 - 0.022 = 0.978$

13. $P(\text{accept}) = P(\text{1st works}) \cdot P(\text{2nd works} \mid \text{1st works})$
$= \dfrac{9}{10} \cdot \dfrac{8}{9} = 0.8$

14. $6! = 720$ different arrangements are possible for the letters in the 'word' LINCEY.

15. $_{29}C_5 = 118{,}755$ different subcommittees are possible.

16. $P(\text{winning Pennsylvania's Cash 5})$
$= \dfrac{1}{_{43}C_5} = \dfrac{1}{962{,}598} \approx 0.00000104$

17. There are 26 ways to select the first character, and 36 ways to select each of the final 7 characters. Therefore, using the Multiplication Rule of Counting, there are
$26 \cdot 36^7 = 2.04 \times 10^{12}$ different passwords that are possible for the local area network.

18. It would take the processor
$\dfrac{26 \cdot 36^7}{3.0 \times 10^6} \approx 679{,}156$ seconds (about 7.9 days)
to compute all the passwords.

19. Subjective probability assignment was used. It is not possible to repeat probability experiments to estimate the probability of life on Mars.

20. Since there are two A's, four C's, four G's, and five T's from which to form a DNA sequence, there are $\dfrac{15!}{2!4!4!5!} = 9,459,450$ distinguishable DNA sequences possible using the 15 letters.

21. There are 40 digits, of which 9 are either a 0 or a I. Therefore,

$$P(\text{guess correctly}) \approx \frac{9}{40} = 0.225.$$

Case Study: The Case of the Body in the Bag

1. Of the 7285 offenders included in the "Victim-Offender Relationship by Age" table, 6327 are known to be at least 18. Thus,

$$P(\text{offender is at least } 18) = \frac{6327}{7285} \approx 0.8685.$$

2. Of the 7285 offenders included in the "Victim-Offender Relationship by Race" table, 3318 are known to be white. Thus,

$$P(\text{offender is white}) = \frac{3318}{7285} \approx 0.4555.$$

3. Of the 7285 offenders included in the "Sex of Victim by Sex of Offender" table, 6495 are known to be male. Thus,

$$P(\text{offender is male}) = \frac{6495}{7285} \approx 0.8916.$$

4. Of the 14,990 victims included in the "Murder Victims by Race and Sex" table, 1883 are known to be white females. Thus,

$$P(\text{victim is white and female}) = \frac{1883}{14,990}$$
$$\approx 0.1256.$$

5. $P(\text{victim is white or female})$

$$= P(\text{victim is white}) + P(\text{victim is female})$$
$$- P(\text{victim is white and female})$$
$$= \frac{6956}{14490} + \frac{3156}{14490} - \frac{1883}{14490} == \frac{8229}{14490} \approx 0.5490$$

6. $P\left(\begin{array}{l}\text{victim and} \\ \text{offender are} \\ \text{from same} \\ \text{age category}\end{array}\right)$

$$= P\left(\begin{array}{l}\text{both are less} \\ \text{than 18}\end{array}\right) + P\left(\begin{array}{l}\text{both are at} \\ \text{least 18}\end{array}\right) + P\left(\begin{array}{l}\text{both are} \\ \text{unknown}\end{array}\right)$$

$$= \frac{124}{7285} + \frac{5628}{7285} + \frac{21}{7285} = \frac{5773}{7285} \approx 0.7925$$

7. $P\left(\begin{array}{l}\text{victim and offender} \\ \text{are from different} \\ \text{age categories}\end{array}\right) = 1 - P\left(\begin{array}{l}\text{victim and} \\ \text{offender are} \\ \text{from same} \\ \text{age categories}\end{array}\right)$

$$\approx 1 - 0.7925 = 0.2075$$

8. $P\left(\begin{array}{l}\text{victim and} \\ \text{offender are} \\ \text{from same} \\ \text{race categories}\end{array}\right)$

$$= P(\text{both are white}) + P(\text{both are black})$$
$$+ P(\text{both are other}) + P(\text{both are unknown})$$
$$= \frac{3026}{7285} + \frac{3034}{7285} + \frac{97}{7285} + \frac{26}{7285}$$
$$= \frac{6183}{7285} \approx 0.8487$$

9. $P\left(\begin{array}{l}\text{victim and offender} \\ \text{are from different} \\ \text{race categories}\end{array}\right)$

$$= 1 - P\left(\begin{array}{l}\text{victim and offender} \\ \text{are from same race} \\ \text{categories}\end{array}\right)$$

$$\approx 1 - 0.8487 = 0.1513$$

10. $P\left(\begin{array}{l}\text{victim and} \\ \text{offender are} \\ \text{of the same} \\ \text{sex categories}\end{array}\right)$

$$= P(\text{both are male}) + P(\text{both are female})$$
$$+ P(\text{both are unknown})$$
$$= \frac{4710}{7285} + \frac{150}{7285} + \frac{26}{7285} = \frac{4886}{7285} \approx 0.6707$$

11. $P\left(\begin{array}{l}\text{victim and} \\ \text{offender are} \\ \text{from different} \\ \text{sex categories}\end{array}\right) = 1 - P\left(\begin{array}{l}\text{victim and} \\ \text{offender are} \\ \text{from same} \\ \text{sex categories}\end{array}\right)$

$$\approx 1 - 0.6707 = 0.3293$$

12. Answers will vary. One possibility is that the victim is more likely to be at least 18, white, and male. The reason is that the probabilities of each outcome is higher than the other outcomes for each category:

$$P(\text{victim is at least 18}) = \frac{6398}{7285} \approx 0.8782 \, ;$$

$$P(\text{victim is white}) = \frac{3709}{7285} \approx 0.5091 \, ;$$

$$P(\text{victim is male}) = \frac{5289}{7285} \approx 0.7260 \, .$$

13. Answers will vary. One possibility is that the offender is more likely to be at least 18, black, and male. The reason is that the probabilities of each outcome is higher than the other outcomes for given category:

$$P(\text{offender is at least 18}) = \frac{6327}{7285} \approx 0.8685 \, ;$$

$$P(\text{offender is black}) = \frac{3662}{7285} \approx 0.5027 \, ;$$

$$P(\text{offender is male}) = \frac{6495}{7285} \approx 0.8916 \, .$$

14. We have no information to relate age of the offender to the race or sex of the victim. The best we can determine is:

$P(\text{offender is at least 18} \mid \text{victim is at least 18})$

$$= \frac{5628}{290 + 5628 + 480} = \frac{5628}{6398} \approx 0.8796$$

15. We have no information to relate race of the offender to the age of the victim. The best we can determine is:

$P(\text{offender is white} \mid \text{victim is white})$

$$= \frac{3026}{3026 + 573 + 53 + 57} = \frac{3026}{3709} \approx 0.8159$$

$P(\text{offender is white} \mid \text{victim is female})$

$$= \frac{1126}{1126 + 700 + 59 + 24} = \frac{1126}{1909} \approx 0.5898$$

Note: We cannot determine the probability that the offender is white given that the victim is a white female because necessary information is not provided.

16. We have no information to relate sex of the offender to the age of the victim. The best we can determine is:

$P(\text{offender is male} \mid \text{victim is white})$

$$= \frac{3322}{3322 + 330 + 57} = \frac{3322}{3709} \approx 0.8957$$

$P(\text{offender is male} \mid \text{victim is female})$

$$= \frac{1735}{1735 + 150 + 24} = \frac{1735}{1909} \approx 0.9089$$

Note: We cannot determine the probability that the offender is male given that the victim is a white female because necessary information is not provided.

17. Since the victim's age category is "at least 18", we find:

$P(\text{offender is at least 18} \mid \text{victim is at least 18})$

$$= \frac{2528}{290 + 5628 + 480} = \frac{2528}{6398} \approx 0.8796$$

18. Since the victim's age category is "at least 18", we find:

$P(\text{offender is not "at least 18"} \mid \text{victim is at least 18})$

$$\approx 1 - 0.8796 = 0.1204$$

19. Since the victim is white, we find:

$P(\text{offender is white} \mid \text{victim is white})$

$$= \frac{3026}{3026 + 573 + 53 + 57} = \frac{3026}{3709} \approx 0.8159$$

20. Since the victim is white, we find:

$P(\text{offender is not white} \mid \text{victim is white})$

$$\approx 1 - 0.8159 = 0.1841$$

21. Since the victim is female, we find:

$P(\text{offender is female} \mid \text{victim is female})$

$$= \frac{150}{1735 + 150 + 24} = \frac{150}{1909} \approx 0.0786$$

22. Since the victim is female, we find:

$P(\text{offender is not female} \mid \text{victim is female})$

$$\approx 1 - 0.0786 = 0.9214$$

23. Answers will vary. One possibility is that, since we know the victim is a 30-year-old, white female, the offender is more likely to be at least 18, white, and male. The reason is that, given the restrictions for the victim, the probabilities of each outcome is higher than the other outcomes for given category:

$P(\text{offender is at least 18} \mid \text{victim is at least 18})$

$$= 0.8796$$

$P(\text{offender is white} \mid \text{victim is female})$

$$= 0.5898$$

$P(\text{offender is male} \mid \text{victim is female})$

$$= 0.9089$$

Copyright © 2013 Pearson Education, Inc.

$P\left(\text{offender is male} \mid \text{victim is white}\right).$

$= 0.8957$

Note: we do not have sufficient information to determine the probability that the offender is a white male who is at least 18 given that the victim is a 30-year-old white female.

24. Answers will vary. In the provided example, the profile of the offender changed from a black male who was at least 18 to a white male who was at least 18 once the age, race, and sex of the victim was known.

Final profiles will vary. One possibility follows: We expect that the offender in this case is a white male who is at least 18 years old. Some key suspects would be any men who have a relationship with the victim, such as husband or boyfriend. We would also want to look at other males who were related to or acquainted with the victim.

Chapter 6

Discrete Probability Distributions

Section 6.1

1. A random variable is a numerical measure of the outcome of a probability experiment, so its value is determined by chance.

3. For a discrete probability distribution, each probability must be between 0 and 1 (inclusive) and the sum of the probabilities must equal one.

5. **(a)** The number of light bulbs that burn out, X, is a discrete random variable because the value of the random variable results from counting.
   Possible values: $x = 0, 1, 2, 3, ..., 20$

   (b) The time it takes to fly from New York City to Los Angeles is a continuous random variable because time is measured. If we let the random variable T represent the time it takes to fly from New York City to Los Angeles, the possible values for T are all positive real numbers; that is $t > 0$.

   (c) The number of hits to a web site in a day is a discrete random variable because the value of the random variable results from counting. If we let the random variable X represent the number of hits, then the possible values of X are $x = 0, 1, 2, ...$.

   (d) The amount of snow in Toronto during the winter is a continuous random variable because the amount of snow is measured. If we let the random variable S represent the amount of snow in Toronto during the winter, the possible values for S are all nonnegative real numbers; that is $s \geq 0$.

7. **(a)** The amount of rain in Seattle during April is a continuous random variable because the amount of rain is measured. If we let the random variable R represent the amount of rain, the possible values for R are all nonnegative real numbers; that is, $r \geq 0$.

   (b) The number of fish caught during a fishing tournament is a discrete random variable because the value of the random variable results from counting. If we let the random variable X represent the number of fish caught, the possible values of X are $x = 0, 1, 2, ...$.

   (c) The number of customers arriving at a bank between noon and 1:00 P.M. is a discrete random variable because the value of the random variable results from counting. If we let the random variable X represent the number of customers arriving at the bank between noon and 1:00 P.M., the possible values of X are $x = 0, 1, 2, ...$

   (d) The time required to download a file from the Internet is a continuous random variable because time is measured. If we let the random variable T represent the time required to download a file, the possible values of T are all positive real numbers; that is, $t > 0$.

9. Yes, because $\sum P(x) = 1$ and $0 \leq P(x) \leq 1$ for all x.

11. No, because $P(50) < 0$.

13. No, because $\sum P(x) = 0.95 \neq 1$.

15. We need the sum of all the probabilities to equal 1. For the given probabilities, we have $0.4 + 0.1 + 0.2 = 0.7$. For the sum of the probabilities to equal 1, the missing probability must be $1 - 0.7 = 0.3$. That is, $P(4) = 0.3$.

17. (a) This is a discrete probability distribution because all the probabilities are between 0 and 1 (inclusive) and the sum of the probabilities is 1.

(b)

Televisions per Household

(c) $\mu_X = \sum[x \cdot P(x)] = 0(0) + 1(0.161) + \ldots + 9(0.010) = 3.210 \approx 3.2$

If we surveyed many households, we would expect the mean number of televisions per household to be about 3.2.

(d) $\sigma_X^2 = \sum\left[(x - \mu_X)^2 \cdot P(x)\right]$

$= (0 - 3.210)^2 (0) + (1 - 3.210)^2 (0.161) + \ldots + (9 - 3.210)^2 (0.010)$

≈ 3.0159

$\sigma_X = \sqrt{\sigma_X^2} = \sqrt{3.0159} \approx 1.7366$ or about 1.7 televisions per household.

(e) $P(3) = 0.176$

(f) $P(3 \text{ or } 4) = P(3) + P(4) = 0.176 + 0.186 = 0.362$

(g) $P(0) = 0$; This is not an impossible event, but it is very unlikely.

19. (a) This is a discrete probability distribution because all the probabilities are between 0 and 1 (inclusive) and the sum of the probabilities is 1.

(b)

Ichiro's Hit Parade

(c) $\mu_X = \sum[x \cdot P(x)] = 0(0.1677) + 1(0.3354) + \ldots + 5(0.0248) = 1.6273 \approx 1.6$

Over many games, Ichiro is expected to average about 1.6 hits per game.

(d) $\sigma_X^2 = \sum\left[(x - \mu_X)^2 \cdot P(x)\right]$

$= (0 - 1.6273)^2 (0.1677) + (1 - 1.6273)^2 (0.3354) + \ldots + (5 - 1.6273)^2 (0.0248)$

≈ 1.389

$\sigma_X = \sqrt{\sigma^2_X} \approx \sqrt{1.389} \approx 1.179$ or about 1.2 hits

(e) $P(2) = 0.2857$

(f) $P(X > 1) = 1 - P(X \le 1) = 1 - P(0 \text{ or } 1) = 1 - (0.1677 + 0.3354) = 1 - 0.5031 = 0.4969$

21. **(a)** Total number of World Series
= 17 + 18 + 19 + 33 = 87.

x (games played)	$P(x)$
4	$\dfrac{17}{87} \approx 0.1954$
5	$\dfrac{18}{87} \approx 0.2069$
6	$\dfrac{19}{87} \approx 0.2184$
7	$\dfrac{33}{87} \approx 0.3793$

(b)

Games Played in the World Series

(c) $\mu_X = \sum \left[x \cdot P(x) \right]$

$\approx 4 \cdot (0.1954) + 5 \cdot (0.2069) + 6 \cdot (0.2184) + 7 \cdot (0.3793) \approx 5.7816$ or about 5.8 games

The World Series, if played many times, would be expected to last about 5.8 games on average.

(d) $\sigma_X^2 = \sum (x - \mu_x)^2 \cdot P(x)$

$\approx (4 - 5.7816)^2 \cdot 0.1954 + (5 - 5.7816)^2 \cdot 0.2069 + (6 - 5.7816)^2 \cdot 0.2184 + (7 - 5.7816)^2 \cdot 0.3793$

≈ 1.3201

$\sigma_X = \sqrt{\sigma_X^2} \approx \sqrt{1.3201} \approx 1.1490$ or about 1.1 games

23. **(a)** Total number of ratings:
= 2752 + 2331 + 4215 + 3902 + 1051 =
23,718.

x (Stars)	$P(x)$
1	$\dfrac{2,752}{23,718} \approx 0.1160$
2	$\dfrac{2,331}{23,718} \approx 0.0983$
3	$\dfrac{4,215}{23,718} \approx 0.1777$
4	$\dfrac{3,902}{23,718} \approx 0.1645$
5	$\dfrac{10,518}{23,718} \approx 0.4435$

(b)

Paper Toss

(c) $\mu_X = \sum \left[x \cdot P(x) \right]$

$\approx 1 \cdot (0.11160) + 2 \cdot (0.0983) + 3 \cdot (0.1777) + 4 \cdot (0.1645) + 5 \cdot (0.4435) \approx 3.7211$ or about 3.7 stars

If we surveyed many Paper Toss players, the mean rating would be 3.7.

(d) $\sigma_X^2 = \sum (x - \mu_x)^2 \cdot P(x)$

$\approx (1 - 3.7211)^2 \cdot 0.1160 + (2 - 3.7211)^2 \cdot 0.0983 + \ldots + (5 - 3.7211)^2 \cdot 0.4435 \approx 1.9808$

$\sigma_X = \sqrt{\sigma_X^2} \approx \sqrt{1.9808} \approx 1.4074$ or about 1.4 stars

25. (a) $P(4) = 0.099$

(b) $P(4 \text{ or } 5) = P(4) + P(5) = 0.099 + 0.054 = 0.153$

(c) $P(X \geq 6) = P(6) + P(7) + P(8) = 0.043 + 0.021 + 0.061 = 0.125$

(d) $E(X) = \mu_X = \sum x \cdot P(x) = 1(0.313) + 2(0.244) + 3(0.165) + \dots + 8(0.061) = 2.855$ or about 2.9

We would expect the mother to have had 2.9 live births, on average.

27. $E(X) = \sum x \cdot P(x)$
$$= (200)(0.999544) + (200 - 250,000)(0.000456) = \$86.00$$

If the company sells many of these policies to 20-year old females, then they will make an average of \$86.00 per policy.

29. Let X = the profit from the investment

Profit, x (\$)	50,000	10,000	−50,000
Probability	0.2	0.7	0.1

$$E(X) = (50,000)(0.2) + (10,000)(0.7) + (-50,000)(0.1) = 12,000$$

The expected profit for the investment is \$12,000.

31. Let X = player winnings for \$5 bet on a single number.

Winnings, x (\$)	175	−5
Probability	$\frac{1}{38}$	$\frac{37}{38}$

$$E(X) = (175)\left(\frac{1}{38}\right) + (-5)\left(\frac{37}{38}\right) = -\$0.26$$

The expected value of the game to the player is a loss of \$0.26. If you played the game 1000 times, you would expect to lose $1000 \cdot (\$0.26) = \$260$.

33. (a) $E(X) = \sum x \cdot P(x)$
$$= (15,000,000)(0.00000000684) + (200,000)(0.00000028)$$
$$+ (10,000)(0.000001711) + (100)(0.000153996) + (7)(0.004778961)$$
$$+ (4)(0.007881463) + (3)(0.01450116) + (0)(0.9726824222)$$
$$\approx 0.30$$

After many \$1 plays, you would expect to win an average of \$0.30 per play. That is, you would lose an average of \$0.70 per \$1 play for a net profit of $-\$0.70$.

(Note: the given probabilities reflect changes made in April 2005 to create larger jackpots that are built up more quickly. It is interesting to note that prior to the change, the expected cash prize was still \$0.30)

(b) We need to find the break-even point. That is, the point where we expect to win the same amount that we pay to play. Let x be the grand prize. Set the expected value equation equal to 1 (the cost for one play) and then solve for x.

$$E(X) = \sum x \cdot P(x)$$
$$1 = (x)(0.00000000684) + (200,000)(0.00000028)$$
$$+ (10,000)(0.000001711) + (100)(0.000153996) + (7)(0.004778961)$$
$$+ (4)(0.007881463) + (3)(0.01450116) + (0)(0.9726824222)$$
$$1 = 0.196991659 + 0.00000000684x$$
$$0.803008341 = 0.00000000684x$$
$$117,398,880.3 = x$$
$$118,000,000 \approx x$$

The grand prize should be at least \$118,000,000 for you to expect a profit after many \$1 plays. (Note: prior to the changes mentioned in part (a), the grand prize only needed to be about \$100 million to expect a profit after many \$1 plays)

(c) No, the size of the grand prize does not affect your chance of winning. Your chance of winning the grand prize is determined by the number of balls that are drawn and the number of balls from which they are drawn. However, the size of the grand prize will impact your expected winnings. A larger grand prize will increase your expected winnings.

35. Answers will vary. The simulations illustrate the Law of Large Numbers.

37. (a) The mean is the sum of the values divided by the total number of observations.

The mean is: $\mu_X = \frac{529}{160} \approx 3.3063$ or about 3.3 credit cards

(b) The standard deviation is computed by subtracting the mean from each value, squaring the result, and summing. Then, to get the population variance, we divide the sum by the number of observations. The square root of the variance is the standard deviation.

$$\sigma_X^2 = \frac{\sum (x - \mu_x)^2}{N} = \frac{(3 - 3.3063)^2 + (2 - 3.3063)^2 + \ldots + (5 - 3.3063)^2}{160} \approx 5.3585$$

$$\sigma_X = \sqrt{\sigma_X^2} \approx \sqrt{5.3585} \approx 2.3148 \text{ or about 2.3}$$

(c)

x (Cards)	$P(x)$	x (Cards)	$P(x)$	x (Cards)	$P(x)$
1	$\frac{23}{160} = 0.14375$	5	$\frac{13}{160} = 0.08125$	9	$\frac{2}{160} = 0.0125$
2	$\frac{44}{160} = 0.275$	6	$\frac{7}{160} = 0.04375$	10	$\frac{2}{160} = 0.0125$
3	$\frac{43}{160} = 0.26875$	7	$\frac{4}{160} = 0.025$	20	$\frac{1}{160} = 0.00625$
4	$\frac{18}{160} = 0.1125$	8	$\frac{3}{160} = 0.01875$		

(d)

Credit Cards

(e) $\mu_X = \sum \left[x \cdot P(x) \right]$

$\approx 1 \cdot (0.14375) + 2 \cdot (0.275) + 3 \cdot (0.26875) + \cdots + 10 \cdot (0.0125) + 20 \cdot (0.00625)$

≈ 3.3063 or about 3.3 credit cards

$\sigma_X^2 = \sum (x - \mu_x)^2 \cdot P(x)$

$= (1 - 3.3063)^2 \cdot 0.14375 + (2 - 3.3063)^2 \cdot 0.275 + \ldots + (20 - 3.3063)^2 \cdot 0.00625$

≈ 5.234961

$\sigma_X = \sqrt{\sigma_X^2} \approx \sqrt{5.234961} \approx 2.3076$ or about 2.3 credit cards

(f) $\mu - 2\sigma = 3.3 - 2 \cdot 2.3 = -1.3$

$\mu + 2\sigma = 3.3 + 2 \cdot 2.3 = 7.9$

So, the probability of selecting an individual whose number of credit cards is more than two standard deviations from the mean is the same as picking someone who has at least 8 credit cards.

$P(X \geq 8) = P(X = 8) + P(X = 9) + P(X = 10 + P(X = 20)$

$= 0.01875 + 0.0125 + 0.0125 + 0.00625$

$= 0.05$

This result is right on the boundary between being unusual and not unusual. We can say that this result is a little unusual.

(g) $P(\text{Choose 2 with exactly 2 cards}) = \left(\dfrac{44}{160} \right) \left(\dfrac{43}{159} \right) = 0.0744$

Section 6.2

1. trial

3. True

5. np

7. This is not a binomial experiment because there are more than two possible values for the random variable 'age'.

9. This is a binomial experiment. There is a fixed number of trials ($n = 100$ where each trial corresponds to administering the drug to one of the 100 individuals), the trials are independent (due to random selection), there are two outcomes (favorable or unfavorable response), and the probability of a favorable response is assumed to be constant for each individual.

11. This is not a binomial experiment because the trials (cards) are not independent and the probability of getting an ace changes for each trial (card). Because the cards are not replaced, the probability of getting an ace on the second card depends on what was drawn first.

13. This is not a binomial experiment because the number of trials is not fixed.

15. This is a binomial experiment. There is a fixed number of trials ($n = 100$ where each trial corresponds to selecting one of the 100 parents), the trials are independent (due to random selection), there are two outcomes (spanked or never spanked), and there is a fixed probability (for a given population) that a parent has ever spanked their child.

17. Using $P(x) = {}_nC_x p^x (1-p)^{n-x}$ with $x = 3$, $n = 10$ and $p = 0.4$:

$$P(3) = {}_{10}C_3 \cdot (0.4)^3 \cdot (1-0.4)^{10-3} = \frac{10!}{3!(10-3)!} \cdot (0.4)^3 \cdot (0.6)^7$$
$$= 120 \cdot (0.064) \cdot (0.0279936) \approx 0.2150$$

19. Using $P(x) = {}_nC_x p^x (1-p)^{n-x}$ with $x = 38$, $n = 40$ and $p = 0.99$:

$$P(38) = {}_{40}C_{38} \cdot (0.99)^{38} \cdot (1-0.99)^{40-38} = \frac{40!}{38!(40-38)!} \cdot (0.99)^{38} \cdot (0.01)^2$$
$$= 780 \cdot (0.6825...) \cdot (0.0001) \approx 0.0532$$

21. Using $P(x) = {}_nC_x p^x (1-p)^{n-x}$ with $x = 3$, $n = 8$ and $p = 0.35$:

$$P(3) = {}_8C_3 \cdot (0.35)^3 \cdot (1-0.35)^{8-3} = \frac{8!}{3!(8-3)!} \cdot (0.35)^3 \cdot (0.65)^5$$
$$= 56 \cdot (0.42875)(0.116029...) \approx 0.2786$$

23. Using $n = 9$ and $p = 0.2$:

$$P(X \le 3) = P(0) + P(1) + P(2) + P(3)$$
$$= {}_9C_0 \cdot (0.2)^0 (0.8)^9 + {}_9C_1 \cdot (0.2)^1 (0.8)^8 + {}_9C_2 \cdot (0.2)^2 (0.8)^7 + {}_9C_3 \cdot (0.2)^3 (0.8)^6$$
$$\approx 0.134218 + 0.301990 + 0.301990 + 0.176161 \approx 0.9144$$

25. Using $n = 7$ and $p = 0.5$:

$$P(X > 3) = P(X \ge 4)$$
$$= P(4) + P(5) + P(6) + P(7)$$
$$= {}_7C_4 \cdot (0.5)^4 (0.5)^3 + {}_7C_5 \cdot (0.5)^5 (0.5)^2 + {}_7C_6 \cdot (0.5)^6 (0.5)^1 + {}_7C_7 \cdot (0.5)^7 (0.5)^0$$
$$= 0.2734375 + 0.1640625 + 0.0546875 + 0.0078125 = 0.5$$

27. Using $n = 12$ and $p = 0.35$:

$$P(X \le 4) = P(0) + P(1) + P(2) + P(3) + P(4)$$
$$= {}_{12}C_0 \cdot (0.35)^0 (0.65)^{12} + {}_{12}C_1 \cdot (0.35)^1 (0.65)^{11} + {}_{12}C_2 \cdot (0.35)^2 (0.65)^{10}$$
$$+ {}_{12}C_3 \cdot (0.35)^3 (0.65)^9 + {}_{12}C_4 \cdot (0.35)^4 (0.65)^8$$
$$\approx 0.005688 + 0.036753 + 0.108846 + 0.195365 + 0.236692 \approx 0.5833$$

29. (a)

Distribution		$x \cdot P(x)$	$(x - \mu_x)^2 \cdot P(x)$
x	$P(x)$		
0	0.1176	0.0000	0.3812
1	0.3025	0.3025	0.1936
2	0.3241	0.6483	0.0130
3	0.1852	0.5557	0.2667
4	0.0595	0.2381	0.2881
5	0.0102	0.0510	0.1045
6	0.0007	0.0044	0.0129
	Σ	1.8000	1.2600

(b) $\mu_X = 1.8$ (from first column in table above and to the right)

$\sigma_X = \sqrt{\sigma_X^2} = \sqrt{1.26} \approx 1.1$ (from second column in table above and to the right)

(c) $\mu_X = n \cdot p = 6 \cdot (0.3) = 1.8$ and $\sigma_X = \sqrt{n \cdot p \cdot (1-p)} = \sqrt{(6) \cdot (0.3) \cdot (0.7)} \approx 1.1$

(d)

The distribution is skewed right.

31. (a)

Distribution		$x \cdot P(x)$	$(x - \mu_x)^2 \cdot P(x)$
x	$P(x)$		
0	0.0000	0.0000	0.0002
1	0.0001	0.0001	0.0034
2	0.0012	0.0025	0.0279
3	0.0087	0.0260	0.1217
4	0.0389	0.1557	0.2944
5	0.1168	0.5840	0.3577
6	0.2336	1.4016	0.1314
7	0.3003	2.1024	0.0188
8	0.2253	1.8020	0.3520
9	0.0751	0.6758	0.3801
Σ		6.7500	1.6876

(b) $\mu_X = 6.75$ (from first column in table above and to the right)

$\sigma_X = \sqrt{\sigma_X^2} = \sqrt{1.6876} \approx 1.3$ (from second column in table above and to the right)

(c) $\mu_X = n \cdot p = 9 \cdot (0.75) = 6.75$ and $\sigma_X = \sqrt{n \cdot p \cdot (1-p)} = \sqrt{(9) \cdot (0.75) \cdot (0.25)} \approx 1.3$

(d)

The distribution is skewed left.

33. (a)

Distribution		$x \cdot P(x)$	$(x - \mu_x)^2 \cdot P(x)$
x	$P(x)$		
0	0.0010	0.0000	0.0250
1	0.0098	0.0098	0.1568
2	0.0439	0.0878	0.3952
3	0.1172	0.3516	0.4690
4	0.2051	0.8204	0.2053
5	0.2461	1.2305	0.0000
6	0.2051	1.2306	0.2049
7	0.1172	e0.8204	0.4686
8	0.0439	0.3512	0.3950
9	0.0098	0.0882	0.1568
10	0.0010	0.0100	0.0250
	Σ	5.0005	2.5016

(b) $\mu_X = 5.0$ (from first column in table above and to the right)

$\sigma_X = \sqrt{\sigma_X^2} = \sqrt{2.5016} \approx 1.6$ (from second column in table above and to the right)

(c) $\mu_X = n \cdot p = 10 \cdot (0.5) = 5$ and $\sigma_X = \sqrt{n \cdot p \cdot (1-p)} = \sqrt{10 \cdot (0.5) \cdot (0.5)} \approx 1.6$

(d)

The distribution is symmetric.

35. (a) This is a binomial experiment because it satisfies each of the four requirements:
   1) There are a fixed number of trials ($n = 15$).
   2) The trials are all independent (randomly selected).
   3) For each trial, there are only two possible outcomes ('on time' and 'not on time').
   4) The probability of "success" (i.e. on time) is the same for all trials ($p = 0.65$).

(b) We have $n = 15$, $p = 0.80$, and $x = 10$. In the binomial table, we go to the section for $n = 15$ and the column that contains $p = 0.80$. Within the $n = 15$ section, we look for the row $x = 10$.

$P(10) = 0.1032$

There is a probability of 0.1032 that in a random sample of 15 such flights, exactly 10 will be on time.

(c) $P(X < 10) = P(X \leq 9)$, so using the cumulative binomial table, we go to the section for $n = 15$ and the column that contains $p = 0.8$. Within the $n = 15$ section, we look for the row $x = 9$. We find $P(X \leq 9) = 0.0611$. In a random sample of 15 such flights, there is a 0.0611 probability that less than 10 flights will be on time.

(d) Here we wish to find $P(X \geq 10)$. Using the complement rule we can write:

$P(X \geq 10) = 1 - P(X < 10) = 1 - P(X \leq 9) = 1 - 0.0611 = 0.9389$ In a random sample of 15 such flights, there is a 0.9389 probability that at least 10 flights will be on time.

(e) Using the binomial probability table we get:
$$P(8 \le X \le 10) = P(8) + P(9) + P(10) = 0.0138 + 0.0430 + 0.1032 = 0.1600$$

Using the cumulative binomial probability table we get:
$$P(8 \le X \le 10) = P(X \le 10) - P(X \le 7) = 0.1642 - 0.0042 = 0.1600$$

In a random sample of 15 such flights, there is a 0.1600 probability that between 8 and 10 flights, inclusive, will be on time.

37. **(a)** We can either use the binomial probability table or compute the probability by hand:
$$P(X = 15) = {}_{25}C_{15} \cdot (0.45)^{15} \cdot (1 - 0.45)^{25-15} = \frac{25!}{15!(25-15)!} \cdot (0.45)^{15} \cdot (0.55)^{10}$$
$$= 3,268,760 \cdot (0.45)^{15} \cdot (0.55)^{10} = 0.0520$$

In a random sample of 25 adult Americans, there is a probability of 0.0520 that exactly 15 will say the state of morals is poor.

(b) Using the cumulative binomial probability table, we find:
$$P(X \le 10) = P(0) + P(1) + P(2) + \cdots + P(10) = 0.3843.$$ We expect that, in a random sample of 25 adult Americans, there is a probability of 0.3843 that no more than 10 will state that they feel the state of morals is poor.

(c) $P(X > 16) = 1 - P(X \le 16) = 1 - 0.9826 = 0.0174$

There is a probability of 0.0174 that more than 16 people in a sample of 25 would state that the condition of morals is poor in the United States.

(d) Using the binomial probability table, we find:
$$P(X = 13 \text{ or } X = 14) = P(13) + P(14) = 0.1236 + 0.0867 = 0.2103$$

There is a probability of 0.2103 that 13 or 14 people in a sample of 25 would state that the condition of morals is poor in the United States.

(e) Using the cumulative binomial probability table, we find that the probability that more than 20 would state that the condition of morals is poor is: $P(X \ge 20) = 1 - P(X \le 19) = 1 - 0.9996 = 0.0004$

Since this profanity is less than 0.05, it would be unusual to find that 20 or more adults in a random sample of 25 adult Americans feel that the condition of morals is poor in the United States.

39. **(a)** Calculating this probability by hand, we get:
$$P(X = 4) = {}_{10}C_4 \cdot (0.267)^4 \cdot (1 - 0.733)^{10-4} = \frac{10!}{4!(10-4)!} \cdot (0.267)^4 \cdot (0.733)^6$$
$$= 210 \cdot (0.267)^4 \cdot (0.733)^6 = 0.1655$$

In a random sample of 10 people, there is a probability of 0.1655 that exactly 4 will cover their mouth when sneezing.

(b) $P(X < 3) = P(0) + P(1) + P(2)$
$$= {}_{10}C_0 \cdot (0.267)^{10} \cdot (0.733)^0 + {}_{10}C_1 \cdot (0.267)^9 \cdot (0.733)^1 + {}_{10}C_2 \cdot (0.267)^8 \cdot (0.733)^2$$
$$= 1 \cdot (0.267)^{10} \cdot (0.733)^0 + 10 \cdot (0.267)^9 \cdot (0.733)^1 + 45 \cdot (0.267)^8 \cdot (0.733)^2$$
$$= 0.0448 + 0.1631 + 0.2673$$
$$= 0.4752$$

We expect that, in a random sample of 10 adult Americans, there is a probability of 0.4752 that less than 3 will cover their mouth when sneezing.

(c) If fewer than half covered their mouth, then at least half did not cover their mouth. In other words, 5 or more out of the 10 people did not cover their mouth.

$$P(X \geq 5) = P(5) + P(6) + P(7) + P(8) + P(9) + P(10)$$
$$= {}_{10}C_5 \cdot (0.267)^5 \cdot (0.733)^5 + {}_{10}C_6 \cdot (0.267)^6 \cdot (0.733)^4 + {}_{10}C_7 \cdot (0.267)^7 \cdot (0.733)^3$$
$$+ {}_{10}C_8 \cdot (0.267)^8 \cdot (0.733)^2 + {}_{10}C_9 \cdot (0.267)^9 \cdot (0.733)^1 + {}_{10}C_{10} \cdot (0.267)^{10} \cdot (0.733)^0$$
$$= 252 \cdot (0.267)^5 \cdot (0.733)^5 + 210 \cdot (0.267)^6 \cdot (0.733)^4 + 120 \cdot (0.267)^7 \cdot (0.733)^3$$
$$+ 45 \cdot (0.267)^8 \cdot (0.733)^2 + 10 \cdot (0.267)^9 \cdot (0.733)^1 + 1 \cdot (0.267)^{10} \cdot (0.733)^0$$
$$= 0.0724 + 0.0220 + 0.0046 + 0.0006 + 0.0001 + 0.0000$$
$$= 0.0996 \text{ or } 0.0997, \text{ depending upon rounding}$$

In a random sample of 10 people, there is a probability of 0.0997 that fewer than half will cover their mouth when sneezing. This is not unusual.

41. (a) The proportion of the jury that is Hispanic is: $\frac{2}{12} = 0.1667$. So, about 16.67% of the jury is Hispanic.

(b) We can compute this by hand or use the binomial probability table. By hand, we have:
$$P(X \leq 2) = P(X = 0) + P(X = 1) + P(X = 2)$$
$$= {}_{12}C_0 \cdot (0.45)^0 \cdot (1-0.45)^{12-0} + {}_{12}C_1 \cdot (0.45)^1 \cdot (1-0.45)^{12-1} + {}_{12}C_2 \cdot (0.45)^2 \cdot (1-0.45)^{12-2}$$
$$= \frac{12!}{0!(12-0)!} \cdot (0.45)^0 \cdot (0.55)^{12} + \frac{12!}{1!(12-1)!} \cdot (0.45)^1 \cdot (0.55)^{11} + \frac{12!}{2!(12-2)!} \cdot (0.45)^2 \cdot (0.55)^{10}$$
$$= 1 \cdot (0.45)^0 \cdot (0.55)^{12} + 12 \cdot (0.45)^1 \cdot (0.55)^{11} + 66 \cdot (0.45)^2 \cdot (0.55)^{10}$$
$$= 0.0008 + 0.0075 + 0.0339$$
$$= 0.0421 \text{ or } 0.0422, \text{ depending on rounding}$$

In a random sample of 12 jurors, there is a probability of 0.421 that 2 or fewer would be Hispanic.

(c) The probability in part (b) is less than 0.05, so this is an unusual event. I would argue that Hispanics are underrepresented on the jury and that the jury selection process was nor fair.

43. (a) We have $n = 100$ and $p = 0.80$.
$$\mu_X = n \cdot p = 100(0.80) = 80 \text{ flights}; \quad \sigma_X = \sqrt{np(1-p)} = \sqrt{100(.80)(1-.80)} = \sqrt{16} = 4 \text{ flights}$$

(b) We expect that, in a random sample of 100 flights from Orlando to Los Angeles, 80 will be on time.

(c) Since $np(1-p) = 16 \geq 10$ the distribution is approximately bell shaped (approximately normal) and we can apply the Empirical Rule.
$$\mu_X - 2\sigma_X = 80 - 2(4) = 72$$
Since 75 is less than 2 standard deviations below the mean, we would conclude that it would not be unusual to observe 75 on-time flights in a sample of 100 flights.

45. (a) We have $n = 500$ and $p = 0.45$.
$$\mu_X = n \cdot p = 500(0.45) = 225 \text{ adult Americans};$$
$$\sigma_X = \sqrt{np(1-p)} = \sqrt{500(.45)(1-.45)} = \sqrt{123.75} \approx 11.1 \text{ adult Americans}$$

(b) In a random sample of 500 adult Americans, we expect 225 to believe that the overall state of moral values in the United States is poor.

(c) Since $np(1-p) = 500(0.45)(0.55) = 123.75 \geq 10$, we can use the Empirical Rule to check for unusual observations.
240 is above the mean, and we have $\mu_X + 2\sigma_X = 225 + 2(11.1) = 247.2 > 240$.
This indicates that 240 is within two standard deviations of the mean. Therefore, it would not be considered unusual to find 240 people who believe the overall state of moral values in the United States is poor in a sample of 500 adult Americans.

47. We have $n = 1030$ and $p = 0.80$.

    $E(X) = \mu_X = n \cdot p = 1030(0.80) = 824$. If attitudes have not changed, we expect 824 of the parents surveyed to spank their children.

    Since $np(1-p) = 164.8 \geq 10$, we can use the Empirical rule to check if 781 would be an unusual observation.

    $\sigma_X = \sqrt{np(1-p)} = \sqrt{1030(0.80)(1-0.80)} = \sqrt{164.8} \approx 12.8$

    781 is below the mean and we have $\mu_X - 2\sigma_X = 824 - 2(12.8) = 798.4$. Since 781 is more than two standard deviations away from the mean, it would be considered unusual if 781 parents from a sample of 1030 said they spank their children. This suggests that parents attitudes have changed since 1995.

49. We have $n = 1000$ and $p = 0.536$.

    $\mu_X = n \cdot p = 1000(0.536) = 536$; $\sigma_X = \sqrt{np(1-p)} = \sqrt{1000(0.536)(1-0.536)} = \sqrt{248.704} \approx 15.8$

    Since $600 > \mu_X + 2\sigma_X = 536 + 2(15.8) = 567.6$, it would be considered unusual to find 600 searches using Google in a random sample of 1000 U.S. internet searches.

51. We would expect $500,000(0.56) = 280,000$ of the stops to be pedestrians who are nonwhite. Because $np(1-p) = 500,000(0.56)(1-0.56) = 123,200 \geq 10$, we can use the Empirical Rule to identify cutoff points for unusual results. The standard deviation number of stops is $\sigma_x = \sqrt{500,000(0.56)(1-0.56)} \approx 351.0$. If the number of stops of nonwhites exceeds $\mu_X + 2\sigma_X = 280,000 + 2(351) = 280,702$, we would say the result is unusual. The actual number of stops is $500,000(0.89) = 445,000$, which is definitely unusual. A potential criticism of this analysis is the use of 0.44 as the proportion of whites, since the actual proportion of whites may be different due to individuals commuting back and forth to the city. Additionally, there could be confounding due to the part of the city where stops are made. If the distribution of nonwhites is not uniform across the city, then location should be taken into account. See Simpson's Paradox in chapter 4.

53. (a) Since $E(X) = np$ and we want $E(X)$ to be at least 11, we solve

    $np = 11$

    $n(0.55) = 11$

    $n = \dfrac{11}{0.55} = 20$

    You would need to randomly select at least 20 high school students.

    (b) We would like the probability, $P(X \geq 12)$, to be 0.99 or greater, with $p = 0.55$ and n to be determined. $P(X \geq 12) = 1 - P(X \leq 11) = 1 - (P(0) + ... + P(11))$ and using technology we find that:

    when $n = 32$, $P(X \geq 12) = 1 - P(X \leq 11) = 1 - 0.0151 = 0.9849$;

    when $n = 33$, $P(X \geq 12) = 1 - P(X \leq 11) = 1 - 0.0100 = 0.9900$.

    Thus the minimum number we would need in our sample is 33 high school students.

55. (a) Using $P(x) = p \cdot (1-p)^{x-1}$ with $p = 0.524$ gives:

    $P(3) = (0.524) \cdot (1-0.524)^{3-1} = 0.1187$

(b)

x	$P(x) = p \cdot (1-p)^{x-1}$
1	0.5240
2	0.2494
3	0.1187
4	0.0565
5	0.0269
6	0.0128
7	0.0061
8	0.0029
9	0.0014
10	0.0007

(c)

$x \cdot P(x)$
0.5240
0.4988
0.3562
0.2261
0.1345
0.0768
0.0427
0.0232
0.0124
0.0066
$\mu_X = \sum xP(x) \approx 1.9$

(d) $\dfrac{1}{p} = \dfrac{1}{0.524} \approx 1.9$ free throws

We would expect Shaq to take about 1.9 free throws before making his first one. This result is the same as that obtained in part (c).

(c) We have $x = 30$ and $r = 3$. $P(30) = {}_{29}C_2 \cdot \left(\dfrac{1}{19}\right)^3 \left(\dfrac{18}{19}\right)^{27} \approx 0.0137$

(d) The expected number of trials before observing $r = 3$ green is $E(X) = \dfrac{r}{p} = \dfrac{3}{1/19} = 3 \cdot 19 = 57$.

57. An experiment is binomial if all of the following conditions are satisfied:
   (1) the experiment consists of a fixed number of trials, where n is the number of trials;
   (2) the trials are independent;
   (3) each trial has two possible, mutually exclusive, outcomes: called success and failure;
   (4) the probability of success, p, remains constant for each trial of the experiment.

59. For large values of n, the shape of the binomial probability histogram is approximately normally distributed. So, as n increases, the shape of the binomial probability histogram becomes more bell shaped.

61. When the sample size, n, is large, the binomial distribution is approximately bell shaped, so about 95% of the outcomes will be in the interval from $\mu - 2\sigma$ to $\mu + 2\sigma$. The sample size, n, is large enough to apply the Empirical Rule when $np(1-p) \geq 10$.

Section 6.3

1. A random variable X, measuring the number of successes in a fixed interval, follows a Poisson process provided that the following conditions are met:
   (1) the probability of two or more successes in any sufficiently small subinterval is 0;
   (2) the probability of success is the same for any two intervals of equal length;
   (3) the number of successes in any interval is independent of the number of successes in any other non-overlapping (disjoint) interval.

3. $\lambda = 10$ hits per minute, $t = 5$ minutes (from 7:30 P.M. to 7:35 P.M.).

5. $\lambda = 0.07$ flaws per linear foot, $t = 20$ linear feet of timber.

7. Use $P(x) = \dfrac{\mu^x}{x!} \cdot e^{-\mu}$ to find the following, assuming $\mu = 5$:

    (a) $P(6) = \dfrac{5^6}{6!} \cdot e^{-5} = 0.1462$

    (b) $P(X < 6) = P(X \le 5) = P(0) + P(1) + P(2) + P(3) + P(4) + P(5)$

$$= \dfrac{5^0}{0!} \cdot e^{-5} + \dfrac{5^1}{1!} \cdot e^{-5} + \dfrac{5^2}{2!} \cdot e^{-5} + \dfrac{5^3}{3!} \cdot e^{-5} + \dfrac{5^4}{4!} \cdot e^{-5} + \dfrac{5^5}{5!} \cdot e^{-5} = 0.6160$$

    (c) $P(X \ge 6) = 1 - P(X < 6) = 1 - 0.6160 = 0.3840$

    (d) $P(2 \le X \le 4) = P(2) + P(3) + P(4) = \dfrac{5^2}{2!} \cdot e^{-5} + \dfrac{5^3}{3!} \cdot e^{-5} + \dfrac{5^4}{4!} \cdot e^{-5} = 0.4001$

9. Use $P(x) = \dfrac{(\lambda t)^x}{x!} \cdot e^{-\lambda t}$ to find the following, given $\lambda = 0.07$, $t = 10$, and $\lambda t = 0.7$:

    (a) $P(4) = \dfrac{0.7^4}{4!} \cdot e^{-0.7} = 0.0050$

    (b) $P(X < 4) = P(X \le 3) = P(X = 0) + P(1) + P(2) + P(3)$

$$= \dfrac{0.7^0}{0!} \cdot e^{-0.7} + \dfrac{0.7^1}{1!} \cdot e^{-0.7} + \dfrac{0.7^2}{2!} \cdot e^{-0.7} + \dfrac{0.7^3}{3!} \cdot e^{-0.7} = 0.9942$$

    (c) $P(X \ge 4) = 1 - P(X < 4) = 1 - 0.9942 = 0.0058$

    (d) $P(4 \le X \le 6) = P(4) + P(5) + P(6) = \dfrac{0.7^4}{4!} \cdot e^{-0.7} + \dfrac{0.7^5}{5!} \cdot e^{-0.7} + \dfrac{0.7^6}{6!} \cdot e^{-0.7} = 0.0057$

    (e) $\mu_X = \lambda t = (0.07)(10) = 0.7$ and $\sigma_X = \sqrt{\lambda t} = \sqrt{0.7} \approx 0.8$

11. We are given that the hits occur at a rate of 1.4 per minute, so $\lambda = 1.4$; and the interval of time we are interested in is 5 minutes (from 7:30 P.M. to 7:35 P.M.), so $t = 5$. Therefore, $\lambda t = (1.4)(5) = 7$.

    (a) $P(7) = \dfrac{7^7}{7!} e^{-7} \approx 0.1490$

    On about 15 out of every 100 days, exactly 7 hits will occur between 7:30 P.M. and 7:35 P.M.

    (b) $P(X < 7) = P(X \le 6) = P(0) + \ldots + P(6)$

$$= \dfrac{7^0}{0!} e^{-7} + \dfrac{7^1}{1!} e^{-7} + \dfrac{7^2}{2!} e^{-7} + \dfrac{7^3}{3!} e^{-7} + \dfrac{7^4}{4!} e^{-7} + \dfrac{7^5}{5!} e^{-7} + \dfrac{7^6}{6!} e^{-7}$$

$$\approx 0.00091 + 0.00638 + 0.02234 + 0.05213 + 0.09123 + 0.12772 + 0.14900$$

$$\approx 0.4497$$

    On about 45 out of every 100 days, fewer than 7 hits will occur between 7:30 P.M. and 7:35 P.M.

    (c)   Use the Complement Rule: $P(X \ge 7) = 1 - P(X < 7) = 1 - 0.4497 = 0.5503$.

    On about 55 out of every 100 days, at least 7 hits will occur between 7:30 P.M. and 7:35 P.M.

13. We are given that FDAL level for insect filth in peanut butter is 0.3 insect fragments per gram, so $\lambda = 0.3$; and the interval we are interested in is 5 grams (the 5-gram sample), so $t = 5$. Therefore, $\lambda t = (0.3)(5) = 1.5$.

(a) $P(2) = \dfrac{(1.5)^2}{2!} e^{-1.5} \approx 0.2510$

In about 25 out of every hundred 5-gram samples of this peanut butter, exactly 2 insect fragments will be found.

(b) $P(X < 2) = P(X \le 1) = P(0) + P(1) \approx 0.22313 + 0.33470 \approx 0.5578$

In about 56 out of every hundred 5-gram samples of this peanut butter, fewer than 2 insect fragments will be found.

(c)   Use the Complement Rule: $P(X \ge 2) = 1 - P(X < 2) = 1 - 0.5578 = 0.4422$.

In about 44 out of every hundred 5-gram samples of this peanut butter, at least 2 insect fragments will be found.

(d)   Use the Complement Rule: $P(X \ge 1) = 1 - P(X < 1) = 1 - P(0) = 1 - 0.2231 = 0.7769$

In about 78 out of every hundred 5-gram samples of this peanut butter, at least 1 insect fragment will be found.

(e) $\begin{aligned} P(X \ge 4) &= 1 - P(X < 4) = 1 - P(X \le 3) \\ &= 1 - \left[P(0) + P(1) + P(2) + P(3) \right] \\ &\approx 1 - \left[0.22313 + 0.33470 + 0.25102 + 0.12551 \right] \\ &= 1 - 0.93436 \approx 0.0656 \end{aligned}$

Since $P(X \ge 4) > 0.05$, it would not be unusual for a 5-gram sample of peanut butter to contain four or more insect fragments. About 7 out of every hundred 5-gram samples of this peanut butter will contain at least 4 fragments.

15. (a)   We are given that fatal accidents occur at the rate of 0.05 per 100 million miles. If we think in terms of 100 million miles, we have $\lambda = 0.05$. The interval we are interested in is 100 million miles, so $t = 1$ (that is, 1 one hundred million mile unit). Therefore, $\lambda t = (0.05)(1) = 0.05$.

$$P(0) = \frac{(0.05)^0}{0!} e^{-0.05} = e^{-0.05} \approx 0.9512$$

In about 95 out of every 100 one hundred million-mile samples, no fatal accidents will occur.

(b) $P(X \ge 1) = 1 - P(X < 1) = 1 - P(0) = 1 - 0.9512 = 0.0488$

In about 5 out of every 100 one hundred million-mile samples, at least one fatal accident will occur.

(c) $P(X > 1) = 1 - P(X \le 1) = 1 - \left[P(0) + P(1) \right]$

$$= 1 - \left[\frac{(0.05)^0}{0!} e^{-0.01} + \frac{(0.05)^1}{1!} e^{-0.01} \right] \approx 1 - 0.9988 = 0.0012$$

In about 1 out of every 1,000 one hundred million-mile samples, more than one fatal accident will occur.

17. We are given that major hurricanes strike Florida at a rate of 29 per 111 year interval. That is, $\lambda = \dfrac{29}{111} \approx 0.2613$ major hurricanes per year. The time interval we are interested in is one year, so $t = 1$.

(a) $P(X = 0) = \dfrac{\left(\frac{29}{111}\right)^0}{0!} e^{-\frac{29}{111}} \approx 0.7701$

(b) $P(X \ge 1) = 1 - P(X = 0) = 1 - 0.7701 = 0.2299$

Since this probability is greater than 0.05 it would not be unusual for Florida to be hit by at least one major hurricane in a single year.

(c) $P(X \geq 3) = 1 - P(X < 3) = 1 - P(X \leq 2)$

$= 1 - \left[P(0) + P(1) + P(2) \right]$

$= 1 - \left[\dfrac{\left(\frac{29}{111}\right)^0}{0!} e^{-\frac{29}{111}} + \dfrac{\left(\frac{29}{111}\right)^1}{1!} e^{-\frac{29}{111}} + \dfrac{\left(\frac{29}{111}\right)^2}{2!} e^{-\frac{29}{111}} \right]$

$\approx 1 - 0.9976$

$= 0.0024$

Since this probability is less than 0.05, it would be unusual for Florida to be hit by at least 3 major hurricanes in a single year.

(d) $P(X \geq 2) = 1 - P(X < 2) = 1 - P(X \leq 1)$

$= 1 - \left[P(0) + P(1) \right]$

$= 1 - \left[\dfrac{\left(\frac{29}{111}\right)^0}{0!} e^{-\frac{29}{111}} + \dfrac{\left(\frac{29}{111}\right)^1}{1!} e^{-\frac{29}{111}} \right]$

$\approx 1 - 0.9713$

$= 0.0287$

Since this probability is less than 0.05, it would be unusual for Florida to be hit by at least 2 major hurricanes in a single year. Thus, the 2005 hurricane season in Florida was unusual.

19. **(a)** and **(b)**

Use technology and the given parameters $\lambda = 0.2$ and $t = 30$ to construct a table:

x	$P(x) = \dfrac{(\lambda t)^x}{x!} \cdot e^{-\lambda t}$	Expected # of restaurants having x arrivals $= 200 \cdot P(x)$	Observed number of restaurants having x arrivals
0	0.0025	0.50	0
1	0.0149	2.97	4
2	0.0446	8.92	5
3	0.0892	17.85	13
4	0.1339	26.77	23
5	0.1606	32.12	25
6	0.1606	32.12	28
7	0.1377	27.54	25
8	0.1033	20.65	27
9	0.0688	13.77	21
10	0.0413	8.26	15
11	0.0225	4.51	5
12	0.0113	2.25	3
13	0.0052	1.04	2
14	0.0022	0.45	2
15	0.0009	0.18	0
16	0.0003	0.07	2

(c) The observed frequencies are lower for fewer cars (x small) and higher for more cars (x large), so the advertising campaign appears to be effective.

21. (a) and (c)

Use technology to construct a table:

x	Observed Frequency	Proportion of years = Frequency ÷ 200	$x \cdot$ proportion of years	$P(x) = \dfrac{(0.61)^x}{x!} \cdot e^{-0.61}$
0	109	0.545	0.000	0.5434
1	65	0.325	0.325	0.3314
2	22	0.110	0.220	0.1011
3	3	0.015	0.045	0.0206
4	1	0.005	0.020	0.0031
		Total =	0.610	

(b) From the data in the table, the mean number of deaths per year is $\mu_X = 0.610$.

(c) Using $\mu = 0.61$ we calculate the Poisson probabilities as shown in the last column.

(d) Answers will vary. These probabilities are very close to the relative frequencies, so the random variable X appears to follow a Poisson distribution.

23. (a) $E(X) = \lambda t = \dfrac{23.6}{100} \cdot (500) = 118$ colds.

(b) – (g) Answers will vary.

Consumer Reports®: Quality Assurance in Customer Relations

(a) Since we are interested in the number of draft responses that contains any number of errors (i.e., multiple errors are counted as one), X is a Binomial random variable. Note that there are only two possible outcomes for the individual drafts (i.e., the draft contains an error or it doesn't), the probability of finding a draft that contains errors is constant (i.e., 0.10 for each draft), and the individual drafts are independent from one another. If we were counting the number of errors in a single draft, the random variable would be Poisson.

(b) In practice, it may not be possible to find an n such that this equation is exactly true. Using Minitab with $p = 0.1$, we find the following:

With $n = 28$: $P(X \geq 1) = 1 - P(X = 0) = 1 - 0.052335 = 0.947665$

With $n = 29$: $P(X \geq 1) = 1 - P(X = 0) = 1 - 0.047101 = 0.952899$.

To be conservative, the manager needs to sample 29 documents to have (at least) a 95% probability of finding at least one document containing an error.

(c) Using Minitab with $p = 0.2$, we find the following:

With $n = 13$: $P(X \geq 1) = 1 - P(X = 0) = 1 - 0.054976 = 0.945024$

With $n = 14$: $P(X \geq 1) = 1 - P(X = 0) = 1 - 0.043980 = 0.956020$

If the error rate is really 20%, the sample size requirement reduces to 14.

(d) Y is a Poisson random variable with $\lambda = 0.3$ errors per document and $t = 1$ (document). Therefore, we have $\mu_Y = \lambda t = (0.3)(1) = 0.3$.

The possible outcomes for Y are $y = 0, 1, 2, ..., \infty$. However, since the expected error rate is low, we would expect the actual number of observed errors to be low.

(e) $P(\text{no errors}) = P(0) = \dfrac{(0.3)^0}{0!} \cdot e^{-0.3} \approx 0.7408$

$P(\text{one error}) = P(1) = \dfrac{(0.3)^1}{1!} \cdot e^{-0.3} \approx 0.2222$

$P(\text{at least 2 errors}) = P(Y \geq 2) = 1 - P(Y \leq 1) = 1 - \left[P(0) + P(1) \right]$
$= 1 - \left[0.7408 + 0.2222 \right] = 1 - 0.9630 = 0.0370$

Section 6.4 (On CD)

1. In both cases, the random variable X represents the number of successes in n trials of an experiment, where each trial results in one of two mutually exclusive (disjoint) outcomes classified as "success" or "failure". The binomial probability distribution requires that each trial be independent of the others and that the probability of success remain the same for each trial. This is not the case for the hypergeometric probability distribution. For the hypergeometric probability distribution, the population size is finite and there are a fixed number of successes in the population. The sample is obtained without replacement, so the trials are not independent and the probability of success does not remain constant. If the population size is infinite, there is no difference between the hypergeometric distribution and the binomial distribution.

3. n is the sample size and there cannot be more successes than trials. Therefore, we must have $x \leq n$. In turn, the population only contains k successes. Therefore, we must also have $x \leq k$. If the sample size n is larger than the number of successes in the population (that is, $n > k$), we are still restricted by the fact that there are only k successes in the population. Therefore, the possible values of the random variable must be less than or equal to the smaller of the sample size (n) and the total number of successes in the population (k).

5. This is a hypergeometric probability experiment because:
    1) The population is finite ($N = 49$ numbers).
    2) There are only two possible outcomes (number matches or does not match).
    3) The sample size is $n = 6$ and is drawn without replacement.
    There are only 6 matching numbers so $k = 6$. The possible values for X are $x = 0, 1, 2, ..., 6$.

7. This is a hypergeometric probability experiment because:
    1) The population is finite ($N = 250$ numbers).
    2) There are only two possible outcomes (chip is defective or chip is not defective).
    3) The sample size is $n = 20$ and is drawn without replacement.
    There are only 12 chips that are defective so $k = 12$. The possible values for X are $x = 0, 1, 2, ..., 12$.

9. Given $N = 150$, $n = 20$, $k = 30$, and $x = 5$:

$P(5) = \dfrac{\left({}_{30}C_5 \right)\left({}_{150-30}C_{20-5} \right)}{{}_{150}C_{20}} = \dfrac{\left({}_{30}C_5 \right)\left({}_{120}C_{15} \right)}{{}_{150}C_{20}} = 0.1856$

11. Given $N = 230$, $n = 15$, $k = 200$, and $x = 12$:

$P(12) = \dfrac{\left({}_{200}C_{12} \right)\left({}_{230-200}C_{15-12} \right)}{{}_{230}C_{15}} = \dfrac{\left({}_{200}C_{12} \right)\left({}_{30}C_3 \right)}{{}_{230}C_{15}} = 0.1939$

13. Given $N = 150$, $n = 20$, and $k = 30$:

$\mu_X = n \cdot \dfrac{k}{N} = 20 \cdot \dfrac{30}{150} = 4$;

$\sigma_X = \sqrt{\left(\dfrac{N-n}{N-1} \right) \cdot n \cdot \dfrac{k}{N} \cdot \dfrac{N-k}{N}} = \sqrt{\left(\dfrac{150-20}{150-1} \right) \cdot 20 \cdot \dfrac{30}{150} \cdot \dfrac{150-30}{150}} \approx 1.67$

15. Given $N = 230$, $n = 15$, and $k = 200$:

$$\mu_X = n \cdot \frac{k}{N} = 15 \cdot \frac{200}{230} \approx 13.0 ;$$

$$\sigma_X = \sqrt{\left(\frac{N-n}{N-1}\right) \cdot n \cdot \frac{k}{N} \cdot \frac{N-k}{N}} = \sqrt{\left(\frac{230-15}{230-1}\right) \cdot 15 \cdot \frac{200}{230} \cdot \frac{230-200}{230}} \approx 1.26$$

17. We are given $N = 47$, $n = 6$, and $k = 6$:

   (a) $P(3) = \dfrac{\left(_6C_3\right)\left(_{47-6}C_{6-3}\right)}{_{47}C_6} = \dfrac{\left(_6C_3\right)\left(_{41}C_3\right)}{_{47}C_6} = 0.01986$

        There is a 0.01986 probability of matching 3 numbers.

   (b) $P(4) = \dfrac{\left(_6C_4\right)\left(_{47-6}C_{6-4}\right)}{_{47}C_6} = \dfrac{\left(_6C_4\right)\left(_{41}C_2\right)}{_{47}C_6} = 0.001146$

        There is a 0.001146 probability of matching 4 numbers.

   (c) $P(5) = \dfrac{\left(_6C_5\right)\left(_{47-6}C_{6-5}\right)}{_{47}C_6} = \dfrac{\left(_6C_5\right)\left(_{41}C_1\right)}{_{47}C_6} = 0.0000229$

        There is a 0.0000229 probability of matching 5 numbers.

   (d) $P(6) = \dfrac{\left(_6C_6\right)\left(_{47-6}C_{6-6}\right)}{_{47}C_6} = \dfrac{\left(_6C_6\right)\left(_{41}C_0\right)}{_{47}C_6} = 0.000000093$

        There is a 0.000000093 probability of matching 6 numbers.

   (e) $P(\text{win}) = P(3) + P(4) + P(5) + P(6)$

                $= 0.01986 + 0.001146 + 0.0000229 + 0.000000093$

                $= 0.02103$

        There is a 0.02103 probability of purchasing a winning ticket. This probability is less than 0.05, so it would be considered unusual to purchase a winning ticket.

   (f) $\mu_X = n \cdot \dfrac{k}{N} = 6 \cdot \dfrac{6}{47} \approx 0.766 ;$

        $\sigma_X = \sqrt{\left(\dfrac{N-n}{N-1}\right) \cdot n \cdot \dfrac{k}{N} \cdot \dfrac{N-k}{N}} = \sqrt{\left(\dfrac{47-6}{47-1}\right) \cdot 6 \cdot \dfrac{6}{47} \cdot \dfrac{47-6}{47}} \approx 0.772$

        $E(X) = \mu_x \approx 0.766$. For a randomly selected ticket, we expect to match 0.766 numbers.

19. We are given $N = 250$, $n = 20$, and $k = 12$:

   (a) $P(4) = \dfrac{\left(_{12}C_4\right)\left(_{250-12}C_{20-4}\right)}{_{250}C_{20}} = \dfrac{\left(_{12}C_4\right)\left(_{238}C_{16}\right)}{_{250}C_{20}} = 0.0087$

        There is a 0.0087 probability of obtaining 4 defective chips.

   (b) $P(3) = \dfrac{\left(_{12}C_3\right)\left(_{250-12}C_{20-3}\right)}{_{250}C_{20}} = \dfrac{\left(_{12}C_3\right)\left(_{238}C_{17}\right)}{_{250}C_{20}} = 0.0507$

        There is a 0.0507 probability of obtaining 3 defective chips.

   (c) $P(0) = \dfrac{\left(_{12}C_0\right)\left(_{250-12}C_{20-0}\right)}{_{250}C_{20}} = \dfrac{\left(_{12}C_0\right)\left(_{238}C_{20}\right)}{_{250}C_{20}} = 0.3590$

        There is a 0.3590 probability of obtaining no defective chips.

(d) $P(14) = 0$ because the random variable cannot take on the value 14. The minimum of n and k is 12. Therefore, the possible values for the random variable are $x = 0, 1, 2, ..., 12$.

(e) $E(X) = \mu_X = n \cdot \dfrac{k}{N} = 20 \cdot \dfrac{12}{250} = 0.96$

In a random sample of 20 chips, we expect to obtain 0.96 defective chips on average.

21. Let the random variable X represent the number of jurors on the jury who will never convict. From the problem statement, we get $N = 30$, $n = 12$, and $k = 2$.

Here it will take less work if we find the probability of avoiding a hung jury and subtracting from 1. To avoid a hung jury, we need $X = 0$.

$$P(X = 0) = \frac{\left({}_2C_0\right)\left({}_{30-2}C_{12-0}\right)}{{}_{30}C_{12}} = \frac{\left({}_2C_0\right)\left({}_{28}C_{12}\right)}{{}_{30}C_{12}} = 0.3517$$

Since, $P(X \geq 1) = 1 - P(X = 0) = 1 - 0.3517 = 0.6483$, there is a 0.6483 probability of having a hung jury.(Note: Here we are assuming that there will never be a hung jury unless one of the two potential jurors who will never convict is on the jury.)

23. Let the random variable X represent the number of cylinders that pass inspection. From the problem statement we have $N = 200$ and $n = 4$. To accept the shipment, we must obtain 0 defective cylinders. That is, we want $X = 0$.

(a) If 10% are defective, we have $k = (0.1)(200) = 20$.

$$P(0) = \frac{\left({}_{20}C_0\right)\left({}_{200-20}C_{4-0}\right)}{{}_{200}C_4} = \frac{\left({}_{20}C_0\right)\left({}_{180}C_4\right)}{{}_{200}C_4} = 0.6539$$

If 10% are defective, there is a 0.6539 probability that the shipment is accepted.

(b) If 20% are defective, we have $k = (0.2)(200) = 40$.

$$P(0) = \frac{\left({}_{40}C_0\right)\left({}_{200-40}C_{4-0}\right)}{{}_{200}C_4} = \frac{\left({}_{40}C_0\right)\left({}_{160}C_4\right)}{{}_{200}C_4} = 0.4065$$

If 20% are defective, there is a 0.4065 probability that the shipment is accepted.

(c) If 40% are defective, we have $k = (0.4)(200) = 80$.

$$P(0) = \frac{\left({}_{80}C_0\right)\left({}_{200-80}C_{4-0}\right)}{{}_{200}C_4} = \frac{\left({}_{80}C_0\right)\left({}_{120}C_4\right)}{{}_{200}C_4} = 0.1270$$

If 40% are defective, there is a 0.1270 probability that the shipment is accepted.

(d) If 60% are defective, we have $k = (0.6)(200) = 120$.

$$P(0) = \frac{\left({}_{120}C_0\right)\left({}_{200-120}C_{4-0}\right)}{{}_{200}C_4} = \frac{\left({}_{120}C_0\right)\left({}_{80}C_4\right)}{{}_{200}C_4} = 0.0245$$

If 60% are defective, there is a 0.0245 probability that the shipment is accepted.

(e) If 80% are defective, we have $k = (0.8)(200) = 160$.

$$P(0) = \frac{\left({}_{160}C_0\right)\left({}_{200-160}C_{4-0}\right)}{{}_{200}C_4} = \frac{\left({}_{160}C_0\right)\left({}_{40}C_4\right)}{{}_{200}C_4} = 0.0014$$

If 80% are defective, there is a 0.0014 probability that the shipment is accepted.

(f)

Chapter 6 Review Exercises

1. (a) The number of students in a randomly selected classroom is a discrete random variable because its value results from counting. If we let the random variable S represent the number of students, the possible values for S are all nonnegative integers (up to the capacity of the room). That is, $s = 0, 1, 2, \ldots$.

(b) The number of inches of snow that falls in Minneapolis during the winter season is a continuous random variable because its value results from a measurement. If we let the random variable S represent the number of inches of snow, the possible values for S are all nonnegative real numbers. That is, $s \geq 0$.

(c) The amount of flight time accumulated is a continuous random variable because its value results from a measurement. If we let the random variable H represent the accumulated flight time, the possible values for H are all nonnegative real numbers. That is, $h \geq 0$.

(d) The number of points scored by the Miami Heat in a game is a discrete random variable because its value results from a count. If we let the random variable X represent the number of points scored by the Miami Heat in a game, the possible values for X are nonnegative integers. That is, $x = 0, 1, 2, \ldots$.

2. (a) This is not a valid probability distribution because the sum of the probabilities does not equal 1. ($\sum P(x) = 0.73$)

(b) This is a valid probability distribution because $\sum P(x) = 1$ and $0 \leq P(x) \leq 1$ for all x.

3. (a) Total number of Stanley Cups
$= 20 + 17 + 19 + 16 = 72.$

x (games played)	$P(x)$
4	$\dfrac{20}{72} \approx 0.2778$
5	$\dfrac{17}{72} = 0.2361$
6	$\dfrac{19}{72} = 0.2639$
7	$\dfrac{16}{72} \approx 0.2222$

(b)

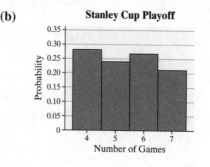

(c) $\mu_X = \sum [x \cdot P(x)] \approx 4 \cdot (0.2778) + 5 \cdot (0.2361) + 6 \cdot (0.2639) + 7 \cdot (0.2222)$
≈ 5.4306 or about 5.4 games

We expect the Stanley Cup to last, on average, about 5.4 games.

(d) $\sigma_X^2 = \sum (x - \mu_x)^2 \cdot P(x)$

$\approx (4 - 5.4306)^2 \cdot 0.2778 + (5 - 5.4306)^2 \cdot 0.2361 + (6 - 5.4306)^2 \cdot 0.2639 + (7 - 5.4306)^2 \cdot 0.2222$

≈ 1.2452

$\sigma_X = \sqrt{\sigma_X^2} \approx \sqrt{1.2452} \approx 1.1159$ or about 1.1 games

4. (a) Let X represent the profit.

$\mu_X = E(X) = \sum [x \cdot P(x)]$

$\approx (200)\left(\dfrac{12}{5525}\right) + (150)\left(\dfrac{1}{425}\right) + (30)\left(\dfrac{36}{1105}\right) + (20)\left(\dfrac{274}{5525}\right) + (5)\left(\dfrac{72}{425}\right) + (-5)\left(\dfrac{822}{1105}\right)$

$\approx -0.1158 \approx -\$0.12$

If this game is played many times, the players can expect to lose about \$0.12 per game, in the long run.

(b) In four hours, at a rate of 35 hands per hour, a total of 140 hands will be played. So, the player can expect to *lose* $(140)(\$0.1158) \approx \$16.21$.

5. (a) This is a binomial experiment. There are a fixed number of trials ($n = 10$) where each trial corresponds to a randomly chosen freshman, the trials are independent, there are only two possible outcomes (graduated or did not graduate within six years), and the probability of a graduation within six years is fixed for all trials ($p = 0.54$).

(b) This is not a binomial experiment because the number of trials is not fixed.

6. (a) We have $n = 10$, $p = 0.05$, and $x = 1$. $P(1) = {}_{10}C_1 \cdot (0.05)^1 (0.95)^9 \approx 0.3151$

In about 31.5% of random samples of 10 visitors to the ER, exactly one will die within a year.

(b) $P(X < 2) = P(X = 0) + P(X = 1) = {}_{25}C_0 \cdot (0.05)^0 (0.95)^{25} + {}_{25}C_1 \cdot (0.05)^1 (0.95)^{24}$

$= 0.2774 + 0.3650 = 0.6424$

In about 64% of random samples of 25 ER patients, fewer than 2 of the twenty-five will die within one year.

(c) $P(X \geq 2) = 1 - P(X < 2) = 1 - 0.6424 = 0.3576$

In about 36% of random samples of 25 ER patients, at least 25 will die within one year.

(d) The probability that at least 8 of 10 will not die is the same as the probability that 2 or fewer out of 10 will die.

$P(X \leq 2) = P(X = 0) + P(X = 1) + P(X = 2)$

$= {}_{10}C_0 \cdot (0.05)^0 (0.95)^{10} + {}_{10}C_1 \cdot (0.05)^1 (0.95)^9 + {}_{10}C_2 \cdot (0.05)^2 (0.95)^8$

$= 0.5987 + 0.3151 + 0.0746 = 0.9885$ or 0.9884, depending on rounding

In about 99% of random samples of 10 ER patients, at least 8 will not die.

(e) $P(X > 3) = 1 - P(X \leq 3)$

$= 1 - [P(X = 0) + P(X = 1) + P(X = 2) + P(X = 3)]$

$= 1 - \left[{}_{30}C_0 \cdot (0.05)^0 (0.95)^{30} + {}_{30}C_1 \cdot (0.05)^1 (0.95)^{29} \right.$

$\left. + {}_{30}C_2 \cdot (0.05)^2 (0.95)^{28} + {}_{30}C_3 \cdot (0.05)^2 (0.95)^{27} \right]$

$= 1 - [0.2146 + 0.3389 + 0.2586 + 0.1270] = 0.0608$ or 0.0609, depending on rounding

Since 0.0608 is greater than 0.05, it would not be unusual for more than 3 of thirty ER patients to die within one year of their visit.

(f) $E(X) = \mu_X = n \cdot p = 1000(0.05) = 50$; $\sigma_X = \sqrt{np(1-p)} = \sqrt{1000(0.05)(0.95)} \approx 6.9$

(g) No. Since $np \cdot (1-p) = 800 \cdot (0.05) \cdot (0.95) = 38 \geq 10$, we can use the Empirical Rule to check for unusual observations. $E(X) = \mu_X = n \cdot p = 800(0.05) = 40$ and

$\sigma_X = \sqrt{np(1-p)} = \sqrt{800(0.05)(0.95)} \approx 6.2$. We have that 51 is above the mean and $\mu + 2\sigma = 40 + 2 \cdot 6.2 = 52.4$. This indicates that 51 is within two standard deviations of the mean, so observing 51 deaths after one year in an ER with 800 visitors would not be unusual.

7. (a) We have $n = 15$ and $p = 0.6$.

Using the binomial probability table, we get: $P(10) = 0.1859$

(b) Using the cumulative binomial probability table, we get: $P(X < 5) = P(X \leq 4) = 0.0093$

(c) Using the complement rule, we get: $P(X \geq 5) = 1 - P(X < 5) = 1 - 0.0093 = 0.9907$

(d) Using the binomial probability table, we get:
$$P(7 \leq X \leq 12) = P(7) + P(8) + P(9) + P(10) + P(11) + P(12)$$
$$= 0.1181 + 0.1771 + 0.2066 + 0.1859 + 0.1268 + 0.0634 = 0.8779$$

Using the cumulative binomial probability table, we get:
$$P(7 \leq X \leq 12) = P(X \leq 12) - P(X \leq 6) = 0.9729 - 0.0950 = 0.8779$$

There is a 0.8779 [Tech: 0.8778] probability that in a random sample of 15 U.S. women 18 years old or older, between 7 and 12, inclusive, would feel that the minimum driving age should be 18.

(e) $E(X) = \mu_X = np = 200(0.6) = 120$ women;

$\sigma_X = \sqrt{np(1-p)} = \sqrt{200 \cdot (0.6) \cdot (0.4)} = \sqrt{48} \approx 6.928$ or about 6.9 women

(f) No; since $np(1-p) = 48 \geq 10$, we can use the Empirical Rule to check for unusual observations.
$\mu - 2\sigma \approx 120 - 13.8 = 106.2$ and $\mu + 2\sigma \approx 120 + 13.8 = 133.8$
Since 110 is within this range of values, it would not be considered an unusual observation.

8. (a)

x	P(x)	$x \cdot P(x)$	$(x - \mu_x)^2 \cdot P(x)$
0	0.00002	0.0000	0.0005
1	0.00037	0.0004	0.0092
2	0.00385	0.0077	0.0615
3	0.02307	0.0692	0.2076
4	0.08652	0.3461	0.3461
5	0.20764	1.0382	0.2076
6	0.31146	1.8688	0.0000
7	0.26697	1.8688	0.2670
8	0.10011	0.8009	0.4005
Σ		6.0000	1.5000

Distribution

(b) Using the formulas from Section 6.1:
$\mu_X = 6$ (from first column in table above and to the right)

$\sigma_X = \sqrt{\sigma_X^2} = \sqrt{1.5} \approx 1.2$ (from second column in table above and to the right)

Using the formulas from Section 6.2:
$\mu_X = np = 8(0.75) = 6$ and $\sigma_X = \sqrt{np \cdot (1-p)} = \sqrt{8(0.75)(0.25)} \approx 1.2$.

(c)

The distribution is skewed left.

9. If $np \cdot (1-p) \geq 10$, then the Empirical Rule can be used to check for unusual observations.

10. In sampling from large populations without replacement, the trials may be assumed to be independent provided that the sample size is small in relation to the size of the population. As a general rule of thumb, this condition is satisfied if the sample size is less than 5% of the population size.

11. A random variable X, the number of successes in a fixed interval, follows a Poisson process provided that following conditions are met:
1) the probability of two or more successes in any sufficiently small subinterval is 0;
2) the probability of success is the same for any two intervals of equal length;
3) the number of successes in any interval is independent of the number of successes in any other non-overlapping interval.

12. Use $P(x) = \dfrac{(\lambda t)^x}{x!} \cdot e^{-\lambda t}$ with $\lambda = 0.05$ and $t = 30$ so that $\lambda t = 1.5$:

(a) $P(2) = \dfrac{1.5^2}{2!} \cdot e^{-1.5} = 0.2510$

(b) $P(X < 2) = P(0) + P(1) = 0.22313 + 0.33470 \approx 0.5578$

(c) $P(X \geq 2) = 1 - P(X < 2) = 1 - 0.5578 = 0.4422$

(d) $P(1 \leq X \leq 3) = P(1) + P(2) + P(3) = 0.33470 + 0.25102 + 0.12551 \approx 0.7112$

(e) $\mu_X = \lambda t = (0.05) \cdot 30 = 1.5$; $\sigma_X = \sqrt{\lambda t} = \sqrt{1.5} \approx 1.2$

13. We need to find $P(X \geq 2)$ where X is the number of defects in a roll. We have $\mu = \lambda \cdot t = \dfrac{1}{500} \cdot (100) = 0.2$

$$P(X \geq 2) = 1 - P(X \leq 1) = 1 - \left[\dfrac{0.2^0}{0!} \cdot e^{-0.2} + \dfrac{0.2^1}{1!} \cdot e^{-0.2} \right] = 1 - 0.9825 = 0.0175$$

About 1.75% of the rolls are rejected.

14. We have that maintenance calls occur at the rate of two per month, so $\lambda = 2$. The time interval of interest is one month, so $t = 1$. Therefore, $\lambda t = (2)(1) = 2$.

$$P(X > 4) = 1 - P(X \leq 4)$$
$$= 1 - \left[\dfrac{2^0}{0!} \cdot e^{-2} + \dfrac{2^1}{1!} \cdot e^{-2} + \dfrac{2^2}{2!} \cdot e^{-2} + \dfrac{2^3}{3!} \cdot e^{-2} + \dfrac{2^4}{4!} \cdot e^{-2} \right]$$
$$= 1 - 0.9473 = 0.0527$$

Since $P(X > 4) > 0.05$, it would not be unusual for the copy machine to require more than 4 maintenance calls in a given month.

15. We have $n = 40$, $p = 0.17$, $x = 12$, and find
$\mu_X = n \cdot p = 40(0.17) = 6.8$. Since 12 is above
the mean, we compute $P(X \geq 12)$. Using
technology we find
$P(X \geq 12) \approx 0.0301 < 0.05$, so the result of
the survey is unusual. This suggests that
emotional abuse may be a factor that increases
the likelihood of self-injurious behavior.

```
1-binomcdf(40,.1
7,11)
        .0300884622
```

Chapter 6 Test

1. (a) The number of days of measurable rainfall in Honolulu during a year is a discrete random variable because its value results from counting. If we let the random variable R represent the number of days for which there is measurable rainfall, the possible values for R are integers between 0 and 365, inclusive. That is, $r = 0, 1, 2, \ldots 365$.

(b) The miles per gallon for a Toyota Prius is a continuous random variable because its value results from a measurement. If we let the random variable M represent the miles per gallon, the possible values for M are all nonnegative real numbers. That is, $m \geq 0$.
(Note: we need to include the possibility that $m = 0$ since the engine could be idling, in which case it would be getting 0 miles to the gallon.)

(c) The number of golf balls hit into the ocean on the famous 18th hole at Pebble Beach is a discrete random variable because its value results from a count. If we let the random variable X represent the number of golf balls hit into the ocean, the possible values for X are nonnegative integers. That is, $x = 0, 1, 2, \ldots$

(d) The weight of a randomly selected robin egg is a continuous random variable because its value results from a measurement. If we let the random variable W represent the weight of a robin's egg, the possible values for W are all nonnegative real numbers. That is, $w > 0$.

2. (a) This is a valid probability distribution because $\sum P(x) = 1$ and $0 \leq P(x) \leq 1$ for all x.

(b) This is not a valid probability distribution because $P(4) < 0$.

3. (a) Total number of Wimbledon Men's Single Finals = 19 + 12 + 13 = 44.

x (sets played)	$P(x)$
3	$\frac{19}{44} \approx 0.4318$
4	$\frac{12}{44} \approx 0.2727$
5	$\frac{13}{44} \approx 0.2955$

(b)

Wimbledon Men's Singles Finals

(c) $\mu_X = \sum [x \cdot P(x)] = 3 \cdot (0.4318) + 4 \cdot (0.2727) + 5 \cdot (0.2955)$
$= 3.8636$ or about 3.9 sets
The Wimbledon men's single finals, if played many times, would be expected to last about 3.9 sets.

(d) $\sigma_X^2 = \sum(x-\mu_x)^2 \cdot P(x)$

$\approx (3-3.8636)^2 \cdot 0.4318 + (4-3.8636)^2 \cdot 0.2727 + (5-3.8636)^2 \cdot 0.2955$

$= 0.7087$

$\sigma_X = \sqrt{\sigma_X^2} \approx \sqrt{0.7087} \approx 0.8418$ or about 0.8 sets

4. $E(X) = \sum x \cdot P(x) = \$200 \cdot (0.998725) + (-\$99,800) \cdot (0.001275) = \$72.50$.

If the company issues many \$100,000 1-year policies to 35-year-old males, then they will make an average profit of \$72.50 per male.

5. A probability experiment is said to be a binomial probability experiment provided:
a) The experiment consists of a fixed number, n, of trials.
b) The trials are all independent.
c) For each trial there are two mutually exclusive (disjoint) outcomes, success or failure.
d) The probability of success, p, is the same for each trial.

6. (a) This is not a binomial experiment because the number of trials is not fixed.

(b) This is a binomial experiment. There are a fixed number of trials ($n = 25$) where each trial corresponding to a randomly chosen property crime, the trials are independent, there are only two possible outcomes (cleared or not cleared), and the probability that the property crime is cleared is fixed for all trials ($p = 0.16$).

7. (a) We have $n = 20$ and $p = 0.80$. We can use the binomial probability table, or compute this by hand.

$P(15) = {}_{20}C_{15} \cdot (0.8)^{15} (0.2)^5 \approx 0.1746$

(b) $P(X \geq 19) = P(X = 19) + P(X = 20) = {}_{20}C_{19} \cdot (0.8)^{19} (0.2)^1 + {}_{20}C_1 \cdot (0.8)^{20} (0.2)^0$

$= 0.0576 + 0.0115 = 0.0692$

(c) Using the complement rule, we get:

$P(X < 19) = 1 - P(X \geq 19) = 1 - 0.0692 = 0.9308$

(d) Using the binomial probability table, we get:

$P(15 \leq X \leq 17) = P(15) + P(16) + P(17) = 0.1746 + 0.2182 + 0.2054 = 0.5981$

Using the cumulative binomial probability table, we get:

$P(15 \leq X \leq 17) = P(X \leq 17) - P(X \leq 14) = 0.7939 - 0.1958 = 0.5981$

There is a probability of 0.5981 that in a random sample of 20 couples, between 15 and 17 inclusive, would hide purchases from their mates.

8. (a) We have $n = 1200$ and $p = 0.5$. $\mu_X = np = 1200 \cdot (0.5) = 600$. We expect that 600 adult Americans would state that their taxes are too high.

(b) Since $np(1-p) = 1200(0.5)(0.5) = 300 \geq 10$, we can use the Empirical Rule to check for unusual observations.

(c) $\mu_X = np = 1200 \cdot (0.5) = 600$ adult Americans

$\sigma_X = \sqrt{np \cdot (1-p)} = \sqrt{1200(0.5)(0.5)} \approx 17.3$

$\mu - 2\sigma = 565.4$ and $\mu + 2\sigma = 634.6$. Since 640 is not between these values, a result of 640 would be unusual. This result contradicts the belief that adult Americans are equally split in their belief that the amount of tax they pay is too high.

9. (a)

Distribution

x	$P(x)$
0	0.3277
1	0.4096
2	0.2048
3	0.0512
4	0.0064
5	0.0003

$x \cdot P(x)$	$(x - \mu_x)^2 \cdot P(x)$
0	0.3277
0.4096	0.0000
0.4096	0.2048
0.1536	0.2048
0.0256	0.0576
0.0016	0.0051
Σ 1.0000	0.8000

(b) Using the formulas from Section 6.1:

$\mu_X = 1$ (from first column in table above and to the right)

$\sigma_X = \sqrt{\sigma_X^2} = \sqrt{0.8} \approx 0.9$ (from second column in table above and to the right)

Using the formulas from Section 6.2:

$\mu_X = np = 5(0.2) = 1$ and $\sigma_X = \sqrt{np \cdot (1-p)} = \sqrt{5(0.2)(0.8)} \approx 0.9$.

(c)

The distribution is skewed right.

10. Use $P(x) = \dfrac{(\lambda t)^x}{x!} \cdot e^{-\lambda t}$ with $\lambda = 0.05$ and $t = 8$ so that $\lambda t = 0.4$:

(a) $P(3) = \dfrac{0.4^3}{3!} \cdot e^{-0.4} = 0.0072$

(b) $P(X < 3) = P(0) + P(1) + P(2) = 0.67032 + 0.26813 + 0.05363 \approx 0.9921$

(c) $P(X \geq 3) = 1 - P(X < 3) = 1 - 0.9921 = 0.0079$

(d) $P(3 \leq X \leq 5) = P(3) + P(4) + P(5) = 0.00715 + 0.00072 + 0.00006 \approx 0.0079$

(e) $\mu_X = \lambda t = (0.05) \cdot 8 = 0.4$; $\sigma_X = \sqrt{\lambda t} = \sqrt{0.4} \approx 0.6$

11. We are given that cars arrive at the rate of 0.41 cars per minute, so $\lambda = 0.41$. The time interval of interest is 10 minutes (4:00 P.M. and 4:10 P.M.), so $t = 10$. Therefore, $\lambda t = 0.41(10) = 4.1$.

(a) $P(4) = \dfrac{4.1^4}{4!} \cdot e^{-4.1} \approx 0.1951$

On about 20 of every 100 randomly chosen days, exactly 4 cars will arrive at the bank's drive-through window between 4:00 P.M. and 4:10 P.M.

(b) $P(X < 4) = P(X \le 3) = P(0) + P(1) + P(2) + P(3)$

$$= \frac{4.1^0}{0!} \cdot e^{-4.1} + \frac{4.1^1}{1!} \cdot e^{-4.1} + \frac{4.1^2}{2!} \cdot e^{-4.1} + \frac{4.1^3}{3!} \cdot e^{-4.1}$$

$$\approx 0.01657 + 0.06795 + 0.13929 + 0.19037$$

$$\approx 0.4142$$

On about 41 of every 100 randomly chosen days, fewer than 4 cars will arrive at the bank's drive-through window between 4:00 P.M. and 4:10 P.M.

(c) Using the complement rule, we get: $P(X \ge 4) = 1 - P(X < 4) = 1 - 0.4142 = 0.5858$

On about 59 of every 100 randomly chosen days, at least 4 cars will arrive at the bank's drive-through window between 4:00 P.M. and 4:10 P.M.

(d) $\mu_X = \lambda t = (0.41) \cdot 10 = 4.1$ cars

(e) $\sigma_X = \sqrt{\lambda t} = \sqrt{4.1} \approx 2.0$ cars

Case Study: The Voyage of the St. Andrew

1. Distribution of the number of families per parish:

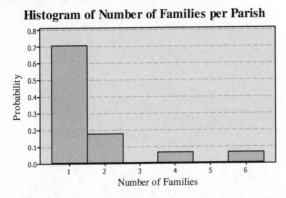

The distribution of the number of families per parish is skewed right with two gaps. The heaviest concentration occurs for 1 family per parish. The expected number of families per parish is $\mu_X \approx 1.65$ with a standard deviation of $\sigma_X \approx 1.33$.

Distribution of the number of freights purchased:

The distribution of the number of freights purchased is skewed right with two gaps. The most frequent purchase was 2 freights. The expected number of freights purchased is $\mu_X \approx 2.54$ with a standard deviation of $\sigma_X \approx 1.00$.

Distribution of the number of family members on board:

Histogram of Number in Family

The distribution of family members is skewed right with no apparent gaps or outliers. The expected number of family members on board is $\mu_X \approx 3.10$ with a standard deviation of $\sigma_X \approx 2.13$.

2. It appears that the *Neuländers* were mildly successful in signing more than one family from a parish. In the vast majority of cases, only one family was signed. However, multiple families were signed in about 30% of the cases. Since villages were likely spread out in the German countryside, it is not likely that many families knew each other prior to undertaking the voyage. When multiple families were signed, it is very likely that families from the same parish knew each other.

3. Using 2.54 as the mean number of freights, the estimated cost per family would be:
   $(2.54)(7.5) = 19.05$ or about 19 pistoles
   $(2.54)(2000) = \$5080$ or about \$5100

4. Answers will vary. It would not be appropriate to estimate the average cost of the voyage from the mean family size because the age of the family members is a major factor. The cost for a family of 6 adults would be significantly more than the cost for a family of 6 with 2 adults and 4 children.

5. Based on the distribution of freights from the *St. Andrew*, we will assume the probability is $p = 0.05$ that a family would purchase more than 4 freights. Using this value, $n = 6$ (families), and assuming each family purchases freights independently of the others, the random variable X is a binomial random variable and represents the number of families purchasing more than 4 freights.
   $$P(X \geq 5) = P(5) + P(6)$$
   $$= {}_6C_5 \cdot (0.05)^5 (0.95)^1 + {}_6C_6 \cdot (0.05)^6 (0.95)^0 \approx 0.0000018$$
   This probability is very small. Therefore, it seems very unlikely that these families came from a population similar to that of the Germans on board the *St. Andrew*.

6. Answers will vary. Topics for discussion could include issues such as independence or whether or not families were disjoint units. For example, did some families send their children with a wealthier family friend to give their children a chance at a better life?

Chapter 7
The Normal Probability Distribution

Section 7.1

1. probability density function

3. True

5. $\mu - \sigma$; $\mu + \sigma$

7. No, the graph cannot represent a normal density function because it is not symmetric.

9. No, the graph cannot represent a normal density function because it crosses below the horizontal axis. That is, it is not always greater than 0.

11. Yes, the graph can represent a normal density function.

13. (a) The figure presents the graph of the density function with the area we wish to find shaded.

    The width of the rectangle is $10 - 5 = 5$ and the height is $\frac{1}{30}$. Thus, the area between 5 and 10 is $5\left(\frac{1}{30}\right) = \frac{1}{6}$. The probability that the friend is between 5 and 10 minutes late is $\frac{1}{6}$.

    (b) 40% of 30 minutes is $0.40 \cdot 30 = 12$ minutes. So, there is a 40% probability your friend will arrive within the next 12 minutes.

15. The figure presents the graph of the density function with the area we wish to find shaded.

The width of the rectangle is $30 - 20 = 10$ and the height is $\frac{1}{30}$. Thus, the area between 20 and 30 is $10\left(\frac{1}{30}\right) = \frac{1}{3}$. The probability that the friend is at least 20 minutes late is $\frac{1}{3}$.

17. (a)

    (b) $P(0 \le X \le 0.2) = 1(0.2 - 0) = 0.2$

    (c) $P(0.25 \le X \le 0.6) = 1(0.6 - 0.25) = 0.35$

    (d) $P(X \ge 0.95) = 1(1 - 0.95) = 0.05$

    (e) Answers will vary.

19. The histogram is symmetrical and bell-shaped, so a normal distribution can be used as a model for the variable.

21. The histogram is skewed to the right, so normal distribution cannot be used as a model for the variable.

23. Graph A matches $\mu = 10$ and $\sigma = 2$, and graph B matches $\mu = 10$ and $\sigma = 3$. We can tell because a higher standard deviation makes the graph lower and more spread out.

25. The center is at 2, so $\mu = 2$. The distance to the inflection points is ± 3, so $\sigma = 3$.

27. The center is at 100, so $\mu = 100$. The distance to the inflection points is ± 15, so $\sigma = 15$.

29.

31. (a)

(b)

(c) Interpretation 1: 15.87% of the cell phone plans in the United States are less than $44.00 per month.

Interpretation 2: The probability is 0.1587 that a randomly selected cell phone plan in the United States is less than $44.00 per month.

33. (a)

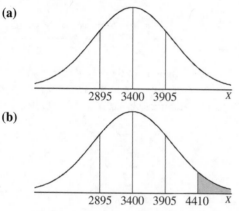

(b)

(c) Interpretation 1: 2.28% of all full-term babies has a birth weight of at least 4410 grams.

Interpretation 2: The probability is 0.0228 that the birth weight of a randomly chosen full-term baby is at least 4410 grams.

35. (a) Interpretation 1: The proportion of human pregnancies that last more than 280 days is 0.1908.

Interpretation 2: The probability is 0.1908 that a randomly selected human pregnancy lasts more than 280 days.

(b) Interpretation 1: The proportion of human pregnancies that last between 230 and 260 days is 0.3416.

Interpretation 2: The probability is 0.3416 that a randomly selected human pregnancy lasts between 230 and 260 days.

37. (a)

(b) The normal density function appears to describe the distance Michael hits a pitching wedge fairly accurately. Looking at the graph, the normal curve is a fairly good approximation to the histogram.

Section 7.2

1. standard normal distribution

3. 0.3085

5. The standard normal table (Table V) gives the area to the left of the z-score. Thus, we look up each z-score and read the corresponding area. We can also use technology. The areas are:

(a) The area to the left of $z = -2.45$ is 0.0071.

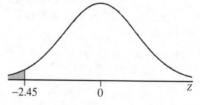

(b) The area to the left of $z = -0.43$ is 0.3336.

(c) The area to the left of $z = 1.35$ is 0.9115.

(d) The area to the left of $z = 3.49$ is 0.9998.

7. The standard normal table (Table V) gives the area to the left of the z-score. Thus, we look up each z-score and read the corresponding area from the table. The area to the right is one minus the area to the left. We can also use technology to find the area. The areas are:

(a) The area to the right of $z = -3.01$ is $1 - 0.0013 = 0.9987$.

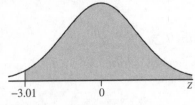

(b) The area to the right of $z = -1.59$ is $1 - 0.0559 = 0.9441$.

(c) The area to the right of $z = 1.78$ is $1 - 0.9625 = 0.0375$.

(d) The area to the right of $z = 3.11$ is $1 - 0.9991 = 0.0009$.

9. To find the area between two z-scores using the standard normal table (Table V), we look up the area to the left of each z-score and then we find the difference between these two. We can also use technology to find the areas:

(a) The area to the left of $z = -2.04$ is 0.0207, and the area to the left of $z = 2.04$ is 0.9793. So, the area between is $0.9793 - 0.0207 = 0.9586$.

(b) The area to the left of $z = -0.55$ is 0.2912, and the area to the left of $z = 0$ is 0.5. So, the area between is $0.5 - 0.2912 = 0.2088$.

(c) The area to the left of $z = -1.04$ is 0.1492, and the area to the left of $z = 2.76$ is 0.9971. So, the area between is $0.9971 - 0.1492 = 0.8479$

11. (a) The area to the left of $z = -2$ is 0.0228, and the area to the right of $z = 2$ is $1 - 0.9772 = 0.0228$. So, the total area is $0.0228 + 0.0228 = 0.0456$. [Tech: 0.0455]

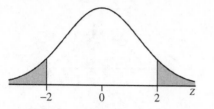

(b) The area to the left of $z = -1.56$ is 0.0594, and the area to the right of $z = 2.56$ is $1 - 0.9948 = 0.0052$. So, the total area is $0.0594 + 0.0052 = 0.0646$.

(c) The area to the left of $z = -0.24$ is 0.4052, and the area to the right of $z = 1.20$ is $1 - 0.8849 = 0.1151$. So, the total area is $0.4052 + 0.1151 = 0.5203$. [Tech: 0.5202]

13. The area in the interior of the standard normal table (Table V) that is closest to 0.1000 is 0.1003, corresponding to $z = -1.28$.

15. The area to the left of the unknown z-score is $1 - 0.25 = 0.75$. The area in the interior of the standard normal table (Table V) that is closest to 0.7500 is 0.7486, corresponding to $z = 0.67$.

17. The z-scores for the middle 99% are the z-scores for the top and bottom 0.5%. The area to the left of z_1 is 0.005, and the area to the left of z_2 is

0.995. The areas in the interior of the standard normal table (Table V) that are closest to 0.0050 are 0.0049 and 0.0051. So, we use the average of their corresponding z-scores: -2.58 and -2.57, respectively.
This gives $z_1 = -2.575$. The areas in the interior of the standard normal table (Table V) that are closest to 0.9950 are 0.9949 and 0.9951. So, we use the average of their corresponding z-scores: 2.57 and 2.58, respectively. This gives $z_2 = 2.575$.

19. The area to the right of the unknown z-score is 0.01, so the area to the left is $1 - 0.01 = 0.99$. The area in the interior of the standard normal table (Table V) that is closest to 0.9900 is 0.9901, so $z_{0.01} = 2.33$.

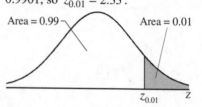

21. The area to the right of the unknown z-score is 0.025, so the area to the left is $1 - 0.025 = 0.975$. The area in the interior of the standard normal table (Table V) that is closest to 0.9750 is 0.9750, so $z_{0.025} = 1.96$.

23. $z = \dfrac{x - \mu}{\sigma} = \dfrac{35 - 50}{7} \approx -2.14$

From Table V, the area to the left is 0.0162, so
$P(X > 35) = 1 - 0.0162 = 0.9838$. [Tech: 0.9839]

25. $z = \dfrac{x - \mu}{\sigma} = \dfrac{45 - 50}{7} \approx -0.71$

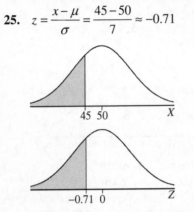

From Table V, the area to the left is 0.2389, so
$P(X \le 45) = 0.2389$. [Tech: 0.2375]

27. $z_1 = \dfrac{x_1 - \mu}{\sigma} = \dfrac{40 - 50}{7} \approx -1.43$;

$z_2 = \dfrac{x_2 - \mu}{\sigma} = \dfrac{65 - 50}{7} \approx 2.14$

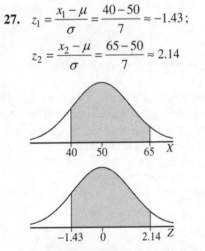

From Table V, the area to the left of
$z_1 = -1.43$ is 0.0764 and the area to the left of
$z_2 = 2.14$ is 0.9838, so $P(40 < X < 65) =$
$0.9838 - 0.0764 = 0.9074$. [Tech: 0.9074]

29. $z_1 = \dfrac{x_1 - \mu}{\sigma} = \dfrac{55 - 50}{7} \approx 0.71$;

$z_2 = \dfrac{x_2 - \mu}{\sigma} = \dfrac{70 - 50}{7} \approx 2.86$

From Table V, the area to the left of $z_1 = 0.71$
is 0.7611 and the area to the left of $z_2 = 2.86$
is 0.9979, so $P(55 \le X \le 70) =$
$0.9979 - 0.7611 = 0.2368$. [Tech: 0.2354]

31. $z_1 = \dfrac{x_1 - \mu}{\sigma} = \dfrac{38 - 50}{7} \approx -1.71$;

$z_2 = \dfrac{x_2 - \mu}{\sigma} = \dfrac{55 - 50}{7} \approx 0.71$

From Table V, the area to the left of $z_1 = -1.71$
is 0.0436 and the area to the left of $z_2 = 0.71$ is
0.7611, so $P(38 < X \le 55) = 0.7611 - 0.0436$
$= 0.7175$. [Tech: 0.7192]

33. The figure below shows the normal curve with
the unknown value of X separating the bottom
9% of the distribution from the top 91% of the
distribution.

From Table V, the area closest to 0.09 is
0.0901. The corresponding z-score is -1.34.
So, the 9th percentile for X is $x = \mu + z\sigma =$
$50 + (-1.34)(7) = 40.62$. [Tech: 40.61]

35. The figure below shows the normal curve with the unknown value of X separating the bottom 81% of the distribution from the top 19% of the distribution.

From Table V, the area closest to 0.81 is 0.8106. The corresponding z-score is 0.88. So, the 81st percentile for X is $x = \mu + z\sigma =$ $50 + 0.88(7) = 56.16$. [Tech: 56.15]

37. (a) $z = \dfrac{x - \mu}{\sigma} = \dfrac{20 - 21}{1} = -1.00$

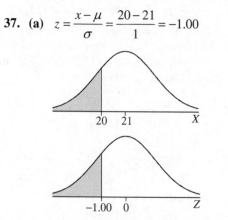

From Table V, the area to the left of $z = -1.00$ is 0.1587, so $P(X < 20) = 0.1587$.

(b) $z = \dfrac{x - \mu}{\sigma} = \dfrac{22 - 21}{1} = 1.00$

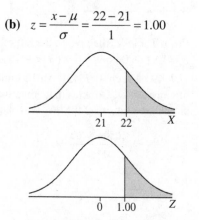

From Table V, the area to the left of $z = 1.00$ is 0.8413, so $P(X > 22) =$ $1 - 0.8413 = 0.1587$.

(c) $z_1 = \dfrac{x_1 - \mu}{\sigma} = \dfrac{19 - 21}{1} = -2.00$;

$z_2 = \dfrac{x_2 - \mu}{\sigma} = \dfrac{21 - 21}{1} = 0$

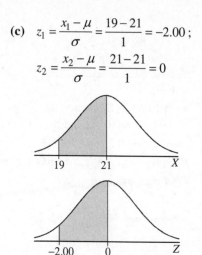

From Table V, the area to the left of $z_1 = -2.00$ is 0.0228 and the area to the left of $z_2 = 0$ is 0.5000, so $P(19 \le X \le 21) =$ $0.5000 - 0.0228 = 0.4772$.

(d) $z = \dfrac{x - \mu}{\sigma} = \dfrac{18 - 21}{1} = -3.00$

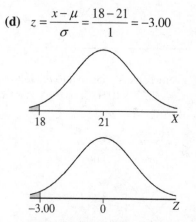

From Table V, the area to the left of $z = -3.00$ is 0.0013, so $P(X < 18) = 0.0013$.

Yes, it would be unusual for an egg to hatch in less than 18 days. Only about 1 egg in 1000 hatches in less than 18 days.

39. (a) $z_1 = \dfrac{x_1 - \mu}{\sigma} = \dfrac{1000 - 1262}{118} \approx -2.22$;

$z_2 = \dfrac{x_2 - \mu}{\sigma} = \dfrac{1400 - 1262}{118} \approx 1.17$

From Table V, the area to the left of $z_1 = -2.22$ is 0.0132 and the area to the left of $z_2 = 1.17$ is 0.8790, so
$P(1000 \le X \le 1400) = 0.8790 - 0.0132$
$= 0.8658$. [Tech: 0.8657]

(b) $z = \dfrac{x - \mu}{\sigma} = \dfrac{1000 - 1262}{118} \approx -2.22$

From Table V, the area to the left of $z = -2.22$ is 0.0132, so $P(X < 1000) = 0.0132$.

(c) $z = \dfrac{x - \mu}{\sigma} = \dfrac{1200 - 1262}{118} \approx -0.53$

From Table V, the area to the left of $Z = -0.53$ is 0.2981, so $P(X > 1200) = 1 - 0.2981 = 0.7019$. [Tech: 0.7004]
So, the proportion of 18-ounce bags of Chip Ahoy! cookies that contains more than 1200 chocolate chips is 0.7019, or 70.19%.

(d) $z = \dfrac{x - \mu}{\sigma} = \dfrac{1125 - 1262}{118} \approx -1.16$.

From Table V, the area to the left of $z = -1.16$ is 0.1230, so $P(X < 1125) = 0.1230$. [Tech: 0.1228] So, the proportion of 18-ounce bags of Chip Ahoy! cookies that contains less than 1125 chocolate chips is 0.1230, or 12.30%.

(e) $z = \dfrac{x - \mu}{\sigma} = \dfrac{1475 - 1262}{118} \approx 1.81$

From Table V, the area to the left of $z = 1.81$ is 0.9649, so $P(X < 1475) = 0.9649$. [Tech: 0.9645] An 18-ounce bag of Chip Ahoy! Cookies that contains 1475 chocolate chips is at the 96th percentile.

(f) $z = \dfrac{x - \mu}{\sigma} = \dfrac{1050 - 1262}{118} \approx -1.80$.

From Table V, the area to the left of $z = -1.80$ is 0.0359, so $P(X < 1050) = 0.0359$. [Tech: 0.0362] An 18-ounce bag of Chip Ahoy! cookies that contains 1050 chocolate chips is at the 4th percentile.

41. (a) $z = \dfrac{x - \mu}{\sigma} = \dfrac{270 - 266}{16} = 0.25$

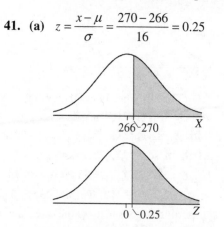

From Table V, the area to the left of $z = 0.25$ is 0.5987, so $P(X > 270) = 1 - 0.5987 = 0.4013$. So, the proportion of human pregnancies that last more than 270 days is 0.4013.

(b) $z = \dfrac{x - \mu}{\sigma} = \dfrac{250 - 266}{16} = -1$

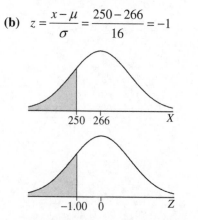

From Table V, the area to the left of $z = -1.00$ is 0.1587, so $P(X < 250) = 0.1587$. So, the proportion of human pregnancies that last less than 250 days is 0.1587, or 15.87%.

(c) $z_1 = \dfrac{x_1 - \mu}{\sigma} = \dfrac{240 - 266}{16} \approx -1.63$;

$z_2 = \dfrac{x_2 - \mu}{\sigma} = \dfrac{280 - 266}{16} \approx 0.88$

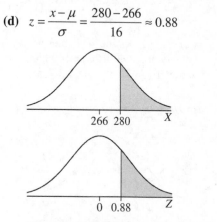

From Table V, the area to the left of $z_1 = -1.63$ is 0.0516 and the area to the left of $z_2 = 0.88$ is 0.8106, so $P(240 \le X \le 280) = 0.8106 - 0.0516 = 0.7590$. [Tech: 0.7571] So, the proportion of human pregnancies lasts between 240 and 280 days is 0.7590, or 75.90%.

(d) $z = \dfrac{x - \mu}{\sigma} = \dfrac{280 - 266}{16} \approx 0.88$

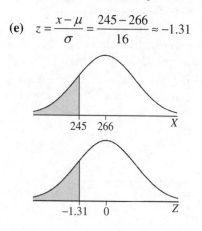

From Table V, the area to the left of $z = 0.88$ is 0.8106, so $P(X > 280) = 1 - 0.8106 = 0.1894$. [Tech: 0.1908]

(e) $z = \dfrac{x - \mu}{\sigma} = \dfrac{245 - 266}{16} \approx -1.31$

From Table V, the area to the left of $z = -1.31$ is 0.0951, so $P(X \le 245) = 0.0951$. [Tech: 0.0947]

(f) $z = \dfrac{x - \mu}{\sigma} = \dfrac{224 - 266}{16} \approx -2.63$

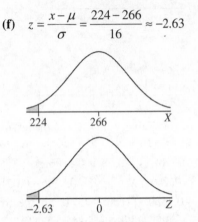

From Table V, the area to the left of $z = -2.63$ is 0.0043, so $P(X < 224) = 0.0043$.

Yes, "very preterm" babies are unusual. Only about 4 births in 1000 are "very preterm."

43. (a) $z = \dfrac{x - \mu}{\sigma} = \dfrac{24.9 - 25}{0.07} \approx -1.43$

From Table V, the area to the left of $z = -1.43$ is 0.0764, so $P(X < 24.9) = 0.0764$. [Tech: 0.0766] So, the proportion of rods that has a length less than 24.9 cm is 0.0764, or 7.64%.

(b) $z_1 = \dfrac{x_1 - \mu}{\sigma} = \dfrac{24.85 - 25}{0.07} \approx -2.14$;

$z_2 = \dfrac{x_2 - \mu}{\sigma} = \dfrac{25.15 - 25}{0.07} \approx 2.14$

From Table V, the area to the left of $z_1 = -2.14$ is 0.0162, so $P(X < 24.85) = 0.0162$. The area to the left of $z_2 = 2.14$ is 0.9838, so $P(X > 25.15) = 1 - 0.9838 = 0.0162$. So, $P(X < 24.85 \text{ or } X > 25.15) = 2(0.0162) = 0.0324$. [Tech: 0.0321] The proportion of rods that will be discarded is 0.0324, or 3.24%.

(c) The manager should expect to discard $5000(0.0324) = 162$ of the 5000 steel rods.

(d) $z_1 = \dfrac{x_1 - \mu}{\sigma} = \dfrac{24.9 - 25}{0.07} \approx -1.43$;

$z_2 = \dfrac{x_2 - \mu}{\sigma} = \dfrac{25.1 - 25}{0.07} \approx 1.43$

From Table V, the area to the left of $z_1 = -1.43$ is 0.0764 and the area to the left of $z_2 = 1.43$ is 0.9236, so $P(24.9 \le X \le 25.1) = 0.9236 - 0.0764 = 0.8472$. [Tech: 0.8469] So, 0.8472, or 84.72%, of the rods manufactured will be between 24.9 and 25.1 cm. Let n represent the number of rods that must be manufactured. Then, $0.8472n = 10,000$, so $n = \dfrac{10,000}{0.8472} \approx 11,803.59$. Increase this to the next whole number: 11,804. To meet the order, the manager should manufacture 11,804 rods. [Tech: 11,808 rods]

45. (a) $z = \dfrac{x-\mu}{\sigma} = \dfrac{5-0}{10.9} \approx 0.46$. From Table V, the area to the left of $z = 0.46$ is 0.6772, so

$$P(X \geq 5) = 1 - P(X < 5)$$
$$= 1 - P(z < 0.46)$$
$$= 1 - 0.6772 = 0.3228.$$

[Tech: 0.3232] The probability that the favored team wins by 5 or more points relative to the spread is 0.3228.

(b) $z = \dfrac{x-\mu}{\sigma} = \dfrac{-2-0}{10.9} \approx -0.18$. From Table V, the area to the left of $z = -0.18$ is 0.4286, so

$$P(X \leq -2) = P(z < -0.18) = 0.4286.$$

[Tech: 0.4272] The probability that the favored team loses by 2 or more points relative to the spread is 0.4286.

(c) If the mean is zero, then the favored team is equally likely to win or lose relative to the spread. Yes, a mean of 0 implies that the spreads are accurate.

47. (a) The figure below shows the normal curve with the unknown value of X separating the bottom 17% of the distribution from the top 83%.

From Table V, the area closest to 0.17 is 0.1711, which corresponds to the z-score -0.95. So, the 17th percentile for incubation times of fertilized chicken eggs is $x = \mu + z\sigma = 21 + (-0.95)(1) \approx 20$ days.

(b) The figure below shows the normal curve with the unknown values of X separating the middle 95% of the distribution from the bottom 2.5% and the top 2.5%.

From Table V, the area 0.0250 corresponds to the z-score -1.96. Likewise, the area $0.0250 + .095 = 0.975$ corresponds to the z-score 1.96. Now,

$x_1 = \mu + z_1\sigma = 21 + (-1.96)(1) \approx 19$ and

$x_2 = \mu + z_2\sigma = 21 + 1.96(1) \approx 23$. Thus, the incubation times that make up the middle 95% of fertilized chicken eggs is between 19 and 23 days.

49. (a) The figure below shows the normal curve with the unknown value of X separating the bottom 30% of the distribution from the top 70%.

From Table V, the area closest to 0.30 is 0.3015, which corresponds to the z-score -0.52. So, the 30th percentile for the number of chocolate chips in an 18-ounce bag of Chips Ahoy! cookies is $x = \mu + z\sigma$
$= 1262 + (-0.52)(118) \approx 1201$ chocolate chips. [Tech: 1200 chocolate chips]

(b) The figure below shows the normal curve with the unknown values of X separating the middle 99% of the distribution from the bottom 0.5% and the top 0.5%.

From Table V, the areas 0.0049 and 0.0051 are equally close to 0.005. We average the corresponding z-scores -2.58 and -2.57 to obtain $z_1 = -2.575$. Likewise, the area $0.005 + 0.99 = 0.995$ is equally close 0. 9949 and 0.9951.
We average the corresponding z-scores 2.57 and 2.58 to obtain $z_2 = 2.575$. Now,

$x_1 = \mu + z_1\sigma = 1262 + (-2.575)(118) \approx 958$

and

$x_2 = \mu + z_2\sigma = 1262 + 2.575(118) \approx 1566$.

The number of chocolate chips that make up the middle 99% of 18-ounce bags of Chips Ahoy! cookies is 958 to 1566 chips.

51. (a) $z = \dfrac{x - \mu}{\sigma} = \dfrac{20 - 17}{2.5} = 1.2$

From Table V, the area to the left of $z = 1.20$ is 0.8849, so $P(X > 20) = 1 - 0.8849 = 0.1151$. So, about 11.51% of Speedy Lube's customers receive the service for half price.

(b) The figure below shows the normal curve with the unknown value of X separating the top 3% of the distribution from the bottom 97%.

From Table V, the area closest to 0.97 is 0.9699, which corresponds to the z-score 1.88. So, $x = \mu + z\sigma = 17 + 1.88(2.5) \approx 22$. In order to discount only about 3% of its customers, Speedy Lube should make the guaranteed time limit 22 minutes.

53. The area under a normal curve can be interpreted as a probability, a proportion, or a percentile.

55. Reporting the probability as > 0.9999 accurately describes the event as highly likely, but not certain. Reporting the probability as 1.0000 might be incorrectly interpreted to mean that the event is certain.

Consumer Reports®: Sunscreens

(a) There tends to be a great deal of variation in the way people respond to treatments such as sunscreen. Hence, we can either test sunscreens on a very large sample of people or block on people, i.e., measuring each person before treatment and then again after treatment. By taking before and after measurements for each person we can determine if there is a treatment effect without testing a very large number of people.

(b) The random assignment of people and application sites to each treatment (the sunscreen) is important to avoid any potential sources of bias.

(c) Product A: $P(X < 15) = P\left(Z < \dfrac{15 - 15.5}{1.5}\right)$

$\approx P(Z < -0.33)$

$= 0.3707$ [Tech: 0.3694]

Product B: $P(X < 15) = P\left(Z < \dfrac{15 - 14.7}{1.2}\right)$

$= P(Z < 0.25)$

$= 0.5987$

(d) Product A:

$P(X > 17.5) = P\left(Z > \dfrac{17.5 - 15.5}{1.5}\right)$

$= P(Z > 1.33)$

$= 0.0918$ [Tech: 0.0912]

Product B:

$P(X > 17.5) = P\left(Z > \dfrac{17.5 - 14.7}{1.2}\right)$

$= P(Z > 2.33)$

$= 0.0099$ [Tech: 0.0098]

(e) Product A:
$P(14.5 < X < 15.5)$

$= P\left(\dfrac{14.5 - 15.5}{1.5} < Z < \dfrac{15.5 - 15.5}{1.5}\right)$

$= P(-0.67 < Z < 0)$

$= 0.5 - 0.2514 = 0.2486$ [Tech: 0.2475]

Product B:
$P(14.5 < X < 15.5)$

$= P\left(\dfrac{14.5 - 14.7}{1.2} < Z < \dfrac{15.5 - 14.7}{1.2}\right)$

$= P(-0.17 < Z < 0.67)$

$= 0.7486 - 0.4325 = 0.3161$ [Tech: 0.3137]

(f) Although Product B has a smaller standard deviation, it also has a smaller mean. Since it is important for a product to meet its advertised claim, it appears as if Product A is slightly superior to Product B. Note that approximately 37% of Product A fails to meet its claim while approximately 60% (or more than half) of Product B fails to meet its claim.

Section 7.3

1. normal probability plot

3. The plotted points do not lie within the provided bounds, so the sample data do not come from a normally distributed population.

5. The plotted points do not lie within the provided bounds, so the sample data do not come from a normally distributed population.

7. The normal probability plot is roughly linear and all the data lie within the provided bounds, so the sample data could come from a normally distributed population.

9. **(a)** The normal probability plot is roughly linear, and all the data lie within the provided bounds, so the sample data could come from a population that is normally distributed.

(b)

$$\sum x = 48,895; \quad \sum x^2 = 62,635,301;$$
$$n = 40$$

$$\bar{x} = \frac{\sum x}{n} = \frac{49,895}{40} \approx 1247.4 \text{ chips };$$

$$s = \sqrt{\frac{\sum x^2 - \frac{(\sum x)^2}{n}}{n-1}}$$

$$= \sqrt{\frac{62,635,301 - \frac{(49,895)^2}{40}}{40-1}}$$

$$\approx 101.0 \text{ chips}$$

(c) $\mu - \sigma \approx \bar{x} - s = 1247.4 - 101.0 = 1146.4$;
$\mu + \sigma \approx \bar{x} + s = 1247.4 + 101.0 = 1348.4$

(d) $z = \frac{x - \mu}{\sigma} = \frac{1000 - 1247.4}{101.0} \approx -2.45$

From Table V, the area to the left of $z = -2.45$ is 0.0071, so $P(X > 1000) = 1 - 0.0071 = 0.9929$. [Tech: 0.9928]

(e) $z_1 = \frac{x_1 - \mu}{\sigma} = \frac{1200 - 1247.4}{101.0} \approx -0.47$;

$z_2 = \frac{x_2 - \mu}{\sigma} = \frac{1400 - 1247.4}{101.0} \approx 1.51$

From Table V, the area to the left of $z_1 = -0.47$ is 0.3192 and the area to the

left of $z_2 = 1.51$ is 0.9345. So, $P(1200 \le X \le 1400) = 0.9345 - 0.3192 = 0.6153$. [Tech: 0.6152] The proportion of 18-ounce bags of Chips Ahoy! that contains between 1200 and 1400 chips is 0.6153, or 61.53%.

11. The normal probability plot is approximately linear, so the sample data could come from a normally distributed population.

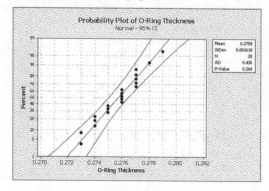

13. The normal probability plot is not approximately linear (points lie outside the provided bounds), so the sample data do not come from a normally distributed population.

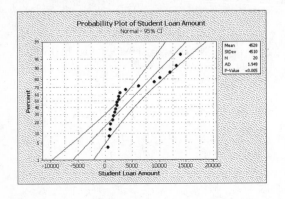

Section 7.4

1. $np(1-p) \geq 10$

3. $P(X \leq 4.5)$; we use the continuity correction and find the area less than or equal to 4.5.

5. Approximate $P(X \geq 40)$ by computing the area under the normal curve to the right of $x = 39.5$.

7. Approximate $P(X = 8)$ by computing the area under the normal curve between $x = 7.5$ and $x = 8.5$.

9. Approximate $P(18 \leq X \leq 24)$ by computing the area under the normal curve between $x = 17.5$ and $x = 24.5$.

11. Approximate $P(X > 20) = P(X \geq 21)$ by computing the area under the normal curve to the right of $x = 20.5$.

13. Approximate $P(X > 500) = P(X \geq 501)$ by computing the area under the normal curve to the right of $x = 500.5$.

15. Using $P(x) = {}_nC_x p^x (1-p)^{n-x}$, with the parameters $n = 60$ and $p = 0.4$, we get

$P(20) = {}_{60}C_{20}(0.4)^{20}(0.6)^{40} \approx 0.0616$. Now $np(1-p) = 60 \cdot 0.4 \cdot (1-0.4) = 14.4 \geq 10$, so the normal approximation can be used, with $\mu_X = np = 60(0.4) = 24$ and $\sigma_X = \sqrt{np(1-p)}$

$= \sqrt{14.4} \approx 3.795$. With continuity correction we calculate:

$P(20) \approx P(19.5 < X < 20.5)$

$= P\left(\dfrac{19.5 - 24}{\sqrt{14.4}} < Z < \dfrac{20.5 - 24}{\sqrt{14.4}}\right)$

$= P(-1.19 < Z < -0.92)$

$= 0.1788 - 0.1170$

$= 0.0618$ [Tech: 0.0603]

17. Using $P(x) = {}_nC_x p^x (1-p)^{n-x}$, with the parameters $n = 40$ and $p = 0.25$, we get

$P(30) = {}_{40}C_{30}(0.25)^{30}(0.75)^{70} \approx 4.1 \times 10^{-11}$. Now $np(1-p) = 40 \cdot 0.25 \cdot (1-0.25) = 7.5$, which is below 10, so the normal approximation cannot be used.

19. Using $P(x) = {}_nC_x p^x (1-p)^{n-x}$, with the parameters $n = 75$ and $p = 0.75$, we get

$P(60) = {}_{75}C_{60}(0.75)^{60}(0.25)^{15} \approx 0.0677$. Now $np(1-p) = 75 \cdot 0.75 \cdot (1-0.75) = 14.0625 \geq 10$, so the normal approximation can be used, with $\mu_X = 75(0.75) = 56.25$ and $\sigma_X = \sqrt{np(1-p)} = \sqrt{14.0625} = 3.75$. With continuity correction we calculate:

$P(60) \approx P(59.5 < X < 60.5)$

$= P\left(\dfrac{59.5 - 56.25}{3.75} < Z < \dfrac{60.5 - 56.25}{3.75}\right)$

$= P(0.87 < Z < 1.13)$

$= 0.8707 - 0.8078$

$= 0.0630$ [Tech: 0.0645]

21. From the parameters $n = 150$ and $p = 0.9$, we get $\mu_X = np = 150 \cdot 0.9 = 135$ and $\sigma_X = \sqrt{np(1-p)} = \sqrt{150 \cdot 0.9 \cdot (1-0.9)} = \sqrt{13.5} \approx 3.674$. Note that $np(1-p) = 13.5 \geq 10$, so the

normal approximation to the binomial distribution can be used.

(a) $P(130) \approx P(129.5 < X < 130.5)$

$$= P\left(\frac{129.5 - 135}{\sqrt{13.5}} < Z < \frac{130.5 - 135}{\sqrt{13.5}}\right)$$

$$= P(-1.50 < Z < -1.22)$$

$$= 0.1112 - 0.0668$$

$$= 0.0444 \quad \text{[Tech: 0.0431]}$$

(b) $P(X \ge 130) \approx P(X \ge 129.5)$

$$= P\left(Z \ge \frac{129.5 - 135}{\sqrt{13.5}}\right)$$

$$= P(Z \ge -1.50)$$

$$= 1 - 0.0668$$

$$= 0.9332 \quad \text{[Tech: 0.9328]}$$

(c) $P(X < 125) = P(X \le 124)$

$$\approx P(X \le 124.5)$$

$$= P\left(Z \le \frac{124.5 - 135}{\sqrt{13.5}}\right)$$

$$= P(Z \le -2.86)$$

$$= 0.0021$$

(d) $P(125 \le X \le 135)$

$$\approx P(124.5 \le X \le 135.5)$$

$$= P\left(\frac{124.5 - 135}{\sqrt{13.5}} < Z < \frac{135.5 - 135}{\sqrt{13.5}}\right)$$

$$= P(-2.86 < Z < 0.14)$$

$$= 0.5557 - 0.0021$$

$$= 0.5536 \quad \text{[Tech: 0.5520]}$$

23. From the parameters $n = 500$ and $p = 0.45$ we get $\mu_X = np = 500 \cdot 0.45 = 225$ and $\sigma_X = \sqrt{np(1-p)} = \sqrt{500 \cdot 0.45 \cdot (1 - 0.45)} = \sqrt{123.75} \approx 11.1243$. Note that $np(1-p) = 123.75 \ge 10$, so the normal approximation to the binomial distribution can be used.

(a) $P(250) \approx P(249.5 \le X \le 250.5)$

$$= P\left(\frac{249.5 - 225}{\sqrt{123.75}} \le Z \le \frac{250.5 - 225}{\sqrt{123.75}}\right)$$

$$= P(2.20 \le Z \le 2.29)$$

$$= 0.9890 - 0.9861$$

$$= 0.0029 \quad \text{[Tech: 0.0029]}$$

(b) $P(X \le 220) \approx P(X \le 220.5)$

$$= P\left(Z \le \frac{220.5 - 225}{\sqrt{123.75}}\right)$$

$$= P(Z \le -0.40)$$

$$= 0.3446 \quad \text{[Tech: 0.3429]}$$

(c) $P(X > 250) \approx P(X \ge 250.5)$

$$= P\left(Z \ge \frac{250.5 - 225}{\sqrt{123.75}}\right)$$

$$= P(Z \ge 2.29) = 1 - P(Z < 2.29)$$

$$= 1 - 0.9890$$

$$= 0.0110 \quad \text{[Tech: 0.0109]}$$

(d) $P(219.5 \le X \le 250.5)$

$$= P\left(\frac{219.5 - 225}{\sqrt{123.75}} \le Z \le \frac{250.5 - 225}{\sqrt{123.75}}\right)$$

$$= P(-0.49 \le Z \le 2.29)$$

$$= 0.9890 - 0.3121$$

$$= 0.6769 \quad \text{[Tech: 0.6785]}$$

(e) $P(X \ge 260) \approx P(X > 259.5)$

$$= 1 - P(X \le 259.5)$$

$$= 1 - P\left(Z \le \frac{259.5 - 225}{\sqrt{123.75}}\right)$$

$$= 1 - P(Z \le 3.10) = 1 - 0.9990$$

$$= 0.0010 \quad \text{[Tech: 0.0010]}$$

This is less than 0.05, so it would be unusual for at least 260 out of 500 adult Americans to indicate that they believe the overall state of moral values is poor.

25. From the parameters $n = 200$ and $p = 0.55$, we get $\mu_X = np = 200 \cdot 0.55 = 110$ and $\sigma_X = \sqrt{np(1-p)} = \sqrt{200 \cdot 0.55 \cdot (1 - 0.55)} = \sqrt{49.5} \approx 7.036$. Note that $np(1-p) = 49.5 \ge 10$, so the normal approximation to the binomial distribution can be used.

(a) $P(X \ge 130) \approx P(X \ge 129.5)$

$$= P\left(Z \ge \frac{129.5 - 110}{\sqrt{49.5}}\right)$$

$$= P(Z \ge 2.77)$$

$$= 1 - 0.9972 = 0.0028$$

(b) Yes, the result from part (a) contradicts the results of the *Current Population Survey* because the result from part (a) is unusual. Fewer than 3 samples in 1000 will result in 130 or more male students living at home if the true percentage is 55%.

27. From the parameters $n = 150$ and $p = 0.37$, we get $\mu_X = np = 150 \cdot 0.37 = 55.5$ and $\sigma_X = \sqrt{np(1-p)} = \sqrt{150 \cdot 0.37 \cdot (1-0.37)} = \sqrt{34.965} \approx 5.913$. Note that $np(1-p) = 34.965 \geq 10$, so the normal approximation to the binomial distribution can be used.

(a) $P(X \geq 75) \approx P(X \geq 74.5)$

$$= P\left(Z \geq \frac{74.5 - 55.5}{\sqrt{34.965}}\right)$$

$$= P(Z \geq 3.21)$$

$$= 1 - 0.9993$$

$$= 0.0007$$

(b) Yes, the result from part (a) contradicts the results of the *Current Population Survey* because the result from part (a) is unusual. Fewer than 1 sample in 1000 will result in 75 or more respondents preferring a male if the true percentage who prefer a male is 37%.

Chapter 7 Review Exercises

1. (a) μ is the center (and peak) of the normal distribution, so $\mu = 60$.

(b) σ is the distance from the center to the points of inflection, so $\sigma = 70 - 60 = 10$.

(c) Interpretation 1: The proportion of values of the random variable to the right of $x = 75$ is 0.0668.
Interpretation 2: The probability that a randomly selected value is greater than $x = 75$ is 0.0668.

(d) Interpretation 1: The proportion of values of the random variable between $x = 50$ and $x = 75$ is 0.7745.
Interpretation 2: The probability that a randomly selected value is between $x = 50$ and $x = 75$ is 0.7745.

2.

Using the standard normal table, the area to the left of $z = -1.04$ is 0.1492.

3.

Using the standard normal table, the area between $z = -0.34$ and $z = 1.03$ is $0.8485 - 0.3669 = 0.4816$.

4. If the area to the right of z is 0.483 then the area to the left z is $1 - 0.483 = 0.5170$.

The closest area to this in the interior of the normal tables is 0.5160, corresponding to $z = 0.04$.

5. The z-scores for the middle 92% are the z-scores for the top and bottom 4%. The area to the left of z_1 is 0.04, and the area to the left of z_2 is 0.96. The area in the interior of the standard normal table (Table V) that is closest to 0.0400 is 0.0401, corresponding to $z_1 = -1.75$. The area in the interior of the standard normal table (Table V) that is closest to 0.9600 is 0.9599, corresponding to $z_2 = 1.75$.

6. The area to the right of the unknown z-score is 0.20, so the area to the left is $1 - 0.20 = 0.80$. The area in the interior of the standard normal table (Table V) that is closest to 0.8000 is 0.7995, corresponding to $z = 0.84$, so $z_{0.20} = 0.84$.

7.

$$z = \frac{55 - 50}{6} \approx 0.83$$

From Table V, the area to the left of $z = 0.83$ is 0.7967, so $P(X > 55) = 1 - 0.7967 = 0.2033$. [Tech: 0.2023]

8.

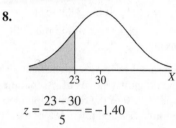

$$z = \frac{23 - 30}{5} = -1.40$$

From Table V, the area to the left of $z = -1.40$ is 0.0808, so $P(X \le 23) = 0.0808$.

9.

$$z_1 = \frac{65 - 70}{10} = -0.50 \; ; \; z_2 = \frac{85 - 70}{10} = 1.50$$

From Table V, the area to the left of $z = -0.50$ is 0.3085 and the area to the left of $z = 1.50$ is 0.9332, so $P(65 < X < 85) = 0.9332 - 0.3085 = 0.6247$.

10. (a)

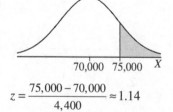

$$z = \frac{75,000 - 70,000}{4,400} \approx 1.14$$

From Table V, the area to the left of $z = 1.14$ is 0.8729, so $P(X \ge 75,000) = 1 - 0.8729 = 0.1271$ [Tech: 0.1279]

So, the proportion of tires that will last at least 75,000 miles is 0.1271, or 12.71%.

(b)

$$z = \frac{60,000 - 70,000}{4,400} \approx -2.27$$

From Table V, the area to the left of $z = -2.27$ is 0.0116, so $P(X \le 60,000) = = 0.0116$ [Tech: 0.0115] So, the proportion of tires that will last 60,000 miles or less is 0.0116, or 1.16%.

(c)

$$z_1 = \frac{65,000 - 70,000}{4,400} \approx -1.14 \; ;$$

$$z_2 = \frac{80,000 - 70,000}{4,400} \approx 2.27$$

From Table V, the area to the left of $z_1 = -1.14$ is 0.1271 and the area to the left of $z_2 = 2.27$ is 0.9884, so $P(65,000 \le X \le 80,000) = 0.9884 - 0.1271 = 0.8613$ [Tech: 0.8606]

(d) The figure below shows the normal curve with the unknown value of X separating the bottom 2% of the distribution from the top 98% of the distribution.

From Table V, 0.0202 is the area closest to 0.02. The corresponding z-score is -2.05. So, $x = \mu + z\sigma = 70,000 + (-2.05)(4,400) = 60,980$. [Tech: 60,964] In order to warrant only 2% of its tires, Dunlop should advertise its warranty mileage as 60,980 miles.

11. (a)

$$z = \frac{125 - 100}{15} \approx 1.67$$

From Table V, the area to the left of $z = 1.67$ is 0.9525, so $P(X > 125) = 1 - 0.9525 = 0.0475$ [Tech: 0.0478].

(b)

$$z = \frac{90 - 100}{15} \approx -0.67$$

From Table V, the area to the left of $z = -0.67$ is 0.2514, so $P(X < 90) = 0.2514$. [Tech: 0.2525]

(c)

$$z_1 = \frac{110 - 100}{15} \approx 0.67 \; ; \; z_2 = \frac{140 - 100}{15} \approx 2.67$$

From Table V, the area to the left of $z_1 = 0.67$ is 0.7486 and the area to the left of $z_2 = 2.67$ is 0.9962, so

$$P(110 < X < 140) = 0.9962 - 0.7486$$

$$= 0.2476 \text{ [Tech: 0.2487]} \text{ So, the}$$

proportion of test takers who score between 110 and 140 is 0.2476, or 24.76%.

(d)

$$z = \frac{150 - 100}{15} \approx 3.33$$

From Table V, the area to the left of $z = 3.33$ is 0.9996, so $P(X > 150) = = 1 - 0.9996 = 0.0004$.

(e) The figure below shows the normal curve with the unknown value of X separating the bottom 98% of the distribution from the top 2% of the distribution.

From Table V, 0.9798 is the area closest to 0.98. The corresponding z-score is 2.05. So, $x = \mu + z\sigma = 100 + 2.05(15) \approx 131$. A score of 131 places a child in the 98th percentile.

(f) The figure below shows the normal curve with the unknown values of X separating the middle 95% of the distribution from the bottom 2.5% and top 2.5% of the distribution.

From Table V, the area 0.0250 corresponds to $z_1 = -1.96$ and the area 0.9750 corresponds to $z_2 = 1.96$. So, $x_1 = \mu + z_1\sigma = 100 + (-1.96)(15) \approx 71$ and $x_2 = \mu + z_2\sigma = 100 + 1.96(15) \approx 129$. Thus, children of normal intelligence scores are between 71 and 129 on the Wechsler Scale.

12. (a)

$$z = \frac{5.25 - 5.11}{0.062} \approx 2.26$$

From Table V, the area to the left of $z = 2.26$ is 0.9881, so $P(X > 5.25) = = 1 - 0.9881 = 0.0119$. [Tech: 0.0120] So, the proportion of baseballs produced by this factory that are too heavy for use by major league baseball is 0.0119, or 1.19%.

(b)

$z = \dfrac{5 - 5.11}{0.062} \approx -1.77$ From Table V, the

area to the left of $z = -1.77$ is 0.0384, so $P(X < 5) = 0.0384$. [Tech: 0.0380] So, the proportion of baseballs produced by this factory that are too light for use by major league baseball is 0.0384, or 3.84%.

(c)

From parts (a) and (b), $z_1 \approx -1.77$ and $z_2 \approx 2.26$. The area to the left of $z_1 = -1.17$ is 0.0384 and the area to the left of $z_2 = 2.26$ is 0.9881, so $P(5 \le X \le 5.25) = 0.9881 - 0.0384 = 0.9497$ [Tech: 0.9500]. So, the proportion of baseballs produced by this factory that can be used by major league baseball is 0.9497, or 94.97%.

(d) From part (c), we know that 94.97% of the baseballs can be used by major league baseball. Let n represent the number of baseballs that must be produced. Then, $0.9497n = 8,000$, so

$n = \dfrac{8,000}{0.9497} \approx 8,423.71$. Increase this to

the next whole number: 8,424. To meet the order, the factory should produce 8,424 baseballs. [Tech: 8421 baseballs]

13. (a) Since $np(1 - p) = 250(0.46)(1 - 0.46) = 62.1 \ge 10$, the normal distribution can be used to approximate the binomial distribution. The parameters are $\mu_X = np$ $= 250(0.46) = 115$ and $\sigma_X = \sqrt{np(1 - p)}$ $= \sqrt{250(0.46)(1 - 0.46)} = \sqrt{62.1} \approx 7.880$

(b) For the normal approximation, we make corrections for continuity 124.5 and 125.5.

$P(125) \approx P(124.5 < X < 125.5)$

$= P\!\left(\dfrac{124.5 - 115}{\sqrt{62.1}} < Z < \dfrac{125.5 - 115}{\sqrt{62.1}} \right)$

$= P(1.21 < Z \le 1.33)$

$= 0.9082 - 0.8869$

$= 0.0213$ [Tech: 0.0226]

Interpretation: Approximately 2 of every 100 random samples of 250 adult Americans will result in exactly 125 who state that they have read at least 6 books within the past year.

(c) $P(X < 120) = P(X \le 119)$

For the normal approximation, we make a correction for continuity to 119.5.

$P(X < 120) = P(X \le 119)$

$\approx P(X < 119.5)$

$= P\!\left(Z < \dfrac{119.5 - 115}{\sqrt{62.1}} \right)$

$= P(Z < 0.57)$

$= 0.7157$ [Tech: 0.7160]

Interpretation: Approximately 72 of every 100 random samples of 250 adult Americans will result in fewer than 120 who state that they have read at least 6 books within the past year.

(d) For the normal approximation, we make a correction for continuity to 139.5.

$P(X \geq 140) \approx P(X > 139.5)$

$$= P\left(Z > \frac{139.5 - 115}{\sqrt{62.1}}\right)$$

$$\approx P(Z > 3.11)$$

$$= 1 - 0.9991 = 0.0009$$

Interpretation: Approximately 1 of every 1000 random samples of 250 adult Americans will result in 140 or more who state that they have read at least 6 books within the past year.

(e) For the normal approximation, we make corrections for continuity 99.5 and 120.5.

$P(100 \leq X \leq 120)$

$\approx P(99.5 < X < 120.5)$

$$= P\left(\frac{99.5 - 115}{\sqrt{62.1}} < Z < \frac{120.5 - 115}{\sqrt{62.1}}\right)$$

$$\approx P(-1.97 < Z < 0.70)$$

$$= 0.7580 - 0.0244$$

$$= 0.7336 \quad \text{[Tech: 0.7328]}$$

Interpretation: Approximately 73 of every 100 random samples of 250 adult Americans will result in between 100 and 120, inclusive, who state that they have read at least 6 books within the past year.

14. The normal probability plot is roughly linear, and all the data lie within the provided bounds, so the sample data could come from a normally distributed population.

15. The plotted points do not appear linear and do not lie within the provided bounds, so the sample data are not from a normally distributed population.

16. The plotted points are not linear and do not lie within the provided bounds, so the data are not from a population that is normally distributed.

17. From the parameters $n = 250$ and $p = 0.20$, we get $\mu_X = np = 250 \cdot 0.20 = 50$ and $\sigma_X = \sqrt{np(1-p)} = \sqrt{250(0.20)(1-0.20)} = \sqrt{40} \approx 6.325$. Note that $np(1-p) = 40 \geq 10$, so the normal approximation to the binomial distribution can be used.

(a) For the normal approximation, we make corrections for continuity to 30.5.

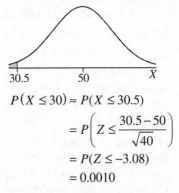

$P(X \leq 30) \approx P(X \leq 30.5)$

$$= P\left(Z \leq \frac{30.5 - 50}{\sqrt{40}}\right)$$

$$= P(Z \leq -3.08)$$

$$= 0.0010$$

(b) Yes, the result from part (a) contradicts the *USA Today* "Snapshot" because the result from part (a) is unusual. About 1 sample in 1000 will result in 30 or fewer who do their most creative thinking while driving, if the true percentage is 20%.

18. (a)

(b) $P(0 \leq X \leq 5) = 0.05 \cdot (5 - 0) = 0.25$

(c) $P(10 \leq X \leq 18) = 0.05 \cdot (18 - 10) = 0.4$

19. The standard normal curve has the following properties:
   (1) It is symmetric about its mean $\mu = 0$ and has standard deviation $\sigma = 1$.
   (2) Its highest point occurs at $\mu = 0$.
   (3) It has inflection points at -1 and 1.
   (4) The area under the curve is 1.
   (5) The area under the curve to the right of $\mu = 0$ equals the area under the curve to the left of $\mu = 0$, which equals 0.5.
   (6) As Z increases, the graph approaches but never equals zero. As Z decreases, the graph approaches but never equals zero.
   (7) It satisfies the Empirical Rule: Approximately 68% of the area under the standard normal curve is between -1 and 1. Approximately 95% of the area under the standard normal curve is between -2 and 2. Approximately 99.7% of the area under the standard normal curve is between -3 and 3.

20. The graph plots actual observations against expected z-scores, assuming that the data are normally distributed. If the plot is not linear, then we have evidence that that data are not from a normal distribution.

Chapter 7 Test

1. (a) μ is the center (and peak) of the normal distribution, so $\mu = 7$.

   (b) σ is the distance from the center to the points of inflection, so $\sigma = 9 - 7 = 2$.

   (c) Interpretation 1: The proportion of values for the random variable to the left of $x = 10$ is 0.9332.
   Interpretation 2: The probability that a randomly selected value is less than $x = 10$ is 0.9332.

   (d) Interpretation 1: The proportion of values for the random variable between $x = 5$ and $x = 8$ is 0.5328.
   Interpretation 2: The probability that a randomly selected value is between $x = 5$ and $x = 8$ is 0.5328.

2.

Using Table V, the area to the left of $z = 2.04$ is 0.9793, so the are to the right of $z = 2.04$ is $1 - 0.9793 = 0.0207$.

3. The z-scores for the middle 88% are the z-scores for the top and bottom 6%.

The area to the left of z_1 is 0.06, and the area to the left of z_2 is 0.94. From the interior of Table V, the areas 0.0606 and 0.0594 are equally close to 0.0600. These areas correspond to $z = -1.55$ and $z = -1.56$, respectively. Splitting the difference we obtain $z_1 = -1.555$. From the interior of Table V, the areas 0.9394 and 0.9406 are equally close to 0.9400. These correspond to $z = 1.55$ and $z = 1.56$, respectively. Splitting the difference, we obtain $z_2 = 1.555$.

4. The area to the right of the unknown z-score is 0.04, so the area to the left is $1 - 0.04 = 0.96$.

The area in the interior of the Table V that is closest to 0.9600 is 0.9599, corresponding to $z = 1.75$, so $z_{0.04} = 1.75$.

5. (a)

(b) $z_1 = \dfrac{22-20}{3} \approx 0.67$; $z_2 = \dfrac{27-20}{3} \approx 2.33$

From Table V, the area to the left of $z_1 = 0.67$ is 0.7486 and the area to the left of $z_2 = 2.33$ is 0.9901, so $P(22 \le X \le 27)$ $= 0.9901 - 0.7486 = 0.2415$. [Tech: 0.2427]

6. (a)

$z = \dfrac{6-7}{0.8} = -1.25$

From Table V, the area to the left of $z = -1.25$ is 0.1956, so $P(X \ge 6) =$ $1 - 0.1056 = 0.8944$. So, the proportion of the time that a fully charged iPhone will last at least 6 hours is 0.8944, or 89.44%.

(b)

$z = \dfrac{5-7}{0.8} = -2.5$

From Table V, the area to the left of $z = -2.50$ is 0.0062, so $P(X < 5) =$ 0.0062. That is, the probability that a fully charged iPhone will last less than 5 hours is 0.0062. This is an unusual result.

(c) The figure below shows the normal curve with the unknown value of X separating the top 5% of the distribution from the bottom 95% of the distribution.

From the interior of Table V, the areas 0.9495 and 0.9505 are equally close to 0.9500. These areas correspond to $z = 1.64$

and $z = 1.65$, respectively. Splitting the difference, we obtain the z-score 1.645. So, $x = \mu + z\sigma = 7 + 1.645(0.8) \approx 8.3$. The cutoff for the top 5% of all talk times is 8.3 hours.

(d)

$z = \dfrac{9-7}{0.8} = 2.5$

From Table V, the area to the left of $z = 2.50$ is 0.9938, so $P(X > 9) =$ $1 - 0.9938 = 0.0062$. So, yes, it would be unusual for the iPhone to last more than 9 hours. Only about 6 out of every 1000 full charges will result in the iPhone lasting more than 9 hours.

7. (a)

$z = \dfrac{100-92.5}{13.7} \approx 0.55$

From Table V, the area to the left of $z = 0.55$ is 0.7088, so $P(X < 6) = 0.7088$. [Tech: 0.7080] So, the proportion of 20- to 29-year-old males whose waist circumferences is less than 100 cm is 0.7088, or 70.88%.

(b)

$z_1 = \dfrac{80-92.5}{13.7} = -0.91$;

$z_2 = \dfrac{100-92.5}{13.7} \approx 0.55$

From Table V, the area to the left of $z_1 = -0.91$ is 0.1814 and the area to the left of $z_2 = 0.55$ is 0.7088, so $P(80 \le X \le 100) = 0.7088 - 0.1814 =$ 0.5274. [Tech: 0.5272] That is, the probability that a randomly selected 20- to 29-year-old male has a waist circumference between 80 and 100 cm is 0.5274.

(c) The figure that follows shows the normal curve with the unknown values of X separating the middle 90% of the distribution from the bottom 5% and the top 5% of the distribution.

From the interior of Table V, the areas 0.0495 and 0.0505 are equally close to 0.0500. These areas correspond to $z = -1.65$ and $z = -1.64$, respectively. Splitting the difference we obtain $z_1 = -1.645$. So, $x_1 = \mu + z_1\sigma =$ $= 92.5 + (-1.645)(13.7) \approx 70$. From the interior of Table V, the areas 0.9495 and 0.9505 are equally close to 0.9500. These areas correspond to $z = 1.64$ and $z = 1.65$, respectively. Splitting the difference we obtain the $z_2 = 1.645$. So, $x_2 = \mu + z_2\sigma =$ $= 92.5 + 1.645(13.7) \approx 115$. Therefore, waist circumferences between 70 and 115 cm make up the middle 90% of all waist circumferences.

(d) The figure below shows the normal curve with the unknown value of X separating the bottom 10% of the distribution from the top 90% of the distribution.

From the interior of Table V, the area closest to 0.10 is 0.1003. The corresponding z-score is -1.28. So, $x = \mu + z\sigma = 92.5 + (-1.28)(13.7) \approx 75$.

Thus, a waist circumference of 75 cm is at the 10th percentile.

8. If the top 6% get an A, then 94% (or 0.9400) of the students will have grades lower than an A. Similarly, 80% (or 0.80) will have grades lower than a B. 20% (or 0.20) will have

grades lower than a C. Finally, 6% (or 0.06) will have a grade lower than a D, that is, an F. To find the cutoff score corresponding to each of these values, we find the z-score for each of the cutoffs, then use the equation $X = \mu + z \cdot \sigma$ to find the point value for each cutoff.

For the cutoff for the low A's: $z_{0.94} = 1.55$, and $X = 64 + 1.55(8) = 76.4$, so everyone who scores at least 76.4 points will get an A.

For the cutoff for the low B's: $z_{0.80} = 0.84$, and $X = 64 + 0.84(8) \approx 70.7$, so everyone who scores at least 70.7 points and less than 76.4 will get a B.

For the cutoff for the low C's: $z_{0.20} = -0.84$, and $X = 64 + (-0.84)(8) \approx 57.3$, so everyone who scores at least 57.3 and less than 70.7 points will get a C.

For the cutoff for the low D's: $z_{0.06} = -1.55$, and $X = 64 + (-1.55)(8) = 51.6$, so everyone who scores at least 51.6 and less than 57.3 points will get a D.

Students who score less than 51.6 points will get an F.

9. $np(1-p) = 500(0.16)(1-0.16) = 67.2 \geq 10$, so the normal distribution can be used to approximate the binomial distribution. The parameters are $\mu_X = np = 500(0.16) = 80$ and $\sigma_X = \sqrt{np(1-p)} = \sqrt{250(0.16)(1-0.16)} = \sqrt{67.2} \approx 8.198$.

(a) For the normal approximation of $P(100)$, we make corrections for continuity 99.5 and 100.5.

$P(100) \approx P(99.5 < X < 100.5)$

$= P\left(\dfrac{99.5 - 80}{\sqrt{67.2}} < Z < \dfrac{100.5 - 80}{\sqrt{67.2}}\right)$

$= P(2.38 < Z \leq 2.50)$

$= 0.9938 - 0.9913$

$= 0.0025$

(b) $P(X < 60) = P(X \le 59)$

For the normal approximation, we make a correction for continuity to 59.5.

$P(X < 60) = P(X \le 59)$

$\approx P(X < 59.5)$

$= P\left(Z < \dfrac{59.5 - 80}{\sqrt{67.2}}\right)$

$= P(Z < -2.50)$

$= 0.0062$

10. The plotted points do not appear linear and one point does not lie within the provided bounds, so the sample data do not likely come from a normally distributed population.

11. **(a)**

(b) $P(20 \le X \le 30) = \dfrac{1}{40}(30 - 20) = 0.25$

(c) $P(X < 15) = P(10 \le X \le 15)$

$= \dfrac{1}{40}(15 - 10)$

$= 0.125$

Case Study: A Tale of Blood Chemistry and Health

1. $\mu \approx \text{midpoint} = \dfrac{35 + 150}{2} = 92.5 \text{ mg/dL}$

$\sigma \approx \dfrac{\text{range}}{4} = \dfrac{150 - 35}{4} = 28.75 \text{ mg/dL}$

These results agree with the entries in the table.

2. White blood cell count:

$P(X \le 5.3) = P\left(Z \le \dfrac{5.3 - 7.25}{1.625}\right)$

$= P(Z \le -1.2)$

$= 0.1151$

Red blood cell count:

$P(X \le 4.62) = P\left(Z \le \dfrac{4.62 - 4.85}{0.375}\right)$

$\approx P(Z \le -0.61)$

$= 0.2709 \quad$ [Tech: 0.2698]

Hemoglobin:

$P(X \le 14.6) = P\left(Z \le \dfrac{14.6 - 14.75}{1.125}\right)$

$\approx P(Z \le -0.13)$

$= 0.4483 \quad$ [Tech: 0.4470]

Hematocrit:

$P(X \le 41.7) = P\left(Z \le \dfrac{41.7 - 43.0}{3.5}\right)$

$\approx P(Z \le -0.37)$

$= 0.3557 \quad$ [Tech: 0.3552]

Glucose, serum:

$P(X \ge 95.0) = P\left(Z \ge \dfrac{95.0 - 87.0}{11.0}\right)$

$\approx P(Z \ge 0.73)$

$= 1 - 0.7673$

$= 0.2327 \quad$ [Tech: 0.2335]

Creatine, serum:

$P(X \le 0.8) = P\left(Z \le \dfrac{0.8 - 1.00}{0.25}\right)$

$= P(Z \le -0.80)$

$= 0.2119$

Sodium, serum:

$P(X \ge 143.0) = P\left(Z \ge \dfrac{143.0 - 141.5}{3.25}\right)$

$\approx P(Z \ge 0.46)$

$= 1 - 0.6772$

$= 0.3228 \quad$ [Tech: 0.3222]

Potassium, serum:

$P(X \ge 5.1) = P\left(Z \ge \dfrac{5.1 - 4.5}{0.5}\right)$

$= P(Z \ge 1.20)$

$= 1 - 0.8849$

$= 0.1151$

Chloride, serum:

$$P(X \le 100.0) = P\left(Z \le \frac{100.0 - 102.5}{3.25}\right)$$
$$\approx P(Z \le -0.77)$$
$$= 0.2206 \quad [\text{Tech: } 0.2209]$$

Carbon dioxide, total:

$$P(X \le 25.0) = P\left(Z \le \frac{25.0 - 26.0}{3.0}\right)$$
$$\approx P(Z \le -0.33)$$
$$= 0.3707 \quad [\text{Tech: } 0.3694]$$

Calcium, serum:

$$P(X \ge 10.1) = P\left(Z \ge \frac{10.1 - 9.55}{0.525}\right)$$
$$\approx P(Z \ge 1.05)$$
$$= 1 - 0.8531$$
$$= 0.1469 \quad [\text{Tech: } 0.1474]$$

Total cholesterol:

$$P(X \ge 253.0) = P\left(Z \ge \frac{253.0 - 149.5}{24.75}\right)$$
$$\approx P(Z \ge 4.18)$$
$$< 1 - 0.9998$$
$$< 0.0002 \quad [\text{Tech: } <0.0001]$$

Triglycerides:

$$P(X \ge 150.0) = P\left(Z \ge \frac{150.0 - 99.5}{49.75}\right)$$
$$\approx P(Z \ge 1.02)$$
$$< 1 - 0.8461$$
$$< 0.1539 \quad [\text{Tech: } 0.1550]$$

HDL cholesterol:

$$P(X \le 42.0) = P\left(Z \le \frac{42.0 - 92.5}{28.75}\right)$$
$$\approx P(Z \le -1.76)$$
$$= 0.0392 \quad [\text{Tech: } 0.0395]$$

LDL cholesterol:

$$P(X \ge 181.0) = P\left(Z \ge \frac{181.0 - 64.5}{32.25}\right)$$
$$\approx P(Z \ge 3.61)$$
$$< 1 - 0.9998$$
$$< 0.0002 \quad [\text{Tech: } 0.0002]$$

LDL/HDL ratio:

$$P(X \ge 4.3) = P\left(Z \ge \frac{4.3 - 1.8}{0.72}\right)$$
$$\approx P(Z \ge 3.47)$$
$$= 1 - 0.9997$$
$$= 0.0003$$

TSH, high sensitivity, serum:

$$P(X \ge 3.15) = P\left(Z \ge \frac{3.15 - 2.925}{1.2875}\right)$$
$$\approx P(Z \ge 0.17)$$
$$= 1 - 0.5675$$
$$= 0.4325 \quad [\text{Tech: } 0.4306]$$

Using Abby's criterion, she should be concerned about her total cholesterol, her LDL cholesterol, and her LDL/HDL ratio.

Chapter 8
Sampling Distributions

Section 8.1

1. sampling distribution

3. standard error; mean

5. False; if the sample is drawn from a population that is not normal, then the distribution of the sample mean will not necessarily be normally distributed.

7. The sampling distribution is normal. The mean of the sampling distribution of $\bar{x}$ is $\mu_{\bar{x}} = \mu = 30$ and the standard deviation is given by $\sigma_{\bar{x}} = \dfrac{\sigma}{\sqrt{n}} = \dfrac{8}{\sqrt{10}}$.

9. $\mu_{\bar{x}} = \mu = 80$; $\sigma_{\bar{x}} = \dfrac{\sigma}{\sqrt{n}} = \dfrac{14}{\sqrt{49}} = \dfrac{14}{7} = 2$

11. $\mu_{\bar{x}} = \mu = 52$; $\sigma_{\bar{x}} = \dfrac{\sigma}{\sqrt{n}} = \dfrac{10}{\sqrt{21}} \approx 2.182$

13. **a.** The sampling distribution is symmetric about 500, so the mean is $\mu_{\bar{x}} = 500$.

   b. The inflection points are at 480 and 520, so $\sigma_{\bar{x}} = 520 - 500 = 20$ (or $500 - 480 = 20$).

   c. Since $n = 16 \leq 30$, the population must be normal so that the sampling distribution of $\bar{x}$ is normal.

   d. $\sigma_{\bar{x}} = \dfrac{\sigma}{\sqrt{n}}$

   $20 = \dfrac{\sigma}{\sqrt{16}}$

   $\sigma = 20\sqrt{16} = 20(4) = 80$

   The standard deviation of the population from which the sample is drawn is 80.

15. **(a)** Since $\mu = 80$ and $\sigma = 14$, the mean and standard deviation of the sampling distribution of $\bar{x}$ are given by:

   $\mu_{\bar{x}} = \mu = 80$; $\sigma_{\bar{x}} = \dfrac{\sigma}{\sqrt{n}} = \dfrac{14}{\sqrt{49}} = \dfrac{14}{7} = 2$.

   We are not told that the population is normally distributed, but we do have a large sample size ($n = 49 \geq 30$). Therefore, we can use the Central Limit Theorem to say

that the sampling distribution of $\bar{x}$ is approximately normal.

   (b) $P(\bar{x} > 83) = P\left(Z > \dfrac{83 - 80}{2}\right)$

   $= P(Z > 1.50)$

   $= 1 - P(Z \leq 1.50)$

   $= 1 - 0.9332 = 0.0668$

   If we take 100 simple random samples of size $n = 49$ from a population with $\mu = 80$ and $\sigma = 14$, then about 7 of the samples will result in a mean that is greater than 83.

   (c) $P(\bar{x} \leq 75.8) = P\left(Z \leq \dfrac{75.8 - 80}{2}\right)$

   $= P(Z \leq -2.10) = 0.0179$

   If we take 100 simple random samples of size $n = 49$ from a population with $\mu = 80$ and $\sigma = 14$, then about 2 of the samples will result in a mean that is less than or equal to 75.8.

   (d) $P(78.3 < \bar{x} < 85.1)$

   $= P\left(\dfrac{78.3 - 80}{2} < Z < \dfrac{85.1 - 80}{2}\right)$

   $= P(-0.85 < Z < 2.55)$

   $= 0.9946 - 0.1977 = 0.7969$ [Tech: 0.7970]

   If we take 100 simple random samples of size $n = 49$ from a population with $\mu = 80$ and $\sigma = 14$, then about 78 of the samples will result in a mean that is between 78.3 and 85.1.

17. **(a)** The population must be normally distributed. If this is the case, then the sampling distribution of $\bar{x}$ is exactly normal. The mean and standard deviation of the sampling distribution are $\mu_{\bar{x}} = \mu = 64$ and

   $\sigma_{\bar{x}} = \dfrac{\sigma}{\sqrt{n}} = \dfrac{17}{\sqrt{12}} \approx 4.907$.

(b) $P(\bar{x} < 67.3) = P\left(Z < \dfrac{67.3 - 64}{17/\sqrt{12}}\right)$

$= P(Z < 0.67)$

$= 0.7486$ [Tech: 0.7493]

If we take 100 simple random samples of size $n = 12$ from a population that is normally distributed with $\mu = 64$ and $\sigma = 17$, then about 75 of the samples will result in a mean that is less than 67.3.

(c) $P(\bar{x} \geq 65.2) = P\left(Z \geq \dfrac{65.2 - 64}{17/\sqrt{12}}\right)$

$= P(Z \geq 0.24)$

$= 1 - P(Z < 0.24)$

$= 1 - 0.5948$

$= 0.4052$ [Tech: 0.4034]

If we take 100 simple random samples of size $n = 12$ from a population that is normally distributed with $\mu = 64$ and $\sigma = 17$ then about 41 of the samples will result in a mean that is greater than or equal to 65.2.

19. (a) $P(X < 260) = P\left(Z < \dfrac{260 - 266}{16}\right)$

$= P(Z < -0.38)$

$= 0.3520$ [Tech: 0.3538]

If we select a simple random sample of $n = 100$ human pregnancies, then about 35 of the pregnancies would last less than 260 days.

(b) Since the length of human pregnancies is normally distributed, the sampling distribution of $\bar{x}$ is normal with $\mu_{\bar{x}} = 266$

and $\sigma_{\bar{x}} = \dfrac{16}{\sqrt{20}} \approx 3.578$.

(c) $P(\bar{x} < 260) = P\left(Z < \dfrac{260 - 266}{16/\sqrt{20}}\right)$

$= P(Z < -1.68)$

$= 0.0465$ [Tech: 0.0468]

If we take 100 simple random samples of size $n = 20$ human pregnancies, then about 5 of the samples will result in a mean gestation period of 260 days or less.

(d) $\mu_{\bar{x}} = \mu = 266$; $\sigma_{\bar{x}} = \dfrac{\sigma}{\sqrt{n}} = \dfrac{16}{\sqrt{50}}$

$P(\bar{x} < 260) = P\left(Z < \dfrac{260 - 266}{16/\sqrt{50}}\right)$

$= P(Z < -2.65)$

$= 0.0040$

If we take 1000 simple random samples of size $n = 50$ human pregnancies, then about 4 of the samples will result in a mean gestation period of 260 days or less.

(e) Answers will vary. Part (d) indicates that this result would be an unusual observation. Therefore, we would conclude that the sample likely came from a population whose mean gestation period is less than 266 days.

(f) $\mu_{\bar{x}} = \mu = 266$; $\sigma_{\bar{x}} = \dfrac{\sigma}{\sqrt{n}} = \dfrac{16}{\sqrt{15}}$

$P(256 \leq \bar{x} \leq 276)$

$= P\left(\dfrac{256 - 266}{16/\sqrt{15}} \leq Z \leq \dfrac{276 - 266}{16/\sqrt{15}}\right)$

$= P(-2.42 \leq Z \leq 2.42)$

$= 0.9922 - 0.0078$

$= 0.9844$ [Tech: 0.9845]

If we take 100 simple random samples of size $n = 15$ human pregnancies, then about 98 of the samples will result in a mean gestation period between 256 and 276 days, inclusive.

21. (a) $P(X > 95) = P\left(Z > \dfrac{95 - 90}{10}\right)$

$= P(Z > 0.5)$

$= 1 - P(Z \leq 0.5)$

$= 1 - 0.6915$

$= 0.3085$

If we select a simple random sample of $n = 100$ second grade students, then about 31 of the students would read more than 95 words per minute.

(b) $\mu_{\bar{x}} = \mu = 90$; $\sigma_{\bar{x}} = \dfrac{\sigma}{\sqrt{n}} = \dfrac{10}{\sqrt{12}}$

$$P(\bar{x} > 95) = P\left(Z > \dfrac{95-90}{10/\sqrt{12}}\right)$$
$$= P(Z > 1.73)$$
$$= 1 - P(Z \le 1.73)$$
$$= 1 - 0.9582$$
$$= 0.0418 \quad [\text{Tech: } 0.0416]$$

If we take 100 simple random samples of size $n = 12$ second grade students, then about 4 of the samples will result in a mean reading rate that is more than 95 words per minute.

(c) $\mu_{\bar{x}} = \mu = 90$; $\sigma_{\bar{x}} = \dfrac{\sigma}{\sqrt{n}} = \dfrac{10}{\sqrt{24}}$

$$P(\bar{x} > 95) = P\left(Z > \dfrac{95-90}{10/\sqrt{24}}\right)$$
$$= P(Z > 2.45)$$
$$= 1 - P(Z \le 2.45)$$
$$= 1 - 0.9929$$
$$= 0.0071 \quad [\text{Tech: } 0.0072]$$

If we take 1000 simple random samples of size $n = 24$ second grade students, then about 7 of the samples will result in a mean reading rate that is more than 95 words per minute.

(d) Increasing the sample size decreases the probability that $\bar{x} > 95$. This happens because $\sigma_{\bar{x}}$ decreases as n increases.

(e) No, this result would not be unusual because, if $\mu_{\bar{x}} = \mu = 90$ and $\sigma_{\bar{x}} = \dfrac{\sigma}{\sqrt{n}} = \dfrac{10}{\sqrt{20}}$, then

$$P(\bar{x} > 92.8) = P\left(Z > \dfrac{92.8-90}{10/\sqrt{20}}\right)$$
$$= P(Z > 1.25)$$
$$= 1 - P(Z \le 1.25)$$
$$= 1 - 0.8944$$
$$= 0.1056 \quad [\text{Tech: } 0.1052]$$

If we take 100 simple random samples of size $n = 20$ second grade students, then about 11 of the samples will result in a mean reading rate that is above 92.8 words per minute. This result does not qualify as unusual.

This means that the new reading program is not abundantly more effective than the old program.

(f) We need to find a number of words per minute c such that $P(\bar{x} > c) = 0.05$. So, we have

$$P(\bar{x} > c) = P\left(Z > \dfrac{c-90}{10/\sqrt{20}}\right) = 0.05.$$

The area to the right of z is 0.05 if $z = 1.645$.

So, $\dfrac{c-90}{10/\sqrt{20}} = 1.645$. Solving for c, we get

$c = 93.7$ [tech 93.7] words per minute.

23. (a) $P(X > 0) = P\left(Z > \dfrac{0-0.007233}{0.04135}\right)$
$$= P(Z > -0.17)$$
$$= 1 - P(Z \le -0.17)$$
$$= 1 - 0.4325$$
$$= 0.5675 \quad [\text{Tech: } 0.5694]$$

If we select a simple random sample of $n = 100$ months, then about 57 of the months would have positive rates of return.

(b) $\mu_{\bar{x}} = \mu = 0.007233$; $\sigma_{\bar{x}} = \dfrac{\sigma}{\sqrt{n}} = \dfrac{0.04135}{\sqrt{12}}$

$$P(\bar{x} > 0) = P\left(Z > \dfrac{0-0.007233}{0.04135/\sqrt{12}}\right)$$
$$= P(Z > -0.61)$$
$$= 1 - P(Z \le -0.61)$$
$$= 1 - 0.2709$$
$$= 0.7291 \quad [\text{Tech: } 0.7277]$$

If we take 100 simple random samples of size $n = 12$ months, then about 73 of the samples will result in a mean monthly rate that is positive.

(c) $\mu_{\bar{x}} = \mu = 0.007233$; $\sigma_{\bar{x}} = \dfrac{\sigma}{\sqrt{n}} = \dfrac{0.04135}{\sqrt{24}}$

$$P(\bar{x} > 0) = P\left(Z > \dfrac{0-0.007233}{0.04135/\sqrt{24}}\right)$$
$$= P(Z > -0.86)$$
$$= 1 - P(Z \le -0.86)$$
$$= 1 - 0.1949$$
$$= 0.8051 \quad [\text{Tech: } 0.8043]$$

If we take 100 simple random samples of size $n = 24$ months, then about 81 of the samples will result in a mean monthly rate that is positive.

(d) $\mu_{\bar{x}} = \mu = 0.007233$; $\sigma_{\bar{x}} = \dfrac{\sigma}{\sqrt{n}} = \dfrac{0.04135}{\sqrt{36}}$

$$P(\bar{x} > 0) = P\left(Z > \dfrac{0 - 0.007233}{0.04135/\sqrt{36}} \right)$$
$$= P(Z > -1.05)$$
$$= 1 - P(Z \leq -1.05)$$
$$= 1 - 0.1469$$
$$= 0.8531 \quad [\text{Tech: } 0.8530]$$

If we take 100 simple random samples of size $n = 36$ months, then about 85 of the samples will result in a mean monthly rate that is positive.

(e) Answers will vary. Based on the results of parts (b)–(d), the likelihood of earning a positive rate of return increases as the investment time horizon increases.

25. (a) Without knowing the shape of the distribution, we would need a sample size of at least 30 so we could apply the Central Limit Theorem.

(b) $\mu_{\bar{x}} = \mu = 11.4$; $\sigma_{\bar{x}} = \dfrac{\sigma}{\sqrt{n}} = \dfrac{3.2}{\sqrt{40}}$

$$P(\bar{x} < 10) = P\left(Z < \dfrac{10 - 11.4}{3.2/\sqrt{40}} \right)$$
$$= P(Z < -2.77)$$
$$= 0.0028$$

If we take 1000 simple random samples of size $n = 40$ oil changes, then about 3 of the samples will result in a mean time less than 10 minutes.

(c) We need to find c such that the probability that the sample mean will be less than c is 10%. So we have:

$$P(\bar{x} \leq c) = P\left(Z \leq \dfrac{c - 11.4}{3.2/\sqrt{40}} \right) = 0.10$$

The value of Z such that 10% of the area is less than or equal to it is $Z = -1.28$. So,

$\dfrac{c - 11.4}{3.2/\sqrt{40}} = -1.28$. Solving for Z, we get:

$$Z = 11.4 + (-1.28)\dfrac{3.2}{\sqrt{40}} \approx 10.8 \text{ minutes.}$$

27. (a) The sampling distribution of $\bar{x}$ is approximately normal because the sample is large, $n = 50 \geq 30$. From the Central Limit Theorem, as the sample size increases, the sampling distribution of the mean becomes more normal.

(b) Assuming that we are sampling from a population that is exactly at the Food Defect Action Level, $\mu_{\bar{x}} = \mu = 3$ and

$$\sigma_{\bar{x}} = \dfrac{\sigma}{\sqrt{n}} = \dfrac{\sqrt{3}}{\sqrt{50}} = \sqrt{\dfrac{3}{50}} \approx 0.245.$$

(c) $P(\bar{x} \geq 3.6) = P\left(Z \geq \dfrac{3.6 - 3}{\sqrt{3}/\sqrt{50}} \right)$

$$= P(Z \geq 2.45) = 1 - 0.9929$$
$$= 0.0071 \quad [\text{Tech: } 0.0072]$$

If we take 1000 simple random samples of size $n = 50$ ten-gram portions of peanut butter, then about 7 of the samples will result in a mean of at least 3.6 insect fragments. This result is unusual. We might conclude that the sample comes from a population with a mean higher than 3 insect fragments per ten-gram portion.

29. (a) No, the variable "weekly time spent watching television" is not likely normally distributed. It is likely skewed right.

(b) Because the sample is large, $n = 40 \geq 30$, the sampling distribution of $\bar{x}$ is approximately normal with $\mu_{\bar{x}} = \mu = 2.35$

and $\sigma_{\bar{x}} = \dfrac{\sigma}{\sqrt{n}} = \dfrac{1.93}{\sqrt{40}} \approx 0.305$.

(c) $P(2 \leq \bar{x} \leq 3)$

$$= P\left(\dfrac{2 - 2.35}{1.93/\sqrt{40}} \leq Z \leq \dfrac{3 - 2.35}{1.93/\sqrt{40}} \right)$$
$$= P(-1.15 \leq Z \leq 2.13)$$
$$= 0.9834 - 0.1251$$
$$= 0.8583 \quad [\text{Tech: } 0.8577]$$

If we take 100 simple random samples of size $n = 40$ adult Americans, then about 86 of the samples will result in a mean time between 2 and 3 hours watching television on a weekday.

(d) $\mu_{\bar{x}} = \mu = 2.35$; $\sigma_{\bar{x}} = \dfrac{\sigma}{\sqrt{n}} = \dfrac{1.93}{\sqrt{35}}$

$$P(\bar{x} \le 1.89) = P\left(Z \le \dfrac{1.89 - 2.35}{1.93/\sqrt{35}}\right)$$

$$= P(Z \le -1.41)$$

$$= 0.0793$$

If we take 100 simple random samples of size $n = 35$ adult Americans, then about 8 of the samples will result in a mean time of 1.89 hours or less watching television on a weekday. This result is not unusual, so this evidence is insufficient to conclude that avid Internet users watch less television.

31. (a) $\mu = \dfrac{\sum x}{N} = \dfrac{278}{6} \approx 46.3$ years

The population mean age is about 46.3 years.

(b) 37, 38; 37, 45; 37, 50; 37, 48; 37, 60;
38, 45; 38, 50; 38, 48; 38, 60; 45, 50;
45, 48; 45, 60; 50, 48; 50, 60; 48, 60

(c) Obtain each sample mean by adding the two ages in a sample and dividing by two.

$\bar{x} = \dfrac{37 + 38}{2} = 37.5$ yr ; $\bar{x} = \dfrac{37 + 45}{2} = 41$ yr ;

$\bar{x} = \dfrac{37 + 50}{2} = 43.5$ yr ; $\bar{x} = \dfrac{37 + 48}{2} = 41.5$ yr ,

etc.

$\bar{x}$	$P(\bar{x})$	$\bar{x}$	$P(\bar{x})$
37.5	$1/15 \approx 0.0667$	46.5	$1/15 \approx 0.0667$
41	$1/15 \approx 0.0667$	47.5	$1/15 \approx 0.0667$
41.5	$1/15 \approx 0.0667$	48.5	$1/15 \approx 0.0667$
42.5	$1/15 \approx 0.0667$	49	$2/15 \approx 0.1333$
43	$1/15 \approx 0.0667$	52.5	$1/15 \approx 0.0667$
43.5	$1/15 \approx 0.0667$	54	$1/15 \approx 0.0667$
44	$1/15 \approx 0.0667$	55	$1/15 \approx 0.0667$

(d) $\mu_{\bar{x}} = (37.5)\left(\dfrac{1}{15}\right) + (41)\left(\dfrac{1}{15}\right) + \ldots + (55)\left(\dfrac{1}{15}\right)$

≈ 46.3 years

Notice that this is the same value we obtained in part (a) for the population mean.

(e) $P(43.3 \le \bar{x} \le 49.3)$

$= P(43.5 \le \bar{x} \le 49)$

$= \dfrac{1}{15} + \dfrac{1}{15} + \dfrac{1}{15} + \dfrac{1}{15} + \dfrac{1}{15} + \dfrac{2}{15} = \dfrac{7}{15} = 0.467$

(f) for part (b):
37, 38, 45;   37, 38, 50;   37, 38, 48;
37, 38, 60;   37, 45, 50;   37, 45, 48;
37, 45, 60;   37, 50, 48;   37, 50, 60;
37, 48, 60;   38, 45, 50;   38, 45, 48;
38, 45, 60;   38, 50, 48;   38, 50, 60;
38, 48, 60;   45, 50, 48;   45, 50, 60;
45, 48, 60;   50, 48, 60

for part (c):
Obtain each sample mean by adding the three ages in a sample and dividing by three.

$\bar{x} = \dfrac{37 + 38 + 45}{3} = 40$ yr ;

$\bar{x} = \dfrac{37 + 38 + 50}{3} \approx 41.7$ yr ;

$\bar{x} = \dfrac{37 + 38 + 48}{3} = 41$ yr ; etc

$\bar{x}$	$P(\bar{x})$	$\bar{x}$	$P(\bar{x})$
40	$1/20 = 0.05$	47.3	$1/20 = 0.05$
41	$1/20 = 0.05$	47.7	$2/20 = 0.1$
41.7	$1/20 = 0.05$	48.3	$1/20 = 0.05$
43.3	$1/20 = 0.05$	48.7	$1/20 = 0.05$
43.7	$1/20 = 0.05$	49	$1/20 = 0.05$
44	$1/20 = 0.05$	49.3	$1/20 = 0.05$
44.3	$1/20 = 0.05$	51	$1/20 = 0.05$
45	$2/20 = 0.1$	51.7	$1/20 = 0.05$
45.3	$1/20 = 0.05$	52.7	$1/20 = 0.05$

for part (d):

$\mu_{\bar{x}} = (40)\left(\dfrac{1}{20}\right) + (41)\left(\dfrac{1}{20}\right) + \ldots + (52.7)\left(\dfrac{1}{20}\right)$

≈ 46.3 years

Notice that this is the same value we obtained previously.

for part (e):

$$P(43.3 \le \bar{x} \le 49.3) = \dfrac{14}{20} = \dfrac{7}{10} = 0.7$$

With the larger sample size, the probability of obtaining a sample mean within 3 years of the population mean increases.

33. (a) Assuming that only one number is selected, the probability distribution will be as follows:

x	$P(x)$
35	$1/38 \approx 0.0263$
-1	$37/38 \approx 0.9737$

(b) $\mu = (35)(0.0263) + (-1)(0.9737) \approx -\$0.05$

$\sigma = \sqrt{(35 - (-.05))^2 (0.0263) + (-1 - (-.05))^2 (0.9737)}$
$\approx \$5.76$

(c) Because the sample size is large, $n = 100 > 30$, the sampling distribution of $\bar{x}$ is approximately normal with $\mu_{\bar{x}} = \mu = -\$0.05$

and $\sigma_{\bar{x}} = \dfrac{\sigma}{\sqrt{n}} = \dfrac{5.76}{\sqrt{100}} \approx \$0.576$.

(d) $P(\bar{x} > 0) = P\left(Z > \dfrac{0 - (-0.05)}{5.76/\sqrt{100}} \right)$

$= P(Z > 0.09)$

$= 1 - P(Z \le 0.09)$

$= 1 - 0.5359$

$= 0.4641$ [Tech: 0.4654]

(e) $\mu_{\bar{x}} = -\$0.05$; $\sigma_{\bar{x}} = \dfrac{5.76}{\sqrt{200}} \approx \$0.407$

$P(\bar{x} > 0) = P\left(Z > \dfrac{0 - (-0.05)}{5.76/\sqrt{200}} \right)$

$= P(Z > 0.12)$

$= 1 - P(Z \le 0.12)$

$= 1 - 0.5478$

$= 0.4522$ [Tech: 0.4511]

(f) $\mu_{\bar{x}} = -\$0.05$; $\sigma_{\bar{x}} = \dfrac{5.76}{\sqrt{1000}} \approx \$0.275$

$P(\bar{x} > 0) = P\left(Z > \dfrac{0 - (-0.05)}{5.76/\sqrt{1000}} \right)$

$= P(Z > 0.27)$

$= 1 - P(Z \le 0.27)$

$= 1 - 0.6064$

$= 0.3936$ [Tech: 0.3918]

(g) The probability of being ahead decreases as the number of games played increases.

35. The Central Limit Theorem states that, regardless of the distribution of the population, the sampling distribution of the sample mean becomes approximately normal as the sample size, n, increases.

37. Probability (b) will be larger. The standard deviation of the sampling distribution of $\bar{x}$ is decreased as the sample size, n, gets larger.

39. (a) We would expect that Jack's distribution would be skewed left, but not as much as the original distribution. Diane's distribution should be bell-shaped and symmetric, i.e., approximately normal.

(b) We would expect both distributions to have a mean of 50. The mean of the distribution of the sample mean is the same as the mean of the distribution from which the data are drawn.

(c) We expect Jack's distribution to have a standard deviation of $\sigma_{\bar{x}} = \dfrac{10}{\sqrt{3}} \approx 5.8$. We expect Diane's distribution to have a standard deviation of $\sigma_{\bar{x}} = \dfrac{10}{\sqrt{30}} \approx 1.8$.

Section 8.2

1. 0.44; $p = \dfrac{220}{500} = 0.44$

3. False; while it is possible for the sample proportion to have the same value as the population proportion, it will not *always* have the same value.

5. The sampling distribution of $\hat{p}$ is approximately normal when $n \le 0.05N$ and $np(1-p) \ge 10$.

7. $25,000(0.05) = 1250$; the sample size, $n = 500$, is less than 5% of the population size and $np(1-p) = 500(0.4)(0.6) = 120 \geq 10$.

The distribution of $\hat{p}$ is approximately normal, with mean $\mu_{\hat{p}} = p = 0.4$ and standard deviation

$$\sigma_{\hat{p}} = \sqrt{\frac{p(1-p)}{n}} = \sqrt{\frac{0.4(1-0.4)}{500}} \approx 0.022 \,.$$

9. $25,000(0.05) = 1250$; the sample size, $n = 1000$, is less than 5% of the population size and $np(1-p) = 1000(0.103)(0.897) = 92.391 \geq 10$.

The distribution of $\hat{p}$ is approximately normal, with mean $\mu_{\hat{p}} = p = 0.103$ and standard deviation

$$\sigma_{\hat{p}} = \sqrt{\frac{p(1-p)}{n}} = \sqrt{\frac{0.103(1-0.103)}{1000}} \approx 0.010 \,.$$

11. (a) $10,000(0.05) = 500$; the sample size, $n = 75$, is less than 5% of the population size and $np(1-p) = 75(0.8)(0.2) = 12 \geq 10$. The distribution of $\hat{p}$ is approximately normal, with mean $\mu_{\hat{p}} = p = 0.8$ and standard deviation

$$\sigma_{\hat{p}} = \sqrt{\frac{p(1-p)}{n}} = \sqrt{\frac{0.8(1-0.8)}{75}} \approx 0.046 \,.$$

(b) $P(\hat{p} \geq 0.84) = P\left(Z \geq \dfrac{0.84 - 0.8}{\sqrt{0.8(0.2)/75}} \right)$

$\phantom{P(\hat{p} \geq 0.84)} = P(Z \geq 0.87)$

$\phantom{P(\hat{p} \geq 0.84)} = 1 - P(Z < 0.87)$

$\phantom{P(\hat{p} \geq 0.84)} = 1 - 0.8078$

$\phantom{P(\hat{p} \geq 0.84)} = 0.1922$ [Tech: 0.1932]

About 19 out of 100 random samples of size $n = 75$ will result in 63 or more individuals (that is, 84% or more) with the characteristic.

(c) $P(\hat{p} \leq 0.68) = P\left(Z \leq \dfrac{0.68 - 0.8}{\sqrt{0.8(0.2)/75}} \right)$

$\phantom{P(\hat{p} \leq 0.68)} = P(Z \leq -2.60) = 0.0047$

About 5 out of 1000 random samples of size $n = 75$ will result in 51 or fewer individuals (that is, 68% or less) with the characteristic.

13. (a) $1,000,000(0.05) = 50,000$; the sample size, $n = 1000$, is less than 5% of the population size and $np(1-p) = 1000(0.35)(0.65) = 227.5 \geq 10$.

The distribution of $\hat{p}$ is approximately normal, with mean $\mu_{\hat{p}} = p = 0.35$ and standard deviation

$$\sigma_{\hat{p}} = \sqrt{\frac{p(1-p)}{n}} = \sqrt{\frac{0.35(1-0.35)}{1000}} \approx 0.015.$$

(b) $\hat{p} = \dfrac{x}{n} = \dfrac{390}{1000} = 0.39$

$P(X \geq 390) = P(\hat{p} \geq 0.39)$

$ = P\left(Z \geq \dfrac{0.39 - 0.35}{\sqrt{0.53(0.65)/1000}} \right)$

$ = P(Z \geq 2.65)$

$ = 1 - P(Z < 2.65)$

$ = 1 - 0.9960 = 0.0040$

About 4 out of 1000 random samples of size $n = 1000$ will result in 390 or more individuals (that is, 39% or more) with the characteristic.

(c) $\hat{p} = \dfrac{x}{n} = \dfrac{320}{1000} = 0.32$

$P(X \leq 320) = P(\hat{p} \leq 0.32)$

$ = P\left(Z \geq \dfrac{0.32 - 0.35}{\sqrt{0.53(0.65)/1000}} \right)$

$ = P(Z \leq -1.99)$

$ = 0.0233$ [Tech: 0.0234]

About 2 out of 100 random samples of size $n = 1000$ will result in 320 or fewer individuals (that is, 32% or less) with the characteristic.

15. (a) The sample size, $n = 200$, is less than 5% of the population size and $np(1-p) = 200(0.47)(0.53) = 49.82 \geq 10$.

The distribution of $\hat{p}$ is approximately normal, with mean $\mu_{\hat{p}} = p = 0.47$ and standard deviation

$$\sigma_{\hat{p}} = \sqrt{\frac{p(1-p)}{n}} = \sqrt{\frac{0.47(1-0.47)}{200}} \approx 0.035 \,.$$

(b) $\hat{p} = \dfrac{x}{n} = \dfrac{100}{200} = 0.5$

$$P(\hat{p} > 0.5) = P\left(Z > \dfrac{0.5 - 0.47}{\sqrt{0.47(0.53)/200}} \right)$$

$$= P(Z > 0.85) = 1 - P(Z \le 0.85)$$

$$= 1 - 0.8023$$

$$= 0.1977 \quad [\text{Tech: } 0.1976]$$

About 20 out of 100 random samples of size $n = 200$ Americans will result in a sample where at least half can order a meal in a foreign language.

(c) $\hat{p} = \dfrac{x}{n} = \dfrac{80}{200} = 0.4$

$$P(X \le 80) = P(\hat{p} \le 0.4)$$

$$= P\left(Z \le \dfrac{0.4 - 0.47}{\sqrt{0.47(0.53)/200}} \right)$$

$$= P(Z \le -1.98)$$

$$= 0.0239 \quad [\text{Tech: } 0.0237]$$

About 2 out of 100 random samples of size $n = 200$ Americans will result in a sample where 80 or fewer can order a meal in a foreign language. This result is unusual.

17. (a) Our sample size, $n = 500$, is less than 5% of the population size and $np(1 - p) = 500(0.39)(0.61) = 118.95 \ge 10$. The distribution of $\hat{p}$ is approximately normal, with mean $\mu_{\hat{p}} = p = 0.39$ and standard deviation

$$\sigma_{\hat{p}} = \sqrt{\dfrac{p(1 - p)}{n}} = \sqrt{\dfrac{0.39(1 - 0.39)}{500}} \approx 0.022 .$$

(b) $P(\hat{p} < 0.38) = P\left(Z < \dfrac{0.38 - 0.39}{\sqrt{0.39(0.61)/500}} \right)$

$$= P(Z < -0.46)$$

$$= 0.3228 \quad [\text{Tech: } 0.3233]$$

About 32 out of 100 random samples of size $n = 500$ adults will result in less than 38% who believe marriage is obsolete.

(c)

$$P(0.40 < \hat{p} < 0.45)$$

$$= P\left(\dfrac{0.40 - 0.39}{\sqrt{\dfrac{0.39(0.61)}{500}}} < Z < \dfrac{0.45 - 0.39}{\sqrt{\dfrac{0.39(0.61)}{500}}} \right)$$

$$= P(0.46 < Z < 2.75)$$

$$= 0.9970 - 0.6772$$

$$= 0.3198 \quad [\text{Tech: } 0.3203]$$

About 32 out of 100 random samples of size $n = 500$ adults will have between 40% and 45% of the respondents who say that marriage is obsolete.

(d)

$$P(X \ge 210)$$

$$= P(\hat{p} \ge 0.42)$$

$$= 1 - P(\hat{p} < 0.42)$$

$$= 1 - P\left(Z < \dfrac{0.42 - 0.39}{\sqrt{\dfrac{0.39(0.61)}{500}}} \right)$$

$$= 1 - P(Z < 1.38)$$

$$= 1 - 0.9162$$

$$= 0.0838 \quad [\text{Tech: } 0.0845]$$

About 8 out of 100 random samples of size $n = 500$ adults will have at least 210 of the respondents who say that marriage is obsolete.

19. $\hat{p} = \dfrac{x}{n} = \dfrac{121}{1100} = 0.11$

$$P(X \ge 121) = P(\hat{p} \ge 0.11)$$

$$= P\left(Z \ge \dfrac{0.11 - 0.10}{\sqrt{0.1(0.9)/1100}} \right)$$

$$= P(Z \ge 1.11)$$

$$= 1 - P(Z < 1.11)$$

$$= 1 - 0.8665$$

$$= 0.1335 \quad [\text{Tech: } 0.1345]$$

This result is not unusual, so this evidence is insufficient to conclude that the proportion of Americans who are afraid to fly has increased above 0.10.

21. $\hat{p} = \dfrac{x}{n} = \dfrac{164}{310} = 0.529$

$P(X \geq 164) = P(\hat{p} \geq 0.529)$

$= P\left(Z \geq \dfrac{0.529 - 0.49}{\sqrt{0.49(0.51)/310}} \right)$

$= P(Z \geq 1.37) = 1 - P(Z < 1.37)$

$= 1 - 0.9147$

$= 0.0853$ [Tech: 0.0846]

This result is not unusual, if the true proportion of voters in favor of the referendum is 0.49. So, it is not surprising that we would get a result as extreme as this result. This illustrates the dangers of using exit polling to call elections. Notice that most people in the town do not favor the referendum, however the exit poll result showed a majority favor the referendum.

23. (a) To say the sampling distribution of $\hat{p}$ is approximately normal, we need $np(1-p) \geq 10$ and $n \leq 0.05N$.

With $p = 0.1$, we need

$n(0.1)(1-0.1) \geq 10$

$n(0.1)(0.9) \geq 10$

$n(0.09) \geq 10$

$n \geq 111.11$

Therefore, we need a sample size of 112, or 62 more adult Americans.

(b) With $p = 0.2$, we need

$n(0.2)(1-0.2) \geq 10$

$n(0.2)(0.8) \geq 10$

$n(0.16) \geq 10$

$n \geq 62.5$

Therefore, we need a sample size of 63, or 13 more, adult Americans.

25. A sample of size $n = 20$ households represents more than 5% of the population size $N = 100$ households in the association. Thus, the results from individuals in the sample are not independent of one another.

Consumer Reports®: Tanning Salons

(a) Selecting a random sample from a large number of facilities increases the likelihood of obtaining an estimate of the population parameter that is close to the true value. Since the reference frame is all American tanning salons, sampling from multiple metropolitan areas ensures representation from a broad spectrum of American tanning salons.

(b) Yes, the sample size, $n = 296$, is much less than 5% of the population (150,000).

(c) The sampling distribution of $\hat{p}$ is approximately normal with a mean of $\mu_{\hat{p}} = 0.25$ and a standard deviation of

$\sigma_{\hat{p}} = \sqrt{\dfrac{(0.25)(1-0.25)}{296}} \approx 0.025.$

$P(\hat{p} < .226) = P\left(Z < \dfrac{0.226 - 0.25}{\sqrt{0.25(0.75)/296}} \right)$

$= P(Z < -0.95)$

$= 0.1711$ [Tech: 0.0.1701]

About 17 out of 100 random samples of size $n = 296$ tanning facilities will result in less than 22.6% (i.e., less than 67) that would state that tanning in a salon is the same as tanning in the sun with respect to causing skin cancer.

(d) The sampling distribution of $\hat{p}$ is approximately normal with a mean of $\mu_{\hat{p}} = 0.18$ and a standard deviation of

$\sigma_{\hat{p}} = \sqrt{\dfrac{(0.18)(1-0.18)}{296}} \approx 0.022.$

$P(\hat{p} \geq 0.142) = P\left(Z \geq \dfrac{0.142 - 0.18}{\sqrt{0.18(0.82)/296}} \right)$

$= P(Z \geq -1.70)$

$= 1 - P(Z < -1.70)$

$= 1 - 0.0446$

$= 0.9554$ [Tech: 0.9556]

About 96 out of 100 random samples of size $n = 296$ tanning facilities will result in at least 14.2% (i.e., at least 42) that would state that tanning does not cause wrinkled skin.

$$\hat{p} = \frac{50}{296} \approx 0.169$$

$$P(X \le 50) = P(\hat{p} \le 0.169)$$

$$= P\left(Z \le \frac{0.169 - 0.18}{\sqrt{0.18(0.82)/296}}\right)$$

$$= P(Z \le -0.49)$$

$$= 0.3121 \quad [\text{Tech: } 0.3111]$$

About 31 out of 100 random samples of size $n = 296$ tanning facilities will result in 50 or fewer facilities (i.e., 16.9% or fewer) that would state that tanning does not cause wrinkled skin. Therefore, it is not unusual for 50 or less facilities to state that tanning in a salon does not cause wrinkled skin.

Chapter 8 Review Exercises

1. Answers will vary. The sampling distribution of a statistic (such as the sample mean) is a probability distribution for all possible values of the statistic computed from a sample of size n.

2. The sampling distribution of $\bar{x}$ is exactly normal if the population distribution is normal. If the population distribution is not normal, we apply the Central Limit Theorem and say that the distribution of $\bar{x}$ is approximately normal for a sufficiently large n (for finite σ). For our purposes, $n \ge 30$ is considered large enough to apply the theorem.

3. The sampling distribution of $\hat{p}$ is approximately normal if $np(1-p) \ge 10$ and $n \le 0.05N$.

4. For $\bar{x}$: $\mu_{\bar{x}} = \mu$ and $\sigma_{\bar{x}} = \frac{\sigma}{\sqrt{n}}$

   For $\hat{p}$: $\mu_{\hat{p}} = p$ and $\sigma_{\hat{p}} = \sqrt{\frac{p(1-p)}{n}}$

5. **(a)** $P(X > 2625) = P\left(Z > \frac{2625 - 2600}{50}\right)$

   $$= P(Z > 0.5)$$

   $$= 1 - P(Z \le 0.5)$$

   $$= 1 - 0.6915$$

   $$= 0.3085$$

   If we select a simple random sample of $n = 100$ pregnant women, then about 31 have energy needs of more than 2625 kcal/day. This result is not unusual.

   (b) Since the population is normally distributed, the sampling distribution of $\bar{x}$ will be normal, regardless of the sample size. The mean of the distribution is $\mu_{\bar{x}} = \mu = 2600$ kcal, and the standard deviation is

   $$\sigma_{\bar{x}} = \frac{\sigma}{\sqrt{n}} = \frac{50}{\sqrt{20}} \approx 11.180 \, \text{kcal.}$$

   (c) $P(\bar{x} > 2625) = P\left(Z > \frac{2625 - 2600}{50/\sqrt{20}}\right)$

   $$= P(Z > 2.24)$$

   $$= 1 - P(Z \le 2.24)$$

   $$= 1 - 0.9875$$

   $$= 0.0125 \quad [\text{Tech: } 0.0127]$$

   If we take 100 simple random samples of size $n = 20$ pregnant women, then about 1 of the samples will result in a mean energy need of more than 2625 kcal/day. This result is unusual.

6. **(a)** Since we have a large sample ($n = 30$), we can use the Central Limit Theorem to say that the sampling distribution of $\bar{x}$ is approximately normal. The mean of the sampling distribution is $\mu_{\bar{x}} = \mu = 0.75$ inch and the standard deviation is

   $$\sigma_{\bar{x}} = \frac{\sigma}{\sqrt{n}} = \frac{0.004}{\sqrt{30}} \approx 0.001 \text{ inch.}$$

   (b) The quality control inspector will determine the machine needs an adjustment if the sample mean is either less than 0.748 inch or more than 0.752 inch.

P(needs adjustment)

$= P(\bar{x} < 0.748) + P(\bar{x} > 0.752)$

$= P\left(Z < \dfrac{0.748 - 0.75}{0.004/\sqrt{30}} \right) + P\left(Z > \dfrac{0.752 - 0.75}{0.004/\sqrt{30}} \right)$

$= P(Z < -2.74) + P(Z > 2.74)$

$= P(Z < -2.74) + \left[1 - P(Z \le 2.74) \right]$

$= 0.0031 + (1 - 0.9969)$

$= 0.0062$

There is a 0.0062 probability that the quality control inspector will conclude the machine needs an adjustment when, in fact, the machine is correctly calibrated.

7. (a) No, the variable "number of television sets" is not likely normally distributed. It is likely skewed right.

(b) $\bar{x} = \dfrac{102}{40} = 2.55$ televisions per household

(c) Because the sample is large, $n = 40 > 30$, the sampling distribution of $\bar{x}$ is approximately normal with $\mu_{\bar{x}} = \mu = 2.24$

and $\sigma_{\bar{x}} = \dfrac{\sigma}{\sqrt{n}} = \dfrac{1.38}{\sqrt{40}} \approx 0.218$.

$P(\bar{x} \ge 2.55) = P\left(Z \ge \dfrac{2.55 - 2.24}{1.38/\sqrt{40}} \right)$

$= P(Z \ge 1.42)$

$= 1 - P(Z < 1.42)$

$= 1 - 0.9222$

$= 0.0778$ [Tech: 0.0777]

If we take 100 simple random samples of size $n = 40$ households, then about 8 of the samples will result in a mean of 2.55 or more televisions. This result is not unusual, so it does not contradict the results reported by A.C. Nielsen.

8. (a) The sample size, $n = 600$, is less than 5% of the population size and
$np(1-p) = 600(0.72)(0.28) = 120.96 \ge 10$.
The sampling distribution of $\hat{p}$ is approximately normal, with mean $\mu_{\hat{p}} = p = 0.72$ and standard deviation

$\sigma_{\hat{p}} = \sqrt{\dfrac{p(1-p)}{n}} = \sqrt{\dfrac{0.72(1-0.72)}{600}} \approx 0.018$.

(b) $P(\hat{p} \le 0.70) = P\left(Z \le \dfrac{0.70 - 0.72}{\sqrt{0.72(0.28)/600}} \right)$

$= P(Z \le -1.09)$

$= 0.1379$ [Tech: 0.1376]

About 14 out of 100 random samples of size $n = 600$ 18- to 29-year-olds will result in no more than 70% who would prefer to start their own business.

(c) $\hat{p} = \dfrac{x}{n} = \dfrac{450}{600} = 0.75$

$P(X \ge 450) = P(\hat{p} \ge 0.75)$

$= P\left(Z \ge \dfrac{0.75 - 0.72}{\sqrt{0.72(0.28)/600}} \right)$

$= P(Z \ge 1.64)$

$= 1 - P(Z < 1.64)$

$= 1 - 0.9495$

$= 0.0505$ [Tech: 0.0509]

About 5 out of 100 random samples of size $n = 600$ 18- to 29-year-olds will result in 450 or more who would prefer to start their own business. Since the probability is approximately 0.05, this result is considered unusual.

9. $\hat{p} = \dfrac{60}{500} = 0.12$; the sample size, $n = 500$, is less than 5% of the population size and $np(1-p) = 500(0.1)(0.9) = 45 \ge 10$, so the sampling distribution of $\hat{p}$ is approximately normal with mean $\mu_{\hat{p}} = p = 0.1$ and standard deviation

$\sigma_{\hat{p}} = \sqrt{\dfrac{p(1-p)}{n}} = \sqrt{\dfrac{0.1(1-0.1)}{500}} \approx 0.013$.

$P(x \ge 60) = P(\hat{p} \ge 0.12)$

$= P\left(Z \ge \dfrac{0.12 - 0.1}{\sqrt{0.1(0.9)/500}} \right)$

$= P(Z \ge 1.49)$

$= 1 - P(Z < 1.49)$

$= 1 - 0.9319$

$= 0.0681$ [Tech: 0.0680]

This result is not unusual, so this evidence is insufficient to conclude that the proportion of adults 25 years of age or older with advanced degrees increased.

10. (a) $np(1-p) = 500(0.28)(0.72) = 100.8 \geq 10$.

The distribution of $\hat{p}$ is approximately normal, with mean $\mu_{\hat{p}} = p = 0.280$ and standard deviation

$$\sigma_{\hat{p}} = \sqrt{\frac{p(1-p)}{n}} = \sqrt{\frac{0.28(1-0.28)}{500}} \approx 0.020.$$

(b) $P(\hat{p} \geq 0.310) = P\left(Z \geq \dfrac{0.310 - 0.280}{\sqrt{0.280(0.720)/500}} \right)$

$= P(Z \geq 1.49)$

$= 1 - P(Z < 1.49)$

$= 1 - 0.9319$

$= 0.0681$ [Tech: 0.0676]

If we take 100 simple random samples of size $n = 500$ at-bats, then about 7 of the samples will result in a batting average of 0.310 or greater. This result is not unusual.

(c) $P(\hat{p} \leq 0.255) = P\left(Z \leq \dfrac{0.255 - 0.280}{\sqrt{0.280(0.720)/500}} \right)$

$= P(Z \leq -1.25)$

$= 0.1056$ [Tech: 0.1066]

If we take 100 simple random samples of size $n = 500$ at-bats, then about 11 of the samples will result in a batting average of 0.255 or lower. This result is not unusual.

(d) $P(0.260 \leq \hat{p} \leq 0.300)$

$= P\left(\dfrac{0.260 - 0.280}{\sqrt{\dfrac{0.280(0.720)}{500}}} \leq Z \leq \dfrac{0.300 - 0.280}{\sqrt{\dfrac{0.280(0.720)}{500}}} \right)$

$= P(-1.00 \leq Z \leq 1.00)$

$= 0.8413 - 0.1587$

$= 0.6827$ [Tech: 0.6808]

Batting averages between 0.260 and 0.300 lie within 1 standard deviation of the mean of the sampling distribution. We expect that about 68% of the time, the sample proportion will be within these bounds.

(e) It is unlikely that a career 0.280 hitter hits exactly 0.280 each season. He will probably have seasons where he bats below 0.280 and other seasons where he bats above 0.280.

$P(\hat{p} \leq 0.260) \approx 0.16$ and

$P(\hat{p} \geq 0.300) \approx 0.16$.

Batting averages as low as 0.260 and as high as 0.300 would not be unusual. Based on a single season, we cannot conclude that a player who hit 0.260 is worse than a player who hit 0.300 because neither result would be unusual for players who had identical career batting averages of 0.280.

Chapter 8 Test

1. The Central Limit Theorem states that, regardless of the shape of the population, the sampling distribution of $\bar{x}$ becomes approximately normal as the sample size n increases.

2. $\mu_{\bar{x}} = \mu = 50$; $\sigma_{\bar{x}} = \dfrac{\sigma}{\sqrt{n}} = \dfrac{24}{\sqrt{36}} = 4$

3. (a) $P(X > 100) = P\left(Z > \dfrac{100 - 90}{35} \right)$

$= P(Z > 0.29)$

$= 1 - P(Z \leq 0.29)$

$= 1 - 0.6141$

$= 0.3859$ [Tech: 0.3875]

If we select a simple random sample of $n = 100$ batteries of this type, then about 39 batteries would last more than 100 minutes. This result is not unusual.

(b) Since the population is normally distributed, the sampling distribution of $\bar{x}$ will be normal, regardless of the sample size. The mean of the distribution is $\mu_{\bar{x}} = \mu = 90$ minutes, and the standard deviation is

$$\sigma_{\bar{x}} = \frac{\sigma}{\sqrt{n}} = \frac{35}{\sqrt{10}} \approx 11.068 \text{ minutes}.$$

(c) $P(\bar{x} > 100) = P\left(Z > \dfrac{100 - 90}{35/\sqrt{10}}\right)$

$\qquad\qquad\quad = P(Z > 0.90)$

$\qquad\qquad\quad = 1 - P(Z \le 0.90)$

$\qquad\qquad\quad = 1 - 0.8159$

$\qquad\qquad\quad = 0.1841$ [Tech: 0.1831]

If we take 100 simple random samples of size $n = 10$ batteries of this type, then about 18 of the samples will result in a mean charge life of more than 100 minutes. This result is not unusual.

(d) $\mu_{\bar{x}} = \mu = 90; \ \sigma_{\bar{x}} = \dfrac{\sigma}{\sqrt{n}} = \dfrac{35}{\sqrt{25}} = 7$

$\quad P(\bar{x} > 100) = P\left(Z > \dfrac{100 - 90}{7}\right)$

$\qquad\qquad\quad = P(Z > 1.43) = 1 - P(Z \le 1.43)$

$\qquad\qquad\quad = 1 - 0.9263$

$\qquad\qquad\quad = 0.0764$ [Tech: 0.0766]

If we take 100 simple random samples of size $n = 25$ batteries of this type, then about 8 of the samples will result in a mean charge life of more than 100 minutes.

(e) The probabilities are different because a change in n causes a change in $\sigma_{\bar{x}}$.

4. (a) Since we have a large sample ($n = 45 \ge 30$), we can use the Central Limit Theorem to say that the sampling distribution of $\bar{x}$ is approximately normal. The mean of the sampling distribution is $\mu_{\bar{x}} = \mu = 2.0$ liters and the standard deviation is

$\quad \sigma_{\bar{x}} = \dfrac{\sigma}{\sqrt{n}} = \dfrac{0.05}{\sqrt{45}} \approx 0.007$ liter.

(b) The quality control manager will determine the machine needs an adjustment if the sample mean is either less than 1.98 liters or greater than 2.02 liters.

$P(\text{needs adjustment})$

$= P(\bar{x} < 1.98) + P(\bar{x} > 2.02)$

$= P\left(Z < \dfrac{1.98 - 2.0}{0.05/\sqrt{45}}\right) + P\left(Z > \dfrac{2.02 - 2.0}{0.05/\sqrt{45}}\right)$

$= P(Z < -2.68) + P(Z > 2.68)$

$= P(Z < -2.68) + \left[1 - P(Z \le 2.68)\right]$

$= 0.0037 + (1 - 0.9963)$

$= 0.0074$ [Tech: 0.0073]

There is a 0.0074 probability that the quality control manager will conclude the machine needs an adjustment even though the machine is correctly calibrated.

5. (a) Our sample size, $n = 300$, is less than 5% of the population size and
$np(1 - p) = 300(0.224)(0.776) \approx 52 \ge 10$.
The distribution of $\hat{p}$ is approximately normal, with mean $\mu_{\hat{p}} = p = 0.224$ and standard deviation

$\quad \sigma_{\hat{p}} = \sqrt{\dfrac{p(1 - p)}{n}} = \sqrt{\dfrac{0.224(1 - 0.224)}{300}} \approx 0.024$.

(b) $\hat{p} = \dfrac{x}{n} = \dfrac{50}{300} = \dfrac{1}{6} \approx 0.167$

$\quad P(X \ge 50) = P\left(\hat{p} \ge \dfrac{1}{6}\right)$

$\qquad\qquad\quad = P\left(Z \ge \dfrac{\dfrac{1}{6} - 0.224}{\sqrt{0.224(0.776)/300}}\right)$

$\qquad\qquad\quad = P(Z \ge -2.38) = 1 - P(Z < -2.38)$

$\qquad\qquad\quad = 1 - 0.0087$

$\qquad\qquad\quad = 0.9913$ [Tech: 0.9914]

About 99 out of 100 random samples of size $n = 300$ adults will result in at least 50 adults (that is, at least 16.7%) who are smokers.

(c) $P(\hat{p} \le 0.18) = P\left(Z \le \dfrac{0.18 - 0.224}{\sqrt{0.224(0.776)/300}}\right)$

$\qquad\qquad\quad = P(Z \le -1.83)$

$\qquad\qquad\quad = 0.0336$ [Tech: 0.0338]

About 3 out of 100 random samples of size $n = 300$ adults will result 18% or less who are smokers. This result would be unusual.

6. (a) For the sample proportion to be normal, the sample size must be large enough to meet the condition $np(1-p) \geq 10$. Since $p = 0.01$, we have:

$$n(0.01)(1-0.01) \geq 10$$

$$0.0099n \geq 10$$

$$n \geq \frac{10}{0.0099} \approx 1010.1$$

Thus, the sample size must be at least 1011 in order to satisfy the condition.

(b) $\hat{p} = \dfrac{x}{n} = \dfrac{9}{1500} = \dfrac{3}{500} = 0.006$

$$P(X < 10) = P(X \leq 9)$$

$$= P(\hat{p} \leq 0.006)$$

$$= P\left(Z \leq \frac{0.006 - 0.01}{\sqrt{0.01(0.99)/1500}}\right)$$

$$= P(Z \leq -1.56)$$

$$= 0.0594 \quad [\text{Tech: } 0.597]$$

About 6 out of 100 random samples of size $n = 1500$ Americans will result in fewer than 10 with peanut or tree nut allergies. This result is not considered unusual.

7. $\hat{p} = \dfrac{82}{1000} = 0.082$; the sample size, $n = 1000$, is less than 5% of the population size and $np(1-p) = 1000(0.07)(0.93) = 65.1 \geq 10$, so the sampling distribution of $\hat{p}$ is approximately normal with mean $\mu_{\hat{p}} = p = 0.07$ and standard deviation

$$\sigma_{\hat{p}} = \sqrt{\frac{p(1-p)}{n}} = \sqrt{\frac{0.07(1-0.07)}{1000}} \approx 0.008.$$

$$P(x \geq 82) = P(\hat{p} \geq 0.082)$$

$$= P\left(Z \geq \frac{0.082 - 0.07}{\sqrt{0.07(0.93)/1000}}\right)$$

$$= P(Z \geq 1.49) = 1 - P(Z < 1.49)$$

$$= 1 - 0.9319$$

$$= 0.0681 \quad [\text{Tech: } 0.0685]$$

This result is not unusual, so this evidence is insufficient to conclude that the proportion of households with a net worth in excess of $1 million has increased above 7%.

Chapter 8 Case Study

1. Open a new MINITAB worksheet, and click on the **Calc** menu. Highlight the **Random Data** option and then **Normal...**. Fill in the information as follows:

Number of rows of data to generate: 250
Sore in Column(s): c1–c80
Mean: 50
Standard deviation: 10
Select the OK button to generate the samples.

2. Select the **Calc** menu and click on the **Row Statistics...** option. Fill in the information as follows:

Click the Mean radio.
Input variables: c1–c10
Store results in: c81
Select the OK button to produce the means.

Note: the values in the left dialogue box fill in automatically as you complete the other information.

3. Select the **Calc** menu and click on the **Row Statistics...** option. Fill in the information as follows:

Click the Median radio.
Input variables: c1–c10
Store results in: c82
Select the OK button to produce the means.

Note: the values in the left dialogue box fill in automatically as you complete the other information.

5. Select the **Stat** menu. Highlight the **Basic Statistics** option and then **Display Descriptive Statistics...**. Scroll to the bottom of the left dialogue box. Double click on "c81   n10mn" to select it as a variable. It should appear in the Variables: dialogue box. Similarly, select "c82   n10med", "c83   n20mn", and so on through "c88   n80med" as variables. Select the OK button to compute the summary statistics.

4. Repeat (2) and (3) for the remaining sample sizes: *n* = 20 (c1–c20), *n* = 40 (c1–c40), and *n* = 80 (c1–c80).

Specific results will vary but should be similar to the following.

Descriptive Statistics: n10mn, n10med, n20mn, n20med, n40mn, n40med, n80mn, ...

Variable	N	N*	Mean	SE Mean	StDev	Minimum	Q1	Median	Q3
n10mn	250	0	50.089	0.189	2.989	42.337	48.151	50.155	51.873
n10med	250	0	49.911	0.234	3.698	40.146	47.315	49.839	52.413
n20mn	250	0	49.936	0.133	2.096	44.011	48.550	49.811	51.355
n20med	250	0	50.006	0.164	2.591	42.746	48.073	49.941	51.761
n40mn	250	0	49.949	0.0930	1.471	46.207	49.055	49.914	50.900
n40med	250	0	49.999	0.119	1.877	45.734	48.794	49.933	51.197
n80mn	250	0	49.947	0.0663	1.048	47.065	49.149	49.936	50.804
n80med	250	0	49.977	0.0845	1.337	45.844	49.047	49.919	50.929

Variable	Maximum
n10mn	58.094
n10med	60.061
n20mn	55.999
n20med	56.150
n40mn	54.141
n40med	54.981
n80mn	52.438
n80med	54.033

6. The averages of the sample means are all very close to the actual value of $\mu = 50$. There does not appear to be much of a difference for the various sample sizes.

The averages of the sample medians are also very close to the actual value of $M = 50$. For smaller sample sizes, the average seems to be slightly more than the true value, while for larger samples the average of the sample medians tends to be slightly less than the actual value.

7. In both cases, the standard deviations decrease as the sample size increases. However, the standard deviation for the sample means is always less than the standard deviation for the sample medians.

8. Answers will vary. It seems that the mean may be a better measure of center since the sampling distribution of the sample mean has less variability (i.e. a smaller standard deviation).

9. We will draw the histograms for the sample means first, and then draw the histograms for the sample medians. We will use a feature in MINITAB that will allow us to draw multiple graphs on the same panel, using the same scale. Select the **Graph** menu and click **Histogram…**. Select the **Simple histogram** type and click on the OK button.

Scroll to the bottom of the left dialogue box. Because we want to draw histograms of the sample means, double click on "c81 n10mn" to select it as a variable. It should appear in the **Graph variables:** dialogue box. Similarly, select "c83 n20mn", "c85 n40mn", and "c87 n80mn" as graph variables.

Click on the "Multiple Graphs…" button. Select the "In Separate panels on the same graph" radio button. Place check marks in the boxes next to "Same Y" and "Same X, including same bins." Select the OK button to close the Multiple Graphs window. Select OK again to draw the graphs.

The histograms for the sample means follow:

The distributions all appear to be roughly bell-shaped. The spread (variability) decreases as the sample size increases, but the center of the histogram remains roughly the same.

Next, we repeat the process to draw histograms for the sample medians:

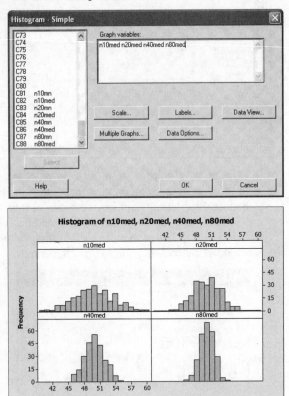

The distributions all appear to be roughly bell-shaped. The spread (variability) decreases as the sample size increases, but the center of the histogram remains roughly the same.

10. Open a new MINITAB worksheet, and click on the **Calc** menu. Highlight the **Random Data** option and then **Exponential…**. Fill in the information as follows:

    Number of rows of data to generate: 250
    Sore in Column(s): c100–c180
    Mean: 50
    Threshold: 10
    Select the OK button to generate the samples.

11. Select the **Calc** menu and click on the **Row Statistics…** option. Fill in the information as follows:

    Click the Mean radio.
    Input variables: c100–c110
    Store results in: c181
    Select the OK button to produce the means.

12. Select the <u>C</u>alc menu and click on the **R<u>o</u>w Statistics...** option. Fill in the information as follows:

    Click the Median radio.
    Input variables: c101–c110
    Store results in: c182
    Select the OK button to produce the means.

Row Statistics

C1	**Statistic**
C2	○ Sum ● Median
C3	○ Mean ○ Sum of squares
C4	○ Standard deviation ○ N total
C5	○ Minimum ○ N nonmissing
C6	○ Maximum ○ N missing
C7	○ Range
C8	
C9	
C10	**Input variables:**
C11	c101-c110

Store result in: c182

Select | Help | OK | Cancel

13. Repeat (11) and (12) for the remaining sample sizes: $n = 20$ (c101–c120), $n = 40$ (c101–c140), and $n = 80$ (c101–c180).

14. Select the **Stat** menu. Highlight the **Basic Statistics** option and then **Display Descriptive Statistics...**. Scroll to the bottom of the left dialogue box. Double click on "c181   e10mn" to select it as a variable. It should appear in the Variables: dialogue box. Similarly, select "c182   e10med", "c183   e20mn", and so on through "c188   e80med" as variables. Select the OK button to compute the summary statistics.

Display Descriptive Statistics

C170	**Variables:**
C171	e10mn e10med e20mn e20med
C172	e40mn e40med e80mn e80med
C173	
C174	
C175	
C176	
C177	
C178	**By variables (optional):**
C179	
C180	
C181 e10mn	
C182 e10med	
C183 e20mn	
C184 e20med	
C185 e40mn	
C186 e40med	
C187 e80mn	Statistics... Graphs...
C188 e80med	

Select | Help | OK | Cancel

Specific results will vary but should be similar to those shown below.

Descriptive Statistics: e10mn, e10med, e20mn, e20med, e40mn, e40med, e80mn, ...

Variable	N	N*	Mean	SE Mean	StDev	Minimum	Q1	Median	Q3
e10mn	250	0	49.454	0.978	15.462	17.347	38.799	48.674	58.552
e10med	250	0	37.66	1.00	15.81	8.81	26.47	35.64	44.79
e20mn	250	0	49.092	0.669	10.572	27.176	41.450	48.512	56.381
e20med	250	0	35.625	0.724	11.443	14.206	27.759	33.494	41.713
e40mn	250	0	49.786	0.482	7.627	32.322	44.789	48.862	54.227
e40med	250	0	35.365	0.494	7.819	19.930	29.199	34.744	40.131
e80mn	250	0	50.081	0.344	5.445	36.059	46.104	49.952	53.103
e80med	250	0	34.891	0.345	5.454	21.648	30.792	34.859	38.212

Variable	Maximum
e10mn	105.964
e10med	98.75
e20mn	82.229
e20med	84.592
e40mn	73.698
e40med	61.172
e80mn	70.860
e80med	58.424

15. The averages of the sample means are all close to the actual value of $\mu = 50$. The sample mean appears to slightly overestimate the true mean for small samples. As the sample size increases, the estimate is much closer.

$$\lambda = \frac{1}{\mu} = \frac{1}{50} = 0.02; \quad M = \frac{\ln 2}{\lambda} = \frac{\ln 2}{0.02} = 34.657$$

The averages of the sample medians are also close to the actual value of $M = 34.657$. For smaller sample sizes, the average of the sample medians was not as close as for larger sample sizes.

16. In both cases, the standard deviations decrease as the sample size increases. However, the standard deviation for the sample medians is always less than the standard deviation for the sample means.

17. Answers will vary. It seems that the median may be a better measure of center since the sampling distribution of the sample median has less variability (i.e. a smaller standard deviation).

18. We will draw the histograms for the sample means first, and then draw the histograms for the sample medians. We will again draw multiple graphs on the same panel, using the same scale. Select the **Graph** menu and click **Histogram…**. Select the **Simple histogram** type and click on the OK button.

Scroll to the bottom of the left dialogue box. Because we want to draw histograms of the sample means, double click on "c181 e10mn" to select it as a variable. It should appear in the **Graph variables:** dialogue box. Similarly, select "c183 e20mn", "c185 e40mn", and "c187 e80mn" as graph variables.

Click on the "Multiple Graphs…" button. Select the "In Separate panels on the same graph" radio button. Place check marks in the boxes next to "Same Y" and "Same X, including same bins." Select the OK button to close the Multiple Graphs window. Select OK again to draw the graphs.

The histograms for the sample means follow:

The distributions all appear to be roughly bell-shaped (though the first graph seems to be slightly skewed right). The spread (variability) decreases as the sample size increases, but the center of the histogram remains roughly the same.

Next, we repeat the process to draw histograms for the sample medians:

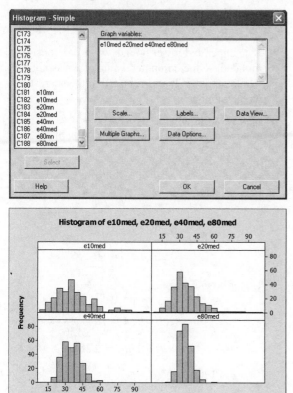

The distributions begin skewed right but become more bell-shaped as the sample size increases. The distributions appear to stay in the same general location, but the variability decreases as the sample size increases.

19. The Central Limit Theorem can be used to explain why the distributions of sample means always become roughly normal as the sample size increases. However, the Central Limit Theorem is for sample *means* only. It *cannot* be used to explain why the sampling distribution of the sample medians appears to approach a normal distribution as the sample size increases.

Chapter 9

Estimating the Value of a Parameter Using Confidence Intervals

Section 9.1

1. point estimate

3. False

5. increases

7. $z_{\alpha/2} = z_{0.05} = 1.645$

9. $z_{\alpha/2} = z_{0.01} = 2.33$

11. The point estimate is in the middle of the upper and lower bounds of the confidence interval. So, the point estimate is the mean of these two values:

$$\hat{p} = \frac{0.249 + 0.201}{2} = \frac{0.45}{2} = 0.225$$

The distance between the upper and lower bounds of the confidence interval is twice the margin of error, so the margin of error, E, is half of the difference between the upper and lower bounds of the confidence interval:

$$E = \frac{0.249 - 0.201}{2} = \frac{0.048}{2} = 0.024$$

Since $\hat{p} = \frac{x}{n}$, we can multiply both sides of the equation by n to solve for x:

$$x = n \cdot \hat{p} = 1200(0.225) = 270$$

13. The point estimate is in the middle of the upper and lower bounds of the confidence interval. So, the point estimate is the mean of these two values:

$$\hat{p} = \frac{0.509 + 0.462}{2} = \frac{0.971}{2} = 0.4855$$

The distance between the upper and lower bounds of the confidence interval is twice the margin of error, so the margin of error, E, is half of the difference between the upper and lower bounds of the confidence interval:

$$E = \frac{0.509 - 0.462}{2} = \frac{0.047}{2} = 0.0235$$

Since $\hat{p} = \frac{x}{n}$, we can multiply both sides of the equation by n to solve for x:

$$x = n \cdot \hat{p} = 1680(0.4855) = 815.64 \rightarrow 816$$

(The upper and lower bounds were previously rounded, so the computed value of x was not a whole number. It was rounded to the nearest whole number.)

15. $\hat{p} = \frac{x}{n} = \frac{30}{150} = 0.20$. For 90% confidence,

$z_{\alpha/2} = z_{0.05} = 1.645$.

Lower bound $= \hat{p} - z_{0.05} \cdot \sqrt{\dfrac{\hat{p}(1-\hat{p})}{n}}$

$$= 0.20 - 1.645 \cdot \sqrt{\frac{0.2(1-0.2)}{150}}$$

$$\approx 0.20 - 0.054 = 0.146$$

Upper bound $= \hat{p} + z_{0.05} \cdot \sqrt{\dfrac{\hat{p}(1-\hat{p})}{n}}$

$$= 0.20 + 1.645 \cdot \sqrt{\frac{0.2(1-0.2)}{150}}$$

$$\approx 0.20 + 0.054 = 0.254$$

17. $\hat{p} = \frac{x}{n} = \frac{120}{500} = 0.24$. For 99% confidence,

$z_{\alpha/2} = z_{0.005} = 2.575$.

Lower bound

$$= \hat{p} - z_{0.005} \cdot \sqrt{\frac{\hat{p}(1-\hat{p})}{n}}$$

$$= 0.24 - 2.575 \cdot \sqrt{\frac{0.24(1-0.24)}{500}}$$

$$\approx 0.24 - 0.049 = 0.191$$

Upper bound $= \hat{p} + z_{0.005} \cdot \sqrt{\dfrac{\hat{p}(1-\hat{p})}{n}}$

$$= 0.24 + 2.575 \cdot \sqrt{\frac{0.24(1-0.24)}{500}}$$

$$\approx 0.24 + 0.049 = 0.289$$

19. $\hat{p} = \frac{x}{n} = \frac{860}{1100} \approx 0.782$. For 94% confidence,

$z_{\alpha/2} = z_{.03} = 1.88$.

Lower bound $= \hat{p} - z_{0.03} \cdot \sqrt{\dfrac{\hat{p}(1-\hat{p})}{n}}$

$$= 0.782 - 1.88 \cdot \sqrt{\frac{0.782(1-0.782)}{1100}}$$

$$\approx 0.782 - 0.023 \approx 0.759$$

21. **(a)** Flawed; no interval has been provided about the population proportion.

(b) Flawed; this interpretation indicates that the level of confidence is varying.

(c) This is the correct interpretation.

(d) Flawed; this interpretation suggest that this interval sets the standard for all the other intervals, which is not true.

23. Rasmussen Reports is 95% confident that the population proportion of adult Americans who dread Valentine's Day is between 0.135 and 0.225.

25. **(a)** $\hat{p} = \dfrac{x}{n} = \dfrac{542}{3611} = 0.150$

(b) The sample size $n = 3611$ is less than 5% of the population, and $n\hat{p}(1-\hat{p}) \approx 460.6 \geq 10$.

(c) For 90% confidence, $z_{\alpha/2} = z_{.05} = 1.645$.
Lower bound

$= \hat{p} - z_{0.05} \cdot \sqrt{\dfrac{\hat{p}(1-\hat{p})}{n}}$

$= 0.150 - 1.645 \cdot \sqrt{\dfrac{0.150(1-0.150)}{3611}}$

$\approx 0.150 - 0.010$

$= 0.140$ [Tech: 0.140]

Upper bound

$= \hat{p} + z_{0.05} \cdot \sqrt{\dfrac{\hat{p}(1-\hat{p})}{n}}$

$= 0.150 + 1.645 \cdot \sqrt{\dfrac{0.150(1-0.150)}{3611}}$

$\approx 0.150 + 0.010$

$= 0.160$ [Tech: 0.160]

(d) We are 90% confident that the proportion of adult Americans age 18 years and older who have used their smartphones to make a purchase is between 0.140 and 0.160.

27. **(a)** $\hat{p} = \dfrac{x}{n} = \dfrac{521}{1003} \approx 0.519$

(b) The sample size $n = 1003$ is less than 5% of the population, and $n\hat{p}(1-\hat{p}) \approx 250.4 \geq 10$.

(c) For 95% confidence, $z_{\alpha/2} = z_{.025} = 1.960$.

Lower bound

$= \hat{p} - z_{0.025} \cdot \sqrt{\dfrac{\hat{p}(1-\hat{p})}{n}}$

$= 0.519 - 1.960 \cdot \sqrt{\dfrac{0.519(1-0.519)}{1003}}$

$\approx 0.519 - 0.031$

$= 0.489$ [Tech: 0.489]

Upper bound

$= \hat{p} + z_{0.025} \cdot \sqrt{\dfrac{\hat{p}(1-\hat{p})}{n}}$

$= 0.519 + 1.960 \cdot \sqrt{\dfrac{0.519(1-0.519)}{1003}}$

$\approx 0.519 + 0.031$

$= 0.550$ [Tech: 0.550]

(d) Yes; it is possible that the population proportion is more than 60%, because it is possible that the true proportion is not captured in the confidence interval. However, it is not likely that the true proportion is greater than 60%.

(e) We subtract the upper and lower limits of the confidence interval from 1 to get a 95% confidence interval for the population proportion of adult Americans who believe that televisions are a necessity: $(0.450, 0.511)$.

29. **(a)** $\hat{p} = \dfrac{x}{n} = \dfrac{944}{1748} = 0.540$

(b) The sample size $n = 1748$ is less than 5% of the population, and $n\hat{p}(1-\hat{p}) = 434.2 \geq 10$.

(c) For 90% confidence, $z_{\alpha/2} = z_{.05} = 1.645$.
Lower bound

$= \hat{p} - z_{0.05} \cdot \sqrt{\dfrac{\hat{p}(1-\hat{p})}{n}}$

$= 0.540 - 1.645 \cdot \sqrt{\dfrac{0.540(1-0.540)}{1748}}$

$\approx 0.540 - 0.020 = 0.520$ [Tech: 0.520]

Upper bound

$= \hat{p} + z_{0.05} \cdot \sqrt{\dfrac{\hat{p}(1-\hat{p})}{n}}$

$= 0.540 + 1.645 \cdot \sqrt{\dfrac{0.540(1-0.540)}{1748}}$

$\approx 0.540 + 0.020 = 0.560$ [Tech: 0.560]

(d) For 99% confidence, $z_{\alpha/2} = z_{.005} = 2.576$.

Lower bound

$$= \hat{p} - z_{0.005} \cdot \sqrt{\frac{\hat{p}(1-\hat{p})}{n}}$$

$$= 0.540 - 2.576 \cdot \sqrt{\frac{0.540(1-0.540)}{1748}}$$

$$\approx 0.540 - 0.031 = 0.509 \quad [\text{Tech: } 0.509]$$

Upper bound

$$= \hat{p} + z_{0.005} \cdot \sqrt{\frac{\hat{p}(1-\hat{p})}{n}}$$

$$= 0.540 + 2.576 \cdot \sqrt{\frac{0.540(1-0.540)}{1748}}$$

$$\approx 0.540 + 0.031 = 0.571 \quad [\text{Tech: } 0.571]$$

(e) Increasing the level of confidence makes the confidence interval wider.

31. Lower bound

$$= \hat{p} - z_{0.05} \cdot \sqrt{\frac{\hat{p}(1-\hat{p})}{n}}$$

$$= 0.256 - 1.645 \cdot \sqrt{\frac{0.256(1-0.256)}{199}}$$

$$\approx 0.256 - 0.051 = 0.205 \quad [\text{Tech: } 0.205]$$

Upper bound

$$= \hat{p} + z_{0.05} \cdot \sqrt{\frac{\hat{p}(1-\hat{p})}{n}}$$

$$= 0.256 + 1.645 \cdot \sqrt{\frac{0.256(1-0.256)}{199}}$$

$$\approx 0.256 + 0.051 = 0.307 \quad [\text{Tech: } 0.307]$$

We are 90% confident that the true proportion of adult Americans who would be willing to pay higher taxes if the revenue went directly toward deficit reduction is between 0.205 and 0.307.

33. For 99% confidence, $z_{\alpha/2} = z_{.005} = 2.575$.

(a) Using $\hat{p} = 0.635$,

$$n = \hat{p}(1-\hat{p})\left(\frac{z_{\alpha/2}}{E}\right)^2$$

$$= 0.635(1-0.635)\left(\frac{2.575}{0.03}\right)^2$$

$$\approx 1708.6$$

which we must increase to 1709. The researcher needs a sample size of 1709.

(b) Without using a prior estimate,

$$n = 0.25\left(\frac{z_{\alpha/2}}{E}\right)^2 = 0.25\left(\frac{2.575}{0.03}\right)^2$$

$$\approx 1843.8$$

which we must increase to 1844. Without the prior estimate, the researcher would need a sample size of 1844.

35. For 98% confidence, $z_{\alpha/2} = z_{.01} = 2.33$.

(a) Using $\hat{p} = 0.15$,

$$n = \hat{p}(1-\hat{p})\left(\frac{z_{\alpha/2}}{E}\right)^2$$

$$= 0.15(1-0.15)\left(\frac{2.33}{0.02}\right)^2$$

$$\approx 1730.5$$

which we must increase to 1731. The researcher needs a sample size of 1731.

(b) Without using a prior estimate,

$$n = 0.25\left(\frac{z_{\alpha/2}}{E}\right)^2 = 0.25\left(\frac{2.33}{0.02}\right)^2$$

$$\approx 3393.1$$

which we must increase to 3394. Without the prior estimate, the researcher would need a sample size of 3394.

37. For 95% confidence, $z_{\alpha/2} = z_{0.025} = 1.96$.

(a) Using $\hat{p} = 0.53$,

$$n = \hat{p}(1-\hat{p})\left(\frac{z_{\alpha/2}}{E}\right)^2$$

$$= 0.53(1-0.53)\left(\frac{1.96}{0.03}\right)^2$$

$$\approx 1063.3$$

which we must increase to 1064. The commentator needs a sample size of 1064.

(b) $n = 0.25\left(\frac{z_{\alpha/2}}{E}\right)^2 = 0.25\left(\frac{1.96}{0.03}\right)^2$

$$\approx 1067.1$$

which we must increase to 1068. Without the prior estimate, the commentator would need a sample size of 1068.

(c) The results are close because $0.48(1-0.48) = 0.2496$ is very close to 0.25. That is, the prior estimate is very close the conservative estimate of 0.5 that we use when no estimate is available.

39. For 95% confidence, $z_{\alpha/2} = z_{0.025} = 1.96$.
Using $E = 0.03$ and $\hat{p} = 0.64$, we get

$$n = \hat{p}(1-\hat{p})\left(\frac{z_{\alpha/2}}{E}\right)^2$$

$$= 0.64(1-0.64)\left(\frac{1.96}{0.03}\right)^2$$

$$\approx 983.4$$

which we must increase to 984. At least 984 people were included in the survey.

41. Answers may vary. One possibility follows: The confidence interval for the percentage intending to vote for George Bush was $(49-3,\ 49+3) = (46\%,\ 52\%)$. Likewise, the confidence interval for the percentage intending to vote for John Kerry was $(47-3,\ 47+3) = (44\%,\ 50\%)$. Since these intervals overlap, it is possible that the true percentage intending to vote for John Kerry could have been greater than the true percentage intending to vote for George Bush, or vice versa. Hence, the result was too close to call.

43. $\tilde{p} = \dfrac{x+2}{n+4} = \dfrac{2+2}{20+4} = \dfrac{4}{24} = 0.167$

For 95% confidence, $z_{\alpha/2} = z_{0.025} = 1.96$.
Lower bound

$$= \tilde{p} - z_{0.05} \cdot \sqrt{\frac{\tilde{p}(1-\tilde{p})}{n}}$$

$$= 0.167 - 1.96 \cdot \sqrt{\frac{0.167(1-0.167)}{24}}$$

$$\approx 0.167 - 0.149 = 0.018$$

Upper bound

$$= \tilde{p} + z_{0.05} \cdot \sqrt{\frac{\tilde{p}(1-\tilde{p})}{n}}$$

$$= 0.167 + 1.96 \cdot \sqrt{\frac{0.167(1-0.167)}{24}}$$

$$\approx 0.167 + 0.149 = 0.316$$

Using Agresti and Coull's method, we are 95% confident that the true proportion of students who eat cauliflower is between 0.018 and 0.316.

45. (a) $\hat{p} = \dfrac{x}{n} = \dfrac{921}{1001} \approx 0.920$

(b) The sample size $n = 1001$ is less than 5% of the population, and $n\hat{p}(1-\hat{p}) \approx 73.7 \geq 10$.

(c) For 95% confidence, $z_{\alpha/2} = z_{0.025} = 1.96$.
Lower bound

$$= \hat{p} - z_{0.05} \cdot \sqrt{\frac{\hat{p}(1-\hat{p})}{n}}$$

$$= 0.920 - 1.96 \cdot \sqrt{\frac{0.920(1-0.920)}{1001}}$$

$$\approx 0.920 - 0.017 = 0.903$$

Upper bound

$$= \hat{p} + z_{0.05} \cdot \sqrt{\frac{\hat{p}(1-\hat{p})}{n}}$$

$$= 0.920 + 1.96 \cdot \sqrt{\frac{0.920(1-0.920)}{1001}}$$

$$\approx 0.920 + 0.017 = 0.937$$

(d) Yes; it is possible that the population proportion is below 85%, because it is possible that the true proportion is not captured in the confidence interval. Since we are 95% confident that the true proportion is between 0.903 and 0.937, it is not likely that the proportion of adults who say they always wash their hands in public rest rooms is less than 85%.

(e) $\hat{p} = \dfrac{x}{n} = \dfrac{4679}{6076} \approx 0.770$

(f) The sample size $n = 6076$ is less than 5% of the population, and $n\hat{p}(1-\hat{p}) \approx 1076 \geq 10$.

(g) For 95% confidence, $z_{\alpha/2} = z_{0.025} = 1.96$.
Lower bound

$$= \hat{p} - z_{0.05} \cdot \sqrt{\frac{\hat{p}(1-\hat{p})}{n}}$$

$$= 0.770 - 1.96 \cdot \sqrt{\frac{0.770(1-0.770)}{6076}}$$

$$\approx 0.770 - 0.011 = 0.759$$

Upper bound

$$= \hat{p} + z_{0.05} \cdot \sqrt{\frac{\hat{p}(1-\hat{p})}{n}}$$

$$= 0.770 + 1.96 \cdot \sqrt{\frac{0.770(1-0.770)}{6076}}$$

$$\approx 0.770 + 0.011 = 0.781$$

(h) Yes; it is possible that the population proportion is greater than 85%, because it is possible that the true proportion is not captured in the confidence interval. Since we are 95% confident that the true proportion is between 0.759 and 0.781, it is not likely that the proportion of adults who wash their hands in public rest rooms is greater than 85%.

(i) The proportion who say they wash their hands in public restrooms is higher than the proportion who actually do. Explanations may vary. One possibility is that people lie about their hand washing habits out of embarrassment.

47. If we let E represent the original margin of error, $E = z_{0.05} \cdot \sqrt{\dfrac{\tilde{p}(1-\tilde{p})}{n}}$ and if we multiply the sample size n by 4, we get:

$$\text{New margin of error} = z_{0.05} \cdot \sqrt{\frac{\tilde{p}(1-\tilde{p})}{4n}}$$

$$= z_{0.05} \cdot \sqrt{\frac{\tilde{p}(1-\tilde{p})}{n}} \cdot \frac{1}{\sqrt{4}}$$

$$= \frac{1}{2} z_{0.05} \cdot \sqrt{\frac{\tilde{p}(1-\tilde{p})}{n}}$$

$$= \frac{1}{2} E$$

So, when the sample size, n, is multiplied by 4, the margin of error is cut in half.

49. The margin of error for Karina's estimate is:

$$z_{0.025} \cdot \sqrt{\frac{\hat{p}(1-\hat{p})}{n}} = 1.96 \cdot \sqrt{\frac{\hat{p}(1-\hat{p})}{100}} \text{ and the}$$

margin of error for Matthew's estimate is:

$$z_{0.005} \cdot \sqrt{\frac{\hat{p}(1-\hat{p})}{n}} = 2.576 \cdot \sqrt{\frac{\hat{p}(1-\hat{p})}{400}}$$

Since the point estimate, $\hat{p}$, is the same for both, we have:

    Matthew: Karina:

$$2.576 \cdot \sqrt{\frac{\hat{p}(1-\hat{p})}{400}} < 1.96 \cdot \sqrt{\frac{\hat{p}(1-\hat{p})}{100}} \text{ since}$$

$$0.1288 = \frac{2.576}{\sqrt{400}} < \frac{1.96}{\sqrt{100}} = 0.196.$$

Matthew's estimate will have the smaller margin of error.

51. The sample is not random. The gender of members of the U.S. Congress is not representative of the human race as a whole.

Section 9.2

1. decreases

3. α

5. False; the population does not need to be normally distributed if the sample size is sufficiently large.

7. (a) From the row with df = 25 and the column headed 0.10, we read $t_{0.10} = 1.316$.

(b) From the row with df = 30 and the column headed 0.05, we read $t_{0.05} = 1.697$.

(c) From the row with df = 18 and the column headed 0.01, we read $t = 2.552$. This is the value with area 0.01 to the *right*. By symmetry, the t-value with an area to the *left* of 0.01 is $t_{0.99} = -2.552$.

(d) For a 90% confidence interval, we want the t-value with an area in the right tail of 0.05. With df = 20, we read from Table VI that $t_{0.05} = 1.725$.

9. No; the data are not normal and contain an outlier.

11. No; the data are not normal.

13. Yes; the data are normal and there are no outliers.

15. The point estimate is in the middle of the upper and lower bounds of the confidence interval. So, the point estimate is the mean of these two values:

$$\bar{x} = \frac{18 + 24}{2} = 21$$

The distance between the upper and lower bounds of the confidence interval is twice the margin of error, so the margin of error, E, is half of the difference between the upper and lower bounds of the confidence interval:

$$E = \frac{24 - 18}{2} = 3$$

17. The point estimate is in the middle of the upper and lower bounds of the confidence interval. So, the point estimate is the mean of these two values:

$$\bar{x} = \frac{5 + 23}{2} = 14$$

The distance between the upper and lower bounds of the confidence interval is twice the margin of error, so the margin of error, E, is half of the difference between the upper and lower bounds of the confidence interval:

$$E = \frac{23 - 5}{2} = 9$$

19. (a) For 96% confidence, $\alpha/2 = 0.02$. Since $n = 25$, then df = 24. The critical value is $t_{0.02} = 2.172$. Then:

Lower bound $= \bar{x} - t_{0.02} \cdot \dfrac{s}{\sqrt{n}}$

$= 108 - 2.172 \cdot \dfrac{10}{\sqrt{25}}$

$\approx 108 - 4.3 = 103.7$

Upper bound $= \bar{x} + t_{0.02} \cdot \dfrac{s}{\sqrt{n}}$

$= 108 + 2.172 \cdot \dfrac{10}{\sqrt{25}}$

$\approx 108 + 4.3 = 112.3$

(b) Since $n = 10$, then df = 9. The critical value is $t_{0.02} = 2.398$. Then:

Lower bound $= \bar{x} - t_{0.02} \cdot \dfrac{s}{\sqrt{n}}$

$= 108 - 2.398 \cdot \dfrac{10}{\sqrt{10}}$

$\approx 108 - 7.6 = 100.4$

Upper bound $= \bar{x} + t_{0.02} \cdot \dfrac{s}{\sqrt{n}}$

$= 108 + 2.398 \cdot \dfrac{10}{\sqrt{10}}$

$\approx 108 + 7.6 = 115.6$

Decreasing the sample size increases the margin of error.

(c) For 90% confidence, $\alpha/2 = 0.05$. With 24 degrees of freedom, $t_{0.05} = 1.711$. Then:

Lower bound $= \bar{x} - t_{0.05} \cdot \dfrac{s}{\sqrt{n}}$

$= 108 - 1.711 \cdot \dfrac{10}{\sqrt{25}}$

$\approx 108 - 3.4 = 104.6$

Upper bound $= \bar{x} + t_{0.05} \cdot \dfrac{s}{\sqrt{n}}$

$= 108 + 1.711 \cdot \dfrac{10}{\sqrt{25}}$

$\approx 108 + 3.4 = 111.4$

Decreasing the level of confidence decreases the margin of error.

(d) No, because in all cases the sample was small ($n < 30$), so the population must be normally distributed.

21. (a) For 95% confidence, $\alpha/2 = 0.025$. Since $n = 35$, then df = 34. The critical value is $t_{0.025} = 2.032$. Then:

Lower bound $= \bar{x} - t_{0.025} \cdot \dfrac{s}{\sqrt{n}}$

$= 18.4 - 2.032 \cdot \dfrac{4.5}{\sqrt{35}}$

$\approx 18.4 - 1.55 = 16.85$

Upper bound $= \bar{x} + t_{0.025} \cdot \dfrac{s}{\sqrt{n}}$

$= 18.4 + 2.032 \cdot \dfrac{4.5}{\sqrt{35}}$

$\approx 18.4 + 1.55 = 19.95$

(b) Since $n = 50$, then df = 49, but since there is no row in the tables for 49 degrees of freedom we use the closest value, df = 50, instead. The critical value is $t_{0.025} = 2.009$. Then:

Lower bound $= \bar{x} - t_{0.025} \cdot \dfrac{s}{\sqrt{n}}$

$= 18.4 - 2.009 \cdot \dfrac{4.5}{\sqrt{50}}$

$\approx 18.4 - 1.28 = 17.12$

Upper bound $= \bar{x} + t_{0.025} \cdot \dfrac{s}{\sqrt{n}}$

$= 18.4 + 2.009 \cdot \dfrac{4.5}{\sqrt{50}}$

$\approx 18.4 + 1.28 = 19.68$

Increasing the sample size decreases the margin of error.

(c) For 99% confidence, $\alpha/2 = 0.005$. With 34 degrees of freedom, $t_{0.005} = 2.728$. Then:

Lower bound $= \bar{x} - t_{0.005} \cdot \dfrac{s}{\sqrt{n}}$

$= 18.4 - 2.728 \cdot \dfrac{4.5}{\sqrt{35}}$

$\approx 18.4 - 2.08 = 16.32$

Upper bound $= \bar{x} + t_{0.005} \cdot \dfrac{s}{\sqrt{n}}$

$= 18.4 + 2.728 \cdot \dfrac{4.5}{\sqrt{35}}$

$\approx 18.4 + 2.08 = 20.48$

$\left[\text{Tech: } (16.33, 20.48) \right]$

Increasing the level of confidence increases the margin of error.

(d) For a small sample ($n = 15 < 30$), the population must be normally distributed.

23. (a) Flawed; this interpretation indicates that the population mean varies rather than the interval.

(b) This is the correct interpretation.

(c) Flawed; this interpretation makes an implication about individuals rather than the mean.

(d) Flawed; the interpretation should be about the mean number of hours worked by adult Americans, not about adults in Idaho.

25. We are 90% confident that the mean drive-through service time of Taco Bell restaurants is between 161.5 and 164.7 seconds.

27. (1) Increase the sample size, and (2) decrease the level of confidence to narrow the confidence interval.

29. (a) Since the distribution of blood alcohol concentrations is not normally distributed (highly skewed right), the sample must be large so that the distribution of the sample mean, $\bar{x}$, will be approximately normal.

(b) The sample is less than 5% of the population.

(c) For 90% confidence, $\alpha/2 = 0.05$. Since $n = 51$, then df = 50 and $t_{0.05} = 1.676$. Then:

Lower bound $= \bar{x} - t_{0.05} \cdot \dfrac{s}{\sqrt{n}}$

$= 0.167 - 1.676 \cdot \dfrac{0.010}{\sqrt{51}}$

$\approx 0.167 - 0.0023 = 0.1647$

Upper bound $= \bar{x} + t_{0.05} \cdot \dfrac{s}{\sqrt{n}}$

$= 0.167 + 1.676 \cdot \dfrac{0.010}{\sqrt{51}}$

$\approx 0.167 + 0.0023 = 0.1693$

We are 90% confident that the mean BAC in fatal crashes where the driver had a positive BAC is between 0.1647 and 0.1693 g/dL.

(d) Yes; it is possible that the mean BAC is less than 0.08 g/dL, because it is possible

that the true mean is not captured in the confidence interval, but it is not likely.

31. For 99% confidence, $\alpha/2 = 0.005$. Since $n = 1006$, then df = 1005. There is no row in the table for 1005 degrees of freedom, so we use the closest value, df = 1000, instead. The critical value is $t_{0.005} = 2.581$. Then:

Lower bound $= \bar{x} - t_{0.005} \cdot \dfrac{s}{\sqrt{n}}$

$= 13.4 - 2.581 \cdot \dfrac{16.6}{\sqrt{1006}}$

$\approx 13.4 - 1.35 = 12.05$ books

Upper bound $= \bar{x} - t_{0.005} \cdot \dfrac{s}{\sqrt{n}}$

$= 13.4 + 2.581 \cdot \dfrac{16.6}{\sqrt{1006}}$

$\approx 13.4 + 1.35 = 14.75$ books

We are 99% confident that the population mean number of books read by Americans during 2005 was between 12.05 and 14.75 books.

33. For 95% confidence, $\alpha/2 = 0.025$. Since $n = 81$, then df = 80 and $t_{0.025} = 1.990$. Then:

Lower bound $= \bar{x} - t_{0.025} \cdot \dfrac{s}{\sqrt{n}}$

$= 4.6 - 1.990 \cdot \dfrac{15.9}{\sqrt{81}}$

$\approx 4.6 - 3.52 = 1.08$ days

Upper bound $= \bar{x} + t_{0.025} \cdot \dfrac{s}{\sqrt{n}}$

$= 4.6 + 1.990 \cdot \dfrac{15.9}{\sqrt{81}}$

$\approx 4.6 + 3.52 = 8.12$ days

We are 95% confident that the population mean incubation period of patients with SARS is between 1.08 and 8.12 days.

35. (a) A point estimate for the population mean price of the memory card is:
$\bar{x} = \$15.382$.

(b) We can reasonably assume that the sample size is small relative to the population size. All the data lie within the bounds of the normal probability plot. The boxplot does not reveal any outliers.

(c) Using technology, we find $s \approx \$2.343$. For 95% confidence, $\alpha/2 = 0.025$.

Since $n = 10$, then df = 9 and $t_{0.025} = 2.262$. Thus:

Lower bound $= \bar{x} - t_{0.025} \cdot \dfrac{s}{\sqrt{n}}$

$$= 15.382 - 2.262 \cdot \dfrac{2.343}{\sqrt{10}}$$

$$\approx 15.382 - 1.676 = 13.706$$

Upper bound $= \bar{x} + t_{0.025} \cdot \dfrac{s}{\sqrt{n}}$

$$= 15.382 + 2.262 \cdot \dfrac{2.343}{\sqrt{10}}$$

$$\approx 15.382 + 1.676 = 17.058$$

We are 95% confident that the population mean price of 4 GB flash memory cards available from online retailers is between $13.706 and $17.058.

(d) For 99% confidence, $\alpha / 2 = 0.005$. So, $t_{0.005} = 3.250$. Thus:

Lower bound $= \bar{x} - t_{0.005} \cdot \dfrac{s}{\sqrt{n}}$

$$= 15.382 - 3.250 \cdot \dfrac{2.343}{\sqrt{10}}$$

$$\approx 15.382 - 2.408 = 12.974$$

Upper bound $= \bar{x} + t_{0.005} \cdot \dfrac{s}{\sqrt{n}}$

$$= 15.382 + 3.250 \cdot \dfrac{2.343}{\sqrt{10}}$$

$$\approx 15.382 + 2.408 = 17.790$$

We are 99% confident that the population mean price of 4 GB flash memory cards available from online retailers is between $12.974 and $17.790.

(e) As the level of confidence increases, the width of the interval increases. A wider interval covers potential values of the parameter that are not included in a smaller interval. In the long run, confidence intervals with a higher level of confidence will contain the population parameter more often than intervals based on a smaller level of confidence.

37. Using technology, we find $\bar{x} = \$2139.9$ and $s \approx \$1114.4$ For 90% confidence, $\alpha / 2 = 0.05$. Since $n = 14$, then df = 13 and $t_{0.05} = 1.771$. Thus:

Lower bound $= \bar{x} - t_{0.05} \cdot \dfrac{s}{\sqrt{n}}$

$$= 2139.9 - 1.771 \cdot \dfrac{1114.4}{\sqrt{14}}$$

$$\approx 2139.9 - 527.5 = 1612.5$$

Upper bound $= \bar{x} + t_{0.05} \cdot \dfrac{s}{\sqrt{n}}$

$$= 2139.9 + 1.771 \cdot \dfrac{1114.4}{\sqrt{14}}$$

$$\approx 2139.9 + 527.5 = 2667.4$$

We are 90% confident that the population mean repair cost is between $1612.5 and 2667.4.

All the data lie within the bounds of the normal probability plot. The boxplot does not reveal any outliers.

39. (a) $\bar{x} = \dfrac{\sum x}{n} = \dfrac{93.48}{40} = 2.337$ million shares

(b) Using technology, we find $s \approx 1.217$ million shares. For 90% confidence, $\alpha / 2 = 0.05$. Since $n = 40$, then df = 39 and $t_{0.05} = 1.685$. Thus:

Lower bound $= \bar{x} - t_{0.05} \cdot \dfrac{s}{\sqrt{n}}$

$$= 2.337 - 1.685 \cdot \dfrac{1.217}{\sqrt{40}}$$

$$\approx 2.337 - 0.324 = 2.013$$

Upper bound $= \bar{x} + t_{0.05} \cdot \dfrac{s}{\sqrt{n}}$

$$= 2.337 + 1.685 \cdot \dfrac{1.217}{\sqrt{40}}$$

$$\approx 2.337 + 0.324 = 2.661$$

We are 90% confident that the population mean number of Harley-Davidson shares traded per day in 2007 was between 2.013 and 2.661 million shares.

(c) Using technology, we find $\bar{x} \approx 2.195$ million shares and $s \approx 0.815$ million shares. For 90% confidence, $\alpha / 2 = 0.05$. Since $n = 40$, then df = 39 and $t_{0.05} = 1.685$. Thus:

Lower bound $= \bar{x} - t_{0.05} \cdot \dfrac{s}{\sqrt{n}}$

$= 2.195 - 1.685 \cdot \dfrac{0.815}{\sqrt{40}}$

$\approx 2.195 - 0.217 = 1.978$

Upper bound $= \bar{x} + t_{0.05} \cdot \dfrac{s}{\sqrt{n}}$

$= 2.195 + 1.685 \cdot \dfrac{0.815}{\sqrt{40}}$

$\approx 2.195 + 0.217 = 2.412$

We are 90% confident that the population mean number of Harley-Davidson shares traded per day in 2007 was between 1.978 and 2.412 million shares.

(d) The confidence intervals are different because of variation in sampling. The samples have different means and standard deviations that lead to different confidence intervals.

41. (a) The histogram is skewed right.

Cell Phone Bills

(b) The boxplot shows three outliers.

Cell Phone Bills

(c) Since the distribution of cell phone bills is not normally distributed (it is skewed right with outliers), the sample size must be large so that the distribution of the sample mean, $\bar{x}$, will be approximately normal.

(d) $\bar{x} = \$104.972$

(e) ($\$95.896$, $\$114.048$)

43. (a) From the output:
$\bar{x} = 22.150$;

$E = \dfrac{22.209 - 22.091}{2} = 0.059$

(b) We are 95% confident that the population mean age when first married is between 22.091 and 22.209 years.

(c) For 95% confidence, $\alpha / 2 = 0.025$. Since $n = 26,540$, then df $= 26,539$. There is no row in the table for 26,539 degrees of freedom, so we use df $= 1000$ instead. and $t_{0.025} = 1.962$.

Lower bound $= \bar{x} - t_{0.025} \cdot \dfrac{s}{\sqrt{n}}$

$= 22.150 - 1.962 \cdot \dfrac{4.885}{\sqrt{26,540}}$

$\approx 22.150 - 0.059 = 22.091$

Upper bound $= \bar{x} + t_{0.025} \cdot \dfrac{s}{\sqrt{n}}$

$= 22.150 + 1.962 \cdot \dfrac{4.885}{\sqrt{26,540}}$

$\approx 22.150 + 0.059 = 22.209$

45. For 99% confidence, $z_{\alpha/2} = 2.576$ and

$n = \left(\dfrac{z_{\alpha/2} \cdot s}{E} \right)^2 = \left(\dfrac{2.576 \cdot 13.4}{2} \right)^2 \approx 297.9$

which we must increase to 298. A total of 298 patients are needed.
For 95% confidence, $z_{\alpha/2} = 1.96$ and

$n = \left(\dfrac{z_{\alpha/2} \cdot s}{E} \right)^2 = \left(\dfrac{1.96 \cdot 13.4}{2} \right)^2 \approx 172.4$

which we must increase to 173. A total of 173 patients are needed.
Decreasing the confidence level decreases the sample size needed.

47. (a) Margin of error = 4:
For 95% confidence, $z_{\alpha/2} = 1.96$ and

$n = \left(\dfrac{z_{\alpha/2} \cdot s}{E} \right)^2 = \left(\dfrac{1.96 \cdot 16.6}{4} \right)^2 \approx 66.2$

which we must increase to 67. So,.67 subjects are needed.

(b) Margin of error = 2:
For 95% confidence, $z_{\alpha/2} = 1.96$ and

$n = \left(\dfrac{z_{\alpha/2} \cdot s}{E} \right)^2 = \left(\dfrac{1.96 \cdot 16.6}{2} \right)^2 \approx 264.6$

which we must increase to 265. A total of 265 subjects are needed.

(c) Note that 265 is about 4 times as large as 67. If we double the required accuracy (i.e., cut the margin of error in half), then the sample size must be quadrupled.

(d) Margin of error = 4:
For 99% confidence, $z_{\alpha/2} = 2.576$ and

$$n = \left(\frac{z_{\alpha/2} \cdot s}{E}\right)^2 = \left(\frac{2.576 \cdot 16.6}{4}\right)^2 \approx 114.3$$

which we must increase to 115. A total of 115 subjects are needed. This is larger than the value in part (a). Increasing the level of confidence increases the sample size. For a fixed margin of error, greater confidence can be achieved with a larger sample size.

49. (a) Data Set I: $\bar{x} = \frac{\sum x}{n} = \frac{793}{8} \approx 99.1$;

Data Set II: $\bar{x} = \frac{\sum x}{n} = \frac{1982}{20} = 99.1$;

Data Set III: $\bar{x} = \frac{\sum x}{n} = \frac{2971}{30} \approx 99.0$

(b) For 95% confidence the critical value is $z_{0.025} = 1.96$.
Set I:

Lower bound $= \bar{x} - z_{0.025} \cdot \frac{\sigma}{\sqrt{n}}$

$= 99.125 - 1.96 \cdot \frac{15}{\sqrt{8}}$

$\approx 99.125 - 10.394$

≈ 88.7

Upper bound $= \bar{x} - z_{0.025} \cdot \frac{\sigma}{\sqrt{n}}$

$= 99.125 + 1.96 \cdot \frac{15}{\sqrt{8}}$

$\approx 99.125 + 10.394$

≈ 109.5

Set II:

Lower bound $= \bar{x} - z_{0.025} \cdot \frac{\sigma}{\sqrt{n}}$

$= 99.1 - 1.96 \cdot \frac{15}{\sqrt{20}}$

$\approx 99.1 - 6.574$

≈ 92.5

Upper bound $= \bar{x} - z_{0.025} \cdot \frac{\sigma}{\sqrt{n}}$

$= 99.1 + 1.96 \cdot \frac{15}{\sqrt{20}}$

$\approx 99.1 + 6.574$

≈ 105.7

Set III:

Lower bound $= \bar{x} - z_{0.025} \cdot \frac{\sigma}{\sqrt{n}}$

$\approx 99.033 - 1.96 \cdot \frac{15}{\sqrt{30}}$

$\approx 99.033 - 5.368$

≈ 93.7

Upper bound $= \bar{x} - z_{0.025} \cdot \frac{\sigma}{\sqrt{n}}$

$\approx 99.033 + 1.96 \cdot \frac{15}{\sqrt{30}}$

$\approx 99.033 + 5.368$

≈ 104.4

(c) As the size of the sample increases, the width of the confidence interval decreases.

(d) Set I: $\bar{x} = \frac{\sum x}{n} = \frac{703}{8} = 87.875$;

Lower bound $= \bar{x} - z_{0.025} \cdot \frac{\sigma}{\sqrt{n}}$

$= 87.875 - 1.96 \cdot \frac{15}{\sqrt{8}}$

$\approx 87.875 - 10.394$

≈ 77.5

Upper bound $= \bar{x} - z_{0.025} \cdot \frac{\sigma}{\sqrt{n}}$

$= 87.875 + 1.96 \cdot \frac{15}{\sqrt{8}}$

$\approx 87.875 + 10.394$

≈ 98.3

Set II: $\bar{x} = \frac{\sum x}{n} = \frac{1892}{20} = 94.6$;

Lower bound $= \bar{x} - z_{0.025} \cdot \dfrac{\sigma}{\sqrt{n}}$

$= 94.6 - 1.96 \cdot \dfrac{15}{\sqrt{8}}$

$\approx 94.6 - 6.574$

≈ 88.0

Upper bound $= \bar{x} - z_{0.025} \cdot \dfrac{\sigma}{\sqrt{n}}$

$= 94.6 + 1.96 \cdot \dfrac{15}{\sqrt{8}}$

$\approx 94.6 + 6.574$

≈ 101.2

Set III: $\bar{x} = \dfrac{\sum x}{n} = \dfrac{2881}{30} \approx 96.033$

Lower bound $= \bar{x} - z_{0.025} \cdot \dfrac{\sigma}{\sqrt{n}}$

$= 96.033 - 1.96 \cdot \dfrac{15}{\sqrt{30}}$

$\approx 96.0 - 5.368$

≈ 90.7

Upper bound $= \bar{x} - z_{0.025} \cdot \dfrac{\sigma}{\sqrt{n}}$

$= 96.033 + 1.96 \cdot \dfrac{15}{\sqrt{30}}$

$\approx 96.0 + 5.368$

≈ 101.4

(e) The confidence intervals for both Data Set II and Data Set III still capture the population mean, 100, with the incorrect entry. The interval from Data Set I does not capture the population mean when the incorrect entry is made. The concept of robustness is illustrated. As the sample size increases the distribution of $\bar{x}$ becomes more normal, making the confidence interval more robust against the effect of the incorrect entry.

51. **(a)-(c)** Answers will vary

(d) We expect that 95% of the confidence intervals will contain the population mean, μ.

53. **(a)** Because all subjects were randomly assigned to the treatments, this is a completely randomized design.

(b) The treatment is the smoking cessation program. There are two levels: internet and phone-based intervention, and self-help booklet.

(c) The response variable was abstinence after 12 months (whether or not the subject quit smoking).

(d) The statistics reported are 22.3% of participants in the experimental group reported abstinence and 13.1% of participants in the control group reported abstinence.

(e) $\dfrac{p(1-q)}{q(1-p)} = \dfrac{0.223(1-0.131)}{0.131(1-0.223)} \approx 1.90$; this indicates that abstinence is more likely to occur using the experimental cessation program.

(f) The authors are 95% confident that the population odds ratio is between 1.12 and 3.26.

(g) Answers will vary. One possibility: Smoking cessation is more likely when the Happy Ending Intervention program is used rather than the control method.

55. The t-distribution has less spread as the degrees of freedom increases because as n increases, s becomes closer to σ by the Law of Large Numbers. As the sample size increases, the t-distribution approaches the standard normal distribution.

57. The degrees of freedom are the number of data values that are free to vary. For example, if we know that $n = 5$ data values have a mean of $\bar{x} = 7.2$, then we are free to choose the value of the first four data values. However, since the sample mean is known, we are not free to choose a value for the last observation.

59. We expect that the margin of error for Population A will be smaller, since there should be less variation in the ages of college students than in the ages of the residents of a town.

Consumer Reports®: Consumer Reports Tests Tires

(a) All the data values lie within the bounds on the normal probability plot, indicating that the data is roughly normally distributed.

(b) The five-number summary for the data is 123.2, 128.8, 131.8, 136.9, 140.3.

$$\text{The lower fence} = Q_1 - 1.5 \cdot \text{IQR}$$
$$= 128.8 - 1.5(136.9 - 128.8)$$
$$= 116.65 \text{ feet}$$
$$\text{The upper fence} = Q_3 + 1.5 \cdot \text{IQR}$$
$$= 136.9 + 1.5(136.9 - 128.8)$$
$$= 149.05 \text{ feet}$$

No data values are below the lower fence or above the upper fence, so the data set does not contain any outliers.

(c) Using technology, $\bar{x} \approx 132.12$ feet and $s \approx 5.47$ feet.
For 95% confidence, $\alpha/2 = 0.025$. Since $n = 9$, then df = 8 and $t_{0.025} = 2.306$.

$$\text{Lower bound} = \bar{x} - t_{0.025} \cdot \frac{s}{\sqrt{n}}$$
$$= 132.12 - 2.306 \cdot \frac{5.47}{\sqrt{9}}$$
$$\approx 132.12 - 4.20 = 127.92 \text{ feet}$$
$$\text{Upper bound} = \bar{x} + t_{0.025} \cdot \frac{s}{\sqrt{n}}$$
$$= 132.12 + 2.306 \cdot \frac{5.47}{\sqrt{9}}$$
$$\approx 132.12 + 4.20 = 136.32 \text{ feet}$$

Based on this test, *Consumer Reports* is 95% confident that the mean dry braking distance for this brand of tires is between 127.92 and 136.32 feet.

Section 9.3

1. False; the chi-square distribution is skewed to the right.

3. False; to construct a confidence interval about a population variance or a standard deviation, the sample must come from a normally distributed population.

5. df $= n - 1 = 19$ and, for 90% confidence, $\alpha/2 = 0.05$. From the table, $\chi^2_{0.05} = 30.144$ and $\chi^2_{1-0.05} = \chi^2_{0.95} = 10.117$.

7. df $= n - 1 = 22$ and, for 98% confidence, $\alpha/2 = 0.01$. From the table, $\chi^2_{0.01} = 40.289$ and $\chi^2_{1-0.01} = \chi^2_{0.99} = 9.542$.

9. (a) df $= n - 1 = 19$ and, for 90% confidence, $\alpha/2 = 0.05$.
From the table, $\chi^2_{0.05} = 30.144$ and $\chi^2_{1-0.05} = \chi^2_{0.95} = 10.117$.
$$\text{Lower bound} = \frac{(n-1)s^2}{\chi^2_{\alpha/2}} = \frac{19 \cdot 12.6}{30.144} \approx 7.94$$
$$\text{Upper bound} = \frac{(n-1)s^2}{\chi^2_{1-\alpha/2}} = \frac{19 \cdot 12.6}{10.117} \approx 23.66$$

(b) df $= n - 1 = 29$ and, for 90% confidence, $\alpha/2 = 0.05$.
From the table, $\chi^2_{0.05} = 42.557$ and $\chi^2_{1-0.05} = \chi^2_{0.95} = 17.708$.
$$\text{Lower bound} = \frac{(n-1)s^2}{\chi^2_{\alpha/2}} = \frac{29 \cdot 12.6}{42.557} \approx 8.59$$
$$\text{Upper bound} = \frac{(n-1)s^2}{\chi^2_{1-\alpha/2}} = \frac{29 \cdot 12.6}{17.708} \approx 20.63$$
Increasing the sample size decreases the width of the confidence interval.

(c) df $= n - 1 = 19$ and, for 98% confidence, $\alpha/2 = 0.01$.
From the table, $\chi^2_{0.05} = 36.191$ and $\chi^2_{1-0.01} = \chi^2_{0.99} = 7.633$.
$$\text{Lower bound} = \frac{(n-1)s^2}{\chi^2_{\alpha/2}} = \frac{19 \cdot 12.6}{36.191} \approx 6.61$$
$$\text{Upper bound} = \frac{(n-1)s^2}{\chi^2_{1-\alpha/2}} = \frac{19 \cdot 12.6}{7.633} \approx 31.36$$
Increasing the level of confidence increases the width of the confidence interval.

11. df $= n - 1 = 10 - 1 = 9$ and, for 95% confidence, $\alpha/2 = 0.025$.
From the table, $\chi^2_{0.025} = 19.023$ and $\chi^2_{1-0.025} = \chi^2_{0.975} = 2.700$.

Lower bound $= \sqrt{\dfrac{(n-1)s^2}{\chi^2_{\alpha/2}}}$

$= \sqrt{\dfrac{9 \cdot (2.343)^2}{19.023}} \approx 1.612$

[Tech: 1.611]

Upper bound $= \sqrt{\dfrac{(n-1)s^2}{\chi^2_{1-\alpha/2}}}$

$= \sqrt{\dfrac{9 \cdot (2.343)^2}{2.700}} \approx 4.278$

[Tech: 4.277]

We are 95% confident that the population standard deviation of the price of a 4 GB flash memory card at online retailers is between $1.611 and $4.277.

13. $df = n-1 = 14-1 = 13$ and for 90% confidence, $\alpha/2 = 0.05$.

From the table, $\chi^2_{0.05} = 22.362$ and

$\chi^2_{1-0.025} = \chi^2_{0.975} = 5.892$.

Lower bound $= \sqrt{\dfrac{(n-1)s^2}{\chi^2_{\alpha/2}}}$

$= \sqrt{\dfrac{13 \cdot (1114.412)^2}{22.362}} \approx 849.7$

Upper bound $= \sqrt{\dfrac{(n-1)s^2}{\chi^2_{1-\alpha/2}}}$

$= \sqrt{\dfrac{13 \cdot (1114.412)^2}{5.892}} \approx 1655.3$

We are 90% confident that the population standard deviation of the repair costs of a low-impact bumper crash on a mini- or micro-car is between $849.7 and $1655.3.

15. (a) From the probability plot shown below, the data appear to be from a population that is normally distributed.

(b) Using technology, we obtain $s \approx 0.349$ oz.

(c) $df = n-1 = 11$ and, for 90% confidence, $\alpha/2 = 0.05$. From the table,

$\chi^2_{1-0.05} = \chi^2_{0.95} = 4.575$ and

$\chi^2_{0.05} = 19.675$.

Lower bound $= \sqrt{\dfrac{(n-1)s^2}{\chi^2_{\alpha/2}}}$

$= \sqrt{\dfrac{11(0.349)^2}{19.675}} \approx 0.261$ oz

Upper bound $= \sqrt{\dfrac{(n-1)s^2}{\chi^2_{1-\alpha/2}}}$

$= \sqrt{\dfrac{11(0.349)^2}{4.575}} \approx 0.541$ oz

The quality-control manager can be 90% confident that the population standard deviation of the number of ounces of peanuts is between 0.261 and 0.541 ounce.

(d) No; we are 90% confident that the population standard deviation is between 0.261 and 0.541 ounces, and so above 0.20 ounces.

17. From the standard normal table, we get $z_{0.975} = -1.96$ and $z_{0.025} = 1.96$. With $v = 100$, we get

$\chi^2_{0.975} \approx \dfrac{\left(z_{0.975} + \sqrt{2v-1}\right)^2}{2}$

$= \dfrac{\left(-1.96 + \sqrt{2 \cdot 100 - 1}\right)^2}{2} = 73.772$

(compared to the table's value of 74.222)

and $\chi^2_{0.025} \approx \dfrac{\left(z_{0.025} + \sqrt{2v-1}\right)^2}{2}$

$= \dfrac{\left(1.96 + \sqrt{2 \cdot 100 - 1}\right)^2}{2} = 129.070$

(compared to the table's value of 129.561).

Section 9.4

1. $\hat{p} = \dfrac{35}{300} \approx 0.117$. The sample size $n = 300$ is less than 5% of the population, and $n\hat{p}(1-\hat{p}) \approx 31.0 \ge 10$, so we can construct a Z-interval. For 99% confidence the critical value is $z_{0.005} = 2.575$. Then:

Lower bound

$$= \hat{p} - z_{0.005} \cdot \sqrt{\dfrac{\hat{p}(1-\hat{p})}{n}}$$

$$= 0.117 - 2.575 \cdot \sqrt{\dfrac{0.117(1-0.117)}{300}}$$

$$\approx 0.117 - 0.048 = 0.069$$

Upper bound

$$= \hat{p} + z_{0.005} \cdot \sqrt{\dfrac{\hat{p}(1-\hat{p})}{n}}$$

$$= 0.117 + 2.575 \cdot \sqrt{\dfrac{0.117(1-0.117)}{300}}$$

$$\approx 0.117 + 0.048 = 0.165 \quad [\text{Tech: } 0.164]$$

3. We construct a *t*-interval because we are estimating the population mean, we do not know the population standard deviation, and the underlying population is normally distributed. For 90% confidence, $\alpha/2 = 0.05$. Since $n = 12$, then df = 11 and $t_{0.05} = 1.796$. Then:

Lower bound $= \bar{x} - t_{0.05} \cdot \dfrac{s}{\sqrt{n}}$

$$= 45 - 1.796 \cdot \dfrac{14}{\sqrt{12}}$$

$$\approx 45 - 7.26 = 37.74$$

Upper bound $= \bar{x} + t_{0.05} \cdot \dfrac{s}{\sqrt{n}}$

$$= 45 + 1.796 \cdot \dfrac{14}{\sqrt{12}}$$

$$\approx 45 + 7.26 = 52.26$$

5. We construct a *t*-interval because we are estimating the population mean, we do not know the population standard deviation, and the underlying population is normally distributed. For 99% confidence, $\alpha/2 = 0.005$. Since $n = 40$, then df = 39 and $t_{0.050} = 2.708$. Then:

Lower bound $= \bar{x} - t_{0.005} \cdot \dfrac{s}{\sqrt{n}}$

$$= 120.5 - 2.708 \cdot \dfrac{12.9}{\sqrt{40}}$$

$$\approx 120.5 - 5.52 = 114.98$$

Upper bound $= \bar{x} + t_{0.005} \cdot \dfrac{s}{\sqrt{n}}$

$$= 120.5 + 2.708 \cdot \dfrac{12.9}{\sqrt{40}}$$

$$\approx 120.5 + 5.52 = 126.02$$

7. df $= n - 1 = 12 - 1 = 11$ and, for 90% confidence, $\alpha/2 = 0.05$.

From the table, $\chi^2_{0.05} = 19.675$ and $\chi^2_{1-0.05} = \chi^2_{0.95} = 4.575$.

Lower bound $= \dfrac{(n-1)s^2}{\chi^2_{\alpha/2}} = \dfrac{11 \cdot 23.7}{19.675} \approx 13.25$

Upper bound $= \dfrac{(n-1)s^2}{\chi^2_{1-\alpha/2}} = \dfrac{11 \cdot 23.7}{4.575} \approx 56.98$

9. We construct a *t*-interval because we are estimating the population mean, we do not know the population standard deviation, and the underlying population is normally distributed. For 95% confidence, $\alpha/2 = 0.025$. Since $n = 40$, then df = 39 and $t_{0.025} = 2.023$. Then:

Lower bound $= \bar{x} - t_{0.025} \cdot \dfrac{s}{\sqrt{n}}$

$$= 54 - 2.023 \cdot \dfrac{8}{\sqrt{40}}$$

$$\approx 54 - 2.6 = 51.4 \text{ months}$$

Upper bound $= \bar{x} + t_{0.025} \cdot \dfrac{s}{\sqrt{n}}$

$$= 54 + 2.023 \cdot \dfrac{8}{\sqrt{40}}$$

$$\approx 54 + 2.6 = 56.6 \text{ months}$$

We can be 95% confident that the population of felons convicted of aggravated assault serve a mean sentence between 51.4 and 56.6 months.

11. We construct a *t*-interval because we are estimating the population mean, we do not know the population standard deviation, and the underlying population is normally distributed. For 90% confidence, $\alpha/2 = 0.05$. Since $n = 100$, then df = 99.

Lower bound $= \bar{x} - t_{0.05} \cdot \dfrac{s}{\sqrt{n}}$

$= 3421 - 1.660 \cdot \dfrac{2583}{\sqrt{210}}$

$\approx 3421 - 428.9$

$= 2992.2$ [Tech: 2992.1]

Upper bound $= \bar{x} + t_{0.05} \cdot \dfrac{s}{\sqrt{n}}$

$= 3421 + 1.660 \cdot \dfrac{2583}{\sqrt{210}}$

$\approx 3421 + 428.9$

$= 3849.8$ [Tech: 3849.9]

The Internal Revenue Service can be 90% confident that the true mean additional tax owed for estate tax returns is between $2,992.2 and $3,849.8.

13. $\hat{p} = \dfrac{526}{1008} \approx 0.522$. The sample size $n = 1008$ is less than 5% of the population, and $n\hat{p}(1-\hat{p}) \approx 251.5 \geq 10$, so we can construct a Z-interval:
For 90% confidence the critical value is $z_{0.05} = 1.645$. Then:
Lower bound

$= \hat{p} - z_{0.05} \cdot \sqrt{\dfrac{\hat{p}(1-\hat{p})}{n}}$

$= 0.522 - 1.645 \cdot \sqrt{\dfrac{0.522(1-0.522)}{1008}}$

$\approx 0.522 - 0.026 = 0.496$

Upper bound

$= \hat{p} + z_{0.05} \cdot \sqrt{\dfrac{\hat{p}(1-\hat{p})}{n}}$

$= 0.522 + 1.645 \cdot \sqrt{\dfrac{0.522(1-0.522)}{1008}}$

$\approx 0.522 + 0.026 = 0.548$

The Gallup Organization can be 90% confident that the population proportion of adult Americans who are worried about having enough money to live comfortably in retirement is between 0.496 and 0.548 (i.e., between 49.6% and 54.8%).

15. The normal probability plot and boxplot show that the data are normal with no outliers, so we can use a *t*-interval. For 95% confidence the critical value is $t_{0.025} = 2.201$. The data give $n = 12$ and $\bar{x} = 267.75$ days. Then:

Lower bound $= \bar{x} - t_{0.025} \cdot \dfrac{s}{\sqrt{n}}$

$= 267.75 - 2.201 \cdot \dfrac{8.49}{\sqrt{12}}$

$\approx 267.75 - 5.392 = 262.4$ days

Upper bound $= \bar{x} + t_{0.025} \cdot \dfrac{s}{\sqrt{n}}$

$= 267.75 + 2.201 \cdot \dfrac{8.49}{\sqrt{12}}$

$\approx 267.75 + 5.392 = 273.1$ days

We can be 95% confident that the mean gestation period for humans is between 262.4 and 273.1 days.

17. The box plot indicates an outlier in the data, so we cannot compute a confidence interval using either method. **18.** The data are not normally distributed, and the sample size is small, so we cannot compute a confidence interval using either method.

Section 9.5

1. Bootstrapping is a method of statistical inference that uses a set of sample data as the population. Many samples of size equal to the original sample size are chosen. The samples are chosen with replacement so that the samples are not necessarily the original sample. The statistic of interest is determined for each sample. The distribution of these statistics is used to make a judgment about the population parameter.

3. Answers will vary. A sample answer generated by StatCrunch is given.

(a) Summary Statistics

Column	Mean	2.5th Percentile	97.5th Percentile
Mean	15.366826	14.1235	16.79

Lower bound: 14.124, upper bound: 16.79

(b)

Histogram of Means

The histogram resembles that of a normal distribution but slightly skewed to the right.

(c) In Exercise 35 of Section 9.2, we found the 95% *t*-interval to be (13.706, 17.058). The confidence interval constructed using the bootstrapping method is narrower.

5. Answers will vary. A sample answer generated by StatCrunch is given.
Summary Statistics

Column	Mean	5th Percentile	95th Percentile
Mean	2159.2126	1691.8928	2632.6072

Lower bound: 1691.8928, upper bound: 2632.6072;
The confidence interval constructed using the bootstrapping method is narrower.

7. (a) A *t*-interval cannot be constructed because there are outliers and the sample size is small.

Home Prices

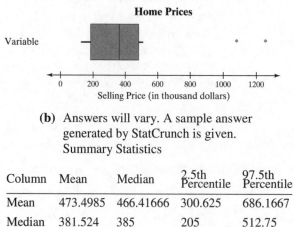

(b) Answers will vary. A sample answer generated by StatCrunch is given.
Summary Statistics

Column	Mean	Median	2.5th Percentile	97.5th Percentile
Mean	473.4985	466.41666	300.625	686.1667
Median	381.524	385	205	512.75

For the mean: lower bound: 300.625, upper bound: 686.167
For the median: lower bound: 205, upper bound: 512.75

(c)

Histogram of Mean

Histogram of Median

The histogram for the means shows a distribution slightly skewed to the right with potential outliers, so we might want to further investigate whether *t*-interval procedures could be used. The histogram for the medians shows a distribution that is skewed to the right and has potential outliers.

(d) The interval for medians better explains the median selling price because the outliers in the original data have had less effect on the sample medians than on the sample means.

9. (a) $\hat{p} = \dfrac{8}{20} = 0.4$

(b) $n\hat{p}(1-\hat{p}) = 4.8 < 10$, so the sample size is not large enough to use the normal model.

(c) Answers will vary. A sample answer generated by StatCrunch is given.
Summary Statistics

Column	Mean	2.5th Percentile	97.5th Percentile
Mean	0.40095	0.2	0.6

Lower bound: 0.2, upper bound: 0.6

10. (a) $\hat{p} = \dfrac{15}{25} = 0.6$

(b) $n\hat{p}(1-\hat{p}) = 6 < 10$, so the sample size is not large enough to use the normal model.

(c) Answers will vary. A sample answer generated by StatCrunch is given. Summary Statistics

Column	Mean	2.5th Percentile	97.5th Percentile
Mean	0.59712	0.4	0.76

Lower bound: 0.4, upper bound: 0.76

11. (a)-(c) Answers will vary.

Chapter 9 Review Exercises

1. (a) For a 99% confidence interval we want the *t*-value with an area in the right tail of 0.005. With df = 17, we read from the table that $t_{0.005} = 2.898$.

(b) For a 90% confidence interval we want the *t*-value with an area in the right tail of 0.05. With df = 26, we read from the table that $t_{0.05} = 1.706$.

2. (a) $df = n - 1 = 21$ and for 95% confidence, $\alpha/2 = 0.025$. From the table,
$$\chi^2_{1-0.025} = \chi^2_{0.975} = 10.283 \text{ and}$$
$$\chi^2_{0.025} = 35.479.$$

(b) $df = n - 1 = 11$ and for 99% confidence, $\alpha/2 = 0.005$. From the table,
$$\chi^2_{1-0.005} = \chi^2_{0.995} = 2.603 \text{ and}$$
$$\chi^2_{0.005} = 26.757.$$

3. In 100 samples, we would expect 95 of the 100 intervals to include the true population mean, 100. Random chance in sampling causes a particular interval to not include the true population mean.

4. In a 95% confidence interval, the 95% represents the proportion of intervals that would contain the parameter of interest (e.g. the population mean, population proportion, or population standard deviation) if a large number of different samples is obtained.

5. If a large number of different samples (of the same size) is obtained, a 90% confidence interval for the population mean will not capture the true value of the population mean 10% of the time.

6. The area to the left of $t = -1.56$ is also 0.0681, because the *t*-distribution is symmetric about 0.

7. The area under the *t*-distribution to the right of $t = 2.32$ is greater than the area under the standard normal distribution to the right of $z = 2.32$ because the *t*-distribution uses *s* to approximate σ, making it more dispersed than the Z-distribution.

8. The properties of the Student's *t*-distribution:

(1) It is symmetric around $t = 0$.

(2) It is different for different sample sizes.

(3) The area under the curve is 1; half the area is to the right of 0 and half the area is to the left of 0.

(4) As *t* gets extremely large, the graph approaches, but never equals, zero. Similarly, as *t* gets extremely small (negative), the graph approaches, but never equals, zero.

(5) The area in the tails of the *t*-distribution is greater than the area in the tails of the standard normal distribution.

(6) As the sample size *n* increases, the distribution (and the density curve) of the *t*-distribution becomes more like the standard normal distribution.

9. (a) For 90% confidence, $\alpha/2 = 0.05$. For a sample size of $n = 30$, $df = n - 1 = 29$, and the critical value is $t_{\alpha/2} = t_{0.05} = 1.699$. Then:

$$\text{Lower bound} = \bar{x} - t_{0.05} \cdot \frac{s}{\sqrt{n}}$$
$$= 54.8 - 1.699 \cdot \frac{10.5}{\sqrt{30}}$$
$$\approx 54.8 - 3.26 = 51.54$$

$$\text{Upper bound} = \bar{x} + t_{0.05} \cdot \frac{s}{\sqrt{n}}$$
$$= 54.8 + 1.699 \cdot \frac{10.5}{\sqrt{30}}$$
$$\approx 54.8 + 3.26 = 58.06$$

(b) For a sample size of $n = 51$, $df = n - 1 = 50$, and the critical value is $t_{\alpha/2} = t_{0.05} = 1.676$. Then:

Lower bound $= \bar{x} - t_{0.05} \cdot \dfrac{s}{\sqrt{n}}$

$= 54.8 - 1.676 \cdot \dfrac{10.5}{\sqrt{51}}$

$\approx 54.8 - 2.49 = 52.34$

Upper bound $= \bar{x} + t_{0.05} \cdot \dfrac{s}{\sqrt{n}}$

$= 54.8 + 1.676 \cdot \dfrac{10.5}{\sqrt{51}}$

$\approx 54.8 + 2.49 = 57.26$

Increasing the sample size decreases the width of the confidence interval.

(c) For 99% confidence, $\alpha / 2 = 0.005$. For a sample size of $n = 30$, $df = n - 1 = 29$, and the critical value is $t_{\alpha/2} = t_{0.05} = 2.756$. Then:

Lower bound $= \bar{x} - t_{0.005} \cdot \dfrac{s}{\sqrt{n}}$

$= 54.8 - 2.756 \cdot \dfrac{10.5}{\sqrt{30}}$

$\approx 54.8 - 5.28 = 49.52$

Upper bound $= \bar{x} + t_{0.005} \cdot \dfrac{s}{\sqrt{n}}$

$= 54.8 + 2.756 \cdot \dfrac{10.5}{\sqrt{30}}$

$\approx 54.8 + 5.28 = 60.08$

Increasing the level of confidence increases the width of the confidence interval.

10. (a) For 90% confidence, $\alpha / 2 = 0.05$. If $n = 15$, then df = 14. The critical value is $t_{0.05} = 1.761$. Then:

Lower bound $= \bar{x} - t_{0.05} \cdot \dfrac{s}{\sqrt{n}}$

$= 104.3 - 1.761 \cdot \dfrac{15.9}{\sqrt{15}}$

$\approx 104.3 - 7.23 = 97.07$

Upper bound $= \bar{x} + t_{0.05} \cdot \dfrac{s}{\sqrt{n}}$

$= 104.3 + 1.761 \cdot \dfrac{15.9}{\sqrt{15}}$

$\approx 104.3 + 7.23 = 111.53$

(b) If $n = 25$, then df = 24. The critical value is $t_{0.05} = 1.711$. Then:

Lower bound $= \bar{x} - t_{0.05} \cdot \dfrac{s}{\sqrt{n}}$

$= 104.3 - 1.711 \cdot \dfrac{15.9}{\sqrt{25}}$

$\approx 104.3 - 5.44 = 98.86$

Upper bound $= \bar{x} + t_{0.05} \cdot \dfrac{s}{\sqrt{n}}$

$= 104.3 + 1.711 \cdot \dfrac{15.9}{\sqrt{25}}$

$\approx 104.3 + 5.44 = 109.74$

Increasing the sample size decreases the width of the confidence interval.

(c) For 95% confidence, $\alpha / 2 = 0.025$. With 14 df, $t_{0.025} = 2.145$. Then:

Lower bound $= \bar{x} - t_{0.025} \cdot \dfrac{s}{\sqrt{n}}$

$= 104.3 - 2.145 \cdot \dfrac{15.9}{\sqrt{15}}$

$\approx 104.3 - 8.81 = 95.49$

Upper bound $= \bar{x} + t_{0.025} \cdot \dfrac{s}{\sqrt{n}}$

$= 104.3 + 2.145 \cdot \dfrac{15.9}{\sqrt{15}}$

$\approx 104.3 + 8.81 = 113.11$

Increasing the level of confidence increases the width of the confidence interval.

(d) df $= n - 1 = 14$ and for 90% confidence, $\alpha / 2 = 0.05$. From the table, $\chi^2_{1-0.05} = \chi^2_{0.95} = 6.571$ and $\chi^2_{0.05} = 23.685$.

Lower bound $= \sqrt{\dfrac{(n-1)s^2}{\chi^2_{\alpha/2}}}$

$= \sqrt{\dfrac{14(15.9)^2}{23.685}} = 12.22$

Upper bound $= \sqrt{\dfrac{(n-1)s^2}{\chi^2_{\alpha/2}}}$

$= \sqrt{\dfrac{14(15.9)^2}{6.571}} = 23.21$

11. (a) The size of the sample ($n = 35$) is sufficiently large to apply the Central Limit Theorem and conclude that the sampling distribution of $\bar{x}$ is approximately normal.

(b) For 95% confidence, $\alpha/2 = 0.025$. If $n = 35$, then df = 34. The critical value is $t_{0.025} = 2.032$. Then:

$$\text{Lower bound} = \bar{x} - t_{0.025} \cdot \frac{s}{\sqrt{n}}$$

$$= 87.9 - 2.032 \cdot \frac{15.5}{\sqrt{35}}$$

$$\approx 87.9 - 5.32 = 82.58$$

$$\text{Upper bound} = \bar{x} + t_{0.025} \cdot \frac{s}{\sqrt{n}}$$

$$= 87.9 + 2.032 \cdot \frac{15.5}{\sqrt{35}}$$

$$\approx 87.9 + 5.32 = 93.22$$

We can be 95% confident that the population mean age people would live to live to is between 82.58 and 93.22 years.

(c) For 95% confidence, $z_{\alpha/2} = 1.96$ and

$$n = \left(\frac{z_{\alpha/2} \cdot s}{E} \right)^2 = \left(\frac{1.96 \cdot 15.5}{2} \right)^2 \approx 230.7$$

which we must increase to 231. A total of 231 subjects are needed.

12. (a) Since the number of emails sent in a day cannot be negative, and one standard deviation to the left of the mean results in a negative number, we expect that the distribution is skewed right.

(b) For 90% confidence, $\alpha/2 = 0.05$. We have df $= n - 1 = 927$. Since there is no row in the table for 927 degrees of freedom, we use df $= 1000$ instead. Thus, $t_{0.05} = 1.646$. Then:

$$\text{Lower bound} = \bar{x} - t_{0.05} \cdot \frac{s}{\sqrt{n}}$$

$$= 10.4 - 1.646 \cdot \frac{28.5}{\sqrt{928}}$$

$$\approx 10.4 - 1.54$$

$$= 8.86 \text{ e-mails}$$

$$\text{Upper bound} = \bar{x} + t_{0.05} \cdot \frac{s}{\sqrt{n}}$$

$$= 10.4 + 1.646 \cdot \frac{28.5}{\sqrt{928}}$$

$$\approx 10.4 + 1.54$$

$$= 11.94 \text{ e-mails}$$

We are 90% confident that the population mean number of e-mails sent per day is between 8.86 and 11.94.

13. (a) The sample size is probably small due to the difficulty and expense of locating highly trained cyclists and gathering the data.

(b) For 95% confidence, $\alpha/2 = 0.025$. We have df $= n - 1 = 15$, and $t_{0.025} = 2.131$. Then:

$$\text{Lower bound} = \bar{x} - t_{0.025} \cdot \frac{s}{\sqrt{n}}$$

$$= 218 - 2.131 \cdot \frac{31}{\sqrt{16}}$$

$$\approx 218 - 16.5$$

$$= 201.5 \text{ kilojoules}$$

$$\text{Upper bound} = \bar{x} + t_{0.025} \cdot \frac{s}{\sqrt{n}}$$

$$= 218 + 2.131 \cdot \frac{31}{\sqrt{16}}$$

$$\approx 218 + 16.5$$

$$= 234.5 \text{ kilojoules}$$

The researchers can be 95% confident that the population mean total work performed for the sports-drink treatment group is between 201.5 and 234.5 kilojoules.

(c) Yes; it is possible that the mean total work performed is less than 198 kilojoules since it is possible that the true mean is not captured in the confidence interval. However, it is not very likely since we are 95% confident the true mean total work performed is between 201.5 and 234.5 kilojoules.

(d) For 95% confidence, $\alpha/2 = 0.025$. We have df $= n - 1 = 15$, and $t_{0.025} = 2.131$. Then:

$$\text{Lower bound} = \bar{x} - t_{0.025} \cdot \frac{s}{\sqrt{n}}$$

$$= 178 - 2.131 \cdot \frac{31}{\sqrt{16}}$$

$$\approx 178 - 16.5$$

$$= 161.5 \text{ kilojoules}$$

$$\text{Upper bound} = \bar{x} + t_{0.025} \cdot \frac{s}{\sqrt{n}}$$

$$= 178 + 2.131 \cdot \frac{31}{\sqrt{16}}$$

$$\approx 178 + 16.5$$

$$= 194.5 \text{ kilojoules}$$

The researchers can be 95% confident that the population mean total work performed for the placebo treatment group is between 161.5 and 194.5 kilojoules.

(e) Yes; it is possible that the mean total work performed is more than 198 kilojoules since it is possible that the true mean is not captured in the confidence interval. However, it is not very likely since we are 95% confident the true mean total work performed is between 161.5 and 194.5 kilojoules.

(f) Yes; our findings support the researchers' conclusion. The confidence intervals do not overlap, so we are confident that the mean total work performed for the sports-drink treatment is greater than the mean total work performed for the placebo treatment.

14. (a) Since the sample size is large ($n \geq 30$), $\bar{x}$ has an approximately normal distribution.

(b) We construct a t-interval because we are estimating a population mean and we do not know the population standard deviation. For 95% confidence, $\alpha/2 = 0.025$. Since $n = 60$, then df = 59. There is no row in the table for 59 degrees of freedom, so we use df = 60 instead. The critical value is $t_{0.025} = 2.000$. Then:

Lower bound $= \bar{x} - t_{0.025} \cdot \dfrac{s}{\sqrt{n}}$

$= 2.27 - 2.000 \cdot \dfrac{1.22}{\sqrt{60}}$

$\approx 2.27 - 0.32$

$= 1.95$ children

Upper bound $= \bar{x} + t_{0.025} \cdot \dfrac{s}{\sqrt{n}}$

$= 2.27 + 2.000 \cdot \dfrac{1.22}{\sqrt{60}}$

$\approx 2.27 + 0.32$

$= 2.59$ children

[Tech: 2.58]

We are 95% confident that couples who have been married for 7 years have a population mean number of children between 1.95 and 2.59.

(c) For 99% confidence, $\alpha/2 = 0.005$. The critical value is $t_{0.005} = 2.660$. Then:

Lower bound $= \bar{x} - t_{0.025} \cdot \dfrac{s}{\sqrt{n}}$

$= 2.27 - 2.660 \cdot \dfrac{1.22}{\sqrt{60}}$

$\approx 2.27 - 0.42$

$= 1.85$ children

Upper bound $= \bar{x} + t_{0.025} \cdot \dfrac{s}{\sqrt{n}}$

$= 2.27 + 2.660 \cdot \dfrac{1.22}{\sqrt{60}}$

$\approx 2.27 + 0.42$

$= 2.69$ children

We are 99% confident that couples who have been married for 7 years have a population mean number of children between 1.85 and 2.69.

15. (a) Using technology, we obtain $\bar{x} \approx 147.3$ cm and $s \approx 28.8$ cm.

(b) Yes. All the data values lie within the bounds on the normal probability plot, indicating that the data should come from a population that is normal. The boxplot does not show any outliers.

(c) We construct a t-interval because we are estimating a population mean and we do not know the population standard deviation. For 95% confidence, $\alpha/2 = 0.025$. Since $n = 12$, then df = 11. The critical value is $t_{0.025} = 2.201$. Then:

Lower bound $= \bar{x} - t_{0.025} \cdot \dfrac{s}{\sqrt{n}}$

$= 147.3 - 2.201 \cdot \dfrac{28.8}{\sqrt{12}}$

$\approx 147.3 - 18.3$

$= 129.0$ cm

Upper bound $= \bar{x} + t_{0.025} \cdot \dfrac{s}{\sqrt{n}}$

$= 147.3 + 2.201 \cdot \dfrac{28.8}{\sqrt{12}}$

$\approx 147.3 + 18.3$

$= 165.6$ cm

We are 95% confident that the population mean diameter of a Douglas fir tree in the western Washington Cascades is between 129.0 and 165.6 centimeters.

(d) From the table, with df = 11 and $\alpha/2 = 0.025$, we find

$$\chi^2_{1-0.025} = \chi^2_{0.975} = 3.816 \text{ and}$$

$$\chi^2_{0.025} = 21.920. \text{ Then:}$$

Lower bound $= \sqrt{\dfrac{(n-1)s^2}{\chi^2_{\alpha/2}}}$

$$= \sqrt{\dfrac{11(28.8)^2}{21.920}} \approx 20.4 \text{ cm}$$

Upper bound $= \sqrt{\dfrac{(n-1)s^2}{\chi^2_{\alpha/2}}}$

$$= \sqrt{\dfrac{11(28.8)^2}{3.816}} \approx 48.9 \text{ cm}$$

We are 95% confident that the population standard deviation diameter of a Douglas fir tree in the western Washington Cascades is between 20.4 and 48.9 centimeters.

16. (a) $\hat{p} = \dfrac{x}{n} = \dfrac{58}{678} \approx 0.086$

(b) For 95% confidence, $z_{\alpha/2} = z_{.025} = 1.96$.

Lower bound

$$= \hat{p} - z_{.025} \cdot \sqrt{\dfrac{\hat{p}(1-\hat{p})}{n}}$$

$$= 0.086 - 1.96 \cdot \sqrt{\dfrac{0.086(1-0.086)}{678}}$$

$$\approx 0.086 - 0.021 = 0.065$$

Upper bound

$$= \hat{p} + z_{.025} \cdot \sqrt{\dfrac{\hat{p}(1-\hat{p})}{n}}$$

$$= 0.086 + 1.96 \cdot \sqrt{\dfrac{0.086(1-0.086)}{678}}$$

$$\approx 0.086 + 0.021 = 0.107$$

The Centers for Disease Control can be 95% confident that the population proportion of adult males aged 20–34 years who have hypertension is between 0.065 and 0.107 (i.e. between 6.5% and 10.7%).

(c) $n = \hat{p}(1-\hat{p})\left(\dfrac{z_{\alpha/2}}{E}\right)^2$

$$= 0.086(1-0.086)\left(\dfrac{1.96}{.03}\right)^2$$

$$\approx 335.5$$

which we must increase to 336. You

would need a sample of size 336 for your estimate to be within 3 percentage points of the true proportion, with 95% confidence.

(d) $n = 0.25\left(\dfrac{z_{\alpha/2}}{E}\right)^2 = 0.25\left(\dfrac{1.96}{0.03}\right)^2$

$$\approx 1067.1$$

which we must increase to 1068. Without the prior estimate, you would need a sample size of 1068.

17. Answers will vary. A sample answer generated by StatCrunch is given.

(a) Summary Statistics

Column	Mean	2.5th Percentile	97.5th Percentile
Mean	147.25333	132.625	162.29167

Lower bound: 132.625, upper bound: 162.292;
The confidence interval constructed using the bootstrapping method is narrower.

(b)

Histogram of Means

The histogram resembles that of a normal distribution, which is not surprising, since the distribution of the original data was approximately normal.

(c) Summary Statistics

Column	Median	2.5th Percentile	97.5th Percentile
Median	151.5	121.5	167.5

Lower bound: 121.5, upper bound: 167.5

Chapter 9 Test

1. Properties of the Student's *t*-distribution:

   (1) It is symmetric around $t = 0$.

   (2) It is different for different sample sizes.

   (3) The area under the curve is 1; half the area is to the right of 0 and half the area is to the left of 0.

   (4) As *t* gets extremely large, the graph approaches, but never equals, zero. Similarly, as *t* gets extremely small (negative), the graph approaches, but never equals, zero.

   (5) The area in the tails of the *t*-distribution is greater than the area in the tails of the standard normal distribution.

   (6) As the sample size *n* increases, the distribution (and the density curve) of the *t*-distribution becomes more like the standard normal distribution.

2. **(a)** For a 96% confidence interval we want the *t*-value with an area in the right tail of 0.02. With df = 25, we read from the table that $t_{0.02} = 2.167$.

   (b) For a 98% confidence interval we want the *t*-value with an area in the right tail of 0.01. With df = 17, we read from the table that $t_{0.01} = 2.567$.

3. **(a)** $df = n - 1 = 15$ and for 98% confidence, $\alpha / 2 = 0.01$. From the table,
   $$\chi^2_{1-0.01} = \chi^2_{0.99} = 5.229 \text{ and } \chi^2_{0.01} = 30.578.$$

   (b) $df = n - 1 = 29$ and for 90% confidence, $\alpha / 2 = 0.05$. From the table,
   $$\chi^2_{1-0.05} = \chi^2_{0.95} = 17.708 \text{ and }$$
   $$\chi^2_{0.05} = 42.557.$$

4. $\bar{x} = \dfrac{125.8 + 152.6}{2} = \dfrac{278.4}{2} = 139.2$

   $E = \dfrac{152.6 - 125.8}{2} = \dfrac{26.8}{2} = 13.4$

5. **(a)** We would expect the distribution to be skewed right. We expect most respondents to have relatively few family members in prison, but there will be some

   with several (possibly many) family members in prison.

   (b) We construct a *t*-interval because we are estimating a population mean and we do not know the population standard deviation. For 99% confidence, $\alpha / 2 = 0.005$. Since $n = 499$, then df = 498 and $t_{0.005} = 2.626$. Then:

   Lower bound $= \bar{x} - t_{0.005} \cdot \dfrac{s}{\sqrt{n}}$

   $= 1.22 - 2.626 \cdot \dfrac{0.59}{\sqrt{499}}$

   $\approx 1.22 - 0.069 = 1.151$

   [Tech: 1.152]

   Upper bound $= \bar{x} + t_{0.005} \cdot \dfrac{s}{\sqrt{n}}$

   $= 1.22 + 2.626 \cdot \dfrac{0.59}{\sqrt{499}}$

   $\approx 1.22 + 0.069 = 1.289$

   [Tech: 1.288]

   We are 99% confident that the population mean number of family members in jail is between 1.151 and 1.289.

6. **(a)** We construct a *t*-interval because we are estimating a population mean and we do not know the population standard deviation. For 90% confidence, $\alpha / 2 = 0.05$. Since $n = 50$, then df = 49. There is no row in the table for 49 degrees of freedom, so we use df = 50 instead. Thus, $t_{0.05} = 1.676$, and

   Lower bound $= \bar{x} - t_{0.05} \cdot \dfrac{s}{\sqrt{n}}$

   $= 4.58 - 1.676 \cdot \dfrac{1.10}{\sqrt{50}}$

   $\approx 4.58 - 0.261 = 4.319$ yrs

   Upper bound $= \bar{x} + t_{0.05} \cdot \dfrac{s}{\sqrt{n}}$

   $= 4.58 + 1.676 \cdot \dfrac{1.10}{\sqrt{50}}$

   $\approx 4.58 + 0.261 = 4.841$ yrs

   We are 90% confident that the population mean time to graduate is between 4.319 years and 4.841 years.

   (b) Yes; we are 90% confident that the mean time to graduate is between 4.319 years and 4.841 years. Since the entire interval is above 4 years, our evidence contradicts the belief that it takes 4 years to complete a bachelor's degree.

7. **(a)** Using technology, we obtain $\bar{x} \approx 57.8$ inches and $s \approx 15.4$ inches.

(b) Yes. The plotted points are generally linear and stay within the bounds of the normal probability plot. The boxplot shows that there are no outliers.

(c) We construct a t-interval because we are estimating a population mean and we do not know the population standard deviation. For 95% confidence, $\alpha/2 = 0.025$. Since $n = 12$, then df = 11 and $t_{0.025} = 2.201$. Then:

Lower bound $= \bar{x} - t_{0.025} \cdot \dfrac{s}{\sqrt{n}}$

$$= 57.8 - 2.201 \cdot \frac{15.4}{\sqrt{12}}$$

$$\approx 57.8 - 9.8 = 48.0 \text{ in.}$$

[Tech: 47.9]

Upper bound $= \bar{x} + t_{0.025} \cdot \dfrac{s}{\sqrt{n}}$

$$= 57.8 + 2.201 \cdot \frac{15.4}{\sqrt{12}}$$

$$\approx 57.8 + 9.8 = 67.6 \text{ in.}$$

The researcher can be 95% confident that the population mean depth of visibility of the Secchi disk is between 48.0 and 67.6 inches.

(d) For 99% confidence, $\alpha/2 = 0.005$. For df = 11, $t_{0.005} = 3.106$. Then:

Lower bound $= \bar{x} - t_{0.005} \cdot \dfrac{s}{\sqrt{n}}$

$$= 57.8 - 3.106 \cdot \frac{15.4}{\sqrt{12}}$$

$$\approx 57.8 - 13.8 = 44.0 \text{ in.}$$

Upper bound $= \bar{x} + t_{0.005} \cdot \dfrac{s}{\sqrt{n}}$

$$= 57.8 + 3.106 \cdot \frac{15.4}{\sqrt{12}}$$

$$\approx 57.8 + 13.8 = 71.6 \text{ in.}$$

The researcher can be 99% confident that the population mean depth of visibility of the Secchi disk is between 44.0 and 71.6 inches.

(e) From the table, with df = 11 and $\alpha/2 = 0.025$, we find

$\chi^2_{1-0.025} = \chi^2_{0.975} = 3.816$ and $\chi^2_{0.025} = 21.920$. Then:

Lower bound $= \sqrt{\dfrac{(n-1)s^2}{\chi^2_{\alpha/2}}} = \sqrt{\dfrac{11(15.45)^2}{21.920}}$

$$\approx 10.9 \text{ in.}$$

Upper bound $= \sqrt{\dfrac{(n-1)s^2}{\chi^2_{\alpha/2}}} = \sqrt{\dfrac{11(15.45)^2}{3.816}}$

$$\approx 26.2 \text{ in.}$$

The researcher can be 99% confident that the population standard deviation of the depth of visibility of the Secchi disk is between 10.9 and 26.2 inches.

8. **(a)** $\hat{p} = \dfrac{x}{n} = \dfrac{1139}{1201} \approx 0.948$

(b) For 99% confidence, $z_{\alpha/2} = z_{.005} = 2.575$.

Lower bound

$= \hat{p} - z_{.005} \cdot \sqrt{\dfrac{\hat{p}(1-\hat{p})}{n}}$

$= 0.948 - 2.575 \cdot \sqrt{\dfrac{0.948(1-0.948)}{1201}}$

$\approx 0.948 - 0.016 = 0.932$

Upper bound

$= \hat{p} + z_{.005} \cdot \sqrt{\dfrac{\hat{p}(1-\hat{p})}{n}}$

$= 0.948 + 2.575 \cdot \sqrt{\dfrac{0.948(1-0.948)}{1201}}$

$\approx 0.948 + 0.016 = 0.964$ [Tech: 0.965]

The EPA can be 99% confident that the population proportion of Americans who live in neighborhoods with acceptable levels of carbon monoxide is between 0.932 and 0.964 (i.e., between 93.2% and 96.4%).

(c) $n = \hat{p}(1-\hat{p})\left(\dfrac{z_{\alpha/2}}{E}\right)^2$

$= 0.948(1-0.948)\left(\dfrac{1.645}{.015}\right)^2$

$= 592.9$

which we must increase to 593. A sample size of 593 would be needed for the estimate to be within 1.5 percentage points with 90% confidence.

(d) $n = 0.25\left(\dfrac{z_{\alpha/2}}{E}\right)^2 = 0.25\left(\dfrac{1.645}{.015}\right)^2$

≈ 3006.7

which we must increase to 3007. Without the prior estimate, a sample size of 3007 would be needed for the estimate to be within 1.5 percentage points with 90% confidence.

9. (a) $\bar{x} = \dfrac{\sum x}{40} = \dfrac{5336}{40} = 133.4$ minutes

(b) Because the population is not normally distributed, a large sample is needed to apply the Central Limit Theorem and say that the $\bar{x}$ distribution is approximately normal.

(c) For 90% confidence, $\alpha/2 = 0.05$. Since $n = 40$, then df = 39 and $t_{0.05} = 1.685$. Then:
Lower bound

$= \bar{x} - t_{0.05} \cdot \dfrac{s}{\sqrt{n}}$

$= 133.4 - 1.685 \cdot \dfrac{43.3}{\sqrt{40}}$

$\approx 133.4 - 11.54 = 121.86$ minutes
Upper bound

$= \bar{x} - t_{0.05} \cdot \dfrac{s}{\sqrt{n}}$

$= 133.4 + 1.685 \cdot \dfrac{43.3}{\sqrt{40}}$

$\approx 133.4 + 11.54 = 144.94$ minutes
. The tennis enthusiast is 99% confident that the population mean length of men's singles matches during Wimbledon is between 121.86 and 144.94 minutes.

(d) For 95% confidence the critical value is $t_{0.025} = 2.023$. Then:

Lower bound $= \bar{x} - t_{0.025} \cdot \dfrac{s}{\sqrt{n}}$

$= 133.4 - 2.023 \cdot \dfrac{43.3}{\sqrt{40}}$

$\approx 133.4 - 13.84$

$= 119.6$ minutes

Upper bound $= \bar{x} + t_{0.025} \cdot \dfrac{s}{\sqrt{n}}$

$= 133.4 + 2.023 \cdot \dfrac{43.3}{\sqrt{40}}$

$\approx 133.4 + 13.84$

$= 147.2$ minutes
The tennis enthusiast is 95% confident that the population mean length of men's singles matches during Wimbledon is between 119.6 and 147.2 minutes.

(e) Decreasing the level of confidence decreases the width of the interval.

(f) No; the tennis enthusiast only sampled Wimbledon matches. Therefore, his results cannot be generalized to all professional tournaments.

10. (a) The distribution is skewed right and includes an outlier, and the sample size is small.

(b) The bootstrapping method uses the given set of sample data as the population. Many samples of size equal to the original sample size are chosen. The samples are chosen with replacement so that the samples are not necessarily the original sample. The mean is determined for each sample. The distribution of these sample means is used to construct the 95% confidence interval.

(c) Answers will vary. A sample answer generated by StatCrunch is given.
Summary Statistics

Column	Mean	2.5th Percentile	97.5th Percentile
Mean	45.220856	25.857143	67.67857

Lower bound: 25.857, upper bound: 67.679

Case Study: When Model Requirements Fail

Note: Results may vary due to the random number generator seed.

(a) Using technology, we have $Q_1 = 0.4$ and $Q_3 = 1.2$. Thus,

$IQR = Q_3 - Q_1 = 1.2 - 0.4 = 0.8$.

Lower fence: $Q_1 - 1.5(IQR) = 0.4 - 1.5(0.8)$
$$= -0.8$$

Upper fence: $Q_3 + 1.5(IQR) = 1.2 + 1.5(0.8)$
$$= 2.4$$

Since $3.1 > 2.4$ (the upper fence), we conclude that 3.1 is an outlier.

(b) Enter data in Excel.

	A	B	C	D
1	1	0.7	0.6	0.5
2	0.3	1.4	3.1	0.2
3				

(c) Enter command in cell A4. Copy to obtain 200 samples.

	A	B	C	D
1	1	0.7	0.6	0.5
2	0.3	1.4	3.1	0.2
3				
4	0.6			
5				

A	B	C	D	E	F	G	H
1	0.7	0.6	0.5				
0.3	1.4	3.1	0.2				
0.5	0.6	0.5	0.7	1.4	0.7	0.6	0.3
0.3	1.4	3.1	0.5	1	0.6	0.6	0.5
1.4	0.2	1	1	0.3	0.3	0.2	1.4
0.3	1	0.6	0.3	1.4	0.6	0.7	0.6
0.3	0.2	0.5	3.1	1.4	0.7	0.6	0.5
0.2	1	3.1	3.1	1	0.2	0.6	0.5
3.1	3.1	1.4	0.5	1	1.4	3.1	1.4
0.6	0.5	1.4	0.5	0.2	0.7	0.3	3.1
1.4	0.6	0.5	0.3	0.3	0.2	0.2	1
0.3	0.6	1.4	3.1	3.1	0.5	0.7	0.3
0.5	0.2	0.5	1	0.5	0.2	1.4	0.5
0.3	1	0.5	0.7	0.3	0.5	0.2	0.6
0.7	3.1	0.2	0.7	0.3	1.4	0.6	1.4
0.3	3.1	0.7	0.6	1.4	0.6	0.7	1.4
0.3	3.1	1	0.5	3.1	1	0.3	0.5
0.5	0.7	0.6	0.2	0.2	1	0.5	3.1
0.5	1.4	1	1	0.5	1.4	0.5	3.1
0.5	1	0.7	1.4	1.4	1.4	1	0.6
0.5	0.2	1.4	1	3.1	0.7	0.5	1
0.6	1	0.3	0.3	3.1	1.4	0.3	0.6
0.7	1.4	1.4	0.5	0.2	0.7	0.7	0.3
0.5	0.3	0.7	1.4	1	0.7	1	0.5
0.6	3.1	3.1	3.1	1	1	1	0.3
3.1	0.6	1.4	3.1	1.4	0.2	1	0.7
0.6	3.1	1	0.6	3.1	0.2	0.6	0.7
0.7	0.6	1.4	0.7	0.5	1.4	3.1	0.3
1	3.1	0.6	0.3	3.1	1	1	1
0.7	1	0.5	0.7	0.2	0.6	0.7	0.6

(d) Find the mean of each sample.

J	K
0.9375	
1.4375	
0.725	
0.9375	
0.6125	
0.8875	
0.8	
1.225	
0.4625	
0.75	
0.675	
1	
0.8125	
1.6	

(e)-(g) Obtain confidence interval and interpret.

	K	L	M
	conf. lev.	0.95	
		0.05	
	samples	200	
35	lower	0.475	
25	upper	1.7625	
.6			
75			
25			

We are 95% confident that the population mean is between 0.475 and 1.7625.

	K	L
	conf. lev.	0.95
		0.05
	samples	200
75	lower	0.4375
25	upper	1.5625
25		
75		

	K	L
	conf. lev.	0.95
		0.05
	samples	200
75	lower	0.425
35	upper	1.6125
75		
3		

With each new interval, the lower and upper bounds fluctuate, but the values do not vary greatly from interval to interval.

(h) The precision of the interval could be improved by drawing larger samples, or decreasing the level of confidence.

(i) *t*-distribution:

For 95% confidence, $\alpha/2 = 0.025$. For df = 7, $t_{0.005} = 2.365$. Then:

Lower bound $= \bar{x} - t_{0.025} \cdot \dfrac{s}{\sqrt{n}}$

$$= 6.19 - 2.365 \cdot \frac{1.16}{\sqrt{8}}$$

$$\approx 6.19 - 0.97 = 5.22$$

Upper bound $= \bar{x} - +_{0.025} \cdot \dfrac{s}{\sqrt{n}}$

$$= 6.19 + 2.365 \cdot \frac{1.16}{\sqrt{8}}$$

$$\approx 6.19 + 0.97 = 7.16$$

Bootstrapping:
(actual interval will vary due to random number generator)

	K	L	M
	conf. lev.	0.95	
		0.05	
	samples	200	
625	lower	5.4375	
375	upper	6.9375	
375			

The two results are very similar.

(j) If $n\hat{p}(1-\hat{p})$ is less than 10, then we cannot say that the distribution of $\hat{p}$ is approximately normal, nor can we compute a confidence interval for *p*.

(k) Simulations will vary. We expect 95 of the samples (95%) to generate confidence intervals that contain the population proportion, 0.15. Unexpected results will likely occur because our assumption for normality has been violated.

(l) **1.** $\tilde{p} = \dfrac{x+2}{n+4} = \dfrac{1+2}{10+4} = \dfrac{3}{14} \approx 0.214$

For 95% confidence we have
$z_{\alpha/2} = z_{0.025} = 1.96$.

Lower bound:

$$= \tilde{p} - z_{\alpha/2} \sqrt{\frac{\tilde{p}(1-\tilde{p})}{n+4}}$$

$$= 0.214 - 1.96 \sqrt{\frac{0.214(1-0.214)}{10+4}}$$

$$\approx 0.214 - 0.215$$

$$= -0.01$$

Upper bound:

$$= \tilde{p} + z_{\alpha/2} \sqrt{\frac{\tilde{p}(1-\tilde{p})}{n+4}}$$

$$= 0.214 + 1.96 \sqrt{\frac{0.214(1-0.214)}{10+4}}$$

$$\approx 0.214 + 0.215$$

$$= 0.429$$

Jane is 95% confident that the proportion of students on her campus who eat cauliflower is between −0.01 (basically 0%) and 0.429 (or 42.9%).

2. $\tilde{p} = \dfrac{x+2}{n+4} = \dfrac{0+2}{12+4} = \dfrac{2}{16} = 0.125$

For 95% confidence we have
$z_{\alpha/2} = z_{0.025} = 1.96$.

Lower bound:

$$= \tilde{p} - z_{\alpha/2} \sqrt{\frac{\tilde{p}(1-\tilde{p})}{n+4}}$$

$$= 0.125 - 1.96 \sqrt{\frac{0.125(1-0.125)}{12+4}}$$

$$\approx 0.125 - 0.162$$

$$= -0.037$$

Upper bound:

$$= \tilde{p} + z_{\alpha/2} \sqrt{\frac{\tilde{p}(1-\tilde{p})}{n+4}}$$

$$= 0.125 + 1.96 \sqrt{\frac{0.125(1-0.125)}{12+4}}$$

$$\approx 0.125 + 0.162$$

$$= 0.287$$

Alan is 95% confident that the proportion of adults who walk to work is between −3.7% (basically 0) and 0.287 (or 28.7%).

Chapter 10

Hypothesis Tests Regarding a Parameter

Section 10.1

1. hypothesis

3. null hypothesis

5. I

7. level of significance

9. Right-tailed since $H_1 : \mu > 5$
   Parameter $= \mu$

11. Two-tailed since $H_1 : \sigma \neq 4.2$
    Parameter $= \sigma$

13. Left-tailed since $H_1 : \mu < 120$
    Parameter $= \mu$

15. **(a)** $H_0 : p = 0.105$, $H_1 : p > 0.105$
    The alternative hypothesis is $p > 0.105$ because the sociologist believes the percent has increased.

    (b) We make a Type I error if the sample evidence leads us to reject H_0 and conclude that the proportion of registered births to teenage mothers has increased above 0.105 when, in fact, it has not increased above 0.105.

    (c) We make a Type II error if the sample evidence does not lead us to conclude that the proportion of registered births to teenage mothers has increased above 0.105 when, in fact, the proportion of registered births to teenage mothers has increased above 0.105.

17. **(a)** $H_0 : \mu = \$218{,}600$; $H_1 : \mu < \$218{,}600$
    The alternative hypothesis is $\mu < \$218{,}600$ because the real estate broker believes the mean price has decreased.

    (b) We make a Type I error if the sample evidence leads us to reject H_0 and conclude that the mean price of a single-family home had decreased below

$218,600 when, in fact, it has not decreased below $218,600.

    (c) We make a Type II error if we do not conclude that the mean price of a single-family home has decreased below $218,600, when, in fact, it has decreased below $218,600.

19. **(a)** $H_0 : \sigma = 0.7$ p.s.i. ; $H_1 : \sigma < 0.7$ p.s.i.
    The alternative hypothesis is $\sigma < 0.7$ p.s.i. because the quality control manager believes the standard deviation of the required pressure has been reduced.

    (b) We make a Type I error if the sample evidence leads us to reject H_0 and conclude that the standard deviation in the pressure required is less than 0.7 p.s.i when, in fact, the true standard deviation is 0.7 p.s.i.

    (c) We make a Type II error if we do not reject the null hypothesis that the standard deviation in the pressure required is 0.7 p.s.i when, in fact, the true standard deviation is less than 0.7 p.s.i.

21. **(a)** $H_0 : \mu = \$47.47$; $H_1 : \mu \neq \$47.47$
    The alternative hypothesis is $\mu \neq \$47.47$ because no direction of change is given. The researcher feels the mean monthly bill has changed but this could mean an increase or a decrease.

    (b) We make a Type I error if the sample evidence leads us to reject H_0 and conclude that the mean monthly cell phone bill is not $47.47 when, in fact, it is $47.47.

    (c) We make a Type II error if we do not reject the null hypothesis that the mean monthly cell phone bill is $47.47 when, in fact, it is different than $47.47.

23. There is sufficient evidence to conclude that the proportion of registered births in the U.S. to teenage mothers has increased above 0.105.

25. There is not sufficient evidence to conclude that the mean price of a single-family home has decreased from $218,600.

27. There is not sufficient evidence to conclude that the standard deviation in the pressure required has been reduced from 0.7 p.s.i.

29. There is sufficient evidence to conclude that the mean monthly cell phone bill is different from $47.47.

31. There is not sufficient evidence to conclude that the proportion of registered births to teenage mothers has increased above 0.105.

33. There is sufficient evidence to conclude that the mean price of a single-family home has decreased from $218,600.

35. (a) $H_0 : \mu = 54$ quarts ; $H_1 : \mu > 54$ quarts

   (b) Answers may vary. One possibility follows: Congratulations to the marketing department at popcorn.org. After a marketing campaign encouraging people to consume more popcorn, our researchers have determined that the mean annual consumption of popcorn is now greater than 54 quarts, the mean consumption prior to the campaign.

   (c) A Type I error has been made, since the true mean consumption has not increased above 54 quarts. The probability of making a Type I error is 0.05.

37. (a) $H_0 : p = 0.067$; $H_1 : p < 0.152$

   (b) There is not sufficient evidence to conclude that changes in the DARE program have resulted in a decrease in the proportion of 12-17 year-olds who used marijuana in the past 6 months.

   (c) Since we failed to reject a false null hypothesis, a Type II error was committed.

39. Let $\mu =$ the mean increase in gas mileage for cars using the Platinum Gasaver. Then the hypotheses would be $H_0 : \mu = 0$ versus $H_1 : \mu > 0$.

41. Answers will vary. One possibility follows: If you are going to accuse a company of wrongdoing, you should have fairly convincing evidence. In addition, you likely do not want to find out down the road that your accusations were unfounded. Therefore, it is likely more serious to make a Type I error. For this reason, we should probably make the level of significance $\alpha = 0.01$.

43. As the level of significance, α, decreases, the probability of making a Type II error, β, increases. As we decrease the probability of rejecting a true null hypothesis, we increase the probability of not rejecting the null hypothesis when the alternative hypothesis is true.

45. Answers will vary. In a court of law in the United States, a defendant is presumed innocent until proven guilty. The prosecution must demonstrate "beyond a reasonable doubt" that the defendant committed the crimes of which they are accused. This means that the prosecution must show that the only logical conclusion when faced with the evidence is that the defendant committed the crime. If a court used the standard "beyond all doubt," a defense team could obtain an acquittal by putting forth any bizarre theory explaining why the defendant could not have committed the crime.

To not reject the null hypothesis would be like accepting the assumption that a person accused of a crime is "innocent." This is the original assumption, so we cannot "prove" it "beyond all doubt" using data.

Section 10.2

1. statistically significant

3. False; when the *P*-value is large, we fail to reject the null hypothesis.

5. We want to find the value of Z such that there is an area of 0.1 to the left of the value. Using Table V, we get -1.28.

7. $np_0(1 - p_0) = 200 \cdot 0.3(1 - 0.3) = 42 \geq 10$, so the requirements of the hypothesis test are satisfied.

   (a) $\hat{p} = \dfrac{75}{200} = 0.375$

   The test statistic is

$$z_0 = \frac{\hat{p} - p_0}{\sqrt{\frac{p_0(1-p_0)}{n}}} = \frac{0.375 - 0.3}{\sqrt{\frac{0.3(1-0.3)}{200}}} \approx 2.31 .$$

This is a right-tailed test, so the critical value is $z_{0.05} = 1.645$. Since $z_0 = 2.31 > z_{0.05} = 1.645$, the test statistic is in the critical region, so we reject the null hypothesis.

(b) P-value $= P(Z > 2.31)$

$= 1 - P(Z \le 2.31)$

$= 1 - 0.9896$

$= 0.0104$ [Tech: 0.0103]

Since P-value $= 0.0104 < \alpha = 0.05$, we reject the null hypothesis.

9. $np_0(1-p_0) = 150 \cdot 0.55(1-0.55) = 37.125 \ge 10$, so the requirements of the hypothesis test are satisfied.

(a) $\hat{p} = \frac{78}{150} = 0.52$

The test statistic is

$$z_0 = \frac{\hat{p} - p_0}{\sqrt{\frac{p_0(1-p_0)}{n}}} = \frac{0.52 - 0.55}{\sqrt{\frac{0.55(1-0.55)}{150}}} \approx -0.74 .$$

This is a left-tailed test so the critical value is $-z_{0.1} = -1.28$. Since $z_0 = -0.74 > -z_{0.1} = -1.28$, the test statistic is not in the critical region, so we do not reject the null hypothesis.

(b) P-value $= P(Z < -0.74)$

$= 0.2296$ [Tech: 0.2301]

Since P-value $= 0.2296 > \alpha = 0.1$, we do not reject the null hypothesis.

11. $np_0(1-p_0) = 500 \cdot 0.9(1-0.9) = 45 \ge 10$, so the requirements of the hypothesis test are satisfied.

(a) $\hat{p} = \frac{440}{500} = 0.88$

The test statistic is

$$z_0 = \frac{\hat{p} - p_0}{\sqrt{\frac{p_0(1-p_0)}{n}}} = \frac{0.88 - 0.9}{\sqrt{\frac{0.9(1-0.9)}{500}}} \approx -1.49$$

This is a two-tailed test so the critical values are $\pm z_{0.025} = \pm 1.96$. Since $z_0 = -1.49$ is between $-z_{0.025} = -1.96$ and $z_{0.025} = 1.96$, the test statistic is not in the critical region, so we do not reject the null hypothesis.

(b) P-value $= P(Z < -1.49) + P(Z > 1.49)$

$= 2 \cdot P(-1.49)$

$= 2(0.0681)$

$= 0.1362$ [Tech: 0.1360]

Since P-value $= 0.1362 > \alpha = 0.05$, we do not reject the null hypothesis.

13. About 27 in 100 samples will give a sample proportion as high as or higher than the one obtained if the population proportion really is 0.5. Because the probability is not small, we do not reject the null hypothesis. There is not sufficient evidence to conclude that the dart-picking strategy resulted in a majority of winners.

15. The hypotheses are $H_0 : p = 0.019$ versus $H_1 : p > 0.019$

$np_0(1-p_0) = 863 \cdot 0.019(1-0.019) = 16.1 \ge 10$, so the requirements of the hypothesis test are satisfied. From the survey, $\hat{p} = \frac{19}{863} = 0.022$.

The test statistic is

$$z_0 = \frac{\hat{p} - p_0}{\sqrt{\frac{p_0(1-p_0)}{n}}} = \frac{0.022 - 0.019}{\sqrt{\frac{0.019(1-0.019)}{863}}} \approx 0.65 .$$

The level of significance is $\alpha = 0.01$.

Classical approach:
This is a right-tailed test, so the critical value is $z_{0.01} = 2.33$. Since $z_0 = 0.65 < z_{0.01} = 2.33$, the test statistic is not in the critical region, so we do not reject the null hypothesis.

P-value approach:
P-value $= P(Z > 0.65)$

$= 1 - P(Z \le 0.65)$

$= 1 - 0.7422$

$= 0.2578$ [Tech: 0.2582]

Since P-value $= 0.2578 > \alpha = 0.01$, we do not reject the null hypothesis.

Conclusion: There is not sufficient evidence to conclude that more than 1.9% of Lipitor users experience flulike symptoms as a side effect.

17. Since a majority would constitute a proportion greater than half, the hypotheses are
$H_0 : p = 0.5$ versus $H_1 : p > 0.5$.
$np_0(1 - p_0) = 676(0.5)(1 - 0.5) = 169 \geq 10$, so the requirements of the hypothesis test are satisfied. From the survey, $\hat{p} = \dfrac{352}{676} = 0.521$.

The test statistic is

$$z_0 = \dfrac{\hat{p} - p_0}{\sqrt{\dfrac{p_0(1 - p_0)}{n}}} = \dfrac{0.521 - 0.5}{\sqrt{\dfrac{0.5(1 - 0.5)}{676}}} \approx 1.08$$

The level of significance is $\alpha = 0.05$.

Classical approach:
This is a right-tailed test, so the critical value is $z_{0.05} = 1.645$. Since $z_0 = 1.08 < z_{0.05} = 1.645$, the test statistic is not in the critical region, so we fail to reject the null hypothesis.

P-value approach:
$P\text{-value} = P(Z > 1.08)$
$= 1 - P(Z \leq 1.08)$
$= 1 - 0.8599$
$= 0.1401$ [Tech: 0.1408]

Since $P\text{-value} = 0.1401 > \alpha = 0.05$, we fail to reject the null hypothesis.

Conclusion: There is not sufficient evidence to conclude that a majority of adults believe they will not have enough money for retirement.

19. The hypotheses are $H_0 : p = 0.56$ versus $H_1 : p > 0.56$.
$np_0(1 - p_0) = 480(0.56)(1 - 0.56) \approx 118.3 \geq 10$, so the requirements of the hypothesis test are satisfied. From the sample, $\hat{p} = \dfrac{297}{480} \approx 0.619$.

$$z_0 = \dfrac{\hat{p} - p_0}{\sqrt{\dfrac{p_0(1 - p_0)}{n}}}$$
$$= \dfrac{0.619 - 0.56}{\sqrt{\dfrac{0.56(1 - 0.56)}{480}}}$$
$$\approx 2.60 \quad [\text{Tech: 2.59}]$$

The level of significance is $\alpha = 0.05$.

Classical approach:
This is a right-tailed test, so the critical value is $z_{0.05} = 1.645$. Since $z_0 = 2.60 > z_{0.05} = 1.645$, the test statistic is in the critical region, so we reject the null hypothesis.

P-value approach:
$P\text{-value} = P(Z > 2.60)$
$= 1 - P(Z \leq 2.60)$
$= 1 - 0.9953$
$= 0.0047$ [Tech: 0.0048]

Since $P\text{-value} = 0.0047 < \alpha = 0.05$, we reject the null hypothesis.

Conclusion: There is sufficient evidence to conclude that the percentage of employed adults who feel basic mathematical skills are critical or very important to their job has increased since August, 2003.

21. The hypotheses are $H_0 : p = 0.52$ versus $H_1 : p \neq 0.52$.
$np_0(1 - p_0) = 800(0.52)(1 - 0.52) \approx 200 \geq 10$, so the requirements of the hypothesis test are satisfied. From the sample, $\hat{p} = \dfrac{256}{800} = 0.32$.

$$z_0 = \dfrac{\hat{p} - p_0}{\sqrt{\dfrac{p_0(1 - p_0)}{n}}} = \dfrac{0.32 - 0.52}{\sqrt{\dfrac{0.52(1 - 0.52)}{800}}} \approx -11.32$$

The level of significance is $\alpha = 0.05$.

Classical approach:
This is a two-tailed test, so the critical values are $\pm z_{\alpha/2} = \pm z_{0.025} = \pm 1.96$. Since $z_0 = -11.32 < -z_{0.025} = -1.96$, the test statistic falls in a critical region, so we reject the null hypothesis.

P-value approach:
The test statistic is so far to the left that $P\text{-value} < 0.0001$. Since this P-value is less

than the $\alpha = 0.05$ level of significance, we reject the null hypothesis.

Conclusion: There is sufficient evidence to conclude that the proportion of parents with children in high school who feel it is a serious problem that high school students are not being taught enough math and science has changed from what it was in 1994.

23. The hypotheses are $H_0 : p = 0.47$ versus $H_1 : p \neq 0.47$.

From the sample, $\hat{p} = \dfrac{437}{1013} \approx 0.431$.

$n\hat{p}(1-\hat{p}) = 1013(0.47)(1-0.47) \approx 252.3 \geq 10$, so the requirements for constructing a confidence interval are satisfied. For a 95% confidence interval, we have $\alpha = 0.05$, so $z_{\alpha/2} = z_{0.025} = 1.96$.

Lower Bound $= \hat{p} - z_{\alpha/2} \cdot \sqrt{\dfrac{\hat{p}(1-\hat{p})}{n}}$

$= 0.431 - 1.96 \cdot \sqrt{\dfrac{0.470(1-0.470)}{1013}}$

≈ 0.401

Upper Bound $= \hat{p} + z_{\alpha/2} \cdot \sqrt{\dfrac{\hat{p}(1-\hat{p})}{n}}$

$= 0.431 + 1.96 \cdot \sqrt{\dfrac{0.470(1-0.470)}{1013}}$

≈ 0.462

Since 0.47 is not contained in this interval, we reject the null hypothesis.

Conclusion: There is sufficient evidence to conclude that the parents' attitudes toward the quality of education in the United States have changed since August, 2002.

25. The hypotheses are $H_0 : p = 0.37$ versus $H_1 : p < 0.37$.

$np_0(1-p_0) = 150(0.37)(1-0.37) \approx 35 \geq 10$, so the requirements of the hypothesis test are satisfied. From the survey, $\hat{p} = \dfrac{54}{150} = 0.36$.

$z_0 = \dfrac{\hat{p} - p_0}{\sqrt{\dfrac{p_0(1-p_0)}{n}}} = \dfrac{0.36 - 0.37}{\sqrt{\dfrac{0.37(1-0.37)}{150}}} \approx -0.25$

The level of significance is $\alpha = 0.05$.

Classical approach:
This is a left-tailed test, so the critical value is $-z_{0.05} = -1.645$. Since $z_0 = -0.25 > -z_{0.05} = -1.645$, the test statistic is not in the critical region, so we do not reject the null hypothesis.

P-value approach:
$P\text{-value} = P(Z < -0.25)$

$\qquad\qquad = 0.4013$ [Tech: 0.3999]

Since $P\text{-value} = 0.4013 > \alpha = 0.05$, we do not reject the null hypothesis.

Conclusion: There is not sufficient evidence to conclude that less than 37% of pet owners speak to their pets on the answering machine or telephone.

27. The hypotheses are $H_0 : p = 0.5$ versus $H_1 : p > 0.5$.

$np_0(1-p_0) = 16 \cdot 0.5(1-0.5) = 4 < 10$, so the requirements to use the normal distribution to complete the hypothesis test are not satisfied. We must use the binomial distribution to do the test. From the survey, $\hat{p} = \dfrac{11}{16} = 0.688$.

The level of significance is $\alpha = 0.05$.

P-value approach:
The P-value is the probability of observing 11 or more successes in 16 independent trials, where the probability of success on each of the trials is 0.5.

$P\text{-value} = P(X \geq 11)$

$\qquad\qquad = P(X = 11) + P(X = 12) + \cdots + P(X = 16)$

$\qquad\qquad = 0.0667 + 0.0278 + \cdots + 0.00002$

$\qquad\qquad = 0.1051$

Since $P\text{-value} = 0.1051 > \alpha = 0.10$, we do not reject the null hypothesis.

Conclusion: There is not sufficient evidence to conclude the experimental course was effective.

29. The hypotheses are $H_0 : p = 0.032$ versus $H_1 : p > 0.021$.

$np_0(1-p_0) = 150 \cdot 0.021(1-0.021) \approx 3.1 < 10$, so we use small sample techniques. This is a right-tailed test, so we calculate the probability of 6 or more successes in 150 binomial trials with $p = 0.021$. Using technology:

P-value $= P(X \geq 6)$

$\qquad = 1 - P(X \leq 5)$

$\qquad = 1 - 0.9024$

$\qquad = 0.0976$

Since this is larger than the $\alpha = 0.05$ level of significance, we do not reject the null hypothesis.

Conclusion: There is not sufficient evidence to conclude that more than 2.1% of Americans now work at home.

31. (a) Answers will vary. The proportion could be changing due to sampling error – different people are in the sample. The proportion could also be changing because people's attitudes are changing, perhaps due to issues such as the economy, taxes, foreign policy, etc.

(b) The hypotheses are $H_0 : p = 0.48$ versus $H_1 : p > 0.48$.

$np_0(1 - p_0) = 1500 \cdot 0.48(1 - 0.48) = 374.4 \geq 10$, so the requirements of the hypothesis test are satisfied. From the poll, $\hat{p} = \dfrac{747}{1500} = 0.498$. The test statistic is

$$z_0 = \frac{\hat{p} - p_0}{\sqrt{\dfrac{p_0(1 - p_0)}{n}}} = \frac{0.498 - 0.48}{\sqrt{\dfrac{0.48(1 - 0.48)}{1500}}} \approx 1.40.$$

P-value $= P(Z > 1.40)$

$\qquad = 0.0808$ [Tech: 0.0814]

Conclusion: The final decision depends on the choice of the level of significance, α. If we assume a level of significance of $\alpha = 0.05$, we would not reject the null hypothesis. The proportion of Americans who approved of the job President Obama was doing in January, 2009 was not significantly greater than the level in December, 2010.

[Note: If we were to choose a level of significance larger than the P-value, such as $\alpha = 0.10$, then we would reject the null hypothesis and conclude that there was a significant increase in the proportion of Americans who approve of the job President Obama is doing.]

33. The hypotheses are $H_0 : p = 0.21$ versus $H_1 : p > 0.21$.

$np_0(1 - p_0) = 199(0.21)(1 - 0.21) \approx 33.01 \geq 10$, so the requirements of the hypothesis test are satisfied. From the survey, $\hat{p} = \dfrac{53}{199} \approx 0.266$.

The test statistic is

$$z_0 = \frac{0.266 - 0.21}{\sqrt{\dfrac{0.21(1 - 0.21)}{199}}} = 1.95.$$

P-value $= P(Z > 1.95)$

$\qquad = 1 - P(X \leq 1.95)$

$\qquad = 1 - 0.9744$

$\qquad = 0.0256$ [Tech: 0.0255]

Conclusion: Since this is smaller than the $\alpha = 0.1$ level of significance, we reject the null hypothesis. There is sufficient evidence at the $\alpha = 0.1$ level of significance that more than 21% of Americans consider themselves to be liberal.

35. The hypotheses are $H_0 : p = 0.5$ versus $H_1 : p \neq 0.5$.

$np_0(1 - p_0) = 45(0.5)(1 - 0.5) \approx 11.25 \geq 10$, so the requirements of the hypothesis test are satisfied. From the survey, $\hat{p} = \dfrac{19}{45} \approx 0.422$.

The test statistic is

$$z_0 = \frac{0.422 - 0.5}{\sqrt{\dfrac{0.5(1 - 0.5)}{45}}} = -1.04.$$

P-value $= P(Z < -1.04 \text{ or } Z > 1.04)$

$\qquad = 0.2983$ [Tech: 0.2967]

Conclusion: The final decision depends on the choice of the level of significance, α. In this case, we have a very large P-value. It indicates that about 30 samples in 100 will yield results similar to those found, *if* the null hypothesis is true. This is a very common result, so we fail to reject the null hypothesis. That is, we conclude that there is not sufficient evidence to suggest that the spreads are not accurate. The data suggest that sport books establish accurate spreads.

37. (a) The P-value is greater than the $\alpha = 0.05$ level of significance for values of p_0 between 0.44 and 0.62, inclusive.

p_0	P-value	p_0	P-value
0.42	0.0258	0.54	0.8410
0.43	0.0434	0.55	0.6877
0.44	0.0698	0.56	0.5456
0.45	0.1078	0.57	0.4191
0.46	0.1602	0.58	0.3110
0.47	0.2293	0.59	0.2225
0.48	0.3169	0.60	0.1530
0.49	0.4236	0.61	0.1010
0.50	0.5485	0.62	0.0637
0.51	0.6891	0.63	0.0383
0.52	0.8414	0.64	0.0219
0.53	1.000		

Each of these values of p_0 represents a possible value of the population proportion at the $\alpha = 0.05$ level of significance.

(b) The lower bound of the 95% confidence interval is:

$$\hat{p} - z_{\alpha/2}\sqrt{\frac{\hat{p}(1-\hat{p})}{n}}$$

$$= 0.530 - 1.96\sqrt{\frac{0.530(1-0.530)}{100}}$$

$$= 0.432$$

The upper bound of the 95% confidence interval is:

$$\hat{p} + z_{\alpha/2}\sqrt{\frac{\hat{p}(1-\hat{p})}{n}}$$

$$= 0.530 + 1.96\sqrt{\frac{0.530(1-0.530)}{100}}$$

$$= 0.628$$

The 95% confidence interval is:
$$(0.432, 0.628)$$

(c) At the $\alpha = 0.01$ level of significance, we would not reject the null hypothesis for any of the values of p_0 given in part (a), so that the range of values of p_0 for which we do not reject the null hypothesis increases. The lower value of α means we need more convincing evidence to reject the null hypothesis. So, we would expect a larger range of possible values for the population proportion.

39. (a) Answers will vary.

(b) At the $\alpha = 0.1$ level of significance, we would expect 10 of the 100 samples to result in a Type I error.

(c) Answers will vary. The sample size is small, so this may account for any discrepancies.

(d) In this situation, we know the population proportion: $p_0 = 0.2$.

41. (a) It is important to randomize the order in which the baby is exposed to the toys. This helps avoid any bias.

(b) The hypotheses are $H_0: p = 0.5$ versus $H_1: p < 0.5$.

(c) $np_0(1-p_0) = 16(0.5)(1-0.5) = 4 < 10$, so we use the binomial distribution to compute the P-value. The P-value is given by:

$$P(X \geq 14) = P(14) + P(15) + P(16)$$
$$= {}_{16}C_{14} \cdot (0.5)^{14}(0.5)^2$$
$$+ {}_{16}C_{15} \cdot (0.5)^{15}(0.5)^1$$
$$+ {}_{16}C_{16} \cdot (0.5)^{16}(0.5)^0$$
$$\approx 0.0018 + 0.0002 + 0.00002$$
$$\approx 0.0021$$

Conclusion: The final decision depends on the choice of the level of significance, α. If we assume a level of significance of $\alpha = 0.05$, we would reject the null hypothesis. There is sufficient evidence to suggest that it is not strictly due to chance that the children were more likely to choose the helper toy.

(d) If the population proportion of babies who choose the helper toy is 0.5, then a sample where all 12 babies choose the helper toy will occur in about 2 out of 10,000 sample of 12 babies.

43. If the P-value for a particular test statistic is 0.23, then we would expect results as extreme as the test statistic in about 23 of 100 similar samples if the null hypothesis is true. Since this event is not unusual, we do not reject the null hypothesis.

45. For the Classical Approach, the calculation of the test statistic is simple but in interpreting the result you need to be very careful when determining the rejecting rejoin in order to interpret the result. For the *P*-value method, the *P*-value can be harder to calculate than the test statistic, but it gives information about the strength of the evidence against the null hypothesis. Most software will calculate both the test statistic and the *P*-value, which eliminates the disadvantages of the *P*-value method. Since the *P*-value is easier to interpret, the *P*-value method is easier to use with technology.

47. Statistical significance means that the results observed in a sample is unusual when the null hypothesis is assumed to be true.

Section 10.3

1. (a) Using Table VI, the critical value is $t_{0.01} = 2.602$.

(b) Using Table VI, the critical value is $-t_{0.05} = -1.729$.

(c) Using Table VI, the critical value is $\pm t_{0.025} = \pm 2.179$.

3. (a) $t_0 = \dfrac{\bar{x} - \mu_0}{s/\sqrt{n}} = \dfrac{47.1 - 50}{10.3/\sqrt{24}} = -1.379$

(b) This is a left-tailed test with $24 - 1 = 23$ degrees of freedom, so the critical value is $-t_{0.05} = -1.714$.

(c)

(d) Since the test statistic is not in the critical region $(-1.379 > -1.714)$, the researcher will not reject the null hypothesis.

5. (a) $t_0 = \dfrac{\bar{x} - \mu_0}{s/\sqrt{n}} = \dfrac{104.8 - 100}{9.2/\sqrt{23}} = 2.502$

(b) This is a two-tailed test with $23 - 1 = 22$ degrees of freedom, so the critical values are $\pm t_{0.005} = \pm 2.819$.

(c)

(d) Since the test statistic is not in the critical region $(-2.819 < 2.502 < 2.819)$, the researcher will not reject the null hypothesis.

(e) Using $\alpha = 0.01$ with 22 degrees of freedom, we have $t_{\alpha/2} = \pm t_{0.005} = \pm 2.819$.

$$\text{Lower bound} = \bar{x} - t_{\alpha/2} \cdot \frac{s}{\sqrt{n}}$$

$$= 104.8 - 2.819 \cdot \frac{9.2}{\sqrt{23}}$$

$$\approx 99.39$$

$$\text{Upper bound} = \bar{x} + t_{\alpha/2} \cdot \frac{s}{\sqrt{n}}$$

$$= 104.8 + 2.819 \cdot \frac{9.2}{\sqrt{23}}$$

$$\approx 110.21$$

Because this confidence interval includes the hypothesized mean of 100, the researcher will not reject the null hypothesis.

7. (a) $t_0 = \dfrac{\bar{x} - \mu_0}{s/\sqrt{n}} = \dfrac{18.3 - 20}{4.3/\sqrt{18}} = -1.677$.

(b)

(c) This is a left-tailed test with $18 - 1 = 17$ degrees of freedom. The *P*-value is the area under the *t*-distribution to the left of the test statistic, $t_0 = -1.677$. Because of symmetry, the area under the distribution to the left of -1.677 equals the area under the distribution to the right of 1.677. From the *t*-distribution table (Table VI) in the row corresponding to 17 degrees of freedom, since $t = 1.677$ is between 1.333 and 1.740, whose right-tail areas are 0.10 and 0.05, respectively. So, $0.05 < P\text{-value} < 0.10$ [Tech: 0.0559].

Interpretation: If we obtain 100 random samples of size 18, we would expect about 6 of the samples to result in a sample mean of 18.3 or less *if* the population mean is $\mu = 20$.

(d) Since $P\text{-value} > \alpha = 0.05$, the researcher will not reject the null hypothesis.

9. (a) No, because this sample is large ($n \geq 30$).

(b) $t_0 = \dfrac{\bar{x} - \mu_0}{s / \sqrt{n}} = \dfrac{101.9 - 105}{5.9 / \sqrt{35}} = -3.108$.

(c)

The sum of the areas is the *P*-value

−3.108   3.108

(d) This is a two-tailed test with $35 - 1 = 34$ degrees of freedom. The *P*-value of this two tailed test is the area to the left of $t_0 = -3.108$, plus the area to the right of 3.108. From the *t*-distribution table (Table VI) in the row corresponding to 34 degrees of freedom, 3.108 falls between 3.002 and 3.348, whose right-tail areas are 0.0025 and 0.001, respectively. We must double these values in order to get the total area in both tails: 0.005 and 0.002. Thus, $0.002 < P\text{-value} < 0.005$ [Tech: $P\text{-value} = 0.0038$].

Interpretation: If we obtain 1000 random samples of size 35, we would expect about 4 of the samples to result in a sample mean as extreme or more extreme than the one observed *if* the population mean is $\mu = 105$.

(e) Since $P\text{-value} < \alpha = 0.010$, we reject the null hypothesis.

11. (a) $H_0 : \mu = \$67$ versus $H_1 : \mu > \$67$

(b) A *P*-value of 0.02 means that there is a 0.02 probability of obtaining a sample mean of $73 or higher from a population whose mean is $67. So, if we obtained 100 simple random samples of size $n = 40$ from a population whose mean is $67, we

would expect about 2 of these samples to result in sample means of $73 or higher.

(c) Since $P\text{-value} = 0.02 < \alpha = 0.05$, we reject the statement in the null hypothesis. There is sufficient evidence to conclude that the mean dollar amount withdrawn from a PayEase ATM is more than the mean amount from a standard ATM. That is, the mean dollar amount withdrawn from a PayEase ATM is more than $67.

13. (a) $H_0 : \mu = 22$ versus $H_1 : \mu > 22$

(b) The sample is random, and the sample size is large: $n > 30$. We can reasonable assume that the sample is small relative to the population, so the observations are independent. The assumptions to conduct the *t*-test appear to be satisfied.

(c) $t_0 = \dfrac{\bar{x} - \mu_0}{s / \sqrt{n}} = \dfrac{22.6 - 22}{3.9 / \sqrt{200}} = 2.176$

$\alpha = 0.05$;

d.f. $= n - 1 = 200 - 1 = 199$

Classical approach:
Because this is a right-tailed test with 199 degrees of freedom, the critical value is $t_{0.05} = 1.660$. Since $t_0 = 2.176 > t_{0.05} = 1.660$, the test statistic falls inside the critical region. We reject the null hypothesis.

P-value approach:
This is a right-tailed test with 199 degrees of freedom. The *P*-value is the area under the *t*-distribution to the right of the test statistic, $t_0 = 2.176$. From the *t*-distribution table (Table VI) in the row corresponding to 100 degrees of freedom (the closest table value to 199,) 2.176 falls between 2.081 and 2.364, , whose right-tail areas are 0.02 and 0.01, respectively. So, $0.01 < P\text{-value} < 0.02$ [Tech: 0.0154]. Since $P\text{-value} < \alpha = 0.05$, we reject the null hypothesis.

(d) Conclusion: There is sufficient evidence to conclude that the mean ACT score is greater than 22 for the students who complete the core curriculum.

15. Hypotheses: $H_0 : \mu = 9.02$ cm^3 versus

$H_1 : \mu < 9.02$ cm^3

Test Statistic:

$$t_0 = \frac{\bar{x} - \mu_0}{s / \sqrt{n}} = \frac{8.10 - 9.02}{0.7 / \sqrt{12}} = -4.553$$

$\alpha = 0.01$; d.f. $= n - 1 = 12 - 1 = 11$

Classical approach:

Because this is a left-tailed test with 11 degrees of freedom, the critical value is $-t_{0.01} = -2.718$. Since $t_0 = -4.553 < -t_{0.01} = -2.718$, the test statistic falls within the critical region. We reject the null hypothesis.

P-value approach:

This is a left-tailed test with 11 degrees of freedom. The P-value is the area under the t-distribution to the left of the test statistic, $t_0 = -4.553$. Because of symmetry, the area under the distribution to the left of -4.553 equals the area under the distribution to the right of 4.553. From the t-distribution table (Table VI) in the row corresponding to 11 degrees of freedom, 4.553 falls to the right of 4.437, whose right-tail area is 0.0005. So, P-value < 0.0005 [Tech: P-value= 0.0004]. Since P-value $< \alpha = 0.01$, we reject the null hypothesis.

Conclusion: There is sufficient evidence to conclude that the mean hippocampal volume in alcoholic adolescents is less than the normal volume of 9.02 cm^3.

17. Hypotheses: $H_0 : \mu = 703.5$ versus

$H_1 : \mu > 703.5$

Test Statistic

$$t_0 = \frac{\bar{x} - \mu_0}{s / \sqrt{n}} = \frac{714.2 - 703.5}{83.2 / \sqrt{40}} = 0.813$$

$\alpha = 0.05$; d.f. $= n - 1 = 40 - 1 = 39$

Classical approach:

Because this is a right-tailed test with 39 degrees of freedom, the critical value is $t_{0.05} = 1.685$. Since $t_0 = 0.813 < t_{0.05} = 1.685$, the test statistic does not fall within the critical region. We do not reject the null hypothesis.

P-value approach:

This is a right-tailed test with 39 degrees of freedom. The P-value is the area under the t-distribution to the right of the test statistic, $t_0 = 0.813$. From the t-distribution table in the row corresponding to 39 degrees of freedom, 0.813 falls between 0.681 and 0.851, whose right-tail areas are 0.25 and 0.20, respectively. So, $0.20 < P$-value < 0.25 [Tech: P-value = 0.2105]. Since P-value $> \alpha = 0.05$, we do not reject the null hypothesis.

Conclusion: There is not sufficient evidence to conclude that the mean FICO score of high-income individuals is greater than that of the general population. In other words, it is not unlikely to obtain a mean credit score of 714.2 even though the true population mean credit score is 703.4.

19. Hypotheses: $H_0 : \mu = 40.7$ years versus

$H_1 : \mu \neq 40.7$ years

Using $\alpha = 0.05$ with $n - 1 = 32 - 1 = 31$ degrees of freedom, we have $t_{\alpha/2} = t_{0.025} = 2.037$.

$$\text{Lower bound} = \bar{x} - t_{\alpha/2} \cdot \frac{s}{\sqrt{n}}$$

$$= 38.9 - 2.037 \cdot \frac{9.6}{\sqrt{32}}$$

$$= 35.44 \text{ years}$$

$$\text{Upper bound} = \bar{x} + t_{\alpha/2} \cdot \frac{s}{\sqrt{n}}$$

$$= 38.9 + 2.037 \cdot \frac{9.6}{\sqrt{32}}$$

$$= 42.36 \text{ years}$$

Because this interval includes the hypothesized mean 40.7 years, we do not reject the null hypothesis. Thus, there is not sufficient evidence to conclude that the mean age of death-row inmates has change since 2002.

21. **(a)** The plotted data are all within the bounds of the normal probability plot, and the boxplot shows that there are no outliers. Therefore, the conditions for a hypothesis test are satisfied.

(b) Hypotheses: $H_0 : \mu = 84.3$ seconds versus $H_1 : \mu < 84.3$ seconds

We compute the sample mean and sample standard deviation to be $\bar{x} = 78$ seconds and $s = 15.21$ seconds.

The test statistic is

$$t_0 = \frac{\bar{x} - \mu_0}{s / \sqrt{n}} = \frac{78 - 84.3}{15.21 / \sqrt{10}} = -1.310$$

Classical approach:

This is a left-tailed test with $n - 1 = 10 - 1 = 9$ degrees of freedom, so the critical value is $-t_{0.10} = -1.383$. Since $t_0 = -1.310 > -t_{0.10} = -1.383$, the test statistic is not in the critical region, so we do not reject the null hypothesis.

P-value approach:

From the t-distribution table (Table VI) in the row corresponding to 9 degrees of freedom, 1.310 falls between 1.100 and 1.383, whose right-tail areas are 0.15 and 0.1, respectively. So, $0.1 < P$-value < 0.15 [Tech: 0.1113]. Since P-value $> \alpha = 0.1$, we do not reject the null hypothesis.

Conclusion: There is not sufficient evidence to conclude that the new system results in a mean wait time that is less than 84.3 seconds. In other words, there is not sufficient evidence to conclude that the new system is effective.

23. (a) The data are all within the confidence bands of the normal probability plot, which also has a generally linear pattern. The boxplot shows that there are no outliers. Therefore, the conditions for a hypothesis test are satisfied.

(b) Hypotheses: $H_0 : \mu = 0.11$ mg/L versus $H_1 : \mu \neq 0.11$ mg/L
We compute the sample mean to be $\bar{x} = 0.1568$ mg/L.

Test Statistic:

$$t_0 = \frac{\bar{x} - \mu_0}{s / \sqrt{n}} = \frac{0.1568 - 0.11}{0.087 / \sqrt{10}} \approx 1.707$$

Classical approach:

This is a two-tailed test, so the critical values are $\pm t_{\alpha/2} = \pm t_{0.025} = \pm 2.262$. Since $t_0 = 1.707$ is between $-t_{0.05} = -2.262$ and $t_{0.05} = 2.262$, the test statistic does not fall in the critical region. Therefore, we do not reject the null hypothesis.

P-value approach:

This is a two-tailed test with $10 - 1 = 9$ degrees of freedom. The P-value of this two tailed test is the area to the right of $t_0 = 1.707$, plus the area to the left of -1.707. From the t-distribution table (Table VI) in the row corresponding to 9 degrees of freedom, 1.7078 falls between 1.383 and 1.833, whose right-tail areas are 0.10 and 0.05, respectively. We must double these values in order to get the total area in both tails: 0.20 and 0.10. Thus, $0.10 < P$-value < 0.20 [Tech: P-value = 0.1220].

Conclusion: There is not sufficient evidence to indicate that the calcium concentration in rainwater in Chautauqua, New York, has changed since 1990.

25. (a) The data are skewed to the right, and there are outliers.

(b) Hypotheses: $H_0 : \mu = 35.1$ versus $H_1 : \mu \neq 35.1$
Test Statistic:

$$t_0 = \frac{\bar{x} - \mu_0}{s / \sqrt{n}} = \frac{23.6 - 35.1}{11.7 / \sqrt{40}} = -6.216$$

Classical approach:

This is a two-tailed test, so the critical values are $\pm t_{\alpha/2} = \pm t_{0.025} = \pm 2.023$. Since $t_0 = -6.216 < -t_{0.025} = -2.023$, the test statistic falls in the critical region. Thus, we reject the null hypothesis.

P-value approach:

This is a two-tailed test with $40 - 1 = 39$ degrees of freedom. The P-value of this two tailed test is the area to the left of $t_0 = -6.216$, plus the area to the right of 6.216. From the t-distribution table (Table VI) in the row corresponding to 39 degrees of freedom, 6.216 falls above 3.558, whose right tail is 0.0005. We must double this value in order to get the bound on the total area in both tails: 0.001. Thus, P-value < 0.001 [Tech: P-value < 0.0001].

Conclusion: There is sufficient evidence to indicate that the mean volume of Apple stock has changed since 2007.

27. Hypotheses: $H_0 : \mu = 0.11$ mg/L versus
$H_1 : \mu \neq 0.11$ mg/L
We have $t_{0.01/2} = t_{0.005} = 3.250$. Using this, we compute the lower and upper bounds of the confidence interval.
Lower bound:

$$\bar{x} - t_0 \frac{s}{\sqrt{n}} = 0.1568 - 3.250 \frac{0.087}{\sqrt{10}} \approx 0.0677$$

Upper bound:

$$\bar{x} + t_0 \frac{s}{\sqrt{n}} = 0.1568 + 3.250 \frac{0.087}{\sqrt{10}} \approx 0.2459$$

So, the 99% confidence interval is:
$(0.0677, 0.2459)$.

Because 0.11 is in the 99% confidence interval, we do not reject the statement in the null hypothesis. There is not sufficient evidence to indicate that the calcium concentration in rainwater in Chautauqua, New York, has changed since 1990.

29. Hypotheses: $H_0 : \mu = 35.1$ versus
$H_1 : \mu \neq 35.1$
We have $t_{0.05/2} = t_{0.025} = 2.023$. Using this, we compute the lower and upper bounds of the confidence interval.
Lower bound:

$$\bar{x} - t_0 \frac{s}{\sqrt{n}} = 23.6 - 2.023 \frac{11.7}{\sqrt{40}} \approx 19.858$$

Upper bound:

$$\bar{x} + t_0 \frac{s}{\sqrt{n}} = 23.6 + 2.023 \frac{11.7}{\sqrt{40}} \approx 27.342$$

So, the 99% confidence interval is:
$(19.858, 27.342)$.

Because 35.1 is not in the 95% confidence interval, we reject the statement in the null hypothesis. The evidence suggests that the volume of Apple stock has changed since 2007.

31. (a) $H_0 : \mu = 515$ versus $H_1 : \mu > 515$

(b) The test statistic is

$$t_0 = \frac{\bar{x} - \mu_0}{s / \sqrt{n}} = \frac{519 - 515}{111 / \sqrt{1800}} = 1.53$$

Classical approach:
This is a right-tailed test. There are $n - 1 = 1800 - 1 = 1799$ degrees of freedom. The closest number of degrees of freedom on Table VI is 1000. We will

use 1000 degrees of freedom. Our critical value is $t_\alpha = t_{0.10} = 1.282$. Since $t_0 = 1.53 > t_{0.10} = 1.282$, the test statistic falls in the critical region. Thus, we reject the null hypothesis.

P-value approach:
$$\begin{aligned} P\text{-value} &= P(t > 1.53) \\ &= 1 - P(t \leq 1.53) \\ &= 1 - 0.9369 \\ &= 0.0631 \ [\text{Tech: } 0.0632] \end{aligned}$$
Since P-value $= 0.0631 < \alpha = 0.10$, we reject the null hypothesis.

Conclusion: There is sufficient evidence to conclude that the mean score of students taking this review is greater than 515.

(c) Answers will vary. In some states this would be regarded as a highly significant increase, although in most states it is not likely to be thought of as an increase that has any practical significance.

(d) The test statistic is now
$$t_0 = \frac{\bar{x} - \mu_0}{s / \sqrt{n}} = \frac{519 - 515}{111 / \sqrt{400}} = 0.72.$$

Classical approach:
This is a right-tailed test. There are $n - 1 = 400 - 1 = 399$ degrees of freedom. The closest number of degrees of freedom on Table VI is 100. We will use 100 degrees of freedom, so our critical value is $t_\alpha = t_{0.10} = 1.290$. Since $t_0 = 0.72 < z_{0.10} = 1.290$, the test statistic does not fall in the critical region. Thus, we do not reject the null hypothesis.

P-value approach:
$$\begin{aligned} P\text{-value} &= P(t > 0.72) \\ &= 1 - P(t \leq 0.72) \\ &= 1 - 0.7640 \\ &= 0.2360 \ [\text{Tech: } 0.2358] \end{aligned}$$
Since P-value $= 0.2360 > \alpha = 0.10$, we do not reject the null hypothesis.

Conclusion: There is not sufficient evidence to conclude that the mean score of students taking this review is greater than 515.

The sample size can dramatically alter the conclusion of a hypothesis test. It is often possible to make even slight changes statistically significant by selecting a large enough sample size.

33. Hypotheses: $H_0 : \mu = 4951$ versus

$H_1 : \mu < 4951$

Test Statistic:

$$t_0 = \frac{\bar{x} - \mu_0}{s / \sqrt{n}} = \frac{2884.32 - 4951}{5908.82 / \sqrt{164}} = -4.479$$

Classical approach:

This is a one-tailed test with $164 - 1 = 163$ degrees of freedom. The closest table value is 100 degrees of freedom, so the critical values are $-t_\alpha = -t_{0.05} = -1.660$.

Since $t_0 = -4.479 < -t_{0.05} = -1.660$, the test statistic falls in the critical region. Thus, we reject the null hypothesis.

P-value approach:

This is a two-tailed test with $164 - 1 = 163$ degrees of freedom. The P-value of this two tailed test is the area to the left of $t_0 = -4.479$. From the t-distribution table (Table VI) in the row corresponding to 100 degrees of freedom, 4.479 falls above 3.390, whose right tail is 0.0005. Thus, P-value < 0.0005 [Tech: P-value < 0.0001].

Conclusion: There is sufficient evidence to indicate that the mean credit card debt is less than \$4951.

35. (a) The Director of Institutional Research would be testing if the mean IQ of JJC students is greater than 100.

Hypotheses: $H_0 : \mu = 100$ versus

$H_1 : \mu > 100$

Test Statistic:

$$t_0 = \frac{\bar{x} - \mu_0}{s / \sqrt{n}} = \frac{103.4 - 100}{13.2 / \sqrt{40}} = 1.629$$

Classical approach:

This is a one-tailed test with $40 - 1 = 39$ The critical values are $t_\alpha = t_{0.05} = 2.023$.

Since $t_0 = 1.629 < t_{0.05} = 2.023$, the test statistic does not fall in the critical region. Thus, we fail to reject the null hypothesis.

P-value approach:

This is a one-tailed test with $40 - 1 = 39$

degrees of freedom. The P-value of this \ test is the area to the right of $t_0 = 1.629$.

From the t-distribution table (Table VI) in the row corresponding to 39 degrees of freedom, 1.629 falls between 1.304 and 1.685, which correspond to areas of .10 and 0.05, respectively. Thus, $0.05 < P$-value < 0.10 [Tech: P-value $= 0.0557$].

Conclusion: There is not sufficient evidence to indicate that the mean IQ of JJC students is greater than 100.

(b) The Director of Institutional Research would be testing if the mean IQ of JJC students is greater than 101.

Hypotheses: $H_0 : \mu = 101$ versus

$H_1 : \mu > 101$

Test Statistic:

$$t_0 = \frac{\bar{x} - \mu_0}{s / \sqrt{n}} = \frac{103.4 - 101}{13.2 / \sqrt{40}} = 1.150$$

Classical approach:

This is a one-tailed test with $40 - 1 = 39$ The critical values are $t_\alpha = t_{0.05} = 2.023$.

Since $t_0 = 1.150 < t_{0.05} = 2.023$, the test statistic does not fall in the critical region. Thus, we fail to reject the null hypothesis.

P-value approach:

This is a one-tailed test with $40 - 1 = 39$ degrees of freedom. The P-value of this test is the area to the right of $t_0 = 1.150$.

From the t-distribution table (Table VI) in the row corresponding to 39 degrees of freedom, 1.150 falls between 1.050 and 1.304, which correspond to areas of .15 and 0.10, respectively. Thus, $0.10 < P$-value < 0.15 [Tech: P-value $= 0.1286$].

Conclusion: There is not sufficient evidence to indicate that the mean IQ of JJC students is greater than 101.

(c) The Director of Institutional Research would be testing if the mean IQ of JJC students is greater than 102.

Hypotheses: $H_0 : \mu = 101$ versus

$H_1 : \mu > 101$

Test Statistic:

$$t_0 = \frac{\overline{x} - \mu_0}{s / \sqrt{n}} = \frac{103.4 - 102}{13.2 / \sqrt{40}} = 0.671$$

Classical approach:
This is a one-tailed test with $40 - 1 = 39$
The critical values are $t_\alpha = t_{0.05} = 2.023$.
Since $t_0 = 0.671 < t_{0.05} = 2.023$, the test statistic does not fall in the critical region. Thus, we fail to reject the null hypothesis.

P-value approach:
This is a one-tailed test with $40 - 1 = 39$ degrees of freedom. The *P*-value of this test is the area to the right of $t_0 = 0.671$
From the *t*-distribution table (Table VI) in the row corresponding to 39 degrees of freedom, 0.671 falls below 0.681, which correspond to an area of 0.25. Thus, $0.25 < P$-value [Tech: *P*-value = 0.2532].

Conclusion: There is not sufficient evidence to indicate that the mean IQ of JJC students is greater than 102.

(d) If we "accept" rather than "not reject" the null hypothesis, then we are saying that the population mean is a specific value such as 100 in part (a) or 101 in part (b), or 102 in part (c), and so we have used the same data to conclude that the population mean is equal to three different values. However, if we do not reject the null hypothesis, then we are saying that the population mean could be 100, 101, 102, or even some other value; we are sampling not ruling them out as the possible value of the population mean. "Accepting" the null hypothesis can lead to contradictory conclusions, whereas "not rejecting" the null hypothesis does not

37. (a) Answers will vary.

(b) Answers will vary.

(c) We expect 5 of the simulations will result in a Type I error.

(d) Answers will vary.
The distribution is skewed right, and the sample size is small. This could account for any discrepancies.

39. Yes. Because the head of institutional research has access to the entire population, inference is unnecessary. He can say with 100% confidence that the mean age decreased because the mean age in the current semester is less than the mean age is 1995.

41. Statistical significance means that the sample statistic likely does not come from the population whose parameter is stated in the null hypothesis. Practical significance refers to whether the difference between the sample statistic and the parameter stated in the null hypothesis is large enough to be considered important in an application. A statistically significant result may be of no practical significance.

Section 10.4

1. (a) Using Table VII, we get $\chi^2_{0.05} = 28.869$.

(b) Using Table VII, we get $\chi^2_{0.9} = 14.041$.

(c) Using Table VII, we get $\chi^2_{0.975} = 16.047$ and $\chi^2_{0.025} = 45.722$.

3. (a) df $= 24 - 1 = 23$

$$\chi^2_0 = \frac{(n-1)s^2}{\sigma_0^2} = \frac{23 \cdot (47.2)^2}{(50)^2} = 20.496$$

(b) This is a left-tailed test with 23 degrees of freedom and $\alpha = 0.05$, so the critical value is $\chi^2_{1-\alpha} = \chi^2_{0.95} = 13.091$.

(c)

Critical Region

13.091 $\chi^2_0 = 20.496$

(d) No; the researcher will not reject the hull hypothesis because the test statistic is not in the critical region.

5. (a) df $= 18 - 1 = 17$

$$\chi^2_0 = \frac{(n-1)s^2}{\sigma_0^2} = \frac{17 \cdot (2.4)^2}{(1.8)^2} = 30.222$$

(b) This is a right-tailed test with 17 degrees of freedom and $\alpha = 0.10$, so the critical value is $\chi^2_{0.10} = 24.769$.

(c)

(d) Yes, the researcher will reject H_0 because the test statistic is in the critical region.

7. (a) $df = 12 - 1 = 11$

$$\chi^2_0 = \frac{(n-1)s^2}{\sigma_0^2} = \frac{11 \cdot (4.8)^2}{(4.3)^2} = 13.707$$

(b) This is a two-tailed test with 11 degrees of freedom and $\alpha = 0.05$, so the critical values are $\chi^2_{0.975} = 3.816$ and $\chi^2_{0.025} = 21.920$.

(c)

(d) No, the researcher will not reject H_0 because the test statistic is not in the critical region.

9. The hypotheses are $H_0 : \sigma = 4.0$ versus $H_1 : \sigma < 4.0$. The test statistic is

$$\chi^2_0 = \frac{(n-1)s^2}{\sigma_0^2} = \frac{(25-1) \cdot (3.01)^2}{(4.0)^2} = 13.590.$$

Classical Approach: This is a left-tailed test with 24 degrees of freedom and $\alpha = 0.05$, so the critical value is $\chi^2_{1-\alpha} = \chi^2_{0.95} = 13.848$. Since the test statistic is in the critical region ($13.590 < 13.848$), we reject the null hypothesis.

P-value approach: The P-value is the area under the χ^2-distribution with 24 degrees of freedom to the left of $\chi^2_0 = 13.590$. From the χ^2-distribution table (Table VII) in the row corresponding to 24 degrees of freedom, 13.590 is between 12.401 and 13.848. The areas to the right of these critical values are 0.975 and 0.95, respectively. This means the area to the left of

these critical values are 0.025 and 0.05, respectively. So, $0.025 < P\text{-value} < 0.05$ [Tech: 0.0446]. Since $P\text{-value} < \alpha = 0.05$, we reject the null hypothesis.

Conclusion: There is sufficient evidence to conclude that the mutual fund has moderate risk.

11. The hypotheses are $H_0 : \sigma = 0.004$ inch versus $H_1 : \sigma < 0.004$ inch. The test statistic is

$$\chi^2_0 = \frac{(n-1)s^2}{\sigma_0^2} = \frac{(25-1) \cdot (0.0025)^2}{(0.004)^2} = 9.375.$$

Classical Approach: This is a left-tailed test with 24 degrees of freedom and $\alpha = 0.01$, so the critical value is $\chi^2_{1-\alpha} = \chi^2_{0.99} = 10.856$. Since the test statistic is in the critical region ($9.375 < 10.856$), we reject the null hypothesis.

P-value approach: The P-value is the area under the χ^2-distribution with 24 degrees of freedom to the left of $\chi^2_0 = 9.375$. From the χ^2-distribution table (Table VII) in the row corresponding to 24 degrees of freedom, 9.375 is to the left of 9.886. The area to the right of this critical value is 0.995, so the area to the left is 0.005. So, $P\text{-value} < 0.005$ [Tech: 0.00339]. Since $P\text{-value} < \alpha = 0.01$, we reject the null hypothesis.

Conclusion: There is sufficient evidence for the manager to conclude that the standard deviation has decreased. The recalibration was effective.

13. The hypotheses are $H_0 : \sigma = 18.0$ versus $H_1 : \sigma < 18.0$. The sample size is $n = 10$. We compute the sample standard deviation to be $s \approx 15.21$. The test statistic is

$$\chi^2_0 = \frac{(n-1)s^2}{\sigma_0^2} = \frac{(10-1) \cdot (15.21)^2}{(18.0)^2} = 6.422$$

Classical Approach: This is a one-tailed test with 9 degrees of freedom and $\alpha = 0.05$, so the critical values are $\chi^2_{1-\alpha} = \chi^2_{0.95} = 3.325$. Since the test statistic is not in the critical region (6.422 is not less than 3.325), we do not reject the null hypothesis.

P-value Approach: Using technology, we find *P*-value = 0.303. Since this is larger than $\alpha = 0.05$, we do not reject the null hypothesis.

Conclusion: There is not sufficient evidence to conclude that the standard deviation of weight times is less than 18.0.

15. The hypotheses are $H_0 : \sigma = 8.3$ points , $H_1 : \sigma < 8.3$ points . The test statistic is

$$\chi_0^2 = \frac{(n-1)s^2}{\sigma_0^2} = \frac{(25-1)\cdot(6.7)^2}{(8.3)^2} = 15.639 .$$

Classical Approach: This is a left-tailed test with 24 degrees of freedom and $\alpha = 0.10$, so the critical value is $\chi_{0.90}^2 = 15.659$. Since the test statistic is in the critical region (15.639 < 15.659), we reject the null hypothesis.

P-value approach: The *P*-value is the area under the χ^2-distribution with 24 degrees of freedom to the left of $\chi_0^2 = 15.639$. From the χ^2-distribution table (Table VII) in the row corresponding to 24 degrees of freedom, 15.639 is between 13.848 and 15.659. The areas to the right of these critical values are 0.95 and 0.90, respectively. This means the area to the left of these critical values are 0.05 and 0.10, respectively. So, $0.05 < P$-value < 0.10 [Tech: 0.0993]. Since *P*-value $< \alpha = 0.10$, we reject the null hypothesis.

Conclusion: There is sufficient evidence to conclude that Allen Iverson is more consistent than other shooting guards in the NBA.

17. (a) The points lie within the bounds of the normal plot and have a generally linear pattern.

(b) Using technology, we compute the sample standard deviation to be $s = 2.059$ inches .

(c) The hypotheses are $H_0 : \sigma = 2.9$ inches versus $H_1 : \sigma < 2.9$ inches . The test statistic is

$$\chi_0^2 = \frac{(n-1)s^2}{\sigma_0^2} = \frac{(20-1)\cdot(2.059)^2}{(2.9)^2} = 9.578 .$$

Classical Approach: This is a left-tailed test with 19 degrees of freedom and $\alpha = 0.01$, so the critical value is $\chi_{0.99}^2 = 7.633$. Since the test statistic is not in the critical region (9.578 > 7.633), we do not reject the null hypothesis.

P-value approach: The *P*-value is the area under the χ^2-distribution with 19 degrees of freedom to the left of $\chi_0^2 = 9.578$.

From the χ^2-distribution table (Table VII) in the row corresponding to 19 degrees of freedom, 9.578 is between 8.907 and 10.117. The areas to the right of these critical values are 0.975 and 0.95, respectively. This means that the areas to the left of these critical values are 0.025 and 0.05, respectively. So, $0.025 < P$-value < 0.05 [Tech: 0.0374]. Since *P*-value $> \alpha = 0.01$, we do not reject the null hypothesis.

Conclusion: There is not sufficient evidence to conclude that the standard deviation of heights of major-league baseball players is less than 2.9 inches.

Section 10.5

1. Hypotheses: $H_0 : \mu = 1.0$ versus $H_1 : \mu < 1.0$

Test statistic: $t_0 = \dfrac{\overline{x} - \mu_0}{s/\sqrt{n}} = \dfrac{0.8 - 1.0}{0.4/\sqrt{19}} = -2.179$

Classical approach:
This is a left-tailed test with $19 - 1 = 18$ degrees of freedom and $\alpha = 0.01$, so the critical value is $-t_{0.01} = -2.552$. Since $-2.179 > -2.552$, the test statistic does not fall in the critical region, so we do not reject the null hypothesis.

P-value approach:
From the *t*-distribution table (Table VI) in the row corresponding to 18 degrees of freedom, 2.179 is between 2.101 and 2.214 whose right-tail areas are 0.025 and 0.02, respectively. So, $0.02 < P\text{-value} < 0.025$ [Tech: 0.0214]. Since $P\text{-value} > \alpha = 0.01$, we do not reject the null hypothesis.

Conclusion:
There is not sufficient evidence to conclude that the population mean is less than 1.0.

3. Hypotheses: $H_0 : \sigma^2 = 95$ versus $H_1 : \sigma^2 < 95$
Test statistic:

$$\chi_0^2 = \frac{(n-1)s^2}{\sigma_0^2} = \frac{(20-1) \cdot 49.3}{95} = 9.86$$

Classical Approach:
This is a left-tailed test with 19 degrees of freedom and $\alpha = 0.1$, so the critical value is $\chi_{1-\alpha}^2 = \chi_{0.90}^2 = 11.651$. Since the test statistic falls in the critical region (9.86 < 11.651), we reject the null hypothesis.

P-value approach:
The *P*-value is the area under the χ^2-distribution with 19 degrees of freedom to the left of $\chi_0^2 = 9.86$. From the χ^2-distribution table (Table VII) in the row corresponding to 19 degrees of freedom, 9.86 is between 8.907 and 10.117. The areas to the right of these critical values are 0.975 and 0.95, respectively. This means the areas to the left of these critical values are 0.025 and 0.05, respectively. So, $0.025 < P\text{-value} < 0.05$ [Tech: 0.0437]. Since $P\text{-value} < \alpha = 0.1$, we reject the null hypothesis.

Conclusion: There is sufficient evidence to conclude that the variance is less than 95.

5. Hypotheses: $H_0 : \sigma^2 = 10$ versus $H_1 : \sigma^2 > 10$
Test statistic:

$$\chi_0^2 = \frac{(n-1)s^2}{\sigma_0^2} = \frac{(16-1) \cdot 13.7}{10} = 20.55$$

Classical Approach:
This is a right-tailed test with 15 degrees of freedom and $\alpha = 0.05$, so the critical value is $\chi_{0.05}^2 = 24.996$. Since the test statistic does

not fall in the critical region (20.55 < 24.996), we do not reject the null hypothesis.

P-value approach:
The *P*-value is the area under the χ^2-distribution with 15 degrees of freedom to the right of $\chi_0^2 = 20.55$. From the χ^2-distribution table (Table VII) in the row corresponding to 15 degrees of freedom, 20.55 is to the left of 22.307. The area to the left of this critical value is 0.10. So, $P\text{-value} > 0.10$ [Tech: 0.1518]. Since $P\text{-value} > \alpha = 0.05$, we do not reject the null hypothesis.

Conclusion:
There is not sufficient evidence to conclude that the population variance is greater than 10.

7. Hypotheses: $H_0 : \mu = 100$ versus $H_1 : \mu > 100$

Test Statistic: $t_0 = \dfrac{\overline{x} - \mu_0}{s / \sqrt{n}} = \dfrac{108.5 - 100}{17.9 / \sqrt{40}} = 3.003$

Classical approach:
This is a right-tailed test with $40 - 1 = 39$ degrees of freedom and $\alpha = 0.05$, so the critical value is $t_{0.05} = 1.685$. Since $t_0 = 3.003 > t_{0.05} = 1.685$, the test statistic falls in the critical region, so we reject the null hypothesis.

P-value approach:
This is a right-tailed test with 39 degrees of freedom. The *P*-value is the area to the right of $t_0 = 3.003$. From the *t*-distribution table (Table VI) in the row corresponding to 39 degrees of freedom, 3.003 falls between 2.976 and 3.313, whose right-tail areas are 0.0025 and 0.001, respectively. So, $0.0025 < P\text{-value} < 0.001$ [Tech: 0.0023]. Since $P\text{-value} < \alpha = 0.05$, we reject the null hypothesis.

Conclusion:
There is sufficient evidence to conclude that the population mean is more than 100.

9. Hypotheses: $H_0 : \mu = 100$ versus $H_1 : \mu > 100$

Test Statistic: $t_0 = \dfrac{\overline{x} - \mu_0}{s / \sqrt{n}} = \dfrac{104.2 - 100}{14.7 / \sqrt{20}} = 1.28$

Classical approach:
This is a right-tailed test with $\alpha = 0.05$, so the

critical values are $t_{0.05} = 1.729$. Since $t_0 = 1.28 < t_{0.05} = 1.729$, the test statistic does not fall in the critical region, so we do not reject the null hypothesis.

P-value approach:
This is a right-tailed test with 19 degrees of freedom. The P-value is the area to the right of $t_0 = 1.28$. From the t-distribution table (Table VI) in the row corresponding to 19 degrees of freedom, 1.28 falls between 1.066 and 1.328, whose right-tail areas are 0.15 and 0.10, respectively. So, $0.10 < P\text{-value} < 0.15$ [Tech: 0.1084]. Since $P\text{-value} > \alpha = 0.05$, we do not reject the null hypothesis.

Conclusion:
There is not sufficient evidence to conclude that mothers who listen to Mozart have children with higher IQs.

11. The hypotheses are $H_0 : \sigma = 7000$ psi versus $H_1 : \sigma > 7000$ psi. The test statistic is

$$\chi_0^2 = \frac{(n-1)s^2}{\sigma_0^2} = \frac{(20-1) \cdot (7500)^2}{(7000)^2} = 21.811.$$

Classical Approach: This is a right-tailed test with 19 degrees of freedom and $\alpha = 0.01$, so the critical value is $\chi_{0.01}^2 = 36.191$. Since the test statistic is not in the critical region $(21.811 < 36.191)$, we do not reject the null hypothesis.

P-value approach: The P-value is the area under the χ^2-distribution with 19 degrees of freedom to the right of $\chi_0^2 = 21.811$. From the χ^2-distribution table (Table VII) in the row corresponding to 19 degrees of freedom, 21.811 falls to the left of 27.204. The area to the right of this critical value is 0.10, so $P\text{-value} > 0.10$ [Tech: 0.2938]. Since $P\text{-value} > \alpha = 0.01$, we do not reject the null hypothesis.

Conclusion: There is not sufficient evidence to conclude that the standard deviation of yield strength exceeds 7000 psi.

13. Hypotheses: $H_0 : p = 0.23$ versus $H_1 : p \neq 0.23$
$np_0(1-p_0) = 1026 \cdot 0.23(1-0.23) \approx 181.7 \geq 10$, so the requirements for the hypothesis test are

satisfied. From the survey, $\hat{p} = \dfrac{254}{1026} \approx 0.248$.
The test statistic is

$$z_0 = \frac{\hat{p} - p_0}{\sqrt{\dfrac{p_0(1-p_0)}{n}}} = \frac{0.248 - 0.23}{\sqrt{\dfrac{0.23(1-0.23)}{1026}}} = 1.37$$

[Tech: 1.34].

Classical approach: This is a two-tailed test with $\alpha = 0.1$, so the critical values are $\pm z_{0.05} = \pm 1.645$. Since $z_0 = 1.37$ is between $-z_{0.05} = -1.645$ and $z_{0.05} = 1.645$, the test statistic does not fall in the critical region, so we do not reject the null hypothesis.

P-value approach:
$$\begin{aligned} P\text{-value} &= P(Z < -1.37) + P(Z > 1.37) \\ &= 2 \cdot P(Z < -1.37) \\ &= 2 \cdot (0.0853) \\ &= 0.1706 \quad [\text{Tech: } 0.1813] \end{aligned}$$

Since $P\text{-value} = 0.1706 > \alpha = 0.1$, we do not reject the null hypothesis.

Conclusion:
There is not sufficient evidence to conclude that the percent of American university undergraduate students with at least one tattoo has changed since 2001.

15. (a) The data appear to be normally distributed and no outliers have been detected. The requirements are satisfied.

(b) The hypotheses are $H_0 : \mu = 10,000$ versus $H_1 : \mu < 10,000$. We compute the sample mean and sample standard deviation to be $\bar{x} \approx 9796.3$ and $s \approx 567.7$. The test statistic is

$$t_0 = \frac{\bar{x} - \mu_0}{s/\sqrt{n}} = \frac{9796.3 - 10,000}{567.7/\sqrt{14}} = -1.343.$$

Classical approach:
This is a left-tailed test with $14 - 1 = 13$ degrees of freedom, so the critical value is $-t_{0.05} = -1.771$. Since $t_0 = -1.343 > -t_{0.005} = -1.771$, the test statistic does not fall in the critical region, so we do not reject the null hypothesis.

P-value approach: This is a left-tailed test with 13 degrees of freedom, so the P-value is the area to the left of $t_0 = -1.343$. From the *t*-distribution table (Table VI) in the row corresponding to 13 degrees of freedom, 1.343 falls between 1.079 and 1.350, whose right-tail areas are 0.15 and 0.10, respectively. So, $0.10 < P\text{-value} < 0.15$ [Tech: 0.1012]. Since $P\text{-value} > \alpha = 0.05$, we do not reject the null hypothesis.

Conclusion:
There is not sufficient evidence to conclude that the toner cartridges have a mean of less than 10,000 copies. The consumer advocate's concerns do not appear to be founded.

17. (a) The data appear to be normally distributed and no outliers have been detected. The requirements are satisfied.

(b) The hypotheses are $H_0 : \mu = 3.2$ days versus $H_1 : \mu > 3.2$ days. We compute the sample mean and sample standard deviation to be $\bar{x} = 4.43$ days and $s = 2.15$ days. The test statistics is

$$t_0 = \frac{\bar{x} - \mu_0}{s/\sqrt{n}} = \frac{4.43 - 3.2}{2.15/\sqrt{12}} = 1.989.$$

Classical approach:
This is a right-tailed test with $12 - 1 = 11$ degrees of freedom and $\alpha = 0.05$, so the critical value is $t_{0.05} = 1.796$. Since $t_0 = 1.989 > t_{0.05} = 1.796$, the test statistic falls in the critical region, so we reject the null hypothesis.

P-value approach:
This is a right-tailed test with 11 degrees of freedom, so the P-value is the area to the right of $t_0 = 1.989$. From the row corresponding to 11 degrees of freedom, 1.989 falls between 1.796 and 2.201, whose right-tail areas are 0.05 and 0.025, respectively. So, $0.025 < P\text{-value} < 0.05$ [Tech: 0.0361]. Since $P\text{-value} < \alpha = 0.05$, we reject the null hypothesis.

Conclusion:
There is sufficient evidence to conclude that the mean time to replace a streetlight for the new contractor is more than 3.2 days. In other words, the new contract is not getting the streetlights replaced as quickly as the previous contractor.

19. Assuming that that congresswoman wants to avoid voting for the tax increase unless she is confident that the majority of her constituents are in favor of it, the level of significance, α, should be small, such as $\alpha = 0.01$ or $\alpha = 0.05$, to avoid committing a Type I error. Let p be the proportion of constituents who are in favor of the tax increase. The hypotheses are $H_0 : p = 0.5$ versus $H_1 : p > 0.5$.

$np_0(1 - p_0) = 8250(0.5)(1 - 0.5) = 2062.5 \geq 10$, so the requirements of the hypothesis test are satisfied. From the survey,

$\hat{p} = \dfrac{4,205}{8,250} \approx 0.510$. The test statistic is

$$z_0 = \frac{\hat{p} - p_0}{\sqrt{\dfrac{p_0(1 - p_0)}{n}}} = \frac{0.510 - 0.5}{\sqrt{\dfrac{0.5(1 - 0.5)}{8250}}} \approx 1.76$$

Classical approach:
If the level of significance is $\alpha = 0.01$, .since this is a right-tailed test, the critical value is $z_{0.01} = 2.326$. Since $z_0 = 1.76 < z_{0.01} = 2.326$,, the test statistic is not in the critical region, so we fail to reject the null hypothesis.

If the level of significance is $\alpha = 0.05$, .since this is a right-tailed test, the critical value is $z_{0.05} = 1.645$. Since $z_0 = 1.76 > z_{0.05} = 1.645$, the test statistic is in the critical region, so we reject the null hypothesis.

P-value approach:
$$P\text{-value} = P(Z > 1.76)$$
$$= 1 - P(Z \leq 1.76)$$
$$= 1 - 0.9608$$
$$= 0.0392 \ [\text{Tech: } 0.0391]$$

If the level of significance is $\alpha = 0.01$, $P\text{-value} = 0.0392 > \alpha = 0.01$, we fail to reject the null hypothesis.

If the level of significance is $\alpha = 0.05$, $P\text{-value} = 0.0392 < \alpha = 0.05$, we reject the null hypothesis.

The *P*-value is greater than 0.01, but less than 0.05. So, the conclusion depends upon the value of α.

Conclusion: If the level of significance is $\alpha = 0.01$, we conclude that there is not sufficient evidence that a majority of the constituents are in favor of the tax increase.

If the level of significance is $\alpha = 0.05$, we conclude that there is sufficient evidence that a majority of the constituents are in favor of the tax increase.

Recommendation: If the level of significance is $\alpha = 0.01$, we would recommend that the congresswoman votes against the tax increase.

If the level of significance is $\alpha = 0.05$, we would recommend that the congresswoman votes in favor of the tax increase.

21. (a) Adding up the frequencies in the table, we find that $n = \Sigma f = 915$. We divide each frequency by this result to obtain the relative frequencies.

Ideal Number of Children	Relative Frequency
0	$\dfrac{15}{915} \approx 0.0164$
1	$\dfrac{31}{915} \approx 0.0339$
2	0.5738
3	0.2798
4	0.0721
5	0.0109
6	0.0044
7	0.0033
8	0.0011
9	0
10	0.0033
11	0.0011

We draw a bar for each group and represent the relative frequencies by the heights of the bars. The distribution is skewed right.

Ideal Number of Children

(b) The mode is 2 because it is the class with the largest frequency.

(c) To find the mean, we use the formula $\bar{x} = \dfrac{\Sigma x_i f_i}{\Sigma f_i}$. To find the standard deviation, we choose to use the computational formula

$$s = \sqrt{s^2} = \sqrt{\dfrac{\Sigma x_i^2 f_i - \dfrac{\left(\Sigma x_i f_i\right)^2}{\Sigma f_i}}{\left(\Sigma f_i\right) - 1}}.$$

We organize our computations of x_i, Σf_i, $\Sigma x_i f_i$, and $\Sigma x_i^2 f_i$ in the table that follows:

x_i	f_i	$x_i f_i$	x_i^2	$x_i^2 f_i$
0	15	0	0	0
1	31	31	1	31
2	525	1050	4	2100
3	256	768	9	2304
4	66	264	16	1056
5	10	50	25	250
6	4	24	36	144
7	3	21	49	147
8	1	8	64	64
9	0	0	81	0
10	3	30	100	300
11	1	11	121	121
$\Sigma f_i = 915$		$\Sigma x_i f_i = 2257$		$\Sigma x_i^2 f_i = 6517$

With the table complete, we compute the sample mean and sample standard deviation:

$$\mu = \dfrac{\Sigma x_i f_i}{\Sigma f_i} = \dfrac{2257}{915} \approx 2.47 \text{ children}$$

$$s = \sqrt{s^2} = \sqrt{\dfrac{\sum x_i^2 f_i - \dfrac{\left(\sum x_i f_i\right)^2}{\sum f_i}}{\left(\sum f_i\right)-1}}$$

$$= \sqrt{\dfrac{6517 - \dfrac{(2257)^2}{915}}{915-1}} \approx 1.02 \text{ children}$$

(d) The distribution is clearly skewed right, not normal. Therefore, a large sample size is necessary in order to apply the Central Limit Theorem.

(e) The hypotheses are $H_0 : \mu = 2.64$ children versus $H_1 : \mu \neq 2.64$ children. The test statistics is

$$t_0 = \dfrac{\overline{x} - \mu_0}{s/\sqrt{n}} = \dfrac{2.47 - 2.64}{1.02/\sqrt{915}} = -5.041$$

[Tech: -5.144]

Classical approach: This is a two-tailed test with $915-1=914$ degrees of freedom and $\alpha = 0.05$. Since our t-distribution table [Table VI] does not contain a row for 914, however, we use 1000 degrees of freedom. The critical values are $\pm t_{0.025} = \pm 1.96$.

Since $t_0 = -5.041 < -t_{0.025} = -1.96$, the test statistic falls within a critical region, so we reject the null hypothesis.

P-value approach: This is a two-tailed test with 914 degrees of freedom, so the P-value is the area to the left of $t_0 = -5.041$, plus the area to the right of 5.041. Again, since our t-distribution table [Table VI] does not contain a row for 914, we use 1000 degrees of freedom. From the row corresponding to 1000 degrees of freedom, 5,041 falls to the right of 3.300, whose right-tail area is 0.0005. We double this to obtain the total area in both tails: $2(0.0005) = 0.001$. So, P-value < 0.001 [Tech: P-value < 0.0001]. Since P-value $< \alpha = 0.05$, we reject the null hypothesis.

Conclusion: There is sufficient evidence to conclude that the mean for the ideal number of children is different from 2.64 children. In other words, the results of the poll indicate that people's beliefs as to the ideal number of children have changed.

Section 10.6

1. A Type II error is made when we fail to reject the null hypothesis when in fact the alternative hypothesis is true.

3. (a) If the null hypothesis is not rejected and the true proportion is less than 0.30, a Type II error has been made.

(b) The null hypothesis will be rejected if $\hat{p}$ is more than 1.645 standard deviations below $p_0 = 0.30$. So, we reject H_0 if

$$\hat{p} < 0.30 - 1.645\sqrt{\tfrac{0.30(1-0.30)}{300}} \approx 0.256.$$

$\beta = P(\text{Type II error})$
$= P(\text{do not reject } H_0 \text{ when } H_1 \text{ is true})$
$= P(\hat{p} > 0.256 \text{ given that } p = 0.28)$

$$= P\left(Z > \dfrac{0.256 - 0.28}{\sqrt{0.28(1-0.28)/300}}\right)$$

$= P(Z > -0.93)$
$= 0.8238$ [Tech: 0.8179]

The power of the test is $1 - \beta = 0.1762$ [Tech: 0.1821].

(c) $\beta = P(\text{Type II error})$
$= P(\text{do not reject } H_0 \text{ when } H_1 \text{ is true})$
$= P(\hat{p} > 0.256 \text{ given that } p = 0.25)$

$$= P\left(Z > \dfrac{0.256 - 0.25}{\sqrt{0.25(1-0.25)/300}}\right)$$

$= P(Z > 0.26)$
$= 0.4052$ [Tech: 0.3977]

The power of the test is $1 - \beta = 0.5948$ [Tech: 0.6023].

5. (a) If the null hypothesis is not rejected and the true proportion is greater than 0.65, a Type II error has been made.

(b) The null hypothesis will be rejected if $\hat{p}$ is more than 2.326 standard deviations above $p_0 = 0.65$. So, we reject H_0 if

$$\hat{p} > 0.65 - 2.326\sqrt{\tfrac{0.65(1-0.65)}{100}} \approx 0.761.$$

$\beta = P(\text{Type II error})$

$= P(\text{do not reject } H_0 \text{ when } H_1 \text{ is true})$

$= P(\hat{p} < 0.761 \text{ given that } p = 0.70)$

$= P\left(Z < \dfrac{0.761 - 0.70}{\sqrt{0.70(1-0.70)/100}}\right)$

$= P(Z < 1.33)$

$= 0.9082 \quad [\text{Tech}: 0.9083]$

The power of the test is $1 - \beta = 0.0918$ [Tech: 0.0917].

(c) $\beta = P(\text{Type II error})$

$= P(\text{do not reject } H_0 \text{ when } H_1 \text{ is true})$

$= P(\hat{p} < 0.761 \text{ given that } p = 0.72)$

$= P\left(Z < \dfrac{0.761 - 0.72}{\sqrt{0.72(1-0.72)/100}}\right)$

$= P(Z < 0.91)$

$= 0.8186 \quad [\text{Tech}: 0.8192]$

The power of the test is $1 - \beta = 0.1814$ [Tech: 0.1808].

7. (a) If the null hypothesis is not rejected and the true proportion is not equal to 0.45, a Type II error has been made.

(b) The null hypothesis will be rejected if $\hat{p}$ is more than 1.645 standard deviations away from $p_0 = 0.45$ (either above or below.) So, we reject H_0 if

$\hat{p} < 0.45 - 1.645\sqrt{\frac{0.45(1-0.45)}{500}} \approx 0.413$ or if

$\hat{p} > 0.45 + 1.645\sqrt{\frac{0.45(1-0.45)}{500}} \approx 0.487$.

$\beta = P(\text{Type II error})$

$= P(\text{do not reject } H_0 \text{ when } H_1 \text{ is true})$

$= P(0.413 < \hat{p} < 0.487 \text{ given } p = 0.49)$

$= P\left(Z < \dfrac{0.487 - 0.49}{\sqrt{0.49(1-0.49)/500}}\right)$

$\quad - P\left(Z < \dfrac{0.413 - 0.49}{\sqrt{0.49(1-0.49)/500}}\right)$

$= P(Z < -0.15) - P(Z < -3.43)$

$= 0.4483 - 0.0003$

$= 0.4480 \quad [\text{Tech}: 0.4392]$

The power of the test is $1 - \beta = 0.5520$ [Tech: 0.5608].

(c) Similarly,

$\beta = P(\text{Type II error})$

$= P(\text{do not reject } H_0 \text{ when } H_1 \text{ is true})$

$= P(0.413 < \hat{p} < 0.487 \text{ given } p = 0.47)$

$= P\left(Z < \dfrac{0.487 - 0.47}{\sqrt{0.47(1-0.47)/500}}\right)$

$\quad - P\left(Z < \dfrac{0.413 - 0.47}{\sqrt{0.47(1-0.47)/500}}\right)$

$= P(Z < 0.74) - P(Z < -2.54)$

$= 0.7764 - 0.0054$

$= 0.7710 \quad [\text{Tech}: 0.7658]$

The power of the test is $1 - \beta = 0.2290$ [Tech: 0.2342].

9. (a) If the null hypothesis is not rejected and the true proportion of adults in the United States who believe they will not have enough money in retirement is greater than 0.5, a Type II error has been made.

(b) The null hypothesis will be rejected if $\hat{p}$ is more than 1.645 standard deviations above $p_0 = 0.5$. So, we reject H_0 if

$\hat{p} > 0.5 + 1.645\sqrt{\frac{0.5(1-0.5)}{676}} \approx 0.532$.

$\beta = P(\text{Type II error})$

$= P(\text{do not reject } H_0 \text{ when } H_1 \text{ is true})$

$= P(\hat{p} < 0.532 \text{ given that } p = 0.53)$

$= P\left(Z < \dfrac{0.532 - 0.53}{\sqrt{0.53(1-0.53)/676}}\right)$

$= P(Z < 0.09)$

$= 0.5398 \quad [\text{Tech}: 0.5339]$

The power of the test is $1 - \beta = 0.4602$ [Tech: 0.4661].

(c) $\beta = P(\text{Type II error})$

$= P(\text{do not reject } H_0 \text{ when } H_1 \text{ is true})$

$= P(\hat{p} < 0.532 \text{ given that } p = 0.55)$

$= P\left(Z < \dfrac{0.532 - 0.55}{\sqrt{0.55(1-0.55)/676}}\right)$

$= P(Z < -0.96)$

$= 0.1736 \quad [\text{Tech}: 0.1685]$

The power of the test is $1 - \beta = 0.8264$ [Tech: 0.8315].

11. (a) If the null hypothesis is not rejected and the true proportion of parents who are satisfied with the quality of education is less than 0.47, a Type II error has been made.

(b) The null hypothesis will be rejected if $\hat{p}$ is more than 1.282 standard deviations below $p_0 = 0.47$. So, we reject H_0 if
$$\hat{p} < 0.47 - 1.645\sqrt{\tfrac{0.47(1-0.47)}{1013}} \approx 0.450.$$

$\beta = P(\text{Type II error})$
$= P(\text{do not reject } H_0 \text{ when } H_1 \text{ is true})$
$= P(\hat{p} > 0.450 \text{ given that } p = 0.42)$
$$= P\left(Z > \frac{0.450 - 0.42}{\sqrt{0.42(1-0.42)/1013}} \right)$$
$= P(Z > 1.93)$
$= 0.0268$ [Tech: 0.0269]

The power of the test is $1 - \beta = 0.9732$ [Tech: 0.9731].

(c) $\beta = P(\text{Type II error})$
$= P(\text{do not reject } H_0 \text{ when } H_1 \text{ is true})$
$= P(\hat{p} > 0.450 \text{ given that } p = 0.46)$
$$= P\left(Z > \frac{0.450 - 0.46}{\sqrt{0.46(1-0.46)/1013}} \right)$$
$= P(Z > -0.64)$
$= 0.7389$ [Tech: 0.7405]

The power of the test is $1 - \beta = 0.2611$ [Tech: 0.2595].

13. (a) If the null hypothesis is not rejected and the true proportion of parents who feel students are not being taught enough math and science is not equal to 0.52, a Type II error has been made.

(b) The null hypothesis will be rejected if $\hat{p}$ is more than 1.96 standard deviations away from $p_0 = 0.52$ (either above or below.) So, we reject H_0 if
$$\hat{p} < 0.52 - 1.96\sqrt{\tfrac{0.52(1-0.52)}{800}} \approx 0.485 \text{ or if}$$
$$\hat{p} > 0.52 + 1.96\sqrt{\tfrac{0.52(1-0.52)}{800}} \approx 0.555.$$

$\beta = P(\text{Type II error})$
$= P(\text{do not reject } H_0 \text{ when } H_1 \text{ is true})$
$= P(0.485 < \hat{p} < 0.555 \text{ given } p = 0.5)$
$$= P\left(Z < \frac{0.555 - 0.5}{\sqrt{0.5(1-0.5)/800}} \right)$$
$$- P\left(Z < \frac{0.485 - 0.5}{\sqrt{0.5(1-0.5)/800}} \right)$$
$= P(Z < 3.09) - P(Z < -0.83)$
$= 0.9991 - 0.1977$
$= 0.8014$ [Tech: 0.7949]

The power of the test is $1 - \beta = 0.1986$ [Tech: 0.2051].

(c) Similarly,
$\beta = P(\text{Type II error})$
$= P(\text{do not reject } H_0 \text{ when } H_1 \text{ is true})$
$= P(0.485 < \hat{p} < 0.555 \text{ given } p = 0.48)$
$$= P\left(Z < \frac{0.555 - 0.48}{\sqrt{0.48(1-0.48)/800}} \right)$$
$$- P\left(Z < \frac{0.485 - 0.48}{\sqrt{0.48(1-0.48)/800}} \right)$$
$= P(Z < 4.22) - P(Z < 0.30)$
$\approx 1.0000 - 0.6103$
$= 0.3897$ [Tech: 0.3803]

The power of the test is $1 - \beta = 0.6103$ [Tech: 0.6197].

15. The null hypothesis will be rejected if $\hat{p}$ is more than 2.326 standard deviations below $p_0 = 0.30$. So, we reject H_0 if
$$\hat{p} < 0.30 - 2.326\sqrt{\tfrac{0.30(1-0.30)}{300}} \approx 0.238.$$

$\beta = P(\text{Type II error})$
$= P(\text{do not reject } H_0 \text{ when } H_1 \text{ is true})$
$= P(\hat{p} > 0.256 \text{ given that } p = 0.28)$
$$= P\left(Z > \frac{0.238 - 0.28}{\sqrt{0.28(1-0.28)/300}} \right)$$
$= P(Z > -1.60)$
$= 0.9474$ [Tech: 0.9455]

The power of the test is $1 - \beta = 0.0526$ [Tech: 0.0545].

When the level of significance was lowered, the power of the test decreased. As we lower the probability of rejecting a true null hypothesis, the probability of not rejecting a false null hypothesis increases, lowering the power of the test.

17. The null hypothesis will be rejected if $\hat{p}$ is more than 1.645 standard deviations below $p_0 = 0.30$. So, we reject H_0 if

$$\hat{p} < 0.30 - 1.645\sqrt{\frac{0.30(1-0.30)}{300}} \approx 0.256 \,.$$

$p = 0.2$:

$\beta = P(\text{Type II error})$
   $= P(\text{do not reject } H_0 \text{ when } H_1 \text{ is true})$
   $= P(\hat{p} > 0.256 \text{ given that } p = 0.2)$

$$= P\left(Z > \frac{0.256 - 0.2}{\sqrt{0.2(1-0.2)/300}} \right)$$

   $= P(Z > 2.45)$
   $= 0.0078$ [Tech: 0.0072]

The power of the test is $1 - \beta = 0.9922$ [Tech: 0.9928].

$p = 0.22$:

$\beta = P(\text{Type II error})$
   $= P(\text{do not reject } H_0 \text{ when } H_1 \text{ is true})$
   $= P(\hat{p} > 0.256 \text{ given that } p = 0.22)$

$$= P\left(Z > \frac{0.256 - 0.22}{\sqrt{0.22(1-0.22)/300}} \right)$$

   $= P(Z > 1.53)$
   $= 0.0655$ [Tech: 0.0636]

The power of the test is $1 - \beta = 0.9345$ [Tech: 0.9364].

$p = 0.24$:

$\beta = P(\text{Type II error})$
   $= P(\text{do not reject } H_0 \text{ when } H_1 \text{ is true})$
   $= P(\hat{p} > 0.256 \text{ given that } p = 0.24)$

$$= P\left(Z > \frac{0.256 - 0.24}{\sqrt{0.24(1-0.24)/300}} \right)$$

   $= P(Z > 0.67)$
   $= 0.2578$ [Tech: 0.2519]

The power of the test is $1 - \beta = 0.7422$ [Tech: 0.7481].

$p = 0.26$:

$\beta = P(\text{Type II error})$
   $= P(\text{do not reject } H_0 \text{ when } H_1 \text{ is true})$
   $= P(\hat{p} > 0.256 \text{ given that } p = 0.26)$

$$= P\left(Z > \frac{0.256 - 0.26}{\sqrt{0.26(1-0.26)/300}} \right)$$

   $= P(Z > -0.14)$
   $= 0.5636$ [Tech: 0.5553]

The power of the test is $1 - \beta = 0.4364$ [Tech: 0.4447].

$p = 0.28$:

$\beta = P(\text{Type II error})$
   $= P(\text{do not reject } H_0 \text{ when } H_1 \text{ is true})$
   $= P(\hat{p} > 0.256 \text{ given that } p = 0.28)$

$$= P\left(Z > \frac{0.256 - 0.28}{\sqrt{0.28(1-0.28)/300}} \right)$$

   $= P(Z > -0.93)$
   $= 0.8238$ [Tech: 0.8179]

The power of the test is $1 - \beta = 0.1762$ [Tech: 0.1821].

$p = 0.29$:

$\beta = P(\text{Type II error})$
   $= P(\text{do not reject } H_0 \text{ when } H_1 \text{ is true})$
   $= P(\hat{p} > 0.256 \text{ given that } p = 0.29)$

$$= P\left(Z > \frac{0.256 - 0.29}{\sqrt{0.29(1-0.29)/300}} \right)$$

   $= P(Z > -1.28)$
   $= 09032$ [Tech: 0.8996]

The power of the test is $1 - \beta = 0.0968$ [Tech: 0.1004].

The values used to create the power curve are:

p	Power
0.2	0.9922
0.22	0.9345
0.24	0.7422
0.26	0.4364
0.28	0.1762
0.29	0.0968

19. (a) The null hypothesis will be rejected if $\hat{p}$ is more than 1.714 standard deviations below $\mu_0 = 50$. So, we reject H_0 if

$$\bar{x} < 50 - 1.714 \frac{6}{\sqrt{24}} \approx 47.9.$$ The sample mean that separates the rejection region from the non-rejection region is 47.9.

(b) $\beta = P(\text{Type II error})$
$= P(\text{do not reject } H_0 \text{ when } H_1 \text{ is true})$
$= P(\bar{x} > 47.9 \text{ given that } \mu = 48.9)$
$= P\left(t > \dfrac{47.9 - 48.9}{6/\sqrt{24}}\right)$
$= P(t > -0.82)$
$= 0.7887 \ [\bar{x} \text{ unrounded: } 0.7885]$

The power of the test is $1 - \beta = 0.2113$
[Unrounded: 0.2115].

21. Increasing the sample size increases the power of the test, assuming all else remains unchanged.

Chapter 10 Review Exercises

1. (a) The hypotheses are $H_0 : \mu = 3173$ versus $H_1 : \mu < 3173$.

(b) We make a Type I error if the sample evidence leads us to reject H_0 and believe that the mean outstanding credit card debt per college undergraduate is less than \$3173 when, in fact, it is \$3173 (that is, we reject H_0 when in fact H_0 is true).

(c) We make a Type II error if we fail to reject the null hypothesis that the mean outstanding credit card debt per college undergraduate is \$3173 when, in fact, it is

less than \$3173 (that is, we do not reject H_0 when in fact H_0 is false).

(d) If the null hypothesis is not rejected, this means there is not enough evidence to support the belief that the mean outstanding credit card debt per college undergraduate is less than \$3173.

(e) If the null hypothesis is rejected, this means there is sufficient evidence to support the belief that the mean outstanding credit card debt per college undergraduate is less than \$3173.

2. (a) The hypotheses are $H_0 : p = 0.13$ versus $H_1 : p \neq 0.13$.

(b) We make a Type I error if the sample evidence leads us to reject H_0 and believe that the percentage of cards that result in default is not different from 13% today, when, in fact, it is 13% (i.e. reject H_0 when H_0 is true).

(c) We make a Type II error if we fail to reject the null hypothesis that the percentage of cards that result in default is 13% today when, in fact, it is different than 13% (i.e. do not reject H_0 when in fact H_0 is false).

(d) If the null hypothesis is not rejected, this means there is not enough evidence to support the credit analyst's suspicion that the percentage of cards that result in default today is different from 13%.

(e) If the null hypothesis is rejected, this means there is sufficient evidence to support the credit analyst's suspicion that the proportion of cards that result in default today is different from 13%.

3. $P(\text{Type I Error}) = \alpha = 0.05$

4. $P(\text{Type II Error}) = \beta = 0.113$;
Power $= 1 - \beta = 1 - 0.113 = 0.887$
The power of the test is the probability of rejecting the null hypothesis when the alternative hypothesis is true.

5. (a) The sample size must be large to use the Central Limit Theorem, because the distribution of the values in the population may not be normal.

(b) Hypotheses: $H_0 : \mu = 100$ versus $H_1 : \mu > 100$

Test Statistic

$$t_0 = \frac{\bar{x} - \mu_0}{s / \sqrt{n}} = \frac{104.3 - 100}{12.4 / \sqrt{35}} = 2.052$$

$\alpha = 0.05$; d.f. $= n - 1 = 35 - 1 = 34$

Classical approach:
Because this is a right-tailed test with 34 degrees of freedom, the critical value is $t_{0.05} = 1.691$. Since $t_0 = 2.052 > t_{0.05} = 1.691$, the test statistic falls within the critical region. We reject the null hypothesis.

P-value approach:
This is a right-tailed test with 34 degrees of freedom. The P-value is the area under the t-distribution to the right of the test statistic, $t_0 = 2.052$. From the t-distribution table in the row corresponding to 34 degrees of freedom, 2.052 falls between 2.032 and 2.136, whose right-tail areas are 0.025 and 0.02, respectively. So, $0.02 < P\text{-value} < 0.025$ [Tech: 0.0240]. Since $P\text{-value} < \alpha = 0.05$, we reject the null hypothesis.

Conclusion: There is not sufficient evidence to indicate that the null hypothesis should be rejected.

6. (a) The distribution must be normal, because the sample size is small. We cannot rely on the Central Limit Theorem to assure that the sample mean is normally distributed.

(b) Hypotheses: $H_0 : \mu = 50$ versus $H_1 : \mu \neq 50$

Test Statistic:

$$t_0 = \frac{\bar{x} - \mu_0}{s / \sqrt{n}} = \frac{48.1 - 50}{4.1 / \sqrt{15}} \approx -1.795$$

There are $n - 1 = 15 - 1 = 14$ degrees of freedom.

Classical approach:
This is a two-tailed test, so the critical values are $\pm t_{\alpha/2} = \pm t_{0.025} = \pm 2.145$. Since $t_0 = -1.795$ is between $-t_{0.05} = -2.145$ and $t_{0.05} = 2.145$, the test statistic does not fall in the critical region. Therefore, we do not reject the null hypothesis.

P-value approach:
This is a two-tailed test. The P-value of this two tailed test is the area to the right of $t_0 = 1.795$, plus the area to the left of -1.795. From the t-distribution table (Table VI) in the row corresponding to 14 degrees of freedom, 1.795 falls between 1.761 and 2.145, whose right-tail areas are 0.05 and 0.025, respectively. We must double these values in order to get the total area in both tails: 0.10 and 0.05. Thus, $0.05 < P\text{-value} < 0.10$ [Tech: $P\text{-value} = 0.0943$].

Conclusion: There is not sufficient evidence to indicate that the null hypothesis should be rejected.

7. $np_0 (1 - p_0) = 250 \cdot 0.6(1 - 0.6) = 60 \geq 10$, so the requirements of the hypothesis test are satisfied.

(a) $\hat{p} = \dfrac{165}{250} = 0.66$

The test statistics is

$$z_0 = \frac{\hat{p} - p_0}{\sqrt{\dfrac{p_0(1 - p_0)}{n}}} = \frac{0.66 - 0.6}{\sqrt{\dfrac{0.6(1 - 0.6)}{250}}} = 1.94.$$

This is a right-tailed test with $\alpha = 0.05$, so the critical value is $z_{0.05} = 1.645$. The test statistic fall in the critical region $(1.94 > 1.645)$, so we reject the null hypothesis.

(b) $P\text{-value} = P(Z > 1.94)$
$= 1 - P(Z < 1.94)$
$= 1 - 0.9738$
$= 0.0262$ [Tech: 0.0264]

Since $P\text{-value} = 0.0262 < \alpha = 0.05$, we reject the null hypothesis.

There is sufficient evidence to conclude that $p > 0.6$.

8. $np_0(1-p_0) = 420 \cdot 0.35(1-0.35) \approx 95.6 \geq 10$, so the requirements of the hypothesis test are satisfied.

(a) $\hat{p} = \dfrac{138}{420} \approx 0.329$

The test statistic is

$z_0 = \dfrac{\hat{p} - p_0}{\sqrt{\dfrac{p_0(1-p_0)}{n}}} = \dfrac{0.329 - 0.35}{\sqrt{\dfrac{0.35(1-0.35)}{420}}} = -0.90$

This is a two-tailed test with $\alpha = 0.01$, so the critical values are $\pm z_{0.005} = \pm 2.575$.

Since $z_0 = -0.90$ is between $-z_{0.005} = -2.575$ and $z_{0.005} = 2.575$, the test statistic does not fall in the critical region, so we do not reject the null hypothesis.

(b) $P\text{-value} = P(Z < -0.90) + P(Z > 0.90)$

$\qquad = 2 \cdot P(Z < -0.90)$

$\qquad = 2(0.1841)$

$\qquad = 0.3682$ [Tech: 0.3572]

Since $P\text{-value} > \alpha = 0.05$, we do not reject the null hypothesis.

There is not sufficient evidence to conclude that $p \neq 0.35$.

9. (a) $\chi_0^2 = \dfrac{(n-1)s^2}{\sigma_0^2} = \dfrac{(18-1) \cdot (4.9)^2}{(5.2)^2} = 15.095$

(b) The hypotheses are $H_0 : \sigma = 5.2$ versus $H_1 : \sigma \neq 5.2$.

Classical Approach: This is a two-tailed test with 17 degrees of freedom and $\alpha = 0.05$, so the critical values are

$\chi_{1-\alpha/2}^2 = \chi_{0.975}^2 = 7.564$ and

$\chi_{\alpha/2}^2 = \chi_{0.025}^2 = 30.191$.

We do not reject the null hypothesis because the test statistic is not in a critical region (15.095 is between 7.564 and 30.191).

P-value Approach: From the chi-square table, we determine that $0.025 < P\text{-value} < 0.05$ Using technology, we find that the *P*-value is 0.0423. Since this is less than $\alpha = 0.05$, we do not reject the null hypothesis.

Conclusion: There is not sufficient evidence to conclude that the standard deviation is not 5.2. In other words, the standard deviation appears to be 5.2.

10. (a) $\chi_0^2 = \dfrac{(n-1)s^2}{\sigma_0^2} = \dfrac{24 \cdot (16.5)^2}{(15.7)^2} = 26.508$

(b) The hypotheses are $H_0 : \sigma = 15.7$ versus $H_1 : \sigma > 15.7$.

Classical Approach: This is a right-tailed test with 24 degrees of freedom and $\alpha = 0.10$, so the critical value is $\chi_{0.10}^2 = 33.196$.

We do not reject the null hypothesis because the test statistic is not in the critical region ($15.095 < 33.196$).

P-value approach: The *P*-value is the area under the χ^2-distribution with 24 degrees of freedom to the right of $\chi_0^2 = 26.508$. From the χ^2-distribution table (Table VII) in the row corresponding to 24 degrees of freedom, 26.508 is between 15.659 and 33.196. The area to the right corresponding to these values is 0.90 and 0.10 respectively. So, $0.10 < P\text{-value} < 0.90$ [Tech: 0.3279]. Since $P\text{-value} > \alpha = 0.05$, we do not reject the null hypothesis.

Conclusion: There is not sufficient evidence to support the representative's belief that the standard deviation is more than 15.7.

11. (a) The hypotheses are $H_0 : p = 0.733$ versus $H_1 : p \neq 0.733$.

(b) $np_0(1-p_0) = 100 \cdot 0.733(1-0.733)$

$$= 19.57 \geq 10$$

so the requirements of the hypothesis test are satisfied.

(c) $\hat{p} = \dfrac{78}{100} = 0.78$

The test statistic is

$$z_0 = \frac{\hat{p}-p_0}{\sqrt{\dfrac{p_0(1-p_0)}{n}}}$$

$$= \frac{0.78-0.733}{\sqrt{\dfrac{0.733(1-0.733)}{100}}} \approx 1.06$$

Classical Approach: This is a two-tailed test so the critical values are $\pm z_{0.025} = \pm 1.96$. Since $z_0 = 1.06$ is between $-z_{0.025} = -1.96$ and $z_{0.025} = 1.96$, the test statistic is not in the critical region, so we do not reject the null hypothesis.

P-value approach: The *P*-value is the area under the standard normal distribution that is left of $-z_0 = -1.06$ or to the right of $z_0 = 1.06$.

$$P\text{-value} = P(Z < -1.06) + P(Z > 1.06)$$

$$= 0.1446 + 0.1446$$

$$= 0.2892 \text{ [Tech: 0.2881]}$$

Since *P*-value $= 0.2892 > \alpha = 0.05$, we do not reject the null hypothesis.

Conclusion: There is not sufficient evidence to contradict Professor's Wilson's findings.

12. The hypotheses are $H_0 : p = 0.05$ versus $H_1 : p > 0.05$.

$np_0(1-p_0) = 250 \cdot 0.05(1-0.05) = 11.88 \geq 10$ so the requirements of the hypothesis test are satisfied.

From the survey, $\hat{p} = \dfrac{17}{250} = 0.068$. The test statistic is

$$z_0 = \frac{\hat{p}-p_0}{\sqrt{\dfrac{p_0(1-p_0)}{n}}} = \frac{0.068-0.50}{\sqrt{\dfrac{0.05(1-0.05)}{250}}} \approx 1.31$$

$$P\text{-value} = P(Z > 1.31)$$

$$= 0.0951 \text{ [Tech: 0.0958]}$$

We would reject the null hypothesis for $\alpha = 0.10$, but we would not reject the null hypothesis for $\alpha = 0.05$ or $\alpha = 0.01$. The administrator should be a little concerned.

13. (a) The sample is random, and the sample size is large: $n > 30$. We can reasonable assume that the sample is small relative to the population, so the observations are independent. The assumptions to conduct the *t*-test appear to be satisfied.

(b) $H_0 : \mu = 0.875$ inch versus
$H_1 : \mu > 0.875$ inch

(c) Recalibrating the machine could be very costly, so the manager wants to avoid the consequences of making a Type I error.

(d) $t_0 = \dfrac{\bar{x}-\mu_0}{s/\sqrt{n}} = \dfrac{0.876-0.875}{0.005/\sqrt{36}} = 1.200$

$\alpha = 0.01$; d.f. $= n-1 = 36-1 = 35$

Classical approach:
Because this is a right-tailed test with 35 degrees of freedom, the critical value is $t_{0.01} = 2.438$. Since $t_0 = 1.200 < t_{0.01} = 2.438$, the test statistic does not fall inside the critical region. We do not reject the null hypothesis.

P-value approach:
This is a right-tailed test with 35 degrees of freedom. The *P*-value is the area under the *t*-distribution to the right of the test statistic, $t_0 = 1.200$. From the *t*-distribution table (Table VI) in the row corresponding to 35 degrees of freedom, 1.200 falls between 1.052 and 1.306, whose right-tail areas are 0.15 and 0.10, respectively. So, $0.10 < P\text{-value} < 0.15$ [Tech: 0.1191]. Since *P*-value $> \alpha = 0.05$, we do not reject the null hypothesis.

(d) There is not sufficient evidence to conclude that the mean distance is greater than 0.875 inch. The machine does not appear to need to be recalibrated.

(e) The quality-control engineer would make a Type I error if he recalibrated the machine when it did not need to be recalibrated. The quality-control engineer would make a Type II error if he did not recalibrate the machine when it actually did need to be recalibrated.

14. Hypotheses: $H_0 : \mu = 98.6°F$ versus $H_1 : \mu < 98.6°F$

Test statistic:

$$t_0 = \frac{\bar{x} - \mu_0}{s / \sqrt{n}} = \frac{98.2 - 98.6}{0.7 / \sqrt{700}} = -15.119.$$

Classical Approach: This is a left-tailed test with $n - 1 =$ 700 – 1 = 699 degrees of freedom. However, since our t-distribution table does not contain a row for 699, we use df = 1000 (the closest one to 699). So, the critical value is $-t_{0.01} = -2.330$. Since $t_0 = -15.119 < \alpha = -2.330$, the test statistic falls in the critical region and we reject the null hypothesis. There is sufficient evidence to conclude that the mean temperature of humans is less than $98.6° F$.

P-value approach: This is a left-tailed test with 699 degrees of freedom. The P-value is the area under the t-distribution to the left of the test statistic $t_0 = -15.119$. Because of symmetry, the area under the distribution to the left of -15.119 equals the area under the distribution to the right of 15.119. From the t-distribution table (Table VI) in the row corresponding to 1000 degrees of freedom (the row closest to df = 699), 15.119 falls to the right of 3.300, whose right-tail area is 0.0005. So, P-value < 0.0005 [Tech: P-value < 0.0001].

Interpretation: There is sufficient evidence to suggest that the mean temperature of humans is less than 98.6 degrees Fahrenheit.

15. (a) The plotted data are all within the bounds of the normal probability plot, which also has a generally linear pattern. The boxplot shows that there are no outliers. Therefore, the conditions for testing the hypothesis are satisfied.

(b) For a 95% confidence interval, we have $\pm t_{\alpha/2} = \pm t_{0.025} = \pm 2.201$.

Lower bound $= \bar{x} - t_{\alpha/2} \cdot \dfrac{s}{\sqrt{n}}$

$$= 1.681 - 2.201 \cdot \frac{0.0045}{\sqrt{12}}$$

$$\approx 1.6782$$

Upper bound $= \bar{x} + t_{\alpha/2} \cdot \dfrac{s}{\sqrt{n}}$

$$= 1.681 + 2.201 \cdot \frac{0.0045}{\sqrt{12}}$$

$$\approx 1.6838$$

Because this confidence interval includes the specified value 1.68, we do not reject the null hypothesis. There is not sufficient evidence to conclude that the mean diameter of Maxfli XS golf balls is different from 1.68 inches. In other words, there is not sufficient evidence to conclude that the golf balls do not conform. Therefore, we will give Maxfli the benefit of the doubt and assume the balls are conforming.

16. Hypotheses: $H_0 : \mu = 480$ versus $H_1 : \mu < 480$

The sample mean is $\bar{x} \approx 419.9$ minutes. The sample standard deviation is $s \approx 126.4$ minutes.

Test statistic:

$$t_0 = \frac{\bar{x} - \mu_0}{s / \sqrt{n}} = \frac{419.9 - 480}{126.4 / \sqrt{11}} = -1.577.$$

Classical approach: This is a left-tailed test with $n - 1 = 11 - 1 = 10$ degrees of freedom. So, the critical value is $-t_{0.05} = -1.812$. Since $t_0 = -1.577 > -t_{0.05} = -1.812$, the test statistic does not fall in the critical region and we do not reject the null hypothesis.

P-value approach: This is a left-tailed test with 10 degrees of freedom. The P-value is the area under the t-distribution to the left of the test statistic $t_0 = -1.577$. From the t-distribution table (Table VI) in the row corresponding to 10 degrees of freedom, 1.577 falls between 1.372 and 1.812, which correspond to right tail areas of 0.10 and 0.05, respectively. So, $0.05 < P$-value < 0.10. [Tech: P-value = 0.0730.]

Conclusion: There is not sufficient evidence to conclude that the mean amount of time students spend studying is less than the recommended 480 minutes per week.

17. Since a majority would constitute a proportion greater than half, the hypotheses are $H_0 : p = 0.5$ versus $H_1 : p > 0.5$.

    $np_0(1 - p_0) = 150(0.5)(1 - 0.5) = 37.5 \geq 10$, so the requirements of the hypothesis test are satisfied. From the sample, $\hat{p} = \dfrac{81}{150} = 0.54$.

    The test statistic is

    $$z_0 = \dfrac{\hat{p} - p_0}{\sqrt{\dfrac{p_0(1 - p_0)}{n}}} = \dfrac{0.54 - 0.5}{\sqrt{\dfrac{0.5(1 - 0.5)}{150}}} \approx 0.98$$

    Classical approach: This is a right-tailed test with $\alpha = 0.05$, so the critical value is $z_{0.05} = 1.645$. Since $z_0 = 0.99 < z_{0.05} = 1.645$, the test statistic does not fall in the critical region, so we do not reject the null hypothesis.

    P-value approach:
    $$\text{P-value} = P(Z > 0.98)$$
    $$= 1 - P(Z \leq 0.98) = 1 - 0.8365$$
    $$= 0.1635 \ [\text{Tech}: 0.1636]$$

    Since $\text{P-value} = 0.1635 > \alpha = 0.05$, we do not reject the null hypothesis.

    Conclusion: There is not sufficient evidence to conclude that a majority of pregnant women nap at least twice each week.

18. **(a)** If we do not reject the null hypothesis and the majority of pregnant women do nap at least twice per week, then we would make a Type II error.

    (b) The null hypothesis will be rejected if $\hat{p}$ is more than 1.645 standard deviations above $p_0 = 0.5$. So, we reject H_0 if
    $$\hat{p} > 0.5 + 1.645\sqrt{\tfrac{0.5(1-0.5)}{150}} \approx 0.567.$$

$\beta = P(\text{Type II error})$
$= P(\text{do not reject } H_0 \text{ when } H_1 \text{ is true})$
$= P(\hat{p} < 0.567 \text{ given that } p = 0.53)$

$= P\left(Z < \dfrac{0.567 - 0.53}{\sqrt{0.53(1 - 0.53)/150}} \right)$

$= P(Z < 0.91)$

$= 0.8186 \ [\text{Tech}: 0.8190]$

The power of the test is $1 - \beta = 0.1814$ [Tech: 0.1810].

19. The hypotheses are $H_0 : \sigma = 505.6$ grams versus $H_1 : \sigma > 505.6$ grams. The test statistic is

    $$\chi_0^2 = \dfrac{(n-1)s^2}{\sigma_0^2} = \dfrac{40 \cdot (840)^2}{(505.6)^2} = 110.409.$$

    Classical Approach: This is a right-tailed test with $n - 1 = 41 - 1 = 40$ degrees of freedom and $\alpha = 0.01$, so the critical value is $\chi_{0.01}^2 = 63.691$. Since the test statistic falls in the critical region ($\chi_0^2 = 110.409 > \chi_{0.01}^2 = 63.691$), we reject the null hypothesis.

    P-value approach: The P-value is the area under the χ^2-distribution with 40 degrees of freedom to the left of $\chi_0^2 = 110.409$. From the χ^2-distribution table (Table VII) in the row corresponding to 40 degrees of freedom, 110.409 is to the right of 66.766. The area to the right of this critical value is 0.005. So, $\text{P-value} < 0.005$ [Tech: < 0.0001]. Since $\text{P-value} < \alpha = 0.01$, we reject the null hypothesis.

    Conclusion: There is sufficient evidence to support the researcher's hypothesis that the variability in the birth weight of preterm babies is more than the variability in the birth weight of full-term babies.

20. The hypotheses are $H_0 : p = 0.52$ versus $H_1 : p > 0.52$.

    $np_0(1 - p_0) = 2843 \cdot 0.52(1 - 0.52)$
    $= 709.6 > 10,$

    so the requirements of the hypothesis test are satisfied. From the sample,

$\hat{p} = \dfrac{1516}{2843} = 0.533$. The test statistic is

$$z_0 = \frac{\hat{p} - p_0}{\sqrt{\dfrac{p_0(1-p_0)}{n}}} = \frac{0.533 - 0.52}{\sqrt{\dfrac{0.52(1-0.52)}{1516}}} \approx 1.41$$

Classical approach: This is a right-tailed test with $\alpha = 0.10$, so the critical value is $z_{0.10} = 1.282$. Since $z_0 = 1.41 > z_{0.10} = 1.282$, the test statistic falls in the critical region, so we reject the null hypothesis.

P-value approach:

P-value $= P(Z > 1.41)$

$\qquad = 0.0793$ [Tech: 0.0788]

Since P-value $= 0.0793 < \alpha = 0.10$, we reject the null hypothesis.

Conclusion: There is sufficient evidence to conclude that a statistically significantly higher proportion of students return for their second year. Yes, the results are statistically significant. The new retention rate was about 53.3%, which is not a huge increase over 52%, so the results might not be considered practically significant. The cost of the new policies would have to be weighed against such a small increase in the retention rate when considering whether other community colleges should implement the policies.

21. The hypotheses are $H_0 : p = 0.4$ versus $H_1 : p > 0.4$.

$np_0(1-p_0) = 40 \cdot 0.4(1-0.4) = 9.6 < 10$, so we use small sample techniques. This is a right-tailed test, so we calculate the probability of 18 or more successes in 40 binomial trials with $p = 0.4$. Using technology:

P-value $= P(X \geq 18)$

$\qquad = 1 - P(X \leq 17)$

$\qquad = 1 - 0.6885$

$\qquad = 0.3115$

Since this is larger than the $\alpha = 0.05$ level of significance, we do not reject the null hypothesis.

Conclusion: There is not sufficient evidence to conclude that the proportion of adolescents who pray daily has increased above 40%.

22. The hypotheses are $H_0 : \mu = 73.2$ versus $H_1 : \mu < 73.2$. The test statistic is

$$t_0 = \frac{\bar{x} - \mu_0}{s/\sqrt{n}} = \frac{72.8 - 73.2}{12.3/\sqrt{3851}} = -2.018.$$

Classical approach: This is a left-tailed test with $n - 1 = 3851 - 1 = 3850$ degrees of freedom and $\alpha = 0.05$. However, since our *t*-distribution table (Table VI) does not have a row for df = 3850, we use df = 1000. The critical value is $-t_{0.05} = -1.646$. Since $t_0 = -2.018 < -t_{0.05} = -1.646$, the test statistic falls in the critical region, so we reject the null hypothesis.

P-value approach: This is a right-tailed test with 3850 degrees of freedom. The *P*-value is the area under the *t*-distribution to the left of the test statistic $t_0 = -2.018$, which is equivalent to the area to the right of 2.018. From the *t*-distribution table (Table VI) in the row corresponding to 1000 degrees of freedom (since the table does not contain a row for df = 3850), 2.018 falls between 1.962 and 2.056, whose right-tail areas are 0.025 and 0.02, respectively. Thus, $0.02 < P$-value < 0.025 [Tech: 0.0218]. Since P-value $< \alpha = 0.05$, we reject the null hypothesis.

Conclusion: There is sufficient evidence to conclude, from a statistical viewpoint, that the scores on the final exam decreased under the new format. The data are statistically significant, but there is no real practical significance. A difference of 0.4 points (less than ½ of a percentage point) is practically insignificant when compared to the savings in resources by the university. For all practical purposes, the average scores would be considered the same.

23. If we accept the null hypothesis, then we are saying it is true. If we do not reject the null hypothesis, then we are saying that we do not have enough evidence to conclude that it is not true. It is the difference between saying H_0 is true and saying that we are not convinced H_0 is false.

24. The hypotheses are $H_0 : \mu = 1.22$ versus $H_1 : \mu < 1.22$.

    The P-value indicates that if the population mean is 1.22 hours as stated in the null hypothesis, then results as low (or lower) than those obtained by the researcher would occur in about 3 out of 100 similar samples.

25. In the Classical Approach, a test statistic is calculated and a rejection rejoin based on the level of significance α is determined within the appropriate distribution for the population parameter stated in the null hypothesis. The null hypothesis is rejected if the test statistic is in the rejection region.

26. In the P-value approach, the probability of getting a result as extreme as the one obtained in the sample is calculated assuming the null hypothesis to be true. If the P-value is less than a predetermined level of significance, α, then we reject the null hypothesis.

Chapter 10 Test

1. (a) The hypotheses are $H_0 : \mu = 42.6$ minutes versus $H_1 : \mu > 46.2$ minutes.

   (b) There is sufficient evidence to conclude that the mean amount of daily time spent on phone calls and answering or writing emails has increased from what it was in 2006.

   (c) Making a Type I error would mean that we rejected the null hypothesis that the mean is 42.6 minutes when, in fact, the mean amount of daily time spent on phone calls and answering or writing emails is 42.6 minutes.

   (d) Making a Type II error would mean that we did not reject the null hypothesis that the mean is 42.6 minutes when, in fact, the mean amount of daily time spent on phone calls and answering or writing emails is greater than 42.6 minutes.

2. (a) The hypotheses are $H_0 : \mu = 167.1$ seconds versus $H_1 : \mu < 167.1$ seconds.

   (b) By choosing a level of significance of 0.01, we make the probability of making a Type I error small. In other words, the probability of rejecting the null hypothesis that the mean is 167.1 seconds in favor of the alternative hypothesis that the mean is less that 167.1 seconds, when in fact the mean is 167.1 seconds, is small.

   (c) The test statistic is
   $$z_0 = \frac{\bar{x} - \mu_0}{\sigma / \sqrt{n}} = \frac{163.9 - 167.1}{15.3 / \sqrt{70}} \approx -1.75.$$

   Classical approach: This is a left-tailed test with $\alpha = 0.01$, so the critical value is $-z_{0.01} = -2.33$. Since $z_0 = -1.75 > -z_{0.01} = -2.33$, the test statistic does not fall in the critical region, so we do not reject the null hypothesis.

   P-value approach:
   P-value $= P(Z < -1.75) = 0.0401$

   Since P-value $= 0.0401 > \alpha = 0.01$, we do not reject the null hypothesis.

   Conclusion: There is not sufficient evidence to conclude that the drive-through service time has decreased.

3. The hypotheses are $H_0 : \mu = 8$ hours versus $H_1 : \mu < 8$ hours. The test statistic is
   $$t_0 = \frac{\bar{x} - \mu_0}{s / \sqrt{n}} = \frac{7.8 - 8}{1.4 / \sqrt{151}} = -1.755.$$

   Classical Approach: This is a left-tailed test with $151 - 1 = 150$ degrees of freedom and $\alpha = 0.05$. However, since our t-distribution table (Table VI) does not have a row for df = 150, we use df = 100. The critical value is $-t_{0.05} = -1.660$. Since $t_0 = -1.755 < -t_{0.05} = -1.660$, the test statistic falls in the critical region, so we reject the null hypothesis.

   P-value approach: This is a left-tailed test with 150 degrees of freedom. The P-value is the area under the t-distribution to the left of the test statistic $t_0 = -1.755$, which is equivalent to the area to the right of 1.755. From the t-distribution table (Table VI) in the row corresponding to 100 degrees of freedom (since the table does not contain a row for df = 150), 1.755 falls between 1.660 and 1.984, whose right-tail areas are 0.05 and 0.025, respectively. Thus, $0.025 < P$-value < 0.05 [Tech: 0.0406]. Since P-value $< \alpha = 0.05$, we reject the null hypothesis.

Conclusion: There is sufficient evidence to conclude that postpartum women get less than 8 hours of sleep each night.

4. The hypotheses are $H_0 : \mu = 1.3825$ inches versus $H_1 : \mu \neq 1.3825$ inches .

We compute the sample mean and sample standard deviation and obtain $\bar{x} = 1.3826$ inches and $s \approx 0.00033$ inch.

Using $\alpha = 0.05$ with $n - 1 = 10 - 1 = 9$ degrees of freedom, we have $t_{\alpha/2} = t_{0.025} = 2.262$.

$$\text{Lower bound} = \bar{x} - t_{\alpha/2} \cdot \frac{s}{\sqrt{n}}$$

$$= 1.3826 - 2.262 \cdot \frac{0.00033}{\sqrt{10}}$$

$$\approx 1.3824 \text{ inches}$$

$$\text{Upper bound} = \bar{x} + t_{\alpha/2} \cdot \frac{s}{\sqrt{n}}$$

$$= 1.3826 + 2.262 \cdot \frac{0.00033}{\sqrt{10}}$$

$$\approx 1.3828 \text{ inches}$$

Because this interval includes the hypothesized mean 1.3825 inches, we do not reject the null hypothesis. There is not sufficient evidence to conclude that the mean is different from 1.3825 inches. Therefore, we presume that the part has been manufactured to specifications.

5. The hypotheses are $H_0 : p = 0.6$ versus $H_1 : p > 0.6$.

$np_0(1 - p_0) = 1561(0.6)(1 - 0.6) = 374.64 \geq 10$, so the requirements of the hypothesis test are satisfied. From the sample, $\hat{p} = \dfrac{954}{1561} \approx 0.611$.

The test statistic is

$$z_0 = \frac{\hat{p} - p_0}{\sqrt{\dfrac{p_0(1 - p_0)}{n}}} = \frac{0.611 - 0.6}{\sqrt{\dfrac{0.6(1 - 0.6)}{1561}}} \approx 0.89$$

[Tech: 0.90]

Classical approach: We use $\alpha = 0.05$. This is a right-tailed test, so the critical value is $z_{0.05} = 1.645$. Since $z_0 = 0.89 < z_{0.05} = 1.645$, the test statistic does not fall in the critical region, so we do not reject the null hypothesis.

P-value approach:

$$P\text{-value} = P(Z > 0.89)$$

$$= 1 - P(Z \leq 0.89)$$

$$= 1 - 0.8133$$

$$= 0.1867 \quad [\text{Tech: } 0.1843]$$

Since $P\text{-value} = 0.1867 > \alpha = 0.05$, we do not reject the null hypothesis.

Conclusion: There is not sufficient evidence to conclude that a supermajority of Americans did not feel that the United States would need to fight Japan in their lifetimes.

6. The hypotheses are $H_0 : \mu = 0$ pounds versus $H_1 : \mu > 0$ pounds . The test statistic is

$$t_0 = \frac{\bar{x} - \mu_0}{s/\sqrt{n}} = \frac{1.6 - 0}{5.4/\sqrt{79}} \approx 2.634 .$$

Classical Approach: This is a right-tailed test with $79 - 1 = 78$ degrees of freedom and $\alpha = 0.05$. However, since our t-distribution table (Table VI) does not have a row for df = 78, we use df = 80. The critical value is $t_{0.05} = 1.664$. Since $t_0 = 2.634 > t_{0.05} = 1.664$, the test statistic falls in the critical region, so we reject the null hypothesis.

P-value approach: This is a right-tailed test with $79 - 1 = 78$ degrees of freedom. The P-value is the area under the t-distribution to the right of the test statistic $t_0 = 2.634$. From the t-distribution table (Table VI) in the row corresponding to 80 degrees of freedom (since the table does not contain a row for df = 78), 2.634 falls between 2.374 and 2.639, whose right-tail areas are 0.01 and 0.005, respectively. Thus, $0.005 < P\text{-value} < 0.01$ [Tech: 0.0051]. Since $P\text{-value} < \alpha = 0.05$, we reject the null hypothesis.

Conclusion: There is sufficient evidence to suggest that the diet is effective. However, loosing 1.6 kg of weight over the course of a year does not seem to have much practical significance.

7. The hypotheses are $H_0 : p = 0.37$ versus $H_1 : p > 0.37$.

$np_0(1 - p_0) = 30 \cdot 0.37(1 - 0.37) = 6.993 < 10$, so we use small sample techniques. This is a right-tailed test, so we calculate the probability of 16 or more successes in 30 binomial trials with $p = 0.37$. Using technology:

$$P\text{-value} = P(X \geq 16)$$
$$= 1 - P(X \leq 15)$$
$$= 1 - 0.9499$$
$$= 0.0501$$

Since this is smaller than the $\alpha = 0.10$ level of significance, we reject the null hypothesis.

Conclusion: There is sufficient evidence to conclude that the proportion of 20- to 24-year-olds who live on their own and do not have a land line is greater than 0.37.

8. The hypotheses are $H_0 : \sigma = 18\%$ versus $H_1 : \sigma < 18\%$. The test statistic is

$$\chi_0^2 = \frac{(n-1)s^2}{\sigma_0^2} = \frac{(10-1)\cdot(16)^2}{(18)^2} \approx 7.111.$$

Classical Approach: This is a left-tailed test with 9 degrees of freedom and $\alpha = 0.05$, so the critical value is $\chi_{1-\alpha}^2 = \chi_{0.95}^2 = 3.325$. Since the test statistic is not in the critical region ($7.111 > 3.325$), we do not reject the null hypothesis.

P-value approach: The *P*-value is the area under the χ^2-distribution with 9 degrees of freedom to the left of $\chi_0^2 = 7.111$. From the χ^2-distribution table (Table VII) in the row corresponding to 9 degrees of freedom, 7.111 is between 4.168 and 14.684. The areas to the right of these critical values are 0.90 and 0.10, respectively. This means the area to the left of these critical values are 0.10 and 0.90, respectively. So, $0.10 < P\text{-value} < 0.90$ [Tech: 0.3744]. Since *P*-value $> \alpha = 0.05$, we do not reject the null hypothesis.

Conclusion: There is not sufficient evidence to conclude that the investment manager's portfolio is less risky than the market.

9. The null hypothesis will be rejected if $\hat{p}$ is more than 1.645 standard deviations

above $p_0 = 0.6$. So, we reject H_0 if

$$\hat{p} > 0.6 + 1.645\sqrt{\tfrac{0.6(1-0.6)}{1561}} \approx 0.620.$$

$$\beta = P(\text{Type II error})$$
$$= P(\text{do not reject } H_0 \text{ when } H_1 \text{ is true})$$
$$= P(\hat{p} < 0.620 \text{ given that } p = 0.63)$$
$$= P\left(Z < \frac{0.620 - 0.63}{\sqrt{0.63(1-0.63)/1561}} \right)$$
$$= P(Z < -0.79)$$
$$= 0.2061 \quad [\text{Tech: } 0.2159]$$

The power of the test is $1 - \beta = 0.7939$ [Tech: 0.7841].

Chapter 10 Case Study

Unshed antler data from ditch:

$H_0 : \mu = 2950$; $H_1 : \mu \neq 2950$
$\bar{x} = 3033.1$, $s = 66.9$, $n = 9$, df $= 8$, $\alpha = 0.05$
$t_0 = \dfrac{3033.1 - 2950}{66.9/\sqrt{9}} = 3.726$; $\pm t_{0.025} = \pm 2.306$;
$P\text{-value} = 0.005817$
Conclusion: Reject H_0. There is enough evidence to indicate that the mean date for construction of the ditch is not 2950 B.C.

Animal bones in ditch terminals:

$H_0 : \mu = 2950$; $H_1 : \mu \neq 2950$
$\bar{x} = 3187.5$, $s = 67.4$, $n = 4$, df $= 3$, $\alpha = 0.05$
$t_0 = \dfrac{3187.5 - 2950}{67.4/\sqrt{4}} = 7.047$; $\pm t_{0.025} = \pm 3.182$;
$P\text{-value} = 0.005872$
Conclusion: Reject H_0. There is enough evidence to indicate that the mean date for construction of the ditch terminals is not 2950 B.C.

Bluestone formations:

$H_0 : \mu = 2950$; $H_0 : \mu \neq 2950$
$\bar{x} = 2193.3$, $s = 104.1$, $n = 3$, df $= 2$, $\alpha = 0.05$
$t_0 = \dfrac{2193.3 - 2950}{104.1/\sqrt{3}} = -12.590$; $\pm t_{0.025} = \pm 4.303$;
$P\text{-value} = 0.006250$
Conclusion: Reject H_0. There is enough evidence to indicate that the mean date for construction of the Bluestone formations is not 2950 B.C.

<u>Y and Z holes:</u>

$H_0 : \mu = 2950$; $H_0 : \mu \neq 2950$

$\bar{x} = 1671.7$, $s = 99.7$, $n = 3$, df $= 2$, $\alpha = 0.05$

$t_0 = \dfrac{1671.7 - 2950}{99.7 / \sqrt{3}} = -22.207$; $\pm t_{0.025} = \pm 4.303$;

P-value $= 0.00202$

Conclusion: Reject H_0. There is enough evidence to indicate that the mean date for construction of the Y and Z holes is not 2950 B.C. So, the dates for the samples from the Y and Z holes is not consistent with 2950 B.C.

Based on the results of the hypothesis tests (including sample data and test statistics), the construction order appears to be: Ditch terminals, Ditch, Bluestone formations, Y and Z holes

Using technology, 95% confidence intervals for the mean date (B.C.) of the various constructions:

Ditch: $(3084.5, 2981.7)$

Ditch terminals: $(3294.7, 3080.3)$

Bluestone formations: $(2451.9, 1934.7)$

Y and Z holes: $(1919.4, 1424.0)$

The only confidence interval that contains 2950 B.C. is the confidence interval for the Bluestone formations. All the others appear to be inconsistent with Corbin's claims.

Based on the results of the confidence intervals, the construction order appears to be: Ditch terminals, Ditch, Bluestone formations, Y and Z holes

The order is somewhat easier to see if we graph the confidence intervals together.

Discussion and reports will vary.

Chapter 11

Inferences on Two Samples

Section 11.1

1. independent

3. Since the members of the two samples are married to each other, the sampling is dependent. The data are quantitative, since a numeric response was given by each subject.

5. Because the two samples were drawn at separate times with two different groups of people, the sampling is independent. The data are qualitative, since each respondent gave a "Yes" or "No" response.

7. Two independent populations are being studied: students who receive the new curriculum and students who receive the traditional curriculum. The people in the two groups are completely independent, so the sampling is independent. The data are quantitative, since it is the test score for each respondent.

9. **(a)** The hypotheses are $H_0 : p_1 = p_2$ versus $H_1 : p_1 > p_2$.

   (b) The two sample estimates are
   $$\hat{p}_1 = \frac{x_1}{n_1} = \frac{368}{541} \approx 0.6802 \text{ and}$$
   $$\hat{p}_2 = \frac{x_2}{n_2} = \frac{351}{593} \approx 0.5919.$$
   The pooled estimate is
   $$\hat{p} = \frac{x_1 + x_2}{n_1 + n_2} = \frac{368 + 351}{541 + 593} \approx 0.6340.$$
   The test statistic is
   $$z_0 = \frac{\hat{p}_1 - \hat{p}_2}{\sqrt{\hat{p}(1 - \hat{p})}\sqrt{\dfrac{1}{n_1} + \dfrac{1}{n_2}}}$$
   $$= \frac{0.6802 - 0.5919}{\sqrt{0.6340(1 - 0.6340)}\sqrt{\dfrac{1}{541} + \dfrac{1}{593}}}$$
   $$\approx 3.08$$

(c) This is a right-tailed test, so the critical value is $z_\alpha = z_{0.05} = 1.645$.

(d) P-value $= P(z_0 \geq 3.08)$
$$= 1 - 0.9990 = 0.0010$$

Since $z_0 = 3.08 > z_{0.05} = 1.645$ and P-value $= 0.0010 < \alpha = 0.05$, we reject H_0. There is sufficient evidence to conclude that $p_1 > p_2$.

11. **(a)** The hypotheses are $H_0 : p_1 = p_2$ versus $H_1 : p_1 \neq p_2$.

(b) The two sample estimates are
$$\hat{p}_1 = \frac{x_1}{n_1} = \frac{28}{254} \approx 0.1102 \text{ and}$$
$$\hat{p}_2 = \frac{x_2}{n_2} = \frac{36}{301} \approx 0.1196.$$
The pooled estimate is
$$\hat{p} = \frac{x_1 + x_2}{n_1 + n_2} = \frac{28 + 36}{254 + 301} \approx 0.1153.$$
The test statistic is
$$z_0 = \frac{\hat{p}_1 - \hat{p}_2}{\sqrt{\hat{p}(1 - \hat{p})}\sqrt{\dfrac{1}{n_1} + \dfrac{1}{n_2}}}$$
$$= \frac{0.1102 - 0.1196}{\sqrt{0.1153(1 - 0.1153)}\sqrt{\dfrac{1}{254} + \dfrac{1}{301}}}$$
$$\approx -0.35 \quad [\text{Tech:} -0.34]$$

(c) This is a two-tailed test, so the critical values are $\pm z_{\alpha/2} = \pm z_{0.025} = \pm 1.96$.

(d) P-value $= 2 \cdot P(z_0 \leq -0.35)$
$$= 2 \cdot 0.3632$$
$$= 0.7264 \quad [\text{Tech: } 0.7307]$$

Since $z_0 = -0.35$ falls between $-z_{0.025} = -1.96$ and $z_{0.025} = 1.96$ and P-value $= 0.7264 > \alpha = 0.05$, we do not reject H_0. There is not sufficient evidence to conclude that $p_1 \neq p_2$.

13. We have $\hat{p}_1 = \dfrac{x_1}{n_1} = \dfrac{368}{541} \approx 0.6802$ and $\hat{p}_2 = \dfrac{x_2}{n_2} = \dfrac{421}{593} \approx 0.7099$. For a 90% confidence level, we use

$\pm z_{0.05} = \pm 1.645$. Then:

Lower Bound: $(\hat{p}_1 - \hat{p}_2) - z_{\alpha/2} \cdot \sqrt{\dfrac{\hat{p}_1(1-\hat{p}_1)}{n_1} + \dfrac{\hat{p}_2(1-\hat{p}_2)}{n_2}}$

$= (0.6802 - 0.7099) - 1.645 \cdot \sqrt{\dfrac{0.6802(1-0.6802)}{541} + \dfrac{0.7099(1-0.7099)}{593}}$

≈ -0.075

Upper Bound: $(\hat{p}_1 - \hat{p}_2) + z_{\alpha/2} \cdot \sqrt{\dfrac{\hat{p}_1(1-\hat{p}_1)}{n_1} + \dfrac{\hat{p}_2(1-\hat{p}_2)}{n_2}}$

$= (0.6802 - 0.7099) + 1.645 \cdot \sqrt{\dfrac{0.6802(1-0.6802)}{541} + \dfrac{0.7099(1-0.7099)}{593}}$

≈ 0.015

15. We have $\hat{p}_1 = \dfrac{x_1}{n_1} = \dfrac{28}{254} \approx 0.1102$ and $\hat{p}_2 = \dfrac{x_2}{n_2} = \dfrac{36}{301} \approx 0.1196$. For a 95% confidence interval we use

$\pm z_{0.025} = \pm 1.96$. Then:

Lower Bound: $(\hat{p}_1 - \hat{p}_2) - z_{\alpha/2} \cdot \sqrt{\dfrac{\hat{p}_1(1-\hat{p}_1)}{n_1} + \dfrac{\hat{p}_2(1-\hat{p}_2)}{n_2}}$

$= (0.1102 - 0.1196) - 1.96 \cdot \sqrt{\dfrac{0.1102(1-0.1102)}{254} + \dfrac{0.1196(1-0.1196)}{301}}$

≈ -0.063

Upper Bound: $(\hat{p}_1 - \hat{p}_2) + z_{\alpha/2} \cdot \sqrt{\dfrac{\hat{p}_1(1-\hat{p}_1)}{n_1} + \dfrac{\hat{p}_2(1-\hat{p}_2)}{n_2}}$

$= (0.1102 - 0.1196) + 1.96 \cdot \sqrt{\dfrac{0.1102(1-0.1102)}{254} + \dfrac{0.1196(1-0.1196)}{301}}$

≈ 0.044

---

17. (a) The hypotheses are $H_0 : p_A = p_B$ versus $H_1 : p_A \neq p_B$.

(b) $\hat{p}_A = \dfrac{45+14}{45+19+14+23} = \dfrac{59}{101} \approx 0.5842$

$\hat{p}_B = \dfrac{45+19}{45+19+14+23} = \dfrac{64}{101} \approx 0.6337$

$z_0 = \dfrac{|f_{12} - f_{21}| - 1}{\sqrt{f_{12} + f_{21}}} = \dfrac{|19-14| - 1}{\sqrt{19+14}} \approx 0.70$

(c) This is a two-tailed test, so the critical values are $\pm z_{\alpha/2} = \pm z_{0.025} = \pm 1.96$.

(d) P-value $= 2 \cdot P(z_0 \geq 0.70)$

$= 2 \cdot \left[1 - P(z_0 \leq 0.70) \right]$

$= 2 \cdot \left[1 - 0.7580 \right]$

$= 2 \cdot \left[0.2420 \right]$

$= 0.4840$

Since $z_0 = 0.70$ falls between $-z_{0.025} = -1.96$ and $z_{0.025} = 1.96$ and since P-value $= 0.4840 > \alpha = 0.05$, we do not reject H_0. There is not sufficient evidence to conclude that $p_A \neq p_B$.

19. We first verify the requirements to perform the hypothesis test: (1) Each sample can be thought of as a simple random sample; (2) We have $x_1 = 107$, $n_1 = 710$, $x_2 = 67$, and $n_2 = 611$, so $\hat{p}_1 = \dfrac{x_1}{n_1} = \dfrac{107}{710} \approx 0.1507$

and $\hat{p}_2 = \dfrac{x_2}{n_2} = \dfrac{67}{611} \approx 0.1097$. Thus,

$n_1 \hat{p}_1 (1 - \hat{p}_1)$

$= 710(0.1507)(1 - 0.1507) \approx 91 \geq 10$ and

$n_2 \hat{p}_2 (1 - \hat{p}_2) = 611(0.1097)(1 - 0.1097) \approx 60$

≥ 10; and (3) Each sample is less than 5% of the population. Thus, the requirements are met, and we can conduct the test.

The hypotheses are $H_0 : p_1 = p_2$ versus $H_1 : p_1 > p_2$. From before, the two sample estimates are $\hat{p}_1 \approx 0.1507$ and $\hat{p}_2 \approx 0.1097$. The pooled estimate is

$\hat{p} = \dfrac{x_1 + x_2}{n_1 + n_2} = \dfrac{107 + 67}{710 + 611} \approx 0.1317$.

The test statistic is

$z_0 = \dfrac{\hat{p}_1 - \hat{p}_2}{\sqrt{\hat{p}(1 - \hat{p})}\sqrt{1/n_1 + 1/n_2}}$

$= \dfrac{0.1507 - 0.1097}{\sqrt{0.1317(1 - 0.1317)}\sqrt{1/710 + 1/611}}$

≈ 2.20 [Tech: 2.19]

Classical approach: This is a right-tailed test, so the critical value is $z_\alpha = z_{0.05} = 1.645$. Since $z_0 > z_{0.05}$ (the test statistic lies within the critical region), we reject H_0.

P-value approach:

P-value = $P(z_0 \geq 2.20) = 1 - 0.9861 = 0.0139$. Since P-value < $\alpha = 0.05$, we reject H_0.

Conclusion: There is sufficient evidence at the $\alpha = 0.05$ level of significance to conclude that a higher proportion of subjects in the treatment group (taking Prevnar) experienced fever as a side effect than in the control (placebo) group.

21. We first verify the requirements to perform the hypothesis test: (1) Each sample is a simple random sample; (2) we have $x_{1947} = 407$, $n_{1947} = 1100$, $x_{2010} = 333$, and $n_{2010} = 1100$, so $\hat{p}_{1947} = \dfrac{x_{1947}}{n_{1947}} = \dfrac{407}{1100} = 0.37$

and $\hat{p}_{2010} = \dfrac{x_{2010}}{n_{2010}} = \dfrac{333}{1100} = 0.30$. Thus,

$n_{1947}\hat{p}_{1947}(1 - \hat{p}_{1947}) = 1100(0.37)(1 - 0.37) \approx$

$256.4 \geq 10$ and $n_{2010}\hat{p}_{2010}(1 - \hat{p}_{2010}) =$

$1100(0.30)(1 - 0.30) \approx 232.2 \geq 10$; and (3) each sample is less than 5% of the population. Thus, the requirements are met, so we can conduct the test.

The hypotheses are $H_0 : p_{1947} = p_{2010}$ versus $H_1 : p_{1947} \neq p_{2010}$. From before, the two sample estimates are $\hat{p}_{1947} = 0.37$ and $\hat{p}_{2010} = 0.30$. The pooled estimate is

$\hat{p} = \dfrac{x_{1947} + x_{2010}}{n_{1947} + n_{2010}} = \dfrac{407 + 333}{1100 + 1100} = 0.336$.

The test statistic is

$z_0 = \dfrac{\hat{p}_{1947} - \hat{p}_{2010}}{\sqrt{\hat{p}(1 - \hat{p})}\sqrt{1/n_{1947} + 1/n_{2010}}}$

$= \dfrac{0.37 - 0.30}{\sqrt{0.336(1 - 0.336)}\sqrt{1/1100 + 1/1100}}$

≈ 3.34

Classical approach: This is a two-tailed test, so the critical values are $\pm z_{\alpha/2} = \pm z_{0.025} = \pm 1.96$. Since the test statistic $z_0 = 3.34$ does not lie between $-z_{0.025} = -1.96$ and $z_{0.025} = 1.96$ (the test statistic falls in the critical region), we reject H_0.

P-value approach: P-value = $2 \cdot P(z_0 \geq 3.34)$ $= 2 \cdot 0.0004 = 0.0008$ [Tech: 0.0008] Since P-value < $\alpha = 0.05$, we reject H_0.

Conclusion: There is sufficient evidence at the $\alpha = 0.05$ level of significance to conclude that the proportion of adult Americans who were abstainers in 1947 is different from the proportion of abstainers in 2010.

23. $H_0 : p_m = p_f$ versus $H_1 : p_m \neq p_f$. We have $x_m = 181$, $n_m = 1205$, $x_f = 143$, and $n_f = 1097$, so

$\hat{p}_m = \dfrac{x_m}{n_m} = \dfrac{181}{1205} \approx 0.1502$ and $\hat{p}_f = \dfrac{x_f}{n_f} = \dfrac{143}{1097} \approx 0.1304$. For a 95% confidence interval we use

$\pm z_{0.025} = \pm 1.96$. Then:

Lower Bound: $(\hat{p}_m - \hat{p}_f) - z_{\alpha/2} \cdot \sqrt{\dfrac{\hat{p}_m (1 - \hat{p}_m)}{n_m} + \dfrac{\hat{p}_f (1 - \hat{p}_f)}{n_f}}$

$= (0.1502 - 0.1304) - 1.96 \cdot \sqrt{\dfrac{0.1502(1 - 0.1502)}{1205} + \dfrac{0.1304(1 - 0.1304)}{1097}} \approx -0.009$

Upper Bound: $(\hat{p}_m - \hat{p}_f) + z_{\alpha/2} \cdot \sqrt{\dfrac{\hat{p}_m (1 - \hat{p}_m)}{n_m} + \dfrac{\hat{p}_f (1 - \hat{p}_f)}{n_f}}$

$= (0.1502 - 0.1304) + 1.96 \cdot \sqrt{\dfrac{0.1502(1 - 0.1502)}{1205} + \dfrac{0.1304(1 - 0.1304)}{1097}} \approx 0.048$

We are 95% confident that the difference in the proportion of males and females that have at least one tattoo is between -0.009 and 0.048. Because the interval includes zero, we do not reject the null hypothesis. There is no significant difference in the proportion of males and females that have tattoos.

---

25. (a) We first verify the requirements to perform the hypothesis test: (1) Each sample can be thought of as a simple random sample; (2) we have $x_C = 50$, $n_C = 1655$, $x_p = 31$,

and $n_p = 1652$, so $\hat{p}_C = \dfrac{50}{1655} \approx 0.0302$ and

$\hat{p}_p = \dfrac{31}{1652} \approx 0.0188$. Thus, $n_C \hat{p}_C (1 - \hat{p}_C) =$

$1655(0.0302)(1 - 0.0302) \approx 48 \geq 10$ and

$n_p \hat{p}_p (1 - \hat{p}_p) = 1652(0.0188)(1 - 0.0188) \approx$

$30 \geq 10$; and (3) each sample is less than 5% of the population. So, the requirements are met, so we can conduct the test.

The hypotheses are $H_0 : p_C = p_p$ versus $H_1 : p_C > p_p$. From before, the two sample estimates are $\hat{p}_C \approx 0.0302$ and

$\hat{p}_p \approx 0.0188$. The pooled estimate is

$\hat{p} = \dfrac{x_C + x_p}{n_C + n_p} = \dfrac{50 + 31}{1655 + 1652} \approx 0.0245.$

The test statistic is

$z_0 = \dfrac{\hat{p}_C - \hat{p}_p}{\sqrt{\hat{p}(1 - \hat{p})} \sqrt{1/n_C + 1/n_p}}$

$= \dfrac{0.0302 - 0.0188}{\sqrt{0.0245(1 - 0.0245)} \sqrt{\dfrac{1}{1655} + \dfrac{1}{1652}}}$

≈ 2.12 [Tech: 2.13]

Classical approach: This is a right-tailed test, so the critical value is $z_{0.05} = 1.645$. Since $z_0 > z_{0.05} = 1.645$ (the test statistic falls in the critical region), we reject H_0.

P-value approach: P-value $= P(z_0 \geq 2.12)$ $= 1 - 0.9830 = 0.0170$ [Tech: 0.0166]. Since this P-value is less than the $\alpha = 0.05$ level of significance, we reject H_0.

Conclusion: There is sufficient evidence at the $\alpha = 0.05$ level of significance to conclude that the proportion of individuals taking Clarinex and experiencing dry mouth is greater than that of those taking a placebo.

(b) No, the difference between the experimental group and the control group is not practically significant. Both Clarinex and the placebo have fairly low proportions of individuals that experience dry mouth.

27. (a) The choices were randomly rotated in order to remove any potential nonsampling error due to the respondent hearing the word "right" or "wrong" first.

(b) We have $x_{2003} = 1086$, $n_{2003} = 1508$, $x_{2010} = 618$, and $n_{2008} = 1508$, so $\hat{p}_{2003} = \dfrac{x_{2003}}{n_{2003}} = \dfrac{1086}{1508} \approx 0.7202$

and $\hat{p}_{2010} = \dfrac{x_{2010}}{n_{2010}} = \dfrac{618}{1508} = 0.4098$. For a 90% confidence interval we use $\pm z_{0.05} = \pm 1.645$. Then:

Lower Bound: $(\hat{p}_{2003} - \hat{p}_{2010}) - z_{\alpha/2} \cdot \sqrt{\dfrac{\hat{p}_{2003}(1 - \hat{p}_{2003})}{n_{2003}} + \dfrac{\hat{p}_{2010}(1 - \hat{p}_{2010})}{n_{2010}}}$

$= (0.7202 - 0.4098) - 1.645 \cdot \sqrt{\dfrac{0.7202(1 - 0.7202)}{1508} + \dfrac{0.4098(1 - 0.4098)}{1508}} \approx 0.282$

Upper Bound: $(\hat{p}_{2003} - \hat{p}_{2010}) + z_{\alpha/2} \cdot \sqrt{\dfrac{\hat{p}_{2003}(1 - \hat{p}_{2003})}{n_{2003}} + \dfrac{\hat{p}_{2010}(1 - \hat{p}_{2010})}{n_{2010}}}$

$= (0.7202 - 0.4098) + 1.645 \cdot \sqrt{\dfrac{0.7202(1 - 0.7202)}{1508} + \dfrac{0.4098(1 - 0.4098)}{1508}} \approx 0.338$

[Tech: 0.339]

We are 90% confident that the difference in the proportion of adult Americans who believe the United States made the right decision to use military force in Iraq from 2003 to 2008 is between 0.282 and 0.338. The attitude regarding the decision to go to war changed substantially.

---

29. (a) We have $x_{\text{males}} = 26$, $n_{\text{males}} = 82$,
$x_{\text{females}} = 25$, and $n_{\text{females}} = 116$, so

$\hat{p}_{\text{males}} = \dfrac{x_{\text{males}}}{n_{\text{males}}} = \dfrac{26}{82} = 0.317$ and

$\hat{p}_{\text{females}} = \dfrac{x_{\text{females}}}{n_{\text{females}}} = \dfrac{25}{116} = 0.216$.

(b) We first verify the requirements to perform the hypothesis test: (1) Each sample is a simple random sample; (2)

$n_{\text{males}} \hat{p}_{\text{males}} (1 - \hat{p}_{\text{males}}) =$
$82(0.317)(1 - 0.317) \approx 17.8 \geq 10$ and

$n_{\text{females}} \hat{p}_{\text{females}} (1 - \hat{p}_{\text{females}}) =$
$116(0.216)(1 - 0.216) \approx 19.6 \geq 10$; and (3)
each sample is less than 5% of the population. Thus, the requirements are met, so we can conduct the test.

The hypotheses are $H_0 : p_{\text{males}} = p_{\text{females}}$
versus $H_1 : p_{\text{males}} \neq p_{\text{females}}$. The pooled estimate of p is

$\hat{p} = \dfrac{x_{\text{males}} + x_{\text{females}}}{n_{\text{males}} + n_{\text{females}}} = \dfrac{26 + 25}{82 + 116} = 0.258$.

The test statistic is

$z_0 = \dfrac{\hat{p}_{\text{males}} - \hat{p}_{\text{females}}}{\sqrt{\hat{p}(1 - \hat{p})} \sqrt{1/n_{\text{males}} + 1/n_{\text{females}}}}$

$= \dfrac{0.317 - 0.216}{\sqrt{0.258(1 - 0.258)} \sqrt{1/82 + 1/116}}$

≈ 1.61 [Tech: 1.61]

Classical approach: This is a two-tailed test, so the critical values are:
$\pm z_{\alpha/2} = \pm z_{0.025} = \pm 1.96$. Since the test statistic $z_0 = 1.61$ lies between
$-z_{0.025} = -1.96$ and $z_{0.025} = 1.96$ (the test statistic does not fall in the critical region), we do not reject H_0.

P-value approach: P-value$= 2 \cdot P(z_0 \geq 1.61)$
$= 2 \cdot 0.0537 = 0.1074$ [Tech: 0.1075]
Since P-value $> \alpha = 0.05$, we do not reject H_0.

Conclusion: There is not sufficient evidence at the $\alpha = 0.05$ level of significance to conclude that the proportion of men and women who are willing to pay higher taxes to reduce the deficit differs.

31. **(a)** In Sentence A, the verbs are "was having" and "was taking." In sentence B, the verbs are "had" and "took."

(b) We first verify the requirements to perform the hypothesis test: (1) Each sample is a simple random sample; (2) we have $x_A = 71$, $n_A = 98$, $x_B = 49$, and $n_B = 98$,

so $\hat{p}_A = \dfrac{x_A}{n_A} = \dfrac{71}{98} = 0.724$ and

$\hat{p}_B = \dfrac{x_B}{n_B} = \dfrac{49}{98} = 0.500$. $n_A \hat{p}_A (1 - \hat{p}_A) =$

$98(0.724)(1 - 0.724) \approx 19.6 \geq 10$ and

$n_B \hat{p}_B (1 - \hat{p}_B) =$

$98(0.500)(1 - 0.216)500 \approx 24.5 \geq 10$; and

(3) each sample is less than 5% of the population. Thus, the requirements are met, so we can conduct the test.

The hypotheses are $H_0 : p_A = p_B$ versus $H_1 : p_A \neq p_B$. The pooled estimate of p is

$\hat{p} = \dfrac{x_A + x_B}{n_A + n_B} = \dfrac{71 + 49}{98 + 98} = 0.612$.

The test statistic is

$z_0 = \dfrac{\hat{p}_A - \hat{p}_B}{\sqrt{\hat{p}(1 - \hat{p})}\sqrt{1/n_A + 1/n_B}}$

$= \dfrac{0.724 - 0.500}{\sqrt{0.612(1 - 0.612)}\sqrt{1/98 + 1/98}}$

≈ 3.22 [Tech: 3.23]

Classical approach: This is a two-tailed test, so the critical values are: $\pm z_{\alpha/2} = \pm z_{0.025} = 1.96$. Since the test statistic $z_0 = 1.61$ does not lie between $-z_{0.025} = -1.96$ and $z_{0.025} = 1.96$ (the test statistic falls in the critical region), we reject H_0.

P-value approach:

P-value $= 2 \cdot P(z_0 \geq 3.22)$

$= 2 \cdot 0.0006 = 0.0012$ [Tech: 0.0013] Since P-value $< \alpha = 0.05$, we reject H_0.

Conclusion: There is sufficient evidence at the $\alpha = 0.05$ level of significance to conclude that the sentence structure makes a difference.

(c) Answers will vary. The wording in Sentence A suggests that the actions were taking place over a period of time and suggests habitual behavior that may be continuing in the present or might be expected to continue at some time in the future. The wording in Sentence B suggests events that are concluded and were possibly brief in duration or one-time occurrences.

33. **(a)** This is a dependent sample because the same person answered both questions.

(b) We first verify the requirements to perform McNemar's test. The samples are dependent and were obtained randomly. The total number of individuals who meet one condition but do not meet the other condition, is $448 + 327 = 775$, which is greater than 10. . Thus, the requirements are met, so we can conduct the test.

The hypotheses are $H_0 : p_{\text{no seat belt}} = p_{\text{smoke}}$ verses $H_1 : p_{\text{no seat belt}} \neq p_{\text{smoke}}$. The test statistic is

$z_0 = \dfrac{|f_{12} - f_{21}| - 1}{\sqrt{f_{12} + f_{21}}} = \dfrac{|448 - 327| - 1}{\sqrt{448 + 327}} \approx 4.31$

Classical approach: The critical value for an $\alpha = 0.05$ level of significance is $z_{\alpha/2} = z_{0.025} = 1.96$. Since $z_0 = 4.31 > z_{0.025} = 1.96$ (the test statistic falls in the critical region), we reject H_0.

P-value approach: The P-value is two times the area under the standard normal distribution to the right of the test statistic, $z_0 = 4.31$. The test statistic is so large that P-value < 0.0001. Since this P-value is less than the $\alpha = 0.05$ level of significance, we reject H_0.

Conclusion: There is sufficient evidence at the $\alpha = 0.05$ level of significance to conclude that there is a difference in the proportion who do not use a seatbelt and the proportion who smoke. The sample proportion of individuals who do not wear a seat belt is

$$\hat{p}_{\text{no seat belt}} = \frac{67 + 327}{3029} \approx 0.13 \text{, while the}$$

proportion of individuals who smoke is

$$\hat{p}_{\text{smoke}} = \frac{67 + 448}{3029} \approx 0.17 \text{. So, smoking}$$

appears to be the more popular hazardous activity.

35. We first verify the requirements to perform McNemar's test. The samples are dependent and were obtained randomly. The total number of words meeting one condition but not meeting the other condition, is $385 + 456 = 841$, which is greater than 10. Thus, the requirements are met, so we can conduct the test.

The hypotheses are $H_0 : p_{\text{NN}} = p_{\text{RN}}$ verses $H_1 : p_{\text{NN}} \neq p_{\text{RN}}$. The test statistic is

$$z_0 = \frac{|f_{12} - f_{21}| - 1}{\sqrt{f_{12} + f_{21}}} = \frac{|385 - 456| - 1}{\sqrt{385 + 456}} \approx 2.41$$

Classical approach: The critical value for $\alpha = 0.05$ is $z_{\alpha/2} = z_{0.025} = 1.96$. Since $z_0 = 2.41 > z_{0.025} = 1.96$ (the test statistic falls in the critical region), we reject H_0.

P-value approach: The P-value is two times the area under the standard normal distribution to the right of the test statistic, $z_0 = 2.41$. P-value $= 2 \cdot P(x > 2.41)$

$$= 2 \cdot [1 - P(x < 2.41)]$$

$$= 2 \cdot (1 - 0.9920)$$

$$= 0.0160$$

Since this P-value is less than the $\alpha = 0.05$ level of significance, we reject H_0.

Conclusion: There is sufficient evidence at the $\alpha = 0.05$ level of significance to conclude that there is a difference in the proportion of words not recognized by the two systems. The sample proportion of words recognized by the neural network is

$$\hat{p}_{\text{NN}} = \frac{9326 + 385}{9326 + 385 + 456 + 29} \approx 0.95 \text{,}$$

while the sample proportion of words recognize by the remapped network is

$$\hat{p}_{\text{RN}} = \frac{9326 + 456}{9326 + 385 + 456 + 29} \approx 0.96 \text{. So, it}$$

appears that the remapped network recognizes more words than the neural network. That is, the neural network appears to make more errors.

37. To answer this question, we test to see if the difference in the proportion of gun owners in October 2010 and the proportion of gun owners in October 2009 is statistically significant. We will use a $\alpha = 0.05$ level of significance.

We first verify the requirements to perform the hypothesis test: (1) Each sample can be thought of as a simple random sample; (2) we have $x_{2010} = 441$, $n_{2010} = 1134$, $x_{2009} = 458$, and $n_{2009} = 1134$, so $\hat{p}_{2010} = \frac{441}{1134} \approx 0.389$

and $\hat{p}_{2009} = \frac{458}{1134} \approx 0.404$. Therefore,

$n_{2010} \hat{p}_{2010} (1 - \hat{p}_{2010}) = 1134(0.389)(1 - 0.389)$

$\approx 269.5 \geq 10$ and $n_{2009} \hat{p}_{2009} (1 - \hat{p}_{2009}) =$

$1134(0.404)(1 - 0.404) \approx 273.0 \geq 10$; and (3) each sample is less than 5% of the population. Thus, the requirements are met, so we can conduct the test. The hypotheses are $H_0 : p_{2009} = p_{2010}$ versus $H_1 : p_{2010} < p_{2009}$. From before, the two sample estimates are $\hat{p}_{2010} \approx 0.389$ and $\hat{p}_{2009} \approx 0.404$. The pooled estimate is

$$\hat{p} = \frac{x_{2010} + x_{2009}}{n_{2010} + n_{2009}} = \frac{441 + 458}{1134 + 1134} \approx 0.396 \text{.}$$

The test statistic is

$$z_0 = \frac{\hat{p}_{2009} - \hat{p}_{2010}}{\sqrt{\hat{p}(1 - \hat{p})}\sqrt{1/n_{2007} + 1/n_{2004}}}$$

$$= \frac{0.404 - 0.389}{\sqrt{0.396(1 - 0.396)}\sqrt{1/1134 + 1/1134}}$$

$$\approx -0.73$$

Classical approach: This is a left-tailed test, so the critical value for a $\alpha = 0.05$ level of significance is $-z_{0.05} = -1.645$. Since $z_0 = -0.73 > -z_{0.05} = -1.645$ (the test statistic does not fall in the critical region), we do not reject H_0.

P-value approach: P-value = $P(z_0 \leq -0.73) =$ 0.2327 [Tech: 0.2328]. Since this P-value is greater than the $\alpha = 0.05$ level of significance, we do not reject H_0.

Conclusion: There is not sufficient evidence at the $\alpha = 0.05$ level of significance to conclude that the proportion of gun owners in 2010 is less than the proportion of gun owners in 2009. Therefore, the headline is not accurate.

39. (a)

$$n = n_1 = n_2$$

$$= \left[\hat{p}_1(1-\hat{p}_1) + \hat{p}_2(1-\hat{p}_2)\right]\left(\frac{z_{\alpha/2}}{E}\right)^2$$

$$= \left[0.219(1-0.219) + 0.197(1-0.197)\right]\left(\frac{1.96}{0.03}\right)^2$$

$$\approx 1405.3$$

We increase this result to 1406.

(b) $n = n_1 = n_2$

$$= 0.5\left(\frac{z_{\alpha/2}}{E}\right)^2 = 0.5\left(\frac{1.96}{0.03}\right)^2 = 2134.2$$

We increase this result to 2135.

41. (a) This is an experiment using a completely randomized design.

(b) The response variable is whether the subject contracted polio or not.

(c) The treatments are the vaccine and the placebo.

(d) A placebo is an innocuous medication that looks, tastes, and smells like the experimental treatment.

(e) Because the incidence rate of polio is low, a large number of subjects is needed so that we are guaranteed a sufficient number of successes.

(f) To answer this question, we test to see if there is a significant difference in the proportion of subjects from the experiment group who contracted polio

and the proportion of subjects from the control group who contracted polio. We will use a $\alpha = 0.01$ level of significance.

We first verify the requirements to perform the hypothesis test: (1) Each sample can be thought of as a simple random sample; (2) we have $x_1 = 33$, $n_1 = 200,000$, $x_2 = 115$, and $n_2 = 200,000$, so

$$\hat{p}_1 = \frac{x_1}{n_1} = \frac{33}{200,000} = 0.000165 \text{ and}$$

$$\hat{p}_2 = \frac{x_2}{n_2} = \frac{115}{200,000} = 0.000575.$$

Therefore, $n_1\hat{p}_1(1-\hat{p}_1) =$

$200,000(0.000165)(1-0.000165) \approx 33 \geq 10$

and $n_2\hat{p}_2(1-\hat{p}_2) =$

$200,000(0.000575)(1-0.000575) \approx 115 \geq 10$;

and (3) each sample is less than 5% of the population. Thus, the requirements are met, so we can conduct the test.

The hypotheses are $H_0 : p_1 = p_2$ versus $H_1 : p_1 < p_2$. From before, the two sample estimates are $\hat{p}_1 = 0.000165$ and $\hat{p}_2 = 0.000575$. The pooled estimate is

$$\hat{p} = \frac{x_1 + x_2}{n_1 + n_2} = \frac{33 + 115}{200,000 + 200,000} = 0.00037.$$

The test statistic is

$$z_0 = \frac{\hat{p}_1 - \hat{p}_2}{\sqrt{\hat{p}(1-\hat{p})}\sqrt{1/n_1 + 1/n_2}}$$

$$= \frac{0.000165 - 0.000575}{\sqrt{0.00037(1-0.00037)}\sqrt{\dfrac{1}{200,000} + \dfrac{1}{200,000}}}$$

$$\approx -6.74$$

Classical approach: This is a left-tailed test, so the critical value is $-z_{0.01} = -2.33$. Since $z_0 = -6.74 < -z_{0.05} = -2.33$ (the test statistic falls in the critical region), we reject H_0.

P-value approach:
P-value = $P(z_0 \leq -6.74) < 0.0001$.
Since this P-value is less than the $\alpha = 0.01$ level of significance, we reject H_0.

Conclusion: There is sufficient evidence at the $\alpha = 0.01$ level of significance to conclude that the proportion of children in the experimental group who contracted polio is less than the proportion of children in the control group who contracted polio.

43. In an independent sample, the individuals in sample A are in no way related to the individuals in sample B; in a dependent sample, the individuals in each sample are somehow related.

Section 11.2

1. $<$

3. (a) We measure differences as $d_i = X_i - Y_i$.

Obs	1	2	3	4	5	6	7
X_i	7.6	7.6	7.4	5.7	8.3	6.6	5.6
Y_i	8.1	6.6	10.7	9.4	7.8	9.0	8.5
d_i	−0.5	1.0	−3.3	−3.7	0.5	−2.4	−2.9

(b) $\sum d_i = (-0.5) + 1.0 + (-3.3) + (-3.7) + 0.5$
$+ (-2.4) + (-2.9) = -11.3$

so $\overline{d} = \dfrac{\sum d_i}{n} = \dfrac{-11.3}{7} \approx -1.614$

$\sum d_i^2 = (-0.5)^2 + (1.0)^2 + (-3.3)^2 + (-3.7)^2$
$+ (0.5)^2 + (-2.4)^2 + (-2.9)^2 = 40.25$

So,

$s_d = \sqrt{\dfrac{\sum d_i^2 - \dfrac{(\sum d_i)^2}{n}}{n-1}}$

$= \sqrt{\dfrac{40.25 - \dfrac{(-11.3)^2}{7}}{7-1}} \approx 1.915$

(c) The hypotheses are $H_0 : \mu_d = 0$ versus $H_1 : \mu_d < 0$. The level of significance is $\alpha = 0.05$. The test statistic is

$t_0 = \dfrac{\overline{d}}{\dfrac{s_d}{\sqrt{n}}} = \dfrac{-1.614}{\dfrac{1.915}{\sqrt{7}}} \approx -2.230$.

Classical approach: Since this is a left-tailed test with 6 degrees of freedom, the

critical value is $-t_{0.05} = -1.943$. Since the test statistic $t_0 \approx -2.230$ is less than the critical value $-t_{0.05} = -1.943$ (i.e., since the test statistic fall within the critical region), we reject H_0.

P-value approach: The P-value for this left-tailed test is the area under the t-distribution with 6 degrees of freedom to the left of the test statistic $t_0 = -2.230$, which by symmetry is equal to the area to the right of $t_0 = 2.230$. From the t-distribution table (Table VI) in the row corresponding to 6 degrees of freedom, 2.230 falls between 1.943 and 2.447 whose right-tail areas are 0.05 and 0.025, respectively. So, $0.025 < P\text{-value} < 0.05$. [Tech: P-value = 0.0336.] Because the P-value is less than the level of significance $\alpha = 0.05$, we reject H_0.

Conclusion: There is sufficient evidence at the $\alpha = 0.05$ level of significance to reject the null hypothesis that $\mu_d = 0$ and conclude that $\mu_d < 0$.

(d) For $\alpha = 0.05$ and df = 6, $t_{\alpha/2} = t_{0.025} = 2.447$. Then:

Lower bound:

$\overline{d} - t_{0.025} \cdot \dfrac{s_d}{\sqrt{n}} = -1.614 - 2.447 \cdot \dfrac{1.915}{\sqrt{7}}$
≈ -3.39

Upper bound:

$\overline{d} + t_{0.025} \cdot \dfrac{s_d}{\sqrt{n}} = -1.614 + 2.447 \cdot \dfrac{1.915}{\sqrt{7}}$
≈ 0.16

We can be 95% confident that the mean difference is between -3.39 and 0.16.

5. (a) The hypotheses are $H_0 : \mu_d = 0$ versus $H_1 : \mu_d > 0$.

(b) The level of significance is $\alpha = 0.05$. The test statistic is $t_0 = \dfrac{\overline{d}}{\dfrac{s_d}{\sqrt{n}}} = \dfrac{1.14}{\dfrac{1.75}{\sqrt{16}}} \approx 2.606$.

Classical approach: Since this is a right-tailed test with 15 degrees of freedom, the critical value is $t_{0.05} = 1.753$. Since the test statistic $t_0 \approx 2.606$ falls beyond the critical value $t_{0.05} = 1.753$ (i.e., since the test statistic falls within the critical region), we reject H_0.

P-value approach: The *P*-value for this one-tailed test is the area under the *t*-distribution with 15 degrees of freedom to the right of the test statistic. From the *t*-distribution table in the row corresponding to 15 degrees of freedom, 2.606 falls between 2.602 and 2.947 whose right-tail areas are 0.01 and 0.005, respectively. So, $0.005 < P\text{-value} < 0.01$. [Tech: *P*-value = 0.0099.] Because the *P*-value is less than the level of significance $\alpha = 0.05$, we reject H_0.

Conclusion: There is sufficient evidence at the $\alpha = 0.05$ level of significance to support the claim that $\mu_d > 0$, suggesting that a baby will watch the climber approach the hinderer toy for a longer time than the baby will watch the climber approach the helper toy.

(c) Answers will vary. The fact that the babies watch the surprising behavior for a longer period of time suggests that they are curious about it.

7. (a) These are matched-pair data because two measurements (A and B) are taken on the same round.

(b) We measure differences as $d_i = A_i - B_i$.

Obs	1	2	3	4	5	6
A	793.8	793.1	792.4	794.0	791.4	792.4
B	793.2	793.3	792.6	793.8	791.6	791.6
d_i	0.6	−0.2	−0.2	0.2	−0.2	0.8

Obs	7	8	9	10	11	12
A	791.4	792.3	789.6	794.4	790.9	793.5
B	791.6	792.4	788.5	794.7	791.3	793.5
d_i	0.1	−0.1	1.1	−0.3	−0.4	0

We compute the mean and standard deviation of the differences and obtain $\bar{d} \approx 0.1167$ feet per second, rounded to four decimal places, and $s_d \approx 0.4745$ feet per second, rounded to four decimal places. The hypotheses are $H_0 : \mu_d = 0$ versus $H_1 : \mu_d \neq 0$. The level of significance is

$\alpha = 0.01$. The test statistic is

$$t_0 = \frac{\bar{d}}{\frac{s_d}{\sqrt{n}}} = \frac{0.1167}{\frac{0.4745}{\sqrt{12}}} \approx 0.852.$$

Classical approach: Since this is a two-tailed test with 11 degrees of freedom, the critical values are $\pm t_{0.005} = \pm 3.106$. Since the test statistic $t_0 \approx 0.852$ falls between the critical values $\pm t_{0.005} = \pm 3.106$ (i.e., since the test statistic does not fall within the critical regions), we do not reject H_0.

P-value approach: The *P*-value for this two-tailed test is the area under the *t*-distribution with 11 degrees of freedom to the right of the test statistic $t_0 = 0.853$ plus the area to the left of $t_0 = -0.853$. From the *t*-distribution table, in the row corresponding to 11 degrees of freedom, 0.852 falls between 0.697 and 0.876 whose right-tail areas are 0.25 and 0.20, respectively.

We must double these values in order to get the total area in both tails: 0.50 and 0.40. So, $0.50 > P\text{-value} > 0.40$. [Tech: *P*-value = 0.4125.] Because the *P*-value is greater than the level of significance $\alpha = 0.01$, we do not reject H_0.

Conclusion: There is not sufficient evidence at the $\alpha = 0.01$ level of significance to conclude that there is a difference in the measurements of velocity between device A and device B.

(c) For $\alpha = 0.01$ and df = 11, $t_{\alpha/2} = t_{0.005} = 3.106$. Then:

Lower bound:

$$\bar{d} - t_{0.005} \cdot \frac{s_d}{\sqrt{n}} = 0.1167 - 3.106 \cdot \frac{0.4745}{\sqrt{12}}$$
$$\approx -0.31$$

Upper bound:

$$\bar{d} + t_{0.005} \cdot \frac{s_d}{\sqrt{n}} = 0.1167 + 3.106 \cdot \frac{0.4745}{\sqrt{12}}$$
$$\approx 0.54$$

We are 99% confident that the population mean difference in measurement is between −0.31 and 0.54 feet per second.

(d)

A–B

Yes. Since a difference of 0 is located in the middle 50%, the boxplot supports that there is no difference in measurements.

9. (a) These are matched pair data because both cars are involved in the same collision.

(b) We measure differences as
$d_i = \text{SUV}_i - \text{Car}_i$.

Obs	1	2	3	4	5	6	7
SUV	1721	1434	850	2329	1415	1470	2884
Car	1274	2327	3223	2058	3095	3386	4560
d_i	447	−893	−2372	271	−1680	−1916	−1676

SUV – Car

The median of the differences is well to the left of 0, suggesting that SUVs do have a lower repair cost.

(c) We compute the mean and standard deviation of the differences and obtain
$\bar{d} = \$-1117.1$ and $s_d \approx \$1100.6$. The hypotheses are $H_0 : \mu_d = 0$ versus $H_1 : \mu_d < 0$. The level of significance is $\alpha = 0.05$. The test statistic is

$$t_0 = \frac{\bar{d}}{\frac{s_d}{\sqrt{n}}} = \frac{-1117.1}{\frac{1100.6}{\sqrt{6}}} \approx -2.685$$

[Tech: $t_0 \approx -2.685$].

Classical approach: Since this is a left-tailed test with 6 degrees of freedom, the critical value is $-t_{0.05} = -1.943$. Since the test statistic $t_0 \approx -2.685$ falls beyond the critical value $-t_{0.05} = -1.943$ (i.e., since the test statistic falls within the critical regions), we reject H_0.

P-value approach: The P-value for this two-tailed test is the area under the *t*-distribution with 6 degrees of freedom to the left of the test statistic $t_0 = -2.685$. From the *t*-distribution table, in the row corresponding to 6 degrees of freedom, 2.685 falls between 2.612 and 3.143 whose right-tail areas are 0.02 and 0.01, respectively. So, $0.01 > P\text{-value} > 0.02$. [Tech: P-value = 0.0181]. Because the P-value is less than the level of significance $\alpha = 0.05$, we reject H_0.

Conclusion: There is sufficient evidence at the $\alpha = 0.01$ level of significance to support the claim that the repair cost for the SUV is lower, or that the repair cost for the car is higher.

11. We measure differences as $d_i = Y_i - X_i$.

Obs	1	2	3	4	5	6	7
X_i	70.3	67.1	70.9	66.8	72.8	70.4	71.8
Y_i	74.1	69.2	66.9	69.2	68.9	70.2	70.4
d_i	3.8	2.1	−4.0	2.4	−3.9	−0.2	−1.4

Obs	8	9	10	11	12	13
X_i	70.1	69.9	70.8	70.2	70.4	72.4
Y_i	69.3	75.8	72.3	69.2	68.6	73.9
d_i	−0.8	5.9	1.5	−1.0	−1.8	1.5

We compute the mean and standard deviation of the differences and obtain $\bar{d} \approx 0.3154$ inches, rounded to four decimal places, and $s_d \approx 2.8971$ inches, rounded to four decimal places. The hypotheses are $H_0 : \mu_d = 0$ versus $H_1 : \mu_d > 0$. The level of significance is $\alpha = 0.1$. The test statistic is

$$t_0 = \frac{\bar{d}}{\frac{s_d}{\sqrt{n}}} = \frac{0.3154}{\frac{2.8971}{\sqrt{13}}} \approx 0.393.$$

Classical approach: Since this is a right-tailed test with 12 degrees of freedom, the critical value is $t_{0.10} = 1.356$. Since the test statistic $t_0 \approx 0.393$ does not fall to the right of the critical value $t_{0.10} = 1.356$ (i.e., since the test statistic falls outside the critical region), we do not reject H_0.

P-value approach: The *P*-value for this right-tailed test is the area under the *t*-distribution with 12 degrees of freedom to the right of the test statistic $t_0 = 0.393$. From the *t*-distribution table, in the row corresponding to 12 degrees of freedom, 0.393 falls to the left of 0.695, whose right-tail area is 0.25. So, *P*-value > 0.25 [Tech: *P*-value = 0.3508]. Because the *P*-value is greater than the level of significance $\alpha = 0.10$, we do not reject H_0.

Conclusion: No, there is not sufficient evidence at the $\alpha = 0.10$ level of significance to conclude that sons are taller than their fathers.

13. We measure differences as $d_i = \text{diamond} - \text{steel}$.

Specimen	1	2	3	4	5	6	7	8	9
Steel ball	50	57	61	71	68	54	65	51	53
Diamond	52	56	61	74	69	55	68	51	56
d_i	2	−1	0	3	1	1	3	0	3

We compute the mean and standard deviation of the differences and obtain $\bar{d} \approx 1.3333$, rounded to four decimal places, and $s_d = 1.5$. The hypotheses are $H_0 : \mu_d = 0$ versus $H_1 : \mu_d \neq 0$. For $\alpha = 0.05$ and df = 8, $t_{\alpha/2} = t_{0.025} = 2.306$. Then:

Lower bound:

$$\bar{d} - t_{0.025} \cdot \frac{s_d}{\sqrt{n}} = 1.3333 - 2.306 \cdot \frac{1.5}{\sqrt{9}} \approx 0.2$$

Upper bound:

$$\bar{d} + t_{0.025} \cdot \frac{s_d}{\sqrt{n}} = 1.3333 + 2.306 \cdot \frac{1.5}{\sqrt{9}} \approx 2.5$$

We can be 95% confident that the population mean difference in hardness reading is between 0.2 and 2.5. This interval does not include 0, so we reject H_0.

Conclusion: There is sufficient evidence to conclude that the two indenters produce different hardness readings.

15. (a) It is important to randomly select whether the student would first be tested with normal or impaired vision in order to control for any "learning" that may occur in using the simulator.

(b) We measure differences as $d_i = Y_i - X_i$.

Subject	1	2	3	4	5
Normal, X_i	4.47	4.24	4.58	4.65	4.31
Impaired, Y_i	5.77	5.67	5.51	5.32	5.83
d_i	1.30	1.43	0.93	0.67	1.52

Subject	6	7	8	9
Normal, X_i	4.80	4.55	5.00	4.79
Impaired, Y_i	5.49	5.23	5.61	5.63
d_i	0.69	0.68	0.61	0.84

We compute the mean and standard deviation of the differences and obtain $\bar{d} \approx 0.9633$ seconds, rounded to four decimal places, and $s_d \approx 0.3576$ seconds, rounded to four decimal places. For $\alpha = 0.05$ and df = 8, $t_{\alpha/2} = t_{0.025} = 2.306$. Lower bound:

$$\bar{d} - t_{0.025} \cdot \frac{s_d}{\sqrt{n}} = 0.9633 - 2.306 \cdot \frac{0.3576}{\sqrt{9}}$$
$$\approx 0.688$$

Upper bound:

$$\bar{d} + t_{0.025} \cdot \frac{s_d}{\sqrt{n}} = 0.9633 + 2.306 \cdot \frac{0.3576}{\sqrt{9}}$$
$$\approx 1.238$$

We can be 95% confident that the population mean difference in reaction time when teenagers are driving impaired from when driving normally is between 0.688 and 1.238 seconds. This interval does not include 0, so we reject H_0.

Conclusion: There is sufficient evidence to conclude that there is a difference in braking time with impaired vision and normal vision.

17. (a) Matching by driver and car for the two different octane levels is important as a means of controlling the experiment. Drivers and cars behave differently, so this matching reduces variability in miles per gallon that is attributable to the driver's driving style.

(b) Conducting the experiment on a closed track allows for control so that all cars and drivers can be put through the exact same driving conditions.

(c) No, neither variable is normally distributed. Both have at least one point outside the bounds of the normal probability plot.

(d) Yes, the differences in mileages appear to the approximately normally distributed since all of the points fall within the boundaries of the probability plot.

(e) From the MINITAB printout, $\bar{d} \approx 5.091$ miles and $s_d \approx 14.876$ miles, where the differences are 92 octane minus 87 octane. The hypotheses are $H_0 : \mu_d = 0$ versus $H_1 : \mu_d > 0$, where $d_i = 92 \text{ Oct} - 87 \text{ Oct}$. The level of significance is $\alpha = 0.05$. The test statistic is $t_0 \approx 1.14$.

Classical approach: Since this is a right-tailed test with 10 degrees of freedom, the critical value is $t_{0.05} = 1.812$. Since the test statistic $t_0 \approx 1.14$ does not fall to the right of the critical value $t_{0.05} = 1.812$ (i.e., since the test statistic does not fall within the critical region), we fail to reject H_0.

P-value approach: From the MINITAB printout, we find that P-value ≈ 0.141. Because the P-value is greater than the level of significance $\alpha = 0.05$, we do not reject H_0.

Conclusion: We would expect to get the results we obtained in about 14 out of 100 samples if the statement in the null hypothesis were true. Our result is not unusual. Thus, there is not sufficient evidence at the $\alpha = 0.05$ level of significance to conclude that cars get better mileage when using 92-octain gasoline than when using 87-octane gasoline.

Section 11.3

1. **(a)** $H_0 : \mu_1 = \mu_2$ versus $H_1 : \mu_1 \neq \mu_2$. The level of significance is $\alpha = 0.05$. Since the sample size of both groups is 15, we use $n_1 - 1 = 14$ degrees of freedom. The test statistic is

$$t_0 = \frac{(\bar{x}_1 - \bar{x}_2) - (\mu_1 - \mu_2)}{\sqrt{\frac{s_1^2}{n_1} + \frac{s_2^2}{n_2}}} = \frac{(15.3 - 14.2) - 0}{\sqrt{\frac{3.2^2}{15} + \frac{3.5^2}{15}}}$$

$$\approx 0.898$$

Classical approach: Since this is a two-tailed test with 14 degrees of freedom, the critical values are $\pm t_{0.025} = \pm 2.145$. Since the test statistic $t_0 \approx 0.898$ is between than the critical values $-t_{0.025} = -2.145$ and $t_{0.025} = 2.145$ (i.e., since the test statistic does not fall within the critical region), we do not reject H_0.

P-value approach: The P-value for this two-tailed test is the area under the t-distribution with 14 degrees of freedom to the right of $t_0 = 0.898$ plus the area to the left of -0.898. From the t-distribution table in the row corresponding to 14 degrees of freedom, 0.898 falls between 0.868 and 1.076 whose right-tail areas are 0.20 and 0.15, respectively. We must double these values in order to get the total area in both tails: 0.40 and 0.30. So, $0.30 < P\text{-value} < 0.40$ [Tech: P-value = 0.3767]. Because the P-value is greater than the level of significance $\alpha = 0.05$, we do not reject H_0.

Conclusion: There is not sufficient evidence at the $\alpha = 0.05$ level of significance to conclude that the population means are different.

(b) For a 95% confidence interval with df = 14, we use $t_{\alpha/2} = t_{0.025} = 2.145$. Then:

Lower bound:

$$(\bar{x}_1 - \bar{x}_2) - t_{\alpha/2} \cdot \sqrt{\frac{s_1^2}{n_1} + \frac{s_2^2}{n_2}}$$

$$= (15.3 - 14.2) - 2.145 \cdot \sqrt{\frac{3.2^2}{15} + \frac{3.5^2}{15}}$$

$$= -1.53 \text{ [Tech: } -1.41]$$

Upper bound:

$$(\bar{x}_1 - \bar{x}_2) + t_{\alpha/2} \cdot \sqrt{\frac{s_1^2}{n_1} + \frac{s_2^2}{n_2}}$$

$$= (15.3 - 14.2) + 2.145 \cdot \sqrt{\frac{3.2^2}{15} + \frac{3.5^2}{15}}$$

$$\approx 3.73 \text{ [Tech: 3.61]}$$

We can be 95% confident that the mean difference is between -1.53 and 3.73 [Tech: between -1.41 and 3.61].

3. (a) $H_0 : \mu_1 = \mu_2$ versus $H_1 : \mu_1 > \mu_2$. The level of significance is $\alpha = 0.10$. Since the smaller sample size is $n_2 = 18$, we use $n_2 - 1 = 18 - 1 = 17$ degrees of freedom. The test statistic is

$$t_0 = \frac{(\bar{x}_1 - \bar{x}_2) - (\mu_1 - \mu_2)}{\sqrt{\dfrac{s_1^2}{n_1} + \dfrac{s_2^2}{n_2}}} = \frac{(50.2 - 42.0) - 0}{\sqrt{\dfrac{6.4^2}{25} + \dfrac{9.9^2}{18}}}$$

$$\approx 3.081$$

Classical approach: Since this is a right-tailed test with 17 degrees of freedom, the critical value is $t_{0.10} = 1.333$.

Since the test statistic $t_0 \approx 3.081$ is to the right of the critical value (i.e., since the test statistic falls within the critical region), we reject H_0.

P-value approach: The *P*-value for this right-tailed test is the area under the *t*-distribution with 17 degrees of freedom to the right of $t_0 = 3.081$. From the *t*-distribution table, in the row corresponding to 17 degrees of freedom, 3.081 falls between 2.898 and 3.222 whose right-tail areas are 0.005 and 0.0025, respectively. Thus, $0.0025 < P$-value < 0.005 [Tech: *P*-value = 0.0024]. Because the *P*-value is less than the level of significance $\alpha = 0.10$, we reject H_0.

Conclusion: There is sufficient evidence at the $\alpha = 0.01$ level of significance to conclude that $\mu_1 > \mu_2$.

(b) For a 90% confidence interval with df = 17, we use $t_{\alpha/2} = t_{0.05} = 1.740$. Then:

Lower bound:

$$(\bar{x}_1 - \bar{x}_2) - t_{\alpha/2} \cdot \sqrt{\frac{s_1^2}{n_1} + \frac{s_2^2}{n_2}}$$

$$= (50.2 - 42.0) - 1.740 \cdot \sqrt{\frac{6.4^2}{25} + \frac{9.9^2}{18}}$$

$$\approx 3.57 \ \text{[Tech: 3.67]}$$

Upper bound:

$$(\bar{x}_1 - \bar{x}_2) + t_{\alpha/2} \cdot \sqrt{\frac{s_1^2}{n_1} + \frac{s_2^2}{n_2}}$$

$$= (50.2 - 42.0) + 1.740 \cdot \sqrt{\frac{6.4^2}{25} + \frac{9.9^2}{18}}$$

$$\approx 12.83 \ \text{[Tech: 12.73]}$$

We can be 90% confident that the mean difference is between 3.57 and 12.83 [Tech: between 3.67 and 12.73].

5. $H_0 : \mu_1 = \mu_2$ versus $H_1 : \mu_1 < \mu_2$. The level of significance is $\alpha = 0.02$. Since the smaller sample size is $n_2 = 25$, we use $n_2 - 1 = 25 - 1 = 24$ degrees of freedom. The test statistic is

$$t_0 = \frac{(\bar{x}_1 - \bar{x}_2) - (\mu_1 - \mu_2)}{\sqrt{\dfrac{s_1^2}{n_1} + \dfrac{s_2^2}{n_2}}} = \frac{(103.4 - 114.2) - 0}{\sqrt{\dfrac{12.3^2}{32} + \dfrac{13.2^2}{25}}}$$

$$\approx -3.158$$

Classical approach: Since this is a left-tailed test with 24 degrees of freedom, the critical value is $-t_{0.02} = -2.172$. Since the test statistic $t_0 \approx -3.158$ is to the left of the critical value (i.e., since the test statistic falls within the critical region), we reject H_0.

P-value approach: The *P*-value for this left-tailed test is the area under the *t*-distribution with 24 degrees of freedom to the left of $t_0 \approx -3.158$, which is equivalent to the area to the right of $t_0 \approx 3.158$. From the *t*-distribution table in the row corresponding to 24 degrees of freedom, 3.158 falls between 3.091 and 3.467 whose right-tail areas are 0.0025 and 0.001, respectively. Thus, $0.001 < P$-value < 0.0025 [Tech: *P*-value = 0.0013]. Because the *P*-value is less than the level of significance $\alpha = 0.02$, we reject H_0.

Conclusion: There is sufficient evidence at the $\alpha = 0.02$ level of significance to conclude that $\mu_1 < \mu_2$.

7. (a) $H_0 : \mu_{\text{CC}} = \mu_{\text{NT}}$ versus $H_1 : \mu_{\text{CC}} > \mu_{\text{NT}}$. The level of significance is $\alpha = 0.01$. Since the smaller sample size is $n_2 = 268$, we use $n_2 - 1 = 268 - 1 = 267$ degrees of freedom. The test statistic is

$$t_0 = \frac{(\overline{x}_1 - \overline{x}_2) - (\mu_1 - \mu_2)}{\sqrt{\frac{s_1^2}{n_1} + \frac{s_2^2}{n_2}}} = \frac{(5.43 - 4.43) - 0}{\sqrt{\frac{1.162^2}{268} + \frac{1.015^2}{1145}}}$$

$$\approx 12.977$$

Classical approach: Since this is a right-tailed test with 267 degrees of freedom. However, since our *t*-distribution table does not contain a row for 267, we use df = 100. Thus, the critical value is $t_{0.01} = 2.364$. Since the test statistic $t_0 \approx 12.977$ is to the right of the critical value (i.e., since the test statistic falls within the critical region), we reject H_0.

P-value approach: The *P*-value for this right-tailed test is the area under the *t*-distribution with 267 degrees of freedom to the right of $t_0 = 12.977$. From the *t*-distribution table, in the row corresponding to 100 degrees of freedom (since the table does not contain a row for df = 267), 12.977 falls to the right of 3.390 whose right-tail area is 0.0005.

Thus, *P*-value < 0.0005 [Tech: *P*-value < 0.0001]. Because the *P*-value is less than the level of significance $\alpha = 0.01$, we reject H_0.

Conclusion: There is sufficient evidence at the $\alpha = 0.01$ level of significance to conclude that $\mu_{CC} > \mu_{NT}$. That is, the evidence suggests that the mean time to graduate for students who first start in community college is longer than the mean time to graduate for those who do not transfer.

(b) For a 95% confidence interval with 100 degrees of freedom (since the table does not contain a row for df = 267), we use $t_{\alpha/2} = t_{0.025} = 1.984$. Then:

Lower bound:

$$(\overline{x}_1 - \overline{x}_2) - t_{\alpha/2} \cdot \sqrt{\frac{s_1^2}{n_1} + \frac{s_2^2}{n_2}}$$

$$= (5.43 - 4.43) - 1.984 \cdot \sqrt{\frac{1.162^2}{268} + \frac{1.015^2}{1145}}$$

$$\approx 0.847 \text{ [Tech: 0.848]}$$

Upper bound:

$$(\overline{x}_1 - \overline{x}_2) + t_{\alpha/2} \cdot \sqrt{\frac{s_1^2}{n_1} + \frac{s_2^2}{n_2}}$$

$$= (5.43 - 4.43) + 1.984 \cdot \sqrt{\frac{1.162^2}{268} + \frac{1.015^2}{1145}}$$

$$\approx 1.153 \text{ [Tech: 1.152]}$$

We can be 95% confident that the mean additional time to graduate for students who start in community college is between 0.847 and 1.153 years [Tech: between 0.848 and 1.152 years].

(c) No, the results from parts (a) and (b) do not imply that community college *causes* one to take extra time to earn a bachelor's degree. This is observational data. Community college students may be working more hours, which does not allow them to take additional classes.

9. (a) This is an observational study, since no treatment is imposed. The researcher did not influence the data.

(b) Though we do not know if the population is normally distributed, the samples are independent with sizes that are sufficiently large ($n_{\text{arrival}} = n_{\text{departure}} = 35$).

(c) $H_0 : \mu_{\text{arrival}} = \mu_{\text{departure}}$ versus $H_1 : \mu_{\text{arrival}} \neq \mu_{\text{departure}}$. The level of significance is $\alpha = 0.05$. Since the sample size of both groups is 35, we use $n_{\text{arrival}} - 1 = 34$ degrees of freedom. The test statistic is

$$t_0 = \frac{(\overline{x}_{\text{arrival}} - \overline{x}_{\text{departure}}) - (\mu_{\text{arrival}} - \mu_{\text{departure}})}{\sqrt{\frac{s_{\text{arrival}}^2}{n_{\text{arrival}}} + \frac{s_{\text{departure}}^2}{n_{\text{departure}}}}}$$

$$= \frac{(269 - 260) - 0}{\sqrt{\frac{53^2}{35} + \frac{34^2}{35}}} \approx 0.846$$

Classical approach: Since this is a two-tailed test with 34 degrees of freedom, the critical values are $\pm t_{0.025} = \pm 2.032$. Since the test statistic $t_0 \approx 0.846$ is between the critical values (i.e., since the test statistic does not fall within the critical regions), we do not reject H_0.

P-value approach: The *P*-value for this two-tailed test is the area under the *t*-distribution with 34 degrees of freedom to the right of $t_0 \approx 0.846$ plus the area to the left of $t_0 \approx -0.846$. From the *t*-distribution table in the row corresponding to 34 degrees of freedom, 0.846 falls between 0.682 and 0.852 whose right-tail areas are 0.25 and 0.20, respectively. We must double these values in order to get the total area in both tails: 0.50 and 0.40. So, $0.40 < P$-value < 0.50 [Tech: *P*-value = 0.4013]. Since the *P*-value is greater than the level of significance $\alpha = 0.05$, we do not reject H_0.

Conclusion: There is not sufficient evidence at the $\alpha = 0.05$ level of significance to conclude that travelers walk at different speeds depending on whether they are arriving or departing an airport.

11. **(a)** Both samples include 200 observations, so there are 199 degrees of freedom. For a 95% confidence interval with df = 199, we use row 100 of Table VI to get $t_{\alpha/2} = t_{0.025} = 1.984$. Then:

Lower bound:

$$(\bar{x}_1 - \bar{x}_2) - t_{\alpha/2} \cdot \sqrt{\frac{s_1^2}{n_1} + \frac{s_2^2}{n_2}}$$

$$= (23.4 - 17.9) - 1.984 \cdot \sqrt{\frac{4.1^2}{200} + \frac{3.9^2}{200}}$$

$$\approx 4.71 \quad \text{[Tech: 4.71]}$$

Upper bound:

$$(\bar{x}_1 - \bar{x}_2) + t_{\alpha/2} \cdot \sqrt{\frac{s_1^2}{n_1} + \frac{s_2^2}{n_2}}$$

$$= (23.4 - 17.9) + 1.984 \cdot \sqrt{\frac{4.1^2}{200} + \frac{3.9^2}{200}}$$

$$\approx 6.29 \quad \text{[Tech: 6.29]}$$

We can be 95% confident that the mean difference in the scores is between students who think about being a professor and student who think about soccer hooligans is between 4.71 and 6.29.

(b) Since the 95% confidence interval does not contain 0, the results suggest that priming does have an effect on scores.

13. **(a)** The five number summaries follow:

Meters Off:
17, 26, 37, 41, 52

Meters On:
25, 31, 42, 48, 56

Based on the boxplots, there does appear to be a difference in the speeds.

(b) $H_0 : \mu_{On} = \mu_{Off}$ versus $H_1 : \mu_{On} > \mu_{Off}$. The level of significance is $\alpha = 0.10$. Both samples involve 15 observations, so we use $n_1 - 1 = 15 - 1 = 14$ degrees of freedom. The test statistic is

$$t_0 = \frac{(\bar{x}_{On} - \bar{x}_{Off}) - (\mu_{On} - \mu_{Off})}{\sqrt{\frac{s_{On}^2}{n_{On}} + \frac{s_{Off}^2}{n_{Off}}}}$$

$$= \frac{(40.67 - 34.53) - 0}{\sqrt{\frac{10.04^2}{15} + \frac{9.56^2}{15}}}$$

$$\approx 1.713$$

Classical approach: This is a right-tailed test with 14 degrees of freedom. Thus, the critical value is $t_{0.10} = 1.345$. Since the test statistic $t_0 \approx 1.713$ is to the right of the critical value (i.e., since the test statistic falls within the critical region), we reject H_0.

P-value approach: The *P*-value for this right-tailed test is the area under the *t*-distribution with 14 degrees of freedom to the right of $t_0 = 1.713$. From the *t*-distribution table in the row corresponding to 14 degrees of freedom, 1.713 falls between 1.345 and 1.761 whose right-tail areas are 0.10 and 0.05, respectively.

Thus, $0.05 < P$-value < 0.10 [Tech: *P*-value = 0.0489]. Because the *P*-value is less than the level of significance $\alpha = 0.10$, we reject H_0.

Conclusion: There is sufficient evidence at the $\alpha = 0.10$ level of significance to conclude that the ram meters are effective in maintaining higher speed on the freeway.

15. $H_0: \mu_{\text{carpeted}} = \mu_{\text{uncarpeted}}$ versus $H_1: \mu_{\text{carpeted}} > \mu_{\text{uncarpeted}}$. The level of significance is $\alpha = 0.05$. The sample statistics for the data are
$\bar{x}_{\text{carpeted}} = 11.2$, $s_{\text{carpeted}} \approx 2.6774$, and $n_{\text{carpeted}} = 8$, and
$\bar{x}_{\text{uncarpeted}} = 9.7875$, $s_{\text{uncarpeted}} \approx 3.2100$, and $n_{\text{uncarpeted}} = 8$. Since the sample size of both groups is 8, we use $n_{\text{carpeted}} - 1 = 7$ degrees of freedom.

The test statistic is

$$t_0 = \frac{(\bar{x}_{\text{carpeted}} - \bar{x}_{\text{uncarpeted}}) - (\mu_{\text{carpeted}} - \mu_{\text{uncarpeted}})}{\sqrt{\dfrac{s_{\text{carpeted}}^2}{n_{\text{carpeted}}} + \dfrac{s_{\text{uncarpeted}}^2}{n_{\text{uncarpeted}}}}}$$

$$= \frac{(11.2 - 9.7875) - 0}{\sqrt{\dfrac{2.6774^2}{8} + \dfrac{3.2100^2}{8}}} \approx 0.956$$

Classical approach: Since this is a right-tailed test with 7 degrees of freedom, the critical value is $t_{0.05} = 1.895$. Since the test statistic $t_0 \approx 0.954$ is not to the right of the critical value (i.e., since the test statistic does not fall within the critical region), we do not reject H_0.

P-value approach: The P-value for this right-tailed test is the area under the t-distribution with 7 degrees of freedom to the right of $t_0 \approx 0.956$. From the t-distribution table in the row corresponding to 7 degrees of freedom, 0.956 falls between 0.896 and 1.119 whose right-tail areas are 0.20 and 0.15, respectively. So, $0.15 < P\text{-value} < 0.20$ [Tech: P-value = 0.1780]. Since the P-value is greater than the level of significance $\alpha = 0.05$, we do not reject H_0.

Conclusion: There is not sufficient evidence at the $\alpha = 0.05$ level of significance to

conclude that carpeted rooms have more bacteria than uncarpeted rooms.

17. Since both sample sizes are sufficiently large ($n_{\text{AL}} = n_{\text{NL}} = 30$), we can use Welch's t-test. The hypotheses are $H_0: \mu_{\text{AL}} = \mu_{\text{NL}}$ versus $H_1: \mu_{\text{AL}} > \mu_{\text{NL}}$. The level of significance is $\alpha = 0.05$. We are given the summary statistics $\bar{x}_{\text{AL}} = 6.0$ and $s_{\text{AL}} = 3.5$, $\bar{x}_{\text{NL}} = 4.3$, and $s_{\text{NL}} = 2.6$. Since the sample size of both groups is 30, we use $n_{\text{AL}} - 1 = 29$ degrees of freedom. The test statistic is

$$t_0 = \frac{(\bar{x}_{\text{AL}} - \bar{x}_{\text{NL}}) - (\mu_{\text{AL}} - \mu_{\text{NL}})}{\sqrt{\dfrac{s_{\text{AL}}^2}{n_{\text{AL}}} + \dfrac{s_{\text{NL}}^2}{n_{\text{NL}}}}} = \frac{(6.0 - 4.3) - 0}{\sqrt{\dfrac{3.5^2}{30} + \dfrac{2.6^2}{30}}}$$

$$\approx 2.136$$

Classical approach: Since this is a right-tailed test with 29 degrees of freedom, the critical value is $t_{0.05} = 1.699$. Since the test statistic $t_0 \approx 2.136$ falls to the right of the critical value 1.699 (i.e., since the test statistic falls within the critical region), we reject H_0.

P-value approach: The P-value for this right-tailed test is the area under the t-distribution with 29 degrees of freedom to the right of $t_0 \approx 2.136$. From the t-distribution table, in the row corresponding to 29 degrees of freedom, 2.136 falls between 2.045 and 2.150 whose right-tail areas are 0.025 and 0.02, respectively. So, $0.02 < P\text{-value} < 0.025$ [Tech: P-value = 0.0187]. Since the P-value is less than the level of significance $\alpha = 0.05$, we reject H_0.

Conclusion: There is sufficient evidence at the $\alpha = 0.05$ level of significance to conclude that games played with a designated hitter result in more runs.

19. Since both sample sizes are sufficiently large ($n_{\text{Male}} = 81$ and $n_{\text{Female}} = 115$), we can use Welch's t-test. The hypotheses are $H_0: \mu_{\text{Male}} = \mu_{\text{Female}}$ versus $H_1: \mu_{\text{Male}} \neq \mu_{\text{Female}}$. The level of significance is $\alpha = 0.05$. Using technology, we get the summary statistics $\bar{x}_{\text{Male}} = 2.35$, $s_{\text{Male}} = 0.78$, $\bar{x}_{\text{Female}} = 2.71$, and $s_{\text{Female}} = 1.01$. Since the sample size of the smaller group is 81, we use $n_{Male} - 1 = 80$ degrees of freedom.

The test statistic is

$$t_0 = \frac{(\bar{x}_{Male} - \bar{x}_{Female}) - (\mu_{Male} - \mu_{Female})}{\sqrt{\dfrac{s_{Male}^2}{n_{Male}} + \dfrac{s_{Female}^2}{n_{Female}}}}$$

$$= \frac{(2.35 - 2.71) - 0}{\sqrt{\dfrac{0.78^2}{81} + \dfrac{1.01^2}{115}}} \approx -2.756$$

Classical approach: Since this is a two-tailed test with 80 degrees of freedom, the critical values are $\pm t_{0.05} = \pm 1.990$. Since the test statistic $t_0 \approx -2.756$ is less than the critical value -1.990 (i.e., since the test statistic falls within the critical regions), we reject H_0.

P-value approach: The P-value for this two-tailed test is the area under the t-distribution with 80 degrees of freedom to the right of $-t_0 \approx 2.756$ plus the area to the left of $t_0 \approx -2.756$. From the t-distribution table in the row corresponding to 80 degrees of freedom, 2.756 falls between 2.639 and 2.887 whose right-tail areas are 0.005 and 0.0025, respectively. We must double these values in order to get that the total area in both tails is between 0.005 and 0.01. So, $0.005 < P\text{-value} < 0.01$ [Tech: P-value = 0.0064]. Since the P-value is less than the level of significance $\alpha = 0.05$, we reject H_0.

Conclusion: There is sufficient evidence at the $\alpha = 0.05$ level of significance to conclude that there is a difference between males and females regarding the ideal number of children.

21. For this 90% confidence level with df $= 40 - 1$ $= 39$, we use $t_{\alpha/2} = t_{0.05} = 1.685$. Then:

Lower bound:

$$(\bar{x}_{no\ children} - \bar{x}_{children}) - t_{\alpha/2} \cdot \sqrt{\frac{s_{no\ children}^2}{n_{no\ children}} + \frac{s_{children}^2}{n_{children}}}$$

$$= (5.62 - 4.10) - 1.685 \cdot \sqrt{\frac{2.43^2}{40} + \frac{1.82^2}{40}}$$

$$\approx 0.71 \ [\text{Tech: } 0.72]$$

Upper bound:

$$(\bar{x}_{no\ children} - \bar{x}_{children}) + t_{\alpha/2} \cdot \sqrt{\frac{s_{no\ children}^2}{n_{no\ children}} + \frac{s_{children}^2}{n_{children}}}$$

$$= (5.62 - 4.10) + 1.685 \cdot \sqrt{\frac{2.43^2}{40} + \frac{1.82^2}{40}}$$

$$\approx 2.33 \ [\text{Tech: } 2.32]$$

We can be 90% confident that the mean difference in daily leisure time between adults without children and those with children is between 0.71 and 2.33 hours [Tech: between 0.72 and 2.33 hours]. Since the confidence interval does not include zero, we can conclude that there is a significance difference in the leisure time of adults without children and those with children.

23. (a) The hypotheses are $H_0 : \mu_{men} = \mu_{women}$ versus $H_1 : \mu_{men} < \mu_{women}$.

(b) From the MINITAB printout, P-value = 0.0051. Because the P-value is less than the level of significance $\alpha = 0.01$, the researcher will reject H_0. Thus, there is sufficient evidence at the $\alpha = 0.01$ level of significance to conclude that the mean step pulse of men is lower than the mean step pulse of women.

(c) From the MINITAB printout, the 95% confidence interval is: Lower bound: -10.7; Upper bound: -1.5 We are 95% confident that the mean step pulse of men is between 1.5 and 10.7 beats per minute lower than the mean step pulse of women

25. (a) This is an experiment using a completely randomized design.

(b) The response variable is scores on the final exam. The treatments are online homework system versus old-fashioned paper-and-pencil homework.

(c) Factors that are controlled in the experiment are the teacher, the text, the syllabus, the tests, the meeting time, the meeting location.

(d) In this case, the assumption is that the students "randomly" enrolled in the course.

(e) The hypotheses are $H_0 : \mu_{\text{fall}} = \mu_{\text{spring}}$ versus $H_1 : \mu_{\text{fall}} > \mu_{\text{spring}}$. The level of significance is $\alpha = 0.05$. We are given the summary statistics $\bar{x}_{\text{fall}} = 73.6$, $s_{\text{fall}} = 10.3$, and $n_{\text{fall}} = 27$, and $\bar{x}_{\text{spring}} = 67.9$, $s_{\text{spring}} = 12.4$, and $n_{\text{spring}} = 25$. Since the smaller sample size is $n_{\text{spring}} = 25$, we use $n_{\text{spring}} - 1 = 24$ degrees of freedom. The test statistic is

$$t_0 = \frac{(\bar{x}_{\text{fall}} - \bar{x}_{\text{spring}}) - (\mu_{\text{fall}} - \mu_{\text{spring}})}{\sqrt{\dfrac{s_{\text{fall}}^2}{n_{\text{fall}}} + \dfrac{s_{\text{spring}}^2}{n_{\text{spring}}}}}$$

$$= \frac{(73.6 - 67.9) - 0}{\sqrt{\dfrac{10.3^2}{27} + \dfrac{12.4^2}{25}}} \approx 1.795$$

Classical approach: Since this is a right-tailed test with 24 degrees of freedom, the critical values are $t_{0.05} = 1.711$. Since the test statistic $t_0 \approx 1.795$ falls to the right of the critical value 1.711 (i.e., since the test statistic falls within the critical region), we reject H_0.

P-value approach: The P-value for this right-tailed test is the area under the t-distribution with 24 degrees of freedom to the right of $t_0 \approx 1.795$.

From the t-distribution table, in the row corresponding to 24 degrees of freedom, 1.795 falls between 1.711 and 2.064 whose right-tail areas are 0.05 and 0.025, respectively. So, $0.025 < P$-value < 0.05 [Tech: P-value = 0.0395]. Since the P-value is less than the level of significance $\alpha = 0.05$, we reject H_0.

Conclusion: There is sufficient evidence at the $\alpha = 0.05$ level of significance to conclude that final exam scores in the fall semester were higher than final exam scores in the spring semester. It would appear to be the case that the online homework system helped in raising final exam scores.

(f) Answers will vary. One possibility follows: One factor that may confound the results is that the weather is pretty lousy at the end of the fall semester, but pretty nice at the end of the spring semester. If "spring fever" kicked in for the spring semester students, then they probably studied less for the final exam.

Consumer Reports®: The High Cost of Convenience

(a) The hypotheses are $H_0 : \mu_{\text{box}} = \mu_{\text{roll}}$ versus $H_1 : \mu_{\text{box}} < \mu_{\text{roll}}$.

(b) The normal probability plots are roughly linear, and all the data values lie within the bounds. We can conclude that the sample data come from a population that is normally distributed. Therefore it is reasonable to conduct a two-sample hypothesis test.

(c) The boxplot does not show any values less than the lower fence ($= Q_1 - 1.5 \cdot \text{IQR}$) or greater than the upper fence ($= Q_3 - 1.5 \cdot \text{IQR}$). Thus, we can conclude that the data set does not contain any outliers. Based on the boxplots, it would seem there does not appear to be much difference in the absorption times of the two paper towel products, although the boxed version may have a slightly lower absorption rate.

(d) The test statistic is $t = -1.09$ and the P-value of the test is 0.150.

Our tests revealed that the absorption rate for the boxed version of paper towels was not significantly lower than the absorption rate for the traditional roll. Therefore, unless the convenience of using boxed towels is worth the extra cost, we recommend using the traditional towels.

Section 11.4

1. $F_{0.05, 9, 10} = 3.02$

3. $\alpha = 0.05$, so $\dfrac{\alpha}{2} = 0.025$. Thus, $F_{0.025, 6, 8} = 4.65$

and $F_{0.975, 6, 8} = \dfrac{1}{F_{0.025, 8, 6}} = \dfrac{1}{5.60} \approx 0.18$.

5. $F_{0.90,25,20} = \dfrac{1}{F_{0.10,20,25}} = \dfrac{1}{1.72} \approx 0.58$

7. $F_{0.05,45,15} = \dfrac{F_{0.05,40,15} + F_{0.05,50,15}}{2}$

$= \dfrac{2.20 + 2.18}{2}$

$= 2.19$

9. The hypotheses are $H_0 : \sigma_1 = \sigma_2$ versus $H_1 : \sigma_1 \neq \sigma_2$. The test statistic is

$F_0 = \dfrac{s_1^2}{s_2^2} = \dfrac{3.2^2}{3.5^2} \approx 0.84$.

Classical approach: This is a two-tailed test with $n_1 = n_2 = 16$ and $\alpha = 0.05$. So, $\dfrac{\alpha}{2} = 0.025$, and the critical values are $F_{0.025,15,15} = 2.86$ and

$F_{0.975,15,15} = \dfrac{1}{F_{0.025,15,15}} = \dfrac{1}{2.86} \approx 0.35$. Since

$F_0 = 0.84$ is between $F_{0.975,15,15} = 0.35$ and $F_{0.025,15,15} = 2.86$ (the test statistic does not fall in the critical regions), we do not reject H_0.

P-value approach: Using technology, we find *P*-value ≈ 0.7330. Since this *P*-value is greater than the $\alpha = 0.05$ level of significance, we do not reject H_0.

Conclusion: There is not sufficient evidence at the $\alpha = 0.05$ level of significance to conclude that $\sigma_1 \neq \sigma_2$.

11. The hypotheses are $H_0 : \sigma_1 = \sigma_2$ versus $H_1 : \sigma_1 > \sigma_2$. The test statistic is

$F_0 = \dfrac{s_1^2}{s_2^2} = \dfrac{9.9^2}{6.4^2} \approx 2.39$.

Classical approach: This is a right-tailed test with $n_1 = 26$, $n_2 = 19$, and $\alpha = 0.01$. So, the critical value is $F_{0.01,25,18}$, which we approximate by $F_{0.01,25,20} = 2.84$. Since $F_0 = 2.39 < F_{0.01,25,20} = 2.84$ the test statistic does not fall in the critical region, we do not reject H_0.

P-value approach: Using technology, we find *P*-value ≈ 0.0303. Since this *P*-value is greater than the $\alpha = 0.01$ level of significance, we do not reject H_0.

Conclusion: There is not sufficient evidence at the $\alpha = 0.01$ level of significance to conclude that $\sigma_1 > \sigma_2$.

13. The hypotheses are $H_0 : \sigma_1 = \sigma_2$ versus $H_1 : \sigma_1 < \sigma_2$. The test statistic is

$F_0 = \dfrac{s_1^2}{s_2^2} = \dfrac{8.3^2}{13.2^2} \approx 0.40$.

Classical approach: This is a left-tailed test with $n_1 = 51$, $n_2 = 26$, and $\alpha = 0.10$. So, the critical value is $F_{0.90,50,25} = \dfrac{1}{F_{0.10,25,50}} = \dfrac{1}{1.53} \approx 0.65$.

Since $F_0 = 0.40 < F_{0.90,50,25} = 0.65$ (the test statistic falls in a critical region), we reject H_0.

P-value approach: Using technology, we find *P*-value ≈ 0.0026. Since this *P*-value is less than the $\alpha = 0.10$ level of significance, we reject H_0.

Conclusion: There is sufficient evidence at the $\alpha = 0.1$ level of significance to conclude that $\sigma_1 < \sigma_2$.

15. The hypotheses are $H_0 : \sigma_1 = \sigma_2$ versus $H_1 : \sigma_1 \neq \sigma_2$. The test statistic is

$F_0 = \dfrac{s_1^2}{s_2^2} = \dfrac{1.162^2}{1.015^2} \approx 1.311$.

Classical approach: This is a two-tailed test with $n_1 = 268$, $n_2 = 1145$, and $\alpha = 0.05$. The critical values are $F_{0.025,267,1144}$ and $F_{0.975,267,1144}$, which we approximate by

$F_{0.025,267,1144} \approx F_{0.025,120,1000} = 1.29$ and

$F_{0.975,267,1144} = \dfrac{1}{F_{0.025,1144,267}} \approx \dfrac{1}{F_{0.025,1000,200}} =$

$\dfrac{1}{1.25} \approx 0.80$. Since $F_0 = 1.311$ does not fall between $F_{0.975,267,1144} \approx 0.80$ and

$F_{0.025,267,1144} \approx 1.29$ (the test statistic falls in the critical region), we reject H_0.

P-value approach: Using technology, we find *P*-value ≈ 0.0036. Since this *P*-value is less than the $\alpha = 0.05$ level of significance, we reject H_0.

Conclusion: There is sufficient evidence at the $\alpha = 0.05$ level of significance to conclude that the standard deviation in time to earn a bachelor's degree is different for students who attend a community college than for students who immediately attend a 4-year institution.

17. The hypotheses are $H_0 : \sigma_1 = \sigma_2$ versus $H_1 : \sigma_1 > \sigma_2$. The test statistic is

$$F_0 = \frac{s_1^2}{s_2^2} = \frac{17.07^2}{6.48^2} \approx 6.94 .$$

Classical approach: This is a right-tailed test with $n_1 = n_2 = 65$ and $\alpha = 0.01$. So, the critical value is $F_{0.01,64,64}$, which we approximate by $F_{0.01,60,50} = 1.91$. Since $F_0 \approx 6.94 > F_{0.01,60,50} = 1.91$ (the test statistic falls in the critical region), we reject H_0.

P-value approach: Using technology, we find *P*-value < 0.0001. Since this *P*-value is less than the $\alpha = 0.01$ level of significance, we reject H_0.

Conclusion: There is sufficient evidence at the $\alpha = 0.01$ level of significance to conclude that the treatment group had a higher standard deviation for serum retinal concentration than did the control group.

19. (a) The hypotheses are $H_0 : \sigma_1 = \sigma_2$ versus $H_1 : \sigma_1 < \sigma_2$, where $\sigma_1 =$ the standard deviation of waiting time for a single line. We compute the sample standard deviations to be $s_1 \approx 0.574$ minutes and $s_2 \approx 1.012$ minutes. The test statistic is

$$F_0 = \frac{s_1^2}{s_2^2} = \frac{0.574^2}{1.012^2} \approx 0.32 .$$

Classical approach: This is a left-tailed test with $n_1 = n_2 = 20$ and $\alpha = 0.05$. The critical value is $F_{0.95,19,19}$. We approximate this by $F_{0.95,20,20} = \dfrac{1}{F_{0.05,20,20}} = \dfrac{1}{2.12} \approx 0.47 .$

Since $F_0 = 0.32 < F_{0.95,20,20} \approx 0.47$ (the test statistic falls in the critical region), we reject H_0.

P-value approach: Using technology, we find *P*-value ≈ 0.0087. Since this *P*-value is less than the $\alpha = 0.05$ level of significance, we reject H_0.

Conclusion: There is sufficient evidence at the $\alpha = 0.05$ level of significance to conclude that the standard deviation of waiting time in a single line is less than the standard deviation for wait time in the multiple lines.

(b) We compute the five-number summary for each variable:
Single: 1.1, 1.9, 2.2, 2.8, 3.1
Multiple: 0.9, 1.8, 2.45, 2.9, 4.6
[Minitab: 0.9, 1.7, 2.45, 2.9, 4.6]

The boxplot indicates that there is much greater variability in the waiting times for multiple lines than for a single line.

21. (a) The normal probability plots are roughly linear, so we conclude that the data could come from normal distributions. Thus, this requirement of the *F*-test is satisfied.

(b) Since *P*-value = 0.0844 is larger than the $\alpha = 0.05$ level of significance, the nursing student should conclude that there is not sufficient evidence to conclude that men have more variability in their systolic blood pressure than women have.

(c) We find the sample standard deviations to be $s_f \approx 10.346$ and $s_m \approx 14.604$. The test statistic is $F_0 = \dfrac{s_m^2}{s_f^2} = \dfrac{14.604^2}{10.346^2} \approx 1.99 .$

We test the hypotheses $H_0 : \sigma_m = \sigma_f$ versus $H_1 : \sigma_m > \sigma_f$.

This is a right-tailed test with df = 19 for the numerator and df = 16 for the denominator. Using Table VIII, we only approximate these by 20 df and 15 df, respectively. Since $F_{0.10,20,15} = 1.92$ and $F_{0.05,20,15} = 2.33$, the P-value should be close to 0.10 which is consistent with the given P-value (0.0844).

Section 11.5

1. We first verify the requirements to perform McNemar's test. The samples are dependent because the different creams were applied to two warts on the same individual. We can think of the sample as a random sample. The total number of individuals for which one treatment was successful while the other was not is 9 + 13 = 22, which is greater than 10. Thus, the requirements are met, so we can conduct the test.

   The hypotheses are $H_0 : p_A = p_B$ verses $H_1 : p_A \neq p_B$. The test statistic is

   $$z_0 = \frac{|f_{12} - f_{21}| - 1}{\sqrt{f_{12} + f_{21}}} = \frac{|9 - 13| - 1}{\sqrt{9 + 13}} \approx 0.64$$

   Classical approach: The critical value for $\alpha = 0.05$ is $z_{\alpha/2} = z_{0.025} = 1.96$. Since $z_0 = 0.64 < z_{0.025} = 1.96$ (the test statistic does not fall in the critical region), we do not reject H_0.

   P-value approach: The P-value is two times the area under the standard normal distribution to the right of the test statistic, $z_0 = 0.64$.
   $$\text{P-value} = 2 \cdot P(Z > 0.64)$$
   $$= 2 \cdot [1 - P(Z < 0.64)]$$
   $$= 2 \cdot (1 - 0.7389)$$
   $$= 0.5222$$
   Since this P-value is greater than $\alpha = 0.05$, we do not reject H_0.

   Conclusion: There is not sufficient evidence at the $\alpha = 0.05$ level of significance to conclude that there is a difference in the proportions.

3. The hypotheses are $H_0 : \sigma_1 = \sigma_2$ versus $H_1 : \sigma_1 > \sigma_2$. We assume that all requirements to conduct the test are satisfied. The test statistic is $F_0 = \frac{s_1^2}{s_2^2} = \frac{12^2}{10^2} = 1.44$.

   Classical approach: This is a right-tailed test with $n_1 = 31$, $n_2 = 51$, and $\alpha = 0.05$. So, the critical value is $F_{0.05,30,50} = 1.69$.

   Since $F_0 = 1.44 < F_{0.05,30,50} = 1.69$ (the test statistic does not fall in the critical region), we do not reject H_0.

   P-value approach: Using technology, we find P-value ≈ 0.1249. Since this P-value is greater than the $\alpha = 0.05$ level of significance, we do not reject H_0.

   Conclusion: There is not sufficient evidence at the $\alpha = 0.05$ level of significance to conclude that $\sigma_1 > \sigma_2$.

5. We verify the requirements to perform the hypothesis test: (1) We are told that the samples are random samples; (2) We have $x_1 = 40$, $n_1 = 135$, $x_2 = 60$, and $n_2 = 150$, so $\hat{p}_1 = \frac{40}{135} \approx 0.2963$ and $\hat{p}_2 = \frac{60}{150} = 0.4$. Thus, $n_1 \hat{p}_1 (1 - \hat{p}_1) = 135(0.2963)(1 - 0.2963) \approx 28 \geq 10$ and $n_2 \hat{p}_2 (1 - \hat{p}_2) = 150(0.4)(1 - 0.4) = 36 \geq 10$; and (3) We assume each sample is less than 5% of the population. So, the requirements are met, and we can conduct the test.

   The hypotheses are $H_0 : p_1 = p_2$ versus $H_1 : p_1 < p_2$. From before, the two sample estimates are $\hat{p}_1 \approx 0.2963$ and $\hat{p}_2 = 0.4$.

   The pooled estimate is $\hat{p} = \frac{40 + 60}{135 + 150} \approx 0.3509$.

   The test statistic is
   $$z_0 = \frac{\hat{p}_1 - \hat{p}_2}{\sqrt{\hat{p}(1 - \hat{p})}\sqrt{\frac{1}{n_1} + \frac{1}{n_2}}}$$
   $$= \frac{0.2963 - 0.4}{\sqrt{0.3509(1 - 0.3509)}\sqrt{\frac{1}{135} + \frac{1}{150}}} \approx -1.83$$

Classical approach:
This is a left-tailed test, so the critical value for $\alpha = 0.05$ is $-z_{0.05} = -1.645$. Since

$z_0 = -1.83 < -z_{0.05} = -1.645$ (the test statistic falls in the critical region), we reject H_0.

P-value approach: The *P*-value is the area under the standard normal distribution to the left of the test statistic, $z_0 = -1.83$.

$$P\text{-value} = P(z_0 \le -1.83)$$
$$= 0.0336 \quad [\text{Tech: } 0.0335]$$

Since the *P*-value $= 0.0336 < \alpha = 0.05$, we reject H_0.

Conclusion: There is sufficient evidence at the $\alpha = 0.05$ level of significance to conclude that $p_1 < p_2$.

7. $H_0 : \mu_1 = \mu_2$ versus $H_1 : \mu_1 \ne \mu_2$. We assume that all requirements to conduct the test are satisfied. The level of significance is $\alpha = 0.01$. The smaller sample size is $n_1 = 41$, so we use $n_1 - 1 = 41 - 1 = 40$ degrees of freedom. The test statistic is

$$t_0 = \frac{(\overline{x}_1 - \overline{x}_2) - (\mu_1 - \mu_2)}{\sqrt{\dfrac{s_1^2}{n_1} + \dfrac{s_2^2}{n_2}}} = \frac{(125.3 - 130.8) - 0}{\sqrt{\dfrac{8.5^2}{41} + \dfrac{7.3^2}{50}}}$$
$$\approx -3.271$$

Classical approach: Since this is a two-tailed test with 40 degrees of freedom, the critical values are $\pm t_{0.005} = \pm 2.704$. Since

$t_0 \approx -3.271 < -t_{0.005} = -2.704$ (the test statistic falls in the critical region), we reject H_0.

P-value approach: The *P*-value for this two-tailed test is the area under the *t*-distribution with 40 degrees of freedom to the left of $t_0 = -3.271$ plus the area to the right of 3.271. From the *t*-distribution table in the row corresponding to 40 degrees of freedom, 3.271 falls between 2.971 and 3.307 whose right-tail areas are 0.0025 and 0.001, respectively. We must double these values in order to get the total area in both tails: 0.005 and 0.002. So, $0.002 < P\text{-value} < 0.005$ [Tech: *P*-value = 0.0016]. Because the *P*-value is less than the level of significance $\alpha = 0.05$, we reject H_0.

Conclusion: There is sufficient evidence at the $\alpha = 0.05$ level of significance to conclude that there is a difference in the population means.

9. These are matched-pair data because two measurements, X_i and Y_i, are taken on the same individual. We measure differences as $d_i = Y_i - X_i$.

Individual	1	2	3	4	5
X_i	40	32	53	48	38
Y_i	38	33	49	48	33
d_i	−2	1	−4	0	−5

We compute the mean and standard deviation of the differences and obtain $\overline{d} = -2$ and $s_d \approx 2.5495$, rounded to four decimal places. The hypotheses are $H_0 : \mu_d = 0$ versus $H_1 : \mu_d < 0$. The level of significance is $\alpha = 0.10$. The test statistic is

$$t_0 = \frac{\overline{d}}{\dfrac{s_d}{\sqrt{n}}} = \frac{-2}{\dfrac{2.5495}{\sqrt{5}}} \approx -1.754.$$

Classical approach: Since this is a left-tailed test with 4 degrees of freedom, the critical value is $-t_{0.10} = -1.533$. Since $t_0 \approx -1.754 < -t_{0.10} = -1.533$ (the test statistic falls within the critical region), we reject H_0.

P-value approach: The *P*-value for this left-tailed test is the area under the *t*-distribution with 4 degrees of freedom to the left of the test statistic $t_0 = -1.754$, which is equivalent to the area to the right of 1.754. From the *t*-distribution table, in the row corresponding to 4 degrees of freedom, 1.754 falls between 1.533 and 2.132 whose right-tail areas are 0.10 and 0.05, respectively. So, $0.10 > P\text{-value} > 0.05$ [Tech: *P*-value = 0.0771]. Because the *P*-value is less than the level of significance $\alpha = 0.10$, we reject H_0.

Conclusion: There is sufficient evidence at the $\alpha = 0.10$ level of significance to conclude that there is a difference in the measure of the variable before and after a treatment. That is, there is sufficient evidence to conclude that the treatment is effective.

11. (a) Collision claims tend to be skewed right because there are a few very large collision claims relative to the majority of claims.

(b) The hypotheses were $H_0 : \mu_{30-59} = \mu_{20-24}$ versus $H_1 : \mu_{30-59} < \mu_{20-24}$. All requirements are satisfied to conduct the test for two independent population means. The level of significance is $\alpha = 0.05$. The sample sizes are both $n = 40$, so we use $n - 1 = 40 - 1 = 39$ degrees of freedom. The test statistic is

$$t_0 = \frac{(\bar{x}_1 - \bar{x}_2) - (\mu_1 - \mu_2)}{\sqrt{\dfrac{s_1^2}{n_1} + \dfrac{s_2^2}{n_2}}}$$

$$= \frac{(3669 - 4586) - 0}{\sqrt{\dfrac{2029^2}{40} + \dfrac{2302^2}{40}}} \approx -1.890$$

Classical approach: Since this is a left-tailed test with 39 degrees of freedom, the critical value is $-t_{0.05} = -1.685$. Since $t_0 = -1.890 < -t_{0.05} = -1.685$ (the test statistic falls in the critical region), we reject H_0.

P-value approach: The P-value for this left-tailed test is the area under the t-distribution with 39 degrees of freedom to the left of $t_0 = -1.890$. From the t-distribution table in the row corresponding to 39 degrees of freedom, 1.890 falls between 1.685 and 2.023 whose right-tail areas are 0.05 and 0.025, respectively. So, $0.025 < P\text{-value} < 0.05$ [Tech: P-value = 0.0313]. Because the P-value is less than the level of significance $\alpha = 0.05$, we reject H_0.

Conclusion: There is sufficient evidence at the $\alpha = 0.05$ level of significance to conclude that the mean collision claim of a 30- to 59-year-old is less than the mean claim of a 20- to 24-year-old. Given that 20- to 24-year-olds tend to claim more for each accident, it makes sense to charge them more for coverage.

13. We first verify the requirements to perform McNemar's test (for dependent population proportions). The samples are dependent because the two questions were asked of the same individual. We can think of the sample as a random sample. The total number of individuals answering favorably for one question while answering unfavorably for the other question is $123 + 70 = 193$, which is greater than 10. Thus, the requirements are met, so we can conduct the test.

The hypotheses are $H_0 : p_{HE} = p_{HA}$ verses $H_1 : p_{HE} \neq p_{HA}$. The test statistic is

$$z_0 = \frac{|f_{12} - f_{21}| - 1}{\sqrt{f_{12} + f_{21}}} = \frac{|123 - 70| - 1}{\sqrt{123 + 70}} \approx 3.74$$

Classical approach: The critical value for $\alpha = 0.05$ is $z_{\alpha/2} = z_{0.025} = 1.96$. Since $z_0 = 3.74 > z_{0.025} = 1.96$ (the test statistic falls in the critical region), we reject H_0.

P-value approach: The P-value is two times the area under the standard normal distribution to the right of the test statistic, $z_0 = 3.74$.

$$\begin{aligned} P\text{-value} &= 2 \cdot P(x > 3.74) \\ &= 2 \cdot [1 - P(x < 3.74)] \\ &= 2 \cdot (1 - 0.9999) \\ &= 0.0002 \end{aligned}$$

Since $P\text{-value} = 0.0002 < \alpha = 0.05$, we reject H_0.

Conclusion: There is sufficient evidence at the $\alpha = 0.05$ level of significance to conclude that there is a difference in the proportions. It seems that the proportion of individuals who are healthy differs from the proportion of individuals who are happy. The data imply that more people are happy even though they are unhealthy.

15. These are matched-pair data because two prices, W_i and T_i, are taken on the same product. We measure differences as $d_i = W_i - T_i$. We are told in the problem to assume the conditions for conducting the test are satisfied.

Product	1	2	3	4	5
W_i	2.94	3.16	4.74	2.84	1.72
T_i	2.94	3.79	4.74	2.49	1.97
d_i	0	−0.63	0	0.35	−0.25

Product	6	7	8	9	10
W_i	1.72	3.77	7.44	3.77	5.98
T_i	1.72	3.99	7.99	3.64	5.99
d_i	0	−0.22	−0.55	0.13	−0.01

We compute the mean and standard deviation of the differences and obtain $\bar{d} = -0.118$ and $s_d \approx 0.3001$, rounded to four decimal places. The hypotheses are $H_0 : \mu_d = 0$ versus $H_1 : \mu_d \neq 0$. The level of significance is $\alpha = 0.05$. The test statistic is

$$t_0 = \frac{\bar{d}}{\frac{s_d}{\sqrt{n}}} = \frac{-0.118}{\frac{0.3001}{\sqrt{10}}} \approx -1.243.$$

Classical approach: Since this is a two-tailed test with 9 degrees of freedom, the critical value is $\pm t_{\alpha/2} = \pm t_{0.025} = \pm 2.262$. Since $t_0 = -1.243$ falls between $-t_{0.025} = -2.262$ and $t_{0.025} = 2.262$ (the test statistic does not fall within the critical region), we do not reject H_0.

P-value approach: The P-value for this two-tailed test is the area under the t-distribution with 9 degrees of freedom to the left of the test statistic $t_0 = -1.243$ plus the area to the right of 1.243. From the t-distribution table in the row corresponding to 9 degrees of freedom, 1.243 falls between 1.100 and 1.383 whose right-tail areas are 0.15 and 0.10, respectively. We must double these values in order to get the total area in both tails: 0.30 and 0.20. So, $0.30 >$ P-value > 0.30. [Tech: P-value = 0.2451]. Because the P-value is greater than the level of significance $\alpha = 0.05$, we do not reject H_0.

Conclusion: There is not sufficient evidence at the $\alpha = 0.05$ level of significance to conclude that there is a difference in the pricing of health and beauty products at Wal-Mart and Target.

17. Since the data could come from a normal population with no outliers, the requirements to conduct the test are satisfied.

Hypotheses: $H_0 : \mu = 3.101$ versus $H_1 : \mu > 3.101$

Test Statistic

$$t_0 = \frac{\bar{x} - \mu_0}{s/\sqrt{n}} = \frac{3.160 - 3.101}{0.066/\sqrt{21}} = 4.064$$

No level of significance was given. We choose $\alpha = 0.05$; d.f. $= n - 1 = 21 - 1 = 20$

Classical approach: Because this is a right-tailed test with 20 degrees of freedom, the critical value is $t_{0.05} = 1.725$. Since $t_0 = 4.064 > t_{0.05} = 1.725$, the test statistic falls within the critical region. We reject the null hypothesis.

P-value approach: This is a right-tailed test with 20 degrees of freedom. The P-value is the area under the t-distribution to the right of the test statistic, $t_0 = 4.064$. From the t-distribution table in the row corresponding to 20 degrees of freedom, 4.064 falls to the right of 3.850, whose right-tail area is 0.0005. So, P-value < 0.0005 [Tech: 0.0003]. Since P-value $< \alpha = 0.05$, we reject the null hypothesis.

Conclusion: There is sufficient evidence to conclude that the mean price of gas in Chicago is more expensive than the national average.

19. The hypotheses were $H_0 : p_{>100K} = p_{<100K}$ versus $H_1 : p_{>100K} \neq p_{<100K}$. We have $x_{>100K} = 710$,

$n_{>100K} = 1205$, $x_{<100K} = 695$, and $n_{<100K} = 1310$, so $\hat{p}_{>100K} = \dfrac{x_{>100K}}{n_{>100K}} = \dfrac{710}{1205} \approx 0.5892$ and

$\hat{p}_{<100K} = \dfrac{x_{<100K}}{n_{<100K}} = \dfrac{695}{1310} = 0.5305$. For a 95% confidence interval we use $\pm z_{0.025} = \pm 1.96$. Then:

Lower Bound: $(\hat{p}_{>100K} - \hat{p}_{<100K}) - z_{\alpha/2} \cdot \sqrt{\dfrac{\hat{p}_{>100K}(1-\hat{p}_{>100K})}{n_{>100K}} + \dfrac{\hat{p}_{<100K}(1-\hat{p}_{<100K})}{n_{<100K}}}$

$= (0.5892 - 0.5305) - 1.96 \cdot \sqrt{\dfrac{0.5892(1-0.5892)}{1205} + \dfrac{0.5305(1-0.5305)}{1310}} \approx 0.020$

Upper Bound: $(\hat{p}_{>100K} - \hat{p}_{<100K}) + z_{\alpha/2} \cdot \sqrt{\dfrac{\hat{p}_{>100K}(1-\hat{p}_{>100K})}{n_{>100K}} + \dfrac{\hat{p}_{<100K}(1-\hat{p}_{<100K})}{n_{<100K}}}$

$= (0.5892 - 0.5305) + 1.96 \cdot \sqrt{\dfrac{0.5892(1-0.5892)}{1205} + \dfrac{0.5305(1-0.5305)}{1310}} \approx 0.097$

We are 95% confident that the difference in the proportion of individuals who believe it is morally wrong for unwed women to have children (for individuals who earned more than \$100,000 versus individuals who earned less than \$100,000) is between 0.020 and 0.097. Because the confidence interval does not include 0, there is sufficient evidence at the $\alpha = 0.05$ level of significance to conclude that there is a difference in the proportions. It appears that a higher proportion of individuals who earn over \$100,000 per year feel it is morally wrong for unwed women to have children.

---

21. (a) Yes, the both plots indicate that the data could come from populations that are normally distributed since all of the points fall within the boundaries of the probability plots.

(b) The five number summaries follow:

Value: 6.02, 6.09, 8.735, 9.54, 12.53
[Minitab: 6.02, 6.073, 8.735, 10.083, 12.53]

Lower fence $= Q_1 - 1.5(Q_3 - Q_1)$
$= 6.09 - 1.5(9.54 - 6.09)$
$= 0.915$ [Tech: -1.063]

Upper fence $= Q_3 + 1.5(Q_3 - Q_1)$
$= 9.54 + 1.5(9.54 - 6.09)$
$= 14.715$ [Tech: 17.966]

There are no outliers in the value funds data.

Growth: 6.93, 7.57, 8.365, 8.63, 9.33
[Minitab: 6.93, 7.56, 8.365, 8.728, 9.33]

Lower fence $= Q_1 - 1.5(Q_3 - Q_1)$
$= 7.57 - 1.5(8.63 - 7.57)$
$= 5.98$ [Tech: 5.808]

Upper fence $= Q_3 - 1.5(Q_3 - Q_1)$
$= 8.63 + 1.5(8.63 - 7.57)$
$= 10.22$ [Tech: 10.480]

There are no outliers in the growth funds data.

So, there are no outliers in either data set.

(c) We verify the requirements to perform Welch's *t*-test. Each sample is a simple random sample. The samples were obtained independently. The normal probability plot indicates that the data could come from a population that is normal, with no outliers. Thus, the requirements are met, so we can conduct the test. The hypotheses are $H_0 : \mu_v = \mu_g$ versus $H_1 : \mu_v \neq \mu_g$. The level of significance is $\alpha = 0.05$.

The sample statistics for the data are
$\overline{x}_v = 8.653$, $s_v \approx 2.2952$, and $n_v = 10$,
$\overline{x}_g = 8.206$, $s_g \approx 0.7256$, and $n_g = 10$.
Since the two sample sizes are equal
$n_v = n_g = 10$, we use $n_v - 1 = 10 - 1 = 9$
degrees of freedom.
The critical t-values are $\pm t_{0.025} = \pm 2.262$.

The test statistic is

$$t_0 = \frac{(\overline{x}_v - \overline{x}_g) - (\mu_v - \mu_g)}{\sqrt{\dfrac{s_v^2}{n_v} + \dfrac{s_g^2}{n_g}}}$$

$$= \frac{(8.653 - 8.206) - 0}{\sqrt{\dfrac{2.2952^2}{10} + \dfrac{0.7256^2}{10}}} \approx 0.587$$

Classical approach: Since this is a two-tailed test with 9 degrees of freedom, the critical values are $\pm t_{0.025} = \pm 2.262$. Since $t_0 = 0.587$ is between $-t_{0.025} = -2.262$ and $t_{0.025} = 2.262$ (the test statistic does not fall within the critical region), we do not reject H_0.

P-value approach: The P-value for this two-tailed test is the area under the t-distribution with 9 degrees of freedom to the right of $t_0 = 0.587$, plus the area to the left of -0.587. From the t-distribution table in the row corresponding to 9 degrees of freedom, 0.587 falls to the left of 0.703 whose right-tail area is 0.25. We must double this value in order to get the total area in both tails: 0.50. Thus, P-value > 0.50 [Tech: P-value $= 0.5872$]. Because the P-value is larger than the level of significance $\alpha = 0.05$, we do not reject H_0.

Conclusion: There is not sufficient evidence at the $\alpha = 0.05$ level of significance to conclude that the mean rate of return of value funds is different from the mean rate of return of growth funds.

(d) The hypotheses are $H_0 : \sigma_v = \sigma_g$ versus $H_1 : \sigma_v \neq \sigma_g$. The test statistic is

$$F_0 = \frac{s_v^2}{s_g^2} = \frac{2.2952^2}{0.7256^2} \approx 10.01.$$

Classical approach: This is a two-tailed test with $n_v = 10$, $n_g = 10$, and $\alpha = 0.05$. The critical values are $F_{0.025,9,9} = 4.03$ and

$$F_{0.975,9,9} = \frac{1}{F_{0.025,9,9}} = \frac{1}{4.03} \approx 0.25. \text{ Since}$$

$F_0 = 10.01 > F_{0.025,9,9} = 4.03$ (the test statistic falls in the critical region), we reject H_0.

P-value approach: Using technology, we find P-value ≈ 0.0021. Since this P-value is less than the $\alpha = 0.05$ level of significance, we reject H_0.

Conclusion: There is sufficient evidence at the $\alpha = 0.05$ level of significance to conclude that the standard deviation of the value funds is different from the standard deviation of the growth funds. In particular, the value funds appear to be more volatile than the growth funds.

(e) Answers may vary. Since there does not appear to be a difference in mean rates of return, risk should be used as a determining factor. Using risk as the criteria for investing, the growth fund appears to be the better option.

23. Hypothesis test for a single mean

25. Matched pairs $t$test on difference of means

27. Confidence interval for a single proportion

Chapter 11 Review Exercises

1. This is dependent since the members of the two samples are matched by diagnosis. The response variable, length of stay, is quantitative.

2. This is independent because the subjects are randomly selected from two distinct populations. The response variable, commute times, is quantitative.

3. **(a)** $F_{0.05,8,9} = 3.23$

(b) The level of significance is $\alpha = 0.05$, so $\frac{\alpha}{2} = 0.025$. Thus, $F_{0.025,10,5} = 6.62$ and

$$F_{0.975,10,5} = \frac{1}{F_{0.025,5,10}} = \frac{1}{4.24} \approx 0.24.$$

4. **(a)** We compute each difference, $d_i = X_i - Y_i$.

Obs	1	2	3	4	5	6
X_i	34.2	32.1	39.5	41.8	45.1	38.4
Y_i	34.9	31.5	39.5	41.9	45.5	38.8
d_i	−0.7	0.6	0	−0.1	−0.4	−0.4

(b) Using technology, $\bar{d} \approx -0.1667$ and $s_d \approx 0.4502$, rounded to four decimal places.

(c) The hypotheses are $H_0 : \mu_d = 0$ versus $H_1 : \mu_d < 0$. The level of significance is $\alpha = 0.05$. The test statistic is

$$t_0 = \frac{\bar{d}}{\frac{s_d}{\sqrt{n}}} = \frac{-0.1667}{\frac{0.4502}{\sqrt{6}}} \approx -0.907.$$

Classical approach: Since this is a left-tailed test with 5 degrees of freedom, the critical value is $-t_{0.05} = -2.015$. Since the test statistic $t_0 = -0.907 > -t_{0.05} = -2.015$ (the test statistic does not fall within the critical region), we do not reject H_0.

P-value approach: The P-value for this left-tailed test is the area under the t-distribution with 5 degrees of freedom to the left of the test statistic $t_0 \approx -0.907$, which is equivalent to the area to the right of 0.907.

From the t-distribution table in the row corresponding to 5 degrees of freedom, 0.907 falls between 0.727 and 0.920 whose right-tail areas are 0.25 and 0.20, respectively. So, $0.20 < P$-value < 0.25 [Tech: P-value = 0.2030]. Because the P-value is greater than the $\alpha = 0.05$ level of significance, we do not reject H_0.

Conclusion: There is not sufficient evidence at the $\alpha = 0.05$ level of significance to conclude that the mean difference is less than zero.

(d) For 98% confidence, we use $\alpha = 0.02$. With 5 degrees of freedom, we have $t_{\alpha/2} = t_{0.01} = 3.365$. Then:

$$\text{Lower bound} = \bar{d} - t_{0.01} \cdot \frac{s_d}{\sqrt{n}}$$

$$= -0.1667 - 3.365 \cdot \frac{0.4502}{\sqrt{6}}$$

$$\approx -0.79$$

$$\text{Upper bound} = \bar{d} + t_{0.01} \cdot \frac{s_d}{\sqrt{n}}$$

$$= -0.167 + 3.365 \cdot \frac{0.450}{\sqrt{6}}$$

$$\approx 0.45$$

We can be 98% confident that the mean difference is between −0.79 and 0.45.

5. **(a)** $H_0 : \mu_1 = \mu_2$ versus $H_1 : \mu_1 \neq \mu_2$. The level of significance is $\alpha = 0.1$. Since the smaller sample size is $n_2 = 8$, we use $n_2 - 1 = 7$ degrees of freedom. The test statistic is

$$t_0 = \frac{(\bar{x}_1 - \bar{x}_2) - (\mu_1 - \mu_2)}{\sqrt{\frac{s_1^2}{n_1} + \frac{s_2^2}{n_2}}} = \frac{(32.4 - 28.2) - 0}{\sqrt{\frac{4.5^2}{13} + \frac{3.8^2}{8}}}$$

$$\approx 2.290$$

Classical approach: Since this is a two-tailed test with 7 degrees of freedom, the critical values are $\pm t_{0.05} = \pm 1.895$. Since the test statistic $t_0 \approx 2.290$ is to the right of the critical value 1.895 (the test statistic falls within a critical region), we reject H_0.

P-value approach: The P-value for this two-tailed test is the area under the t-distribution with 7 degrees of freedom to the right of $t_0 = 2.290$ plus the area to the left of $-t_0 = -2.290$. From the t-distribution table in the row corresponding to 7 degrees of freedom, 2.290 falls between 1.895 and 2.365 whose right-tail areas are 0.05 and 0.025, respectively. We must double these values in order to get the total area in both tails: 0.10 and 0.05. Thus, $0.05 < P$-value < 0.10 [Tech: P-value = 0.0351]. Because the *P*-value is less than the $\alpha = 0.10$ level of significance, we reject H_0.

Conclusion: There is sufficient evidence at the $\alpha = 0.1$ level of significance to conclude that $\mu_1 \neq \mu_2$.

(b) For a 90% confidence interval with df = 7, we use $t_{\alpha/2} = t_{0.05} = 1.895$. Then:

Lower bound

$$= (\bar{x}_1 - \bar{x}_2) - t_{\alpha/2} \cdot \sqrt{\frac{s_1^2}{n_1} + \frac{s_2^2}{n_2}}$$

$$= (32.4 - 28.2) - 1.895 \cdot \sqrt{\frac{4.5^2}{13} + \frac{3.8^2}{8}}$$

$$\approx 0.73 \quad [\text{Tech: } 1.01]$$

Upper bound

$$= (\bar{x}_1 - \bar{x}_2) + t_{\alpha/2} \cdot \sqrt{\frac{s_1^2}{n_1} + \frac{s_2^2}{n_2}}$$

$$= (32.4 - 28.2) + 1.895 \cdot \sqrt{\frac{4.5^2}{13} + \frac{3.8^2}{8}}$$

$$\approx 7.67 \quad [\text{Tech: } 7.39]$$

We are 90% confident that the population mean difference is between 0.73 and 7.67. [Tech: between 1.01 and 7.39].

(c) The hypotheses are $H_0 : \sigma_1 = \sigma_2$ versus $H_1 : \sigma_1 \neq \sigma_2$. The test statistic is

$$F_0 = \frac{s_1^2}{s_2^2} = \frac{4.5^2}{3.8^2} \approx 1.40.$$

Classical approach: This is a two-tailed test with $n_1 = 13$, $n_2 = 8$, and $\alpha = 0.05$. The critical values are $F_{0.025,12,7}$, which we approximate by $F_{0.025,10,7} = 4.76$ and

$$F_{0.975,12,7} = \frac{1}{F_{0.025,7,12}} = \frac{1}{3.61} \approx 0.28. \text{ Since}$$

$F_0 = 1.40$ falls between $F_{0.975,12,7} \approx 0.28$ and $F_{0.025,12,7} \approx 4.76$ (the test statistic does not fall in the critical region), we do not reject H_0.

P-value approach: Using technology, we find *P*-value ≈ 0.6734. Since this *P*-value

is greater than the $\alpha = 0.05$ level of significance, we do not reject H_0.

Conclusion: There is not sufficient evidence at the $\alpha = 0.05$ level of significance to conclude that the standard deviation in population 1 is different from the standard deviation in population 2.

6. (a) $H_0 : \mu_1 = \mu_2$ versus $H_1 : \mu_1 > \mu_2$. The level of significance is $\alpha = 0.01$. Since the smaller sample size is $n_2 = 41$, we use $n_2 - 1 = 40$ degrees of freedom. The test statistic is

$$t_0 = \frac{(\bar{x}_1 - \bar{x}_2) - (\mu_1 - \mu_2)}{\sqrt{\frac{s_1^2}{n_1} + \frac{s_2^2}{n_2}}}$$

$$= \frac{(48.2 - 45.2) - 0}{\sqrt{\frac{8.4^2}{45} + \frac{10.3^2}{41}}} \approx 1.472$$

Classical approach: Since this is a right-tailed test with 40 degrees of freedom, the critical value is $t_{0.01} = 2.423$. Since the test statistic $t_0 \approx 1.472 < t_{0.01} = 2.423$ (the test statistic does not fall within the critical region), we do not reject H_0.

P-value approach: The *P*-value for this right-tailed test is the area under the *t*-distribution with 40 degrees of freedom to the right of $t_0 \approx 1.472$. From the *t*-distribution table in the row corresponding to 40 degrees of freedom, 1.472 falls between 1.303 and 1.684 whose right-tail areas are 0.10 and 0.05, respectively.

Thus, $0.10 > P\text{-value} > 0.05$ [Tech: *P*-value = 0.0726]. Because the *P*-value is greater than the $\alpha = 0.01$ level of significance, we do not reject H_0.

Conclusion: There is not sufficient evidence at the $\alpha = 0.01$ level of significance to conclude that the mean of population 1 is larger than the mean of population 2.

(b) The hypotheses are $H_0 : \sigma_1 = \sigma_2$ versus $H_1 : \sigma_1 < \sigma_2$. The test statistic is

$$F_0 = \frac{s_1^2}{s_2^2} = \frac{8.4^2}{10.3^2} \approx 0.67 .$$

Classical approach: This is a left-tailed test with $n_1 = 45$, $n_2 = 41$, and $\alpha = 0.01$.

The critical value is $F_{0.99, 44, 40} = \dfrac{1}{F_{0.01, 40, 44}}$

$\approx \dfrac{1}{F_{0.01, 40, 50}} \approx \dfrac{1}{2.01} \approx 0.50$. Since

$F_0 = 0.67 > F_{0.99, 44, 40} \approx 0.50$ (the test statistic does not fall in the critical region), we do not reject H_0.

P-value approach: Using technology, we find P-value ≈ 0.0940. Since this P-value is greater than the $\alpha = 0.01$ level of significance, we do not reject H_0.

Conclusion: There is not sufficient evidence at the $\alpha = 0.01$ level of significance to conclude that the standard deviation in population 1 is less than the standard deviation in population 2.

7. The hypotheses are $H_0 : p_1 = p_2$ versus $H_1 : p_1 \neq p_2$. The level of significance is $\alpha = 0.05$. The two sample estimates are

$\hat{p}_1 = \dfrac{451}{555} \approx 0.8126$ and $\hat{p}_2 = \dfrac{510}{600} = 0.85$. The

pooled estimate is $\hat{p} = \dfrac{451 + 510}{555 + 600} \approx 0.8320.$

The test statistic is

$$z_0 = \frac{\hat{p}_1 - \hat{p}_2}{\sqrt{\hat{p}(1 - \hat{p})}\sqrt{\dfrac{1}{n_1} + \dfrac{1}{n_2}}}$$

$$= \frac{0.8126 - 0.85}{\sqrt{0.8320(1 - 0.8320)}\sqrt{\dfrac{1}{555} + \dfrac{1}{600}}} \approx -1.70$$

Classical approach: This is a two-tailed test, so the critical values for $\alpha = 0.05$ are $\pm z_{\alpha/2} = \pm z_{0.025} = \pm 1.96$. Since $z_0 = -1.70$ is between $-z_{0.025} = -1.96$ and $z_{0.025} = 1.96$ (the test statistic does not fall in the critical region), we do not reject H_0.

P-value approach: The P-value is two times the area under the standard normal distribution to the left of the test statistic, $z_0 = -1.70$.
P-value $= 2 \cdot P(z \leq -1.70) = 2(0.0446)$
$\qquad = 0.0892$ [Tech: 0.0895]
Since the P-value is greater than the $\alpha = 0.05$ level of significance, we do not reject H_0.

Conclusion: There is not sufficient evidence at the $\alpha = 0.05$ level of significance to conclude that the proportion in population 1 is different from the proportion in population 2.

8. (a) Since the same individual is used for both measurements, the sampling method is dependent.

(b) We measure differences as $d_i = A_i - H_i$.

Student	1	2	3	4	5
Height, H_i	59.5	69	77	59.5	74.5
Arm span, A_i	62	65.5	76	63	74
$d_i = A_i - H_i$	2.5	−3.5	−1	3.5	−0.5

Student	6	7	8	9	10
Height, H_i	63	61.5	67.5	73	69
Arm span, A_i	66	61	69	70	71
$d_i = A_i - H_i$	3	−0.5	1.5	−3	2

We compute the mean and standard deviation of the differences and obtain $\bar{d} = 0.4$ inches and $s_d \approx 2.4698$ inches, rounded to four decimal places. The hypotheses are $H_0 : \mu_d = 0$ versus $H_1 : \mu_d \neq 0$. The level of significance is $\alpha = 0.05$. The test statistic is

$$t_0 = \frac{\bar{d}}{\dfrac{s_d}{\sqrt{n}}} = \frac{0.4}{\dfrac{2.4698}{\sqrt{10}}} \approx 0.512 .$$

Classical approach: Since this is a two-tailed test with 9 degrees of freedom, the critical values are $\pm t_{0.025} = \pm 2.262$. Since $t_0 = 0.512$ falls between $-t_{0.025} = -2.262$ and $t_{0.025} = 2.262$ (the test statistic does not fall within the critical regions), we do not reject H_0.

P-value approach: The P-value for this two-tailed test is the area under the *t*-

distribution with 9 degrees of freedom to the right of the test statistic $t_0 = 0.512$ plus the area to the left of $-t_0 = -0.512$. From the *t*-distribution table in the row corresponding to 9 degrees of freedom, 0.512 falls to the left of 0.703 whose right-tail area is 0.25. We must double this value in order to get the total area in both tails: 0.50. So, *P*-value > 0.50 [Tech: *P*-value = 0.6209]. Because the *P*-value is greater than the $\alpha = 0.05$ level of significance, we do not reject H_0.

Conclusion: There is not sufficient evidence at the $\alpha = 0.05$ level of significance to conclude that an individual's arm span is different from the individual's height. That is, the sample evidence does not contradict the belief that arm span and height are the same.

9. **(a)** The sampling method is independent since the cars selected for the McDonald's sample has no bearing on the cars chosen for the Wendy's sample.

(b) The hypotheses are $H_0 : \mu_{\text{McD}} = \mu_{\text{W}}$ versus $H_1 : \mu_{\text{McD}} \neq \mu_{\text{W}}$. We compute the means and standard deviations, rounded to four decimal places when necessary, of both samples and obtain $\bar{x}_{\text{McD}} = 133.601$, $s_{\text{McD}} \approx 39.6110$, $n_{\text{McD}} = 30$, and $\bar{x}_{\text{W}} \approx 219.1444$, $s_{\text{W}} \approx 102.8459$, and $n_{\text{W}} = 27$. Since the smaller sample size is $n_{\text{W}} = 27$, we use $n_{\text{W}} - 1 = 26$ degrees of freedom.

The test statistic is

$$t_0 = \frac{(\bar{x}_{\text{McD}} - \bar{x}_{\text{W}}) - (\mu_{\text{McD}} - \mu_{\text{W}})}{\sqrt{\dfrac{s_{\text{McD}}^2}{n_{\text{McD}}} + \dfrac{s_{\text{W}}^2}{n_{\text{W}}}}}$$

$$= \frac{(133.601 - 219.1444) - 0}{\sqrt{\dfrac{39.6110^2}{30} + \dfrac{102.8459^2}{27}}} \approx -4.059$$

Classical approach: Since this is a two-tailed test with 26 degrees of freedom, the

critical values are $\pm t_{0.05} = \pm 1.706$. Since the test statistic $t_0 \approx -4.059$ is to the left of -1.706 (the test statistic falls within a critical region), we reject H_0.

P-value approach: The *P*-value for this two-tailed test is the area under the *t*-distribution with 26 degrees of freedom to the left of $t_0 \approx -4.059$ plus the area to the right of 4.059. From the *t*-distribution table in the row corresponding to 26 degrees of freedom, 4.059 falls to the right of 3.707 whose right-tail area is 0.0005. We must double this value in order to get the total area in both tails: 0.001. So, *P*-value < 0.001 [Tech: *P*-value = 0.0003]. Since the *P*-value is less than $\alpha = 0.1$, we reject H_0.

Conclusion: There is sufficient evidence at the $\alpha = 0.1$ level of significance to conclude that the population mean wait times are different for McDonald's and Wendy's drive-through windows.

(c) We compute the five-number summary for each restaurant:

McDonald's: 71.37, 111.84, 127.13, 147.28, 246.59 [Minitab: 71.37, 109.14, 127.13, 148.23, 246.59]

Wendy's: 71.02, 133.56, 190.91, 281.90, 471.62 [Minitab: 71.0, 133.6, 190.9, 281.9, 471.6]

Yes, the boxplots support the results from part (b). Based on the boxplot, it would appear to be the case that the wait time at McDonald's is less than the wait time at Wendy's drive-through windows.

10. (a) We first verify the requirements to perform the hypothesis test: (1) We assume the data represent a simple random sample; (2) We have $x_{exp} = 27$, $n_{exp} = 696$, $x_{control} = 49$, and $n_{control} = 678$, so

$$\hat{p}_{exp} = \frac{27}{696} \approx 0.0388 \text{ and } \hat{p}_{control} = \frac{49}{678} \approx 0.0723. \text{ So,}$$

$$n_{exp}\hat{p}_{exp}\left(1 - \hat{p}_{exp}\right) = 696(0.0388)(1 - 0.0388) \approx 26 \geq 10 \text{ and}$$

$$n_{control}\hat{p}_{control}\left(1 - \hat{p}_{control}\right) = 678(0.0723)(1 - 0.0723) \approx 45 \geq 10; \text{ and (3) Each sample is less than 5\% of}$$

the population. Thus, the requirements are met, and we can conduct the test. The hypotheses are $H_0 : p_{exp} = p_{control}$ versus $H_1 : p_{exp} < p_{control}$. From before, the two sample estimates are $\hat{p}_{exp} \approx 0.0388$

and $\hat{p}_{control} \approx 0.0723$. The pooled estimate is $\hat{p} = \dfrac{x_{exp} + x_{control}}{n_{exp} + n_{control}} = \dfrac{27 + 49}{696 + 678} \approx 0.0553$. The test statistic

is $z_0 = \dfrac{\hat{p}_{exp} - \hat{p}_{control}}{\sqrt{\hat{p}(1 - \hat{p})}\sqrt{\dfrac{1}{n_{exp}} + \dfrac{1}{n_{control}}}} = \dfrac{0.0388 - 0.0723}{\sqrt{0.0553(1 - 0.0553)}\sqrt{\dfrac{1}{696} + \dfrac{1}{678}}} \approx -2.71$.

Classical approach: This is a left-tailed test, so the critical value is $-z_{0.01} = -2.33$. Since $z_0 = -2.71 < -z_{0.01} = -2.33$ (the test statistic falls in the critical region), we reject H_0.

P-value approach: P-value $= P(z_0 \leq -2.71) = 0.0034$ [Tech: 0.0033]. Since this P-value is less than the $\alpha = 0.01$ level of significance, we reject H_0.

Conclusion: There is sufficient evidence at the $\alpha = 0.01$ level of significance to conclude that a lower proportion of women in the experimental group experienced a bone fracture than in the control group.

(b) For a 95% confidence interval we use $\pm z_{0.025} = \pm 1.96$. Then:

$$\text{Lower bound} = (\hat{p}_{exp} - \hat{p}_{control}) - z_{\alpha/2} \cdot \sqrt{\frac{\hat{p}_{exp}(1 - \hat{p}_{exp})}{n_{exp}} + \frac{\hat{p}_{control}(1 - \hat{p}_{control})}{n_{control}}}$$

$$= (0.0388 - 0.0723) - 1.96 \cdot \sqrt{\frac{0.0388(1 - 0.0388)}{696} + \frac{0.0723(1 - 0.0723)}{678}} \approx -0.06$$

$$\text{Upper bound} = (\hat{p}_{exp} - \hat{p}_{control}) + z_{\alpha/2} \cdot \sqrt{\frac{\hat{p}_{exp}(1 - \hat{p}_{exp})}{n_{exp}} + \frac{\hat{p}_{control}(1 - \hat{p}_{control})}{n_{control}}}$$

$$= (0.0388 - 0.0723) + 1.96 \cdot \sqrt{\frac{0.0388(1 - 0.0388)}{696} + \frac{0.0723(1 - 0.0723)}{678}} \approx -0.01$$

We can be 95% confident that the difference in the proportion of women who experienced a bone fracture between the experimental and control group is between -0.06 and -0.01.

(c) This is a completely randomized design with two treatments: 5 mg of Actonel versus the placebo.

(d) A double-blind experiment is one in which neither the subject nor the individual administering the treatment knows to which group (experimental or control) the subject belongs.

11. We first verify the requirements to perform McNemar's test. The samples are dependent because the different question were asked of the same individual. The sample is a random sample. The total number of individuals for which one response was favorable while the other was not favorable is $65 + 45 = 110$, which is greater than 10. Thus, the requirements are met, so we can conduct the test.

The hypotheses are $H_0 : p_J = p_V$ verses $H_1 : p_J \neq p_V$. The test statistic is

$$z_0 = \frac{|f_{12} - f_{21}| - 1}{\sqrt{f_{12} + f_{21}}} = \frac{|65 - 45| - 1}{\sqrt{65 + 45}} \approx 1.81$$

Classical approach: The critical value for $\alpha = 0.05$ is $z_{\alpha/2} = z_{0.025} = 1.96$. Since $z_0 = 1.81 < z_{0.025} = 1.96$ (the test statistic does not fall within the critical region), we do not reject H_0.

P-value approach: The P-value is two times the area under the standard normal distribution to the right of the test statistic, $z_0 = 1.81$.

$$P\text{-value} = 2 \cdot P(Z > 1.81)$$
$$= 2 \cdot [1 - P(Z < 1.81)]$$
$$= 2 \cdot (1 - 0.9649) = 0.0702$$

Since this P-value is greater than $\alpha = 0.05$, we do not reject H_0.

Conclusion: There is not sufficient evidence at the $\alpha = 0.05$ level of significance to conclude that there is a difference in the proportions. It seems that the proportion of individuals who feel serving on a jury is a civic duty is the same as the proportion who feel voting is a civic duty.

12. (a) $n = n_1 = n_2$

$$= \left[\hat{p}_1(1 - \hat{p}_1) + \hat{p}_2(1 - \hat{p}_2) \right] \left(\frac{z_{\alpha/2}}{E} \right)^2$$

$$= \left[0.188(1 - 0.188) + 0.205(1 - 0.205) \right] \left(\frac{1.645}{0.02} \right)^2$$

$$\approx 2135.3, \text{ which we must increase to } 2136.$$

(b) $n = n_1 = n_2 = 0.5 \left(\frac{z_{\alpha/2}}{E} \right)^2 = 0.5 \left(\frac{1.645}{0.02} \right)^2$

$$\approx 3382.5 \text{ which we increase to } 3383.$$

13. The hypotheses are $H_0 : \sigma_{McD} = \sigma_W$ versus $H_1 : \sigma_{McD} < \sigma_W$. From problem 9, we know $s_{McD} \approx 39.6110$ and $s_W \approx 102.8459$. The test statistic is

$$F_0 = \frac{s_{McD}^2}{s_W^2} = \frac{39.6110^2}{102.8459^2} \approx 0.15.$$

Classical approach: This is a left-tailed test with $n_{McD} = 30$ and $n_W = 27$. The critical values is $F_{0.95, 29, 26}$, which we approximate by

$$F_{0.95, 25, 25} = \frac{1}{F_{0.05, 25, 25}} = \frac{1}{1.96} \approx 0.51.$$ Since $F_0 = 0.15 < F_{0.95, 29, 26} \approx 0.51$ (the test statistic falls in a critical region), we reject H_0.

P-value approach: Using technology, we find P-value < 0.0001. Since this P-value is less than the $\alpha = 0.05$ level of significance, we reject H_0.

Conclusion: There is sufficient evidence at the $\alpha = 0.05$ level of significance to conclude that the standard deviation in wait time at Wendy's is more than the standard deviation in wait time at McDonald's.

14. From problem 8, we have $\bar{d} = 0.4$ and $s_d \approx 2.4698$. For 95% confidence, we use $\alpha = 0.05$, so with 9 degrees of freedom, we have $t_{\alpha/2} = t_{0.025} = 2.262$. Then:

$$\text{Lower bound} = \bar{d} - t_{0.025} \cdot \frac{s_d}{\sqrt{n}}$$

$$= 0.4 - 2.262 \cdot \frac{2.4698}{\sqrt{10}} \approx -1.37$$

$$\text{Upper bound} = \bar{d} + t_{0.025} \cdot \frac{s_d}{\sqrt{n}}$$

$$= 0.4 + 2.262 \cdot \frac{2.4698}{\sqrt{10}} \approx 2.17$$

We can be 95% confident that the mean difference between height and arm span is between −1.37 and 2.17. The interval contains zero, so we conclude that there is not sufficient evidence at the $\alpha = 0.05$ level of significance to reject the claim that arm span and height are equal.

15. From problem 9, we have $\bar{x}_{McD} = 133.601$, $s_{McD} \approx 39.6110$, $n_{McD} = 30$, $\bar{x}_{W} \approx 219.1444$, $s_{W} \approx 102.8459$, and $n_{W} = 27$. For a 95% confidence interval with 26 degrees of freedom, we use $t_{\alpha/2} = t_{0.025} = 2.056$. Then:

$$\text{Lower bound} = (\bar{x}_{McD} - \bar{x}_{W}) - t_{\alpha/2} \cdot \sqrt{\frac{s_{McD}^2}{n_{McD}} + \frac{s_{W}^2}{n_{W}}} = (133.601 - 219.1444) - 2.056 \cdot \sqrt{\frac{39.6110^2}{30} + \frac{102.8459^2}{27}}$$

$$\approx -128.869 \quad [\text{Tech:} -128.4]$$

$$\text{Upper bound} = (\bar{x}_{McD} - \bar{x}_{W}) + t_{\alpha/2} \cdot \sqrt{\frac{s_{McD}^2}{n_{McD}} + \frac{s_{W}^2}{n_{W}}} = (133.601 - 219.1444) + 2.056 \cdot \sqrt{\frac{39.6110^2}{30} + \frac{102.8459^2}{27}}$$

$$\approx -42.218 \quad [\text{Tech:} -42.67]$$

We can be 95% confident that the mean difference in wait times between McDonald's and Wendy's drive-through windows is between -128.869 and -42.218 seconds [Tech: between -128.4 and -42.67].

Answer will vary. One possibility is that a marketing campaign could be initiated by McDonald's touting the fact that wait times are up to 2 minutes less at McDonald's.

Chapter 11 Test

1. This is independent because the subjects are randomly selected from two distinct populations.

2. This is dependent since the members of the two samples are matched by type of crime committed.

3. **(a)** We compute $d_i = X_i - Y_i$.

Obs	1	2	3	4	5	6	7
X_i	18.5	21.8	19.4	22.9	18.3	20.2	23.1
Y_i	18.3	22.3	19.2	22.3	18.9	20.7	23.9
d_i	0.2	–0.5	0.2	0.6	–0.6	–0.5	–0.8

(b) Using technology, $\bar{d} = -0.2$ and $s_d \approx 0.5260$, rounded to four decimal places.

(c) The hypotheses are $H_0 : \mu_d = 0$ versus $H_1 : \mu_d \neq 0$. The level of significance is $\alpha = 0.01$. The test statistic is

$$t_0 = \frac{\bar{d}}{\frac{s_d}{\sqrt{n}}} = \frac{-0.2}{\frac{0.5260}{\sqrt{7}}} \approx -1.006.$$

<u>Classical approach</u>: Since this is a two-tailed test with 6 degrees of freedom, the critical values are $\pm t_{0.005} = \pm 3.707$. Since $t_0 \approx -1.006$ falls between $-t_{0.005} = -3.707$ and $t_{0.005} = 3.707$ (the test statistic does

not fall within the critical regions), we do not reject H_0.

<u>P-value approach</u>: The P-value for this two-tailed test is the area under the t-distribution with 6 degrees of freedom to the left of the test statistic $t_0 = -1.006$ plus the area to the right of $t_0 = 1.006$. From the t-distribution table in the row corresponding to 6 degrees of freedom, 1.006 falls between 0.906 and 1.134 whose right-tail areas are 0.20 and 0.15, respectively. We must double these values in order to get the total area in both tails: 0.40 and 0.30. So, $0.30 < P\text{-value} < 0.40$ [Tech: P-value = 0.3532]. Because the P-value is greater than the $\alpha = 0.01$ level of significance, we do not reject H_0.

<u>Conclusion</u>: There is not sufficient evidence at the $\alpha = 0.01$ level of significance to conclude that the population mean difference is different from zero.

(d) For $\alpha = 0.05$ and 7 degrees of freedom, $t_{\alpha/2} = t_{0.025} = 2.447$. Then:

$$\text{Lower bound} = \bar{d} - t_{0.025} \cdot \frac{s_d}{\sqrt{n}}$$

$$= -0.2 - 2.447 \cdot \frac{0.5260}{\sqrt{7}} \approx -0.69$$

Upper bound $= \bar{d} - t_{0.025} \cdot \dfrac{s_d}{\sqrt{n}}$

$$= -0.2 + 2.447 \cdot \dfrac{0.5260}{\sqrt{7}} \approx 0.29$$

We are 95% confident that the population mean difference is between -0.69 and 0.29.

4. (a) $H_0 : \mu_1 = \mu_2$ versus $H_1 : \mu_1 \neq \mu_2$. The level of significance is $\alpha = 0.1$. Since the smaller sample size is $n_1 = 24$, we use $n_1 - 1 = 23$ degrees of freedom. The test statistic is

$$t_0 = \dfrac{(\bar{x}_1 - \bar{x}_2) - (\mu_1 - \mu_2)}{\sqrt{\dfrac{s_1^2}{n_1} + \dfrac{s_2^2}{n_2}}}$$

$$= \dfrac{(104.2 - 110.4) - 0}{\sqrt{\dfrac{12.3^2}{24} + \dfrac{8.7^2}{27}}} \approx -2.054$$

Classical approach: Since this is a two-tailed test with 23 degrees of freedom, the critical values are $\pm t_{0.05} = \pm 1.714$. Since the test statistic $t_0 \approx -2.054$ is to the left of the critical value -1.714 (the test statistic falls within a critical region), we reject H_0.

P-value approach: The P-value for this two-tailed test is the area under the t-distribution with 23 degrees of freedom to the left of $t_0 = -2.054$ plus the area to the right of $t_0 = 2.054$. From the t-distribution table in the row corresponding to 23 degrees of freedom, 2.054 falls between 1.714 and 2.069 whose right-tail areas are 0.05 and 0.025, respectively. We must double these values in order to get the total area in both tails: 0.10 and 0.05. Thus, $0.05 < P\text{-value} < 0.10$ [Tech: $P\text{-value} = 0.0464$]. Because the P-value is less than the $\alpha = 0.1$ level of significance, we reject H_0.

Conclusion: There is sufficient evidence at the $\alpha = 0.1$ level of significance to conclude that the means are different.

(b) For a 95% confidence interval with 23 degrees of freedom, we use $t_{\alpha/2} = t_{0.025} = 2.069$. Then:

Lower bound

$$= (\bar{x}_1 - \bar{x}_2) - t_{\alpha/2} \cdot \sqrt{\dfrac{s_1^2}{n_1} + \dfrac{s_2^2}{n_2}}$$

$$= (104.2 - 110.4) - 2.069 \cdot \sqrt{\dfrac{12.3^2}{24} + \dfrac{8.7^2}{27}}$$

$$\approx -12.44 \quad [\text{Tech}: -12.3]$$

Upper bound:

$$= (\bar{x}_1 - \bar{x}_2) + t_{\alpha/2} \cdot \sqrt{\dfrac{s_1^2}{n_1} + \dfrac{s_2^2}{n_2}}$$

$$= (104.2 - 110.4) + 2.069 \cdot \sqrt{\dfrac{12.3^2}{24} + \dfrac{8.7^2}{27}}$$

$$\approx 0.04 \quad [\text{Tech}: -0.10]$$

We are 95% confident that the population mean difference is between -12.44 and 0.04 [Tech: between -12.3 and -0.10].

(c) The hypotheses are $H_0 : \sigma_1 = \sigma_2$ versus $H_1 : \sigma_1 > \sigma_2$. The test statistic is

$$F_0 = \dfrac{s_1^2}{s_2^2} = \dfrac{12.3^2}{8.7^2} \approx 2.00.$$

Classical approach: This is a right-tailed test with $n_1 = 24$, $n_2 = 27$, and $\alpha = 0.10$. The critical value is $F_{0.10, 23, 26}$, which we approximate by $F_{0.10, 25, 25} = 1.68$. Since $F_0 = 2.00 > F_{0.10, 23, 26} \approx 1.68$ (the test statistic falls in the critical region), we reject H_0.

P-value approach: Using technology, we find $P\text{-value} \approx 0.0448$. Since this P-value is less than the $\alpha = 0.05$ level of significance, we reject H_0.

Conclusion: There is sufficient evidence at the $\alpha = 0.1$ level of significance to conclude that the standard deviation in population 1 is greater than the standard deviation in population 2.

5. (a) $H_0 : \mu_1 = \mu_2$ versus $H_1 : \mu_1 < \mu_2$. The level of significance is $\alpha = 0.05$. Since the smaller sample size is $n_2 = 8$, we use $n_2 - 1 = 7$ degrees of freedom. The test statistic is

$$t_0 = \frac{(\bar{x}_1 - \bar{x}_2) - (\mu_1 - \mu_2)}{\sqrt{\frac{s_1^2}{n_1} + \frac{s_2^2}{n_2}}}$$

$$= \frac{(96.6 - 98.3) - 0}{\sqrt{\frac{3.2^2}{13} + \frac{2.5^2}{8}}} \approx -1.357$$

Classical approach: Since this is a left-tailed test with 7 degrees of freedom, the critical value is $-t_{0.05} = -1.895$. Since $t_0 = -1.375 > -t_{0.05} = -1.895$ (the test statistic does not fall within the critical region), we do not reject H_0.

P-value approach: The P-value for this left-tailed test is the area under the t-distribution with 7 degrees of freedom to the left of $t_0 = -1.375$, which is equivalent to the area to the right of $t_0 = 1.375$. From the t-distribution table in the row corresponding to 7 degrees of freedom, 1.375 falls between 1.119 and 1.415 whose right-tail areas are 0.15 and 0.10, respectively. Thus, $0.10 < P\text{-value} < 0.15$ [Tech: P-value = 0.0959]. Because the P-value is greater than the $\alpha = 0.05$ level of significance, we do not reject H_0.

Conclusion: There is not sufficient evidence at the $\alpha = 0.05$ level of significance to conclude that the mean of population 1 is less than the mean of population 2.

(b) The hypotheses are $H_0 : \sigma_1 = \sigma_2$ versus $H_1 : \sigma_1 \neq \sigma_2$. The test statistic is

$$F_0 = \frac{s_1^2}{s_2^2} = \frac{3.2^2}{2.5^2} \approx 1.64.$$

Classical approach: This is a two-tailed test with $n_1 = 13$, $n_2 = 8$, and $\alpha = 0.05$. The critical values are $F_{0.025,12,7}$, which we approximate by $F_{0.025,10,7} = 4.76$ and

$F_{0.975,12,7} = \frac{1}{F_{0.025,7,12}} = \frac{1}{3.61} \approx 0.28$. Since $F_0 = 1.64$ is between $F_{0.025,12,7} \approx 4.76$ and $F_{0.975,12,7} \approx 0.28$ (the test statistic does not fall in the critical region), we do not reject H_0.

P-value approach: Using technology, we find P-value ≈ 0.5243. Since this P-value is greater than the $\alpha = 0.05$ level of significance, we do not reject H_0.

Conclusion: There is not sufficient evidence at the $\alpha = 0.05$ level of significance to conclude that the standard deviation in population 1 is different from the standard deviation in population 2.

6. The hypotheses are $H_0 : p_1 = p_2$ versus $H_1 : p_1 < p_2$. The level of significance is $\alpha = 0.05$. The two sample estimates are $\hat{p}_1 = \frac{156}{650} = 0.24$ and $\hat{p}_2 = \frac{143}{550} = 0.26$. The pooled estimate is $\hat{p} = \frac{156+143}{650+550} \approx 0.2492$. The test statistic is

$$z_0 = \frac{\hat{p}_1 - \hat{p}_2}{\sqrt{\hat{p}(1-\hat{p})}\sqrt{\frac{1}{n_1} + \frac{1}{n_2}}}$$

$$= \frac{0.24 - 0.26}{\sqrt{0.2492(1-0.2492)}\sqrt{\frac{1}{650} + \frac{1}{550}}} \approx -0.80$$

Classical approach: This is a left-tailed test, so the critical value is $-z_{0.05} = -1.645$. Since $z_0 = -0.80 > -z_{0.05} = -1.645$ (the test statistic does not fall in the critical region), we do not reject H_0.

P-value approach: The P-value is the area under the standard normal distribution to the left of the test statistic, $z_0 = -0.80$. P-value = $P(z \leq -0.80) = 0.2119$ [Tech: 0.2124]. Since the P-value is greater than the $\alpha = 0.05$ level of significance, we do not reject H_0.

Conclusion: There is not sufficient evidence at the $\alpha = 0.05$ level of significance to conclude that the proportion in population 1 is less than the proportion in population 2.

7. (a) Since the dates selected for the Texas sample have no bearing on the dates chosen for the Illinois sample, the testing method is independent.

(b) Because the sample sizes are small, both samples must come from populations that are normally distributed.

(c) We compute the five-number summary for each restaurant:

Texas: 4.11, 4.555, 4.735, 5.12, 5.22
[Minitab: 4.11, 4.508, 4.735, 5.13, 5.22]

Illinois: 4.22, 4.40, 4.505, 4.64, 4.75
[Minitab: 4.22, 4.39, 4.505, 4.6525, 4.75]

The boxplots that follow indicate that rain in Chicago, Illinois has a lower pH than rain in Houston, Texas.

(d) The hypotheses are $H_0 : \mu_{\text{Texas}} = \mu_{\text{Illinois}}$ versus $H_1 : \mu_{\text{Texas}} \neq \mu_{\text{Illinois}}$. We compute the summary statistics, rounded to four decimal places when necessary: $\bar{x}_{\text{Texas}} = 4.77$, $s_{\text{Texas}} \approx 0.3696$, $n_{\text{Texas}} = 12$, $\bar{x}_{\text{Illinois}} \approx 4.5093$, $s_{\text{Illinois}} \approx 0.1557$, and $n_{\text{Illinois}} = 14$. Since the smaller sample size is $n_{\text{Texas}} = 12$, we use $n_{\text{Texas}} - 1 = 11$ degrees of freedom. The test statistic is

$$t_0 = \frac{(\bar{x}_{\text{Texas}} - \bar{x}_{\text{Illinois}}) - (\mu_{\text{Texas}} - \mu_{\text{Illinois}})}{\sqrt{\dfrac{s_{\text{Texas}}^2}{n_{\text{Texas}}} + \dfrac{s_{\text{Illinois}}^2}{n_{\text{Illinois}}}}}$$

$$= \frac{(4.77 - 4.5093) - 0}{\sqrt{\dfrac{0.3696^2}{12} + \dfrac{0.1557^2}{14}}} \approx 2.276$$

Classical approach: Since this is a two-tailed test with 11 degrees of freedom, the critical values are $\pm t_{0.025} = \pm 2.201$. Since the test statistic $t_0 \approx 2.276$ is to the right of 2.201 (the test statistic falls within a critical region), we reject H_0.

P-value approach: The *P*-value for this two-tailed test is the area under the *t*-distribution with 11 degrees of freedom to the right of $t_0 \approx 2.276$ plus the area to the left of $-t_0 \approx -2.276$. From the *t*-distribution table in the row corresponding to 11 degrees of freedom, 2.276 falls between 2.201 and 2.328 whose right-tail areas are 0.025 and 0.02, respectively. We must double these values in order to get the total area in both tails: 0.05 and 0.04. So, $0.04 < P\text{-value} < 0.05$ [Tech: *P*-value = 0.0387]. Since the *P*-value is less than $\alpha = 0.05$, we reject H_0.

Conclusion: There is sufficient evidence at the $\alpha = 0.05$ level of significance to conclude that the acidity of the rain near Houston, Texas is different from the acidity of rain near Chicago, Illinois.

8. (a) The response variable is the student's GPA. The explanatory variable is whether the student has a sleep disorder or not.

(b) The hypotheses are:
$H_0 : \mu_{\text{Sleep Disorder}} = \mu_{\text{No Sleep Disorder}}$ versus
$H_1 : \mu_{\text{Sleep Disorder}} < \mu_{\text{No Sleep Disorder}}$. The level of significance is $\alpha = 0.05$. Since the smaller sample size is $n_{\text{Sleep Disorder}} = 503$, we use $n_{\text{Sleep Disorder}} - 1 = 503 - 1 = 502$ degrees of freedom. The test statistic is

$$t_0 = \frac{(\bar{x}_{\text{Sleep Disorder}} - \bar{x}_{\text{No Sleep Disorder}}) - (0)}{\sqrt{\dfrac{s_{\text{Sleep Disorder}}^2}{n_{\text{Sleep Disorder}}} + \dfrac{s_{\text{No Sleep Disorder}}^2}{n_{\text{No Sleep Disorder}}}}}$$

$$= \frac{(2.65 - 2.82) - 0}{\sqrt{\dfrac{0.87^2}{503} + \dfrac{0.83^2}{1342}}}$$

$$\approx -3.784$$

Classical approach: Since this is a left-tailed test with 502 degrees of freedom, we use the closest table value, 100 degrees of freedom, and the critical value is $t_{0.05} = 1.660$. Since the test statistic $t_0 \approx -3.784$ is to the left of the critical value (i.e., since the test statistic falls within the critical region), we reject H_0.

P-value approach: The P-value for this left-tailed test is the area under the t-distribution with 502 degrees of freedom to the left of $t_0 \approx -3.784$, which is equivalent to the area to the right of $-t_0 \approx 3.784$. From the t-distribution table, we use the row corresponding to 100 degrees of freedom, 3.784 falls to the right of 3.390 whose right-tail area is are 0.0005. Thus, P-value < 0.0005. [Tech: P-value < 0.0001]. Because the P-value is less than the level of significance, $\alpha = 0.05$, we reject H_0.

Conclusion: There is sufficient evidence at the $\alpha = 0.05$ level of significance to conclude that sleep disorders adversely affect a student's GPA.

9. We measure differences as $d_i = SUV_i - Car_i$.

SUV_i	2091	1338	1053	1872
Car_i	1510	2559	4921	4555
d_i	581	−1221	−3868	−2683

SUV_i	1428	2208	6015
Car_i	5114	5203	3852
d_i	−3686	−2995	2163

We compute the mean and standard deviation of the differences and obtain $\bar{d} \approx -1672.71$ minutes, rounded to two decimal places, and $s_d \approx 2296.29$ minutes, rounded to two decimal places. The hypotheses are $H_0 : \mu_d = 0$ versus $H_1 : \mu_d < 0$. The level of significance is $\alpha = 0.05$. The test statistic is

$$t_0 = \frac{\bar{d}}{\frac{s_d}{\sqrt{n}}} = \frac{-1672.71}{\frac{2296.29}{\sqrt{7}}} \approx -1.927.$$

Classical approach: Since this is a left-tailed test with $7 - 1 = 6$ degrees of freedom, the critical value is $-t_{0.10} = -1.440$. Since the test statistic $t_0 \approx -1.927$ falls to the left of the critical value $-t_{0.10} = -1.440$ (i.e., since the

test statistic falls in the critical region), we reject H_0.

P-value approach: The P-value for this left-tailed test is the area under the t-distribution with $7 - 1 = 6$ degrees of freedom to the left of the test statistic $t_0 \approx -1.927$, which is equivalent to the area to the right of 1.927. From the t-distribution table, in the row corresponding to 6 degrees of freedom, 1.927 falls between 1.440 and 1.943, whose right-tail areas are 0.10 and 0.05, respectively. So, $0.05 < P\text{-value} < 0.10$ [Tech: P-value = 0.0511]. Because the P-value is less than the level of significance, $\alpha = 0.10$, we reject H_0.

Conclusion: There is sufficient evidence at the $\alpha = 0.10$ level of significance to conclude that the repair cost for the car is higher.

10. (a) This is a completely randomized design. The subjects were randomly divided into two groups.

   (b) The response variable is whether the subjects get dry mouth or not.

   (c) We verify the requirements to perform the hypothesis test: (1) We treat the sample as if it was a simple random sample; (2) We have $x_{\exp} = 77$, $n_{\exp} = 553$, $x_{\text{control}} = 34$, and $n_{\text{control}} = 373$, so

$$\hat{p}_{\exp} = \frac{x_{\exp}}{n_{\exp}} = \frac{77}{553} \approx 0.1392 \text{ and}$$

$$\hat{p}_{\text{control}} = \frac{x_{\text{control}}}{n_{\text{control}}} = \frac{34}{373} \approx 0.0912. \text{ So,}$$

$n_{\exp}\hat{p}_{\exp}\left(1 - \hat{p}_{\exp}\right) = 553(0.1392)(1 - 0.1392) \approx 66 \geq 10$ and $n_{\text{control}}\hat{p}_{\text{control}}\left(1 - \hat{p}_{\text{control}}\right) = 373(0.0912)(1 - 0.0912) \approx 31 \geq 10$; and (3) Each sample is less than 5% of the population. Thus, the requirements are met, and we can conduct the test. The hypotheses are $H_0 : p_{\exp} = p_{\text{control}}$ and $H_1 : p_{\exp} > p_{\text{control}}$. From before, the two sample estimates are $\hat{p}_{\exp} \approx 0.1392$ and $\hat{p}_{\text{control}} \approx 0.0912$. The pooled estimate is

$$\hat{p} = \frac{x_{\exp} + x_{\text{control}}}{n_{\exp} + n_{\text{control}}} = \frac{77 + 34}{553 + 373} \approx 0.1199.$$

The test statistic is

$$z_0 = \frac{\hat{p}_{\text{exp}} - \hat{p}_{\text{control}}}{\sqrt{\hat{p}(1-\hat{p})}\sqrt{\dfrac{1}{n_{\text{exp}}} + \dfrac{1}{n_{\text{control}}}}}$$

$$= \frac{0.1392 - 0.0912}{\sqrt{0.1199(1-0.1199)}\sqrt{\dfrac{1}{553} + \dfrac{1}{373}}}$$

$$\approx 2.21$$

Classical approach: This is a right-tailed test, so the critical value is $z_{0.05} = 1.645$. Since $z_0 = 2.21 > z_{0.05} = 1.645$ (the test statistic falls in the critical region), we reject H_0.

P-value approach: The P-value is the area under the standard normal curve to the right of $z_0 = 2.21$:

$$P\text{-value} = P(z_0 \geq 2.21)$$
$$= 1 - P(z_0 < 2.21)$$
$$= 1 - 0.9864$$
$$= 0.0136$$

Since this P-value is less than the $\alpha = 0.05$ level of significance, we reject H_0.

Conclusion: There is sufficient evidence at the $\alpha = 0.05$ level of significance to conclude that a higher proportion of subjects in the experimental group experienced dry mouth than in the control group.

11. We have $x_F = 644$, $n_F = 2800$, $x_M = 840$, and $n_M = 2800$, so $\hat{p}_F = \dfrac{x_F}{n_F} = \dfrac{644}{2800} = 0.23$ and

$\hat{p}_M = \dfrac{x_M}{n_M} = \dfrac{840}{2800} = 0.3$. The hypotheses are $H_0 : p_M = p_F$ and $H_1 : p_M \neq p_F$. For a 90% confidence interval we use $\pm z_{0.05} = \pm 1.645$. Then:

Lower bound:

$$(\hat{p}_M - \hat{p}_F) - z_{\alpha/2} \cdot \sqrt{\frac{\hat{p}_M(1-\hat{p}_M)}{n_M} + \frac{\hat{p}_F(1-\hat{p}_F)}{n_F}} = (0.3 - 0.23) - 1.645 \cdot \sqrt{\frac{0.3(1-0.3)}{2800} + \frac{0.23(1-0.23)}{2800}} \approx 0.051$$

Upper bound:

$$(\hat{p}_M - \hat{p}_F) + z_{\alpha/2} \cdot \sqrt{\frac{\hat{p}_M(1-\hat{p}_M)}{n_M} + \frac{\hat{p}_F(1-\hat{p}_F)}{n_F}} = (0.3 - 0.23) + 1.645 \cdot \sqrt{\frac{0.3(1-0.3)}{2800} + \frac{0.23(1-0.23)}{2800}} \approx 0.089$$

We are 90% confident that the difference in the proportions for males and females for which hypnosis led to quitting smoking is between 0.051 and 0.089. Because the confidence interval does not include 0, we reject the null hypothesis. There is sufficient evidence at the $\alpha = 0.1$ level of significance to conclude that the proportion of males and females for which hypnosis led to quitting smoking is different.

12. We first verify the requirements to perform McNemar's test. The samples are dependent because the different questions were posed to the same individual. The sample is a random sample. The total number of individuals for which one response was favorable while the other was not favorable is 2,549 + 11,685 = 14,234, which is greater than 10. Thus, the requirements are met, so we can conduct the test.

The hypotheses are $H_0 : p_A = p_{DP}$ verses $H_1 : p_A \neq p_{DP}$. The test statistic is

$$z_0 = \frac{|f_{12} - f_{21}| - 1}{\sqrt{f_{12} + f_{21}}} = \frac{|2{,}549 - 11{,}685| - 1}{\sqrt{2{,}549 + 11{,}685}} \approx 76.6$$

Classical approach: The critical value for $\alpha = 0.01$ is $z_{\alpha/2} = z_{0.005} = 2.575$. Since $z_0 = 76.6 > z_{0.005} = 2.575$ (the test statistic falls within the critical region), we reject H_0.

P-value approach: The *P*-value is two times the area under the standard normal distribution to the right of the test statistic, $z_0 = 76.6$. The test statistic is so large that $P\text{-value} < 0.0001$.

Since this *P*-value is less than $\alpha = 0.01$, we reject H_0.

Conclusion: There is sufficient evidence at the $\alpha = 0.01$ level of significance to conclude that the proportion of individuals who favor the death penalty is different from the proportion who favor abortion.

13. (a) $n = n_1 = n_2$

$$= \left[\hat{p}_1(1-\hat{p}_1) + \hat{p}_2(1-\hat{p}_2)\right]\left(\frac{z_{\alpha/2}}{E}\right)^2$$

$$= \left[0.322(1-0.322) + 0.111(1-0.111)\right]\left(\frac{1.96}{0.04}\right)^2$$

$$\approx 761.1, \text{ which we must increase to } 762.$$

(b) $n = n_1 = n_2 = 0.5\left(\frac{z_{\alpha/2}}{E}\right)^2 = 0.5\left(\frac{1.96}{0.04}\right)^2$

$$= 1200.5$$

which we must increase to 1201.

14. We have $n_p = 72$, $\overline{x}_p = 1.9$, $s_p = 0.22$, $n_n = 75$, $\overline{x}_n = 0.8$, and $s_n = 0.21$. The smaller sample size is $n_p = 72$, so we have $n_p - 1 = 71$ degrees of freedom. Since our *t*-distributions table does not have a row for df = 71, we will use df = 70, For a 95% confidence interval with 70 degrees of freedom, we use $t_{\alpha/2} = t_{0.025} = 1.994$. Then:

Lower bound

$$= (\overline{x}_p - \overline{x}_n) - t_{\alpha/2} \cdot \sqrt{\frac{s_p^2}{n_p} + \frac{s_n^2}{n_n}}$$

$$= (1.9 - 0.8) - 1.994 \cdot \sqrt{\frac{0.22^2}{72} + \frac{0.21^2}{75}}$$

$$\approx 1.03$$

Upper bound

$$= (\overline{x}_p - \overline{x}_n) + t_{\alpha/2} \cdot \sqrt{\frac{s_p^2}{n_p} + \frac{s_n^2}{n_n}}$$

$$= (1.9 - 0.8) + 1.994 \cdot \sqrt{\frac{0.22^2}{72} + \frac{0.21^2}{75}}$$

$$\approx 1.17$$

We can be 95% confident that the mean difference in weight of the placebo group versus the Naltrexone group is between 1.03 and 1.17 pounds. Because the confidence interval does not contain 0, we reject H_0. There is sufficient evidence to conclude that Naltrexone is effective in preventing weight gain among individuals who quit smoking. Answers will vary regarding practical significance, but one must ask, "Do I want to take a drug so that I can keep about 1 pound off ?" Probably not.

15. The hypotheses are $H_0 : \sigma_{\text{Texas}} = \sigma_{\text{Illinois}}$ versus $H_1 : \sigma_{\text{Texas}} \neq \sigma_{\text{Illinois}}$. From problem 7, we know $s_{\text{Texas}} \approx 0.3696$, $n_{\text{Texas}} = 12$, $s_{\text{Illinois}} \approx 0.1557$, and $n_{\text{Illinois}} = 14$. The test statistic is

$$F_0 = \frac{s_{\text{Texas}}^2}{s_{\text{Illinois}}^2} = \frac{0.3696^2}{0.1557^2} \approx 5.63.$$

Classical approach: This is a two-tailed test with $n_{\text{Texas}} = 12$ and $n_{\text{Illinois}} = 14$. The critical values are $F_{0.025,12,14} \approx F_{0.025,12,15} = 2.96$ and

$$F_{0.975,12,14} = \frac{1}{F_{0.025,14,12}} \approx \frac{1}{F_{0.025,15,12}} = \frac{1}{3.18} \approx 0.31.$$

Since $F_0 = 5.63 > F_{0.025,12,14} \approx 2.96$ (the test statistic falls in a critical region), we reject H_0.

P-value approach: Using technology, we find $P\text{-value} \approx 0.0044$. Since this *P*-value is less than the $\alpha = 0.05$ level of significance, we reject H_0.

Conclusion: There is sufficient evidence at the $\alpha = 0.05$ level of significance to conclude that the standard deviation in the acidity of rain near Houston, Texas is different from the standard deviation in the acidity of rain near Chicago, Illinois.

Case Study: Control in the Design of an Experiment

Since the individuals selected for the ESRD sample have no bearing on the individuals selected for the control group, the testing method is independent.

Hypothesis Tests for each clinical and biochemical parameter:

For each test of parameters, the hypotheses are $H_0 : \mu_{ESRD} = \mu_{control}$ versus $H_1 : \mu_{ESRD} \neq \mu_{control}$. We use technology to find each P-value. The sample sizes are $n_{ESRD} = 75$ and $n_{control} = 57$, so we use $n_{control} - 1 = 56$ degrees of freedom.

Clinical Characteristics

Age: P-value ≈ 0.0024

Sex: Sample statistics indicate no difference.

Body surface area: P-value ≈ 0.00002

Body mass index: P-value ≈ 0.0034

Systolic BP: P-value ≈ 0.3437

Diastolic BP: P-value ≈ 0.0636

Mean BP: P-value ≈ 0.7442

Pulse pressure: P-value ≈ 0.0092

Heart rate: P-value ≈ 0.000006

Biological Findings

Total cholesterol: P-value ≈ 0.0469

HDL cholesterol: P-value ≈ 0.00001

Triglycerides: P-value ≈ 0.0006

Serum albumin: P-value $\approx 2.6 \times 10^{-17}$

Plasma fibrinogen: P-value $\approx 4.0 \times 10^{-17}$

Plasma creatinine: P-value $\approx 2.5 \times 10^{-61}$

Blood urea: P-value $\approx 1.5 \times 10^{-97}$

Calcium: P-value ≈ 0.5675

Phosphates: P-value $\approx 2.5 \times 10^{-32}$

Thus, the parameters for age, body surface area, body mass index, pulse pressure, heart rate, HDL cholesterol, Triglycerides, serum albumin, plasma fibrinogen, plasma creatinine, blood urea, and phosphates all have P-values less than 0.01. In addition, the parameter for total cholesterol has a P-value less than 0.05. For each of these parameters, we would reject the null hypothesis and conclude that $\mu_{ESRD} \neq \mu_{control}$. In other words, for each of these parameters, the ESRD group and the control group are different.

The parameters for sex, systolic BP, diastolic BP, mean BP, and calcium each had P-values above 0.05. For each of these parameters, we would not reject the null hypothesis and conclude that $\mu_{ESRD} = \mu_{control}$. In other words, for each of these parameters, the ESRD group and the control group have equivalent means.

Assumptions will vary. Additional information and statistical procedures desired for analysis will vary.

Based on the above findings, it does not appear that the ESRD and control groups have similar initial clinical characteristics and biochemical findings. It does not appear that the authors of the article were successful in reducing the likelihood that a confounding effect might obscure the results.

Using the nonrandom process, the authors of the article were able to sufficiently control for sex and blood pressure as they desired, but they were not able to control for the other parameters. In fact, one could argue that the authors, by using the nonrandom process, actually caused potential confounding effects to be more prevalent. Thus, Dr. Nicholls can determine that the effort to control confounding effects by using the nonrandom technique did not work well. She would likely be better off using a random process to assign patients to the treatment and control groups For example, a stratified random assignment would allow for control of some variables while maintaining randomization for the remaining variables.

Reports and recommendations will vary.

Chapter 12

Inference on Categorical Data

Section 12.1

1. True

3. expected counts; $n \cdot p_i$

5. Each expected count is $n \cdot p_i$ where $n = 500$.
This gives expected counts of $500(0.2) = 100$, $500(0.1) = 50$, $500(0.45) = 225$, and $500(0.25) = 125$.

p_i	0.2	0.1	0.45	0.25
E_i	100	50	225	125

7. (a)

O_i	E	$(O_i - E_i)^2$	$\dfrac{(O_i - E_i)^2}{E_i}$
30	25	25	1
20	25	25	1
28	25	9	0.36
22	25	9	0.36
		$\chi_0^2 =$	2.72

(b) df = 4 – 1 = 3

(c) With 3 degrees of freedom, the critical value is $\chi_{0.05}^2 = 7.815$.

(d) The test statistic is not in the (right-tailed) critical region (that is, $\chi_0^2 < \chi_{0.05}^2$), so we do not reject H_0. There is not sufficient evidence at the 5% level of significance to conclude that any one of the proportions is different from the others.

9. (a)

O_i	E_i	$(O_i - E_i)^2$	$\dfrac{(O_i - E_i)^2}{E_i}$
1	1.6	0.36	0.225
38	25.6	153.76	6.006
132	153.6	466.56	3.038
440	409.6	924.16	2.256
389	409.6	424.36	1.036
		$\chi_0^2 =$	12.561

(b) df = 5 – 1 = 4

(c) With 4 degrees of freedom, the critical value is $\chi_{0.05}^2 = 9.488$.

(d) The test statistic is in the (right-tailed) critical region (that is, $\chi_0^2 > \chi_{0.05}^2$), so we reject H_0. There is sufficient evidence at the 5% level of significance to conclude that the random variable X is not binomial with $n = 4$ and $p = 0.8$.

11. Using $\alpha = 0.05$, we want to test
H_0: The color of plain M&Ms follows the manufacturer's stated distribution
vs.
H_1: The color of plain M&Ms follows a different distribution.
There are 400 candies in the bag. To determine the expected counts, multiply 400 by the given percentage for each color. For example, the expected count of brown M&Ms would be $400(0.13) = 52$. We summarize the observed and expected counts in the following table:

	O_i	E_i	$(O_i - E_i)^2$	$\dfrac{(O_i - E_i)^2}{E_i}$
Brown	61	52	81	1.558
Yellow	64	56	64	1.143
Red	54	52	4	0.077
Blue	61	96	1225	12.760
Orange	96	80	256	3.2
Green	64	64	0	0
			$\chi_0^2 =$	18.738

Since all the expected cell counts are greater than or equal to 5, the requirements for the goodness-of-fit test are satisfied.

Classical approach: The critical value, with df = 6 – 1 = 5, is $\chi_{0.05}^2 = 11.071$. Since the test statistic is in the critical region ($\chi_0^2 > \chi_{0.05}^2$), we reject the null hypothesis.

P-value approach: Using the chi-square table, we find the row that corresponds to 5 degrees of freedom. The value of 18.738 is greater than 16.750, which has an area under the chi-square distribution of 0.005 to the right. Therefore, we have P-value < 0.005. Since P-value $< \alpha$, we reject the null hypothesis. [Note: using technology, we find P-value $= 0.0022 < \alpha$ so the null hypothesis is rejected.]

Conclusion: There is enough evidence to conclude that the distribution of colors of plain M&Ms differs from that claimed by M&M/Mars.

13. (a) Answers will vary. Because of the seriousness of the situation, we would want to have strong evidence that there is fraudulent activity. Therefore, we would want to select a small value for α, such as $\alpha = 0.01$.

(b) There are 200 digits being checked. To determine the expected counts, multiply 200 by the percentage for each digit given by Benford's Law. For example, the expected count of the digit 1 would be $200(0.301) = 60.2$. We summarize the observed and expected counts in the following table:

	O_i	E_i	$\left(O_i - E_i\right)^2$	$\dfrac{\left(O_i - E_i\right)^2}{E_i}$
1	36	60.2	585.64	9.728
2	32	35.2	10.24	0.291
3	28	25	9	0.360
4	26	19.4	43.56	2.245
5	23	15.8	51.84	3.281
6	17	13.4	12.96	0.967
7	15	11.6	11.56	0.997
8	16	10.2	33.64	3.298
9	7	9.2	4.84	0.526
			$\chi_0^2 =$	21.693

Since all the expected cell counts are greater than or equal to 5, the requirements for the goodness-of-fit test are satisfied. Our hypotheses are:

H_0 : the digits obey Benford's law

H_1 : the digits do not obey Benford's law

Classical approach: The critical value, with df $= 9 - 1 = 8$, is $\chi_{0.01}^2 = 20.090$. Since the test statistic is in the critical

region ($\chi_0^2 > \chi_{0.01}^2$), we reject the null hypothesis.

P-value approach: Using the chi-square table, we find the row that corresponds to 8 degrees of freedom. The value of 21.693 is greater than 20.090, which has an area under the chi-square distribution of 0.01 to the right. Therefore, we have P-value < 0.01. Since P-value $< \alpha$, we reject the null hypothesis. [Note: using technology, we find P-value $= 0.006 < \alpha$ and reject H_0]

Conclusion: There is enough evidence to reject the null hypothesis. The digits do not appear to obey Benford's law.

(c) Answers will vary. Based on the results from part (b), it appears that the employee is guilty of embezzlement at the $\alpha = 0.01$ level of significance.

15. (a) To determine the expected counts, multiply 2068 by the given percentages. For example, the expected count for Multiple Locations would be $2068(0.57) = 1178.76$. We summarize the observed and expected counts in the following table:

	O_i	E_i	$\left(O_i - E_i\right)^2$	$\dfrac{\left(O_i - E_i\right)^2}{E_i}$
Mult. Locations	1036	1178.76	20,380.4176	17.2897
Head	864	641.08	49,693.3264	77.5150
Neck	38	62.04	577.9216	9.3153
Thorax	83	124.08	1687.5664	13.6006
Ab/Lum/ Spine	47	62.04	226.2016	3.6461
			$\chi_0^2 \approx$	121.367

Since all the expected cell counts are greater than or equal to 5, the requirements for the goodness-of-fit test are satisfied. Our hypotheses are

H_0 : the distribution of fatal injuries for those not wearing helmets is the same as for all riders.

H_1 : the distributions are not the same

Classical approach: The critical value, with df $= 5 - 1 = 4$, is $\chi_{0.05}^2 = 9.488$. Since the test statistic is in the critical region ($\chi_0^2 > \chi_{0.05}^2$), we reject the null hypothesis.

P-value approach: Using the chi-square table, we find the row that corresponds to 4 degrees of freedom. The value of 121.367 is greater than 14.860, which has an area under the chi-square distribution of 0.005 to the right. Therefore, we have *P*-value < 0.005 [Tech: *P*-value < 0.001]. Since *P*-value < α, we reject the null hypothesis.

Conclusion: There is sufficient evidence to conclude that the distribution of fatal injuries for those not wearing helmets is different than for all riders.

(b) The observed count for head injuries is much higher than expected, while the observed count for all other fatal injuries are lower than expected. We might conclude that motorcycle fatalities from head injuries occur more frequently for riders not wearing helmets.

17. (a) We obtain the number of students from each group attending by multiplying each percent by 100:

Group 1: $100(0.84) = 84$

Group 2: $100(0.84) = 84$

Group 3: $100(0.84) = 84$

Group 4: $100(0.81) = 81$

If seat location has no effect, we expect 83% of each group to attend. Since there were 100 students in each group, we expect 83 of the 100 students in each group to attend. We summarize the observed and expected counts in the following table:

	O_i	E_i	$(O_i - E_i)^2$	$\dfrac{(O_i - E_i)^2}{E_i}$
1	84	83	1	0.0120
2	84	83	1	0.0120
3	84	83	1	0.0120
4	81	83	4	0.0482
			$\chi_0^2 \approx$	0.084

Since all the expected cell counts are greater than or equal to 5, the requirements for the goodness-of-fit test

are satisfied. Our hypotheses are

H_0 : there is no difference in attendance patterns among the 4 groups

H_1 : there are differences in attendance patterns among the 4 groups.

Classical approach: The critical value, with df = 4 − 1 = 3, is $\chi_{0.05}^2 = 7.815$. Since the test statistic is not in the critical region ($\chi_0^2 < \chi_{0.05}^2$), we do not reject the null hypothesis.

P-value approach: Using the chi-square table, we find the row that corresponds to 3 degrees of freedom. The value of 0.084 [Tech: 0.081] is less than 0.115, which has an area under the chi-square distribution of 0.99 to the right. Therefore, we have *P*-value > 0.99 [Tech: *P*-value = 0.994]. Since *P*-value > α, we do not reject the null hypothesis.

Conclusion: There is not sufficient evidence to conclude that there are differences among the groups in terms of attendance patterns.

(b) We obtain the number of students from each group attending by multiplying each percent by 100:

Group 1: $100(0.84) = 84$

Group 2: $100(0.81) = 81$

Group 3: $100(0.78) = 78$

Group 4: $100(0.76) = 76$

If seat location has no effect, we expect 80% of each group to attend. Since there were 100 students in each group, we expect 80 of the 100 students in each group to attend. We summarize the observed and expected counts in the following table:

	O_i	E_i	$(O_i - E_i)^2$	$\dfrac{(O_i - E_i)^2}{E_i}$
1	84	80	16	0.2
2	81	80	1	0.0125
3	78	80	4	0.05
4	76	80	16	0.2
			$\chi_0^2 \approx$	0.463

Since all the expected cell counts are greater than or equal to 5, the requirements for the goodness-of-fit test are satisfied. Our hypotheses are

H_0 : there is no difference in attendance patterns among the 4 groups

H_1 : there is a difference in attendance patterns among the 4 groups.

Classical approach: The critical value, with df $= 4 - 1 = 3$, is $\chi^2_{0.05} = 7.815$. Since the test statistic is not in the critical region ($\chi^2_0 < \chi^2_{0.05}$), we do not reject the null hypothesis.

P-value approach: Using the chi-square table, we find the row that corresponds to 3 degrees of freedom. The value of 0.463 [Tech: 0.461] is less than 0.584, which has an area under the chi-square distribution of 0.90 to the right. Therefore, we have P-value > 0.90 [Tech: P-value $= 0.927$]. Since P-value $> \alpha$, we do not reject the null hypothesis.

Conclusion: There is not sufficient evidence to conclude that there are differences among the groups in terms of attendance patterns.

It is curious that Group 1's attendance rate stayed the same, and the farther a group's original position is located from the front of the room, the more the attendance rate for the group decreases.

(c) We obtain the number of students from each group in the top 20% by multiplying each percent by 100:

Group 1: $100(0.25) = 25$

Group 2: $100(0.20) = 20$

Group 3: $100(0.15) = 15$

Group 4: $100(0.19) = 19$

If seat location has no effect, we expect 20% of each group to be in the top 20%. Since there were 100 students in each group, we expect 20 of the 100 students in each group to be in the top 20%. We summarize the observed and expected counts in the following table:

	O_i	E_i	$(O_i - E_i)^2$	$\dfrac{(O_i - E_i)^2}{E_i}$
1	25	20	25	1.25
2	20	20	0	0
3	15	20	25	1.25
4	19	20	1	0.05
			$\chi^2_0 =$	2.55

Since all the expected cell counts are greater than or equal to 5, the requirements for the goodness-of-fit test are satisfied. Our hypotheses are

H_0 : there is no difference in the number of students in the top 20% of the class by group.

H_1 : there is a difference in the number of students in the top 20% of the class by group.

Classical approach: The critical value, with df $= 4 - 1 = 3$, is $\chi^2_{0.05} = 7.815$. Since the test statistic is not in the critical region ($\chi^2_0 < \chi^2_{0.05}$), we do not reject the null hypothesis.

P-value approach: Using the chi-square table, we find the row that corresponds to 3 degrees of freedom. The value of 2.55 [Tech: 2.57] is less than 6.251, which has an area under the chi-square distribution of 0.10 to the right. Therefore, we have P-value > 0.10 [Tech: P-value $= 0.463$]. Since P-value $> \alpha$, we do not reject the null hypothesis.

Conclusion: There is not sufficient evidence to conclude that there is a difference in the number of students in the top 20% of the class by group.

(d) Though not statistically significant, the group located in the front had both better attendance and a larger number of students in the top 20%. Given a choice, one should choose to sit in the front (though other factors, such as study habits, also affect student success).

19. We summarize the observed and expected counts in the following table:

	O_i	E_i	$(O_i - E_i)^2$	$\dfrac{(O_i - E_i)^2}{E_i}$
Jan-Mar	63	$\dfrac{181}{4}$	315.0625	6.9627
Apr-Jun	56	$\dfrac{181}{4}$	115.5625	2.5539
Jul-Sep	28	$\dfrac{181}{4}$	297.5625	6.5760
Oct-Dec	34	$\dfrac{181}{4}$	126.5625	2.7970
			$\chi_0^2 =$	18.8895

Since all the expected cell counts are greater than or equal to 5, the requirements for the goodness-of-fit test are satisfied. Our hypotheses are:

H_0 : distribution of birth months is uniform

H_1 : distribution of birth months is not uniform

A level of significance was not specified. We can choose our own level of significance. Suppose $\alpha = 0.05$.

Classical approach: The critical value, with df $= 4 - 1 = 3$, is $\chi_{0.05}^2 = 7.815$. Since the test statistic is in the critical region ($\chi_0^2 > \chi_{0.01}^2$), we reject the null hypothesis.

P-value approach: Using the chi-square table, we find the row that corresponds to $4 - 1 = 3$ degrees of freedom. The value of 18.8895 is greater than 12.838, which has an area under the chi-square distribution of 0.005 to the right. Therefore, we have P-value < 0.005. Since P-value $< \alpha$, we reject the null hypothesis.
[Note: using technology, we find P-value $= 0.003 < \alpha = 0.05$ so the null hypothesis is not rejected.]

Conclusion: There is enough evidence to reject the null hypothesis. The birth months of players in the 2010 NHL draft does not appear to occur with equal frequency. Players birthdays are not distributed equally throughout the year. Note that this same conclusion would be drawn if we had chosen a level of significance of $\alpha = 0.01$ or $\alpha = 0.10$.

21. We summarize the observed and expected counts in the following table:

	O_i	E_i	$(O_i - E_i)^2$	$\dfrac{(O_i - E_i)^2}{E_i}$
Sun	39	$\dfrac{300}{7} = 42.857$	14.878	0.347
Mon	40	42.857	8.163	0.190
Tue	30	42.857	165.306	3.857
Wed	40	42.857	8.163	0.190
Thu	41	42.857	3.449	0.080
Fri	49	42.857	37.735	0.880
Sat	61	42.857	329.163	7.680
			$\chi_0^2 =$	13.227

Since all the expected cell counts are greater than or equal to 5, the requirements for the goodness-of-fit test are satisfied. Our hypotheses are:

H_0 : the distribution of pedestrian fatalities is uniformly distributed during the week

H_1 : the distribution of pedestrian fatalities is not uniformly distributed during the week

Classical approach: The critical value, with df $= 7 - 1 = 6$, is $\chi_{0.05}^2 = 12.592$. Since the test statistic is in the critical region ($\chi_0^2 > \chi_{0.05}^2$), we reject the null hypothesis.

P-value approach: Using the chi-square table, we find the row that corresponds to 6 degrees of freedom. The value of 13.227 is greater than 12.592, which has an area under the chi-square distribution of 0.05 to the right. Therefore, we have P-value < 0.05
[Tech: P-value $= 0.040$]. Since P-value $< \alpha$, we reject the null hypothesis.

Conclusion: There is enough evidence to indicate that pedestrian deaths are not uniformly distributed over the days of the week. We might conclude that fewer pedestrian deaths occur on Tuesdays, and more on Saturdays.

23. (a)

	Rel. Freq.	Obs. Freq.	Exp. Freq.
K-3	0.329	15	$25(0.329) = 8.225$
4-8	0.393	7	$25(0.393) = 9.825$
9-12	0.278	3	$25(0.278) = 6.95$

(b) Since all the expected cell counts are greater than or equal to 5, the requirements for the goodness-of-fit test are satisfied. Our hypotheses are:

H_0: grade distribution is the same as the national distribution

H_1: grade distribution is not the same as the national distribution.

$$\chi_0^2 = \frac{(15-8.225)^2}{8.225} + \frac{(7-9.825)^2}{9.825}$$
$$+ \frac{(3-6.95)^2}{6.95}$$
$$\approx 8.638$$

Classical approach: The critical value, with df = 3 – 1 = 2, is $\chi_{0.05}^2 = 5.991$. Since the test statistic is in the critical region ($\chi_0^2 > \chi_{0.05}^2$), we reject the null hypothesis.

P-value approach: Using the chi-square table, we find the row that corresponds to 2 degrees of freedom. The value of 8.638 is greater than 7.378, which has an area under the chi-square distribution of 0.025 to the right. Therefore, we have P-value < 0.025 [Tech: P-value = 0.013]. Since P-value < α, we reject the null hypothesis.

Conclusion: There is enough evidence to indicate that the grade distribution of home-schooled students in the social worker's district is different from the national distribution. We might conclude that a greater proportion of students in grades K-3 are home-schooled in this district than nationally.

25. (a) Northeast: 950; Midwest: 1145; South: 1780; West: 1125

(b) Northeast: 0.18; Midwest: 0.218; South: 0.369; West: 0.233
There have been some changes in the proportions since 2000.

(c) We summarize the observed and expected counts in the following table:

	O_i	E_i	$(O_i - E_i)^2$	$\frac{(O_i - E_i)^2}{E_i}$
Northeast	900	950	2500	2.632
Midwest	1089	1145	3136	2.739
South	1846	1780	4356	2.447
West	1165	1125	1600	1.422
			$\chi_0^2 =$	9.240

Since all the expected cell counts are greater than or equal to 5, the requirements for the goodness-of-fit test are satisfied. Our hypotheses are:

H_0: the distribution of residents has not changed since 2000

H_1: the distribution of residents has changed since 2000

We assume a level of significance of $\alpha = 0.05$.

Classical approach: The critical value, with df = 4 – 1 = 3, is $\chi_{0.05}^2 = 7.815$. Since the test statistic is in the critical region ($\chi_0^2 > \chi_{0.05}^2$), we reject the null hypothesis.

P-value approach: Using the chi-square table, we find the row that corresponds to 4 – 1 = 3 degrees of freedom. The value of 9.240 is between 7.815 and 9.348, which correspond to areas under the chi-square distribution of 0.05 and 0.025 to the right. Therefore, we have 0.025 < P-value < 0.5 [Tech: P-value = 0.0263]. Since P-value < α, we reject the null hypothesis.

Conclusion: There is enough evidence to indicate that the distribution of residents in the United States has changed since 2000.

(d) If the sample size is larger, the test will be more sensitive to small differences in the observed and expected counts. So, higher sample sizes require less significant evidence.

27. (a) For 240 randomly selected births, we expect $240(0.071) = 17.04$ to result in low birth weight, and $240(1-0.071) = 222.96$ to not result in low birth weight.

(b) We summarize the observed and expected counts

	O_i	E_i	$(O_i - E_i)^2$	$\dfrac{(O_i - E_i)^2}{E_i}$
Low	22	17.04	24.6016	1.4438
Not Low	218	222.96	24.6016	0.1103
	160		$\chi_0^2 \approx$	1.554

Our hypotheses are:

$H_0 : p = 0.071$

$H_1 : p \neq 0.071$

Classical approach: The critical value, with df = 2 – 1 = 1, is $\chi_{0.05}^2 = 3.841$. Since the test statistic is not in the critical region ($\chi_0^2 < \chi_{0.05}^2$), we do not reject H_0.

P-value approach: Using the chi-square table, we find the row that corresponds to 1 degree of freedom. The value of 1.554 is less than 2.706, which has an area under the chi-square distribution of 0.10 to the right. Therefore, we have P-value > 0.10 [Tech: P-value = 0.213]. Since P-value > α, we do not reject the null hypothesis.

Conclusion: There is not enough evidence to conclude that the percentage of low-birth-weight babies is higher for mothers 35–39 years old.

(c) Note that

$$n \cdot p_0 (1 - p_0) = 240(0.071)(1 - 0.071)$$
$$\approx 15.83 \geq 10$$

so the requirements of the hypothesis test are satisfied.

The hypotheses are $H_0 : p = 0.071$,

$H_1 : p > 0.071$. From the sample data,

$\hat{p} = \dfrac{22}{240} \approx 0.0917$. The test statistic is

$$z_0 = \frac{\hat{p} - p_0}{\sqrt{p_0(1 - p_0)/n}}$$
$$= \frac{0.092 - 0.071}{\sqrt{0.071(1 - 0.071)/240}}$$
$$= 1.27$$

Classical approach: This is a right-tailed test so the critical value is $z_{0.05} = 1.645$. The test statistic is not in the critical region ($z_0 < z_{0.05}$) so we do not reject the null hypothesis.

P-value approach: Since this is a right-tailed test,

P-value $= P(Z > 1.27) = 0.1020$

[Tech: P-value = 0.1063]. Since P-value > α, we do not reject the null hypothesis.

Conclusion: There is not enough evidence to conclude that the percentage of low-birth-weight babies is higher for mothers 35–39 years old.

29. (a) There are $229 + 211 + ... + 0 + 1 = 576$ rockets that landed in London. We obtain the probabilities by dividing each frequency by 576 (obtaining the relative frequencies).

# of rockets, x	$P(x)$
0	$\dfrac{229}{576} \approx 0.3976$
1	$\dfrac{211}{576} \approx 0.3663$
2	$\dfrac{93}{576} \approx 0.1615$
3	$\dfrac{35}{576} \approx 0.0608$
4	$\dfrac{7}{576} \approx 0.0122$
5	$\dfrac{0}{576} = 0$
6	$\dfrac{0}{576} = 0$
7	$\dfrac{1}{576} \approx 0.0017$

$$\mu = 0\left(\tfrac{229}{576}\right) + 1\left(\tfrac{211}{576}\right) + ... + 6(0) + 7\left(\tfrac{1}{576}\right)$$
$$\approx 0.9323$$

The mean number of hits in a region was approximately 0.9323.

(b) The requirements for conducting a goodness-of-fit test are not satisfied because all the expected frequencies will not be greater than or equal to 1, and more than 20% of the expected frequencies are less than 5 (37.5%).

(c) Using the model $P(x) = \dfrac{0.9323^x}{x!} e^{-0.9323}$, the probability distribution is given by

x	$P(x)$
0	0.3936
1	0.3670
2	0.1711
3	0.0532
4 or more	0.0151

(d) We determine the expected number of regions that will be hit by x rockets by multiplying the number of rockets (576) by the probability of a region being hit by x rockets, $P(x) = \dfrac{0.9323^x}{x!} e^{-0.9323}$.

x	Expected number of rocket hits
0	$576 \cdot \dfrac{0.9323^0}{0!} e^{-0.9323} = 226.741$
1	$576 \cdot \dfrac{0.9323^1}{1!} e^{-0.9323} = 211.390$
2	$576 \cdot \dfrac{0.9323^2}{2!} e^{-0.9323} = 98.540$
3	$576 \cdot \dfrac{0.9323^3}{3!} e^{-0.9323} = 30.623$
4 or more	$576 \cdot \left(\dfrac{0.9323^4}{4!} + ... + \dfrac{0.9323^7}{7!} \right) e^{-0.9323} = 8.706$

(e) We summarize the observed and expected counts:

	O_i	E_i	$(O_i - E_i)^2$	$\dfrac{(O_i - E_i)^2}{E_i}$
0	229	226.741	5.1031	0.0225
1	211	211.390	0.1521	0.0007
2	93	98.540	30.6916	0.3115
3	35	30.623	19.1581	0.6256
4+	8	8.706	0.4984	0.0573
			$\chi_0^2 \approx$	1.018

Our hypotheses are:
H_0 : the rockets can be modeled by a Poisson random variable.
H_1 : the rockets cannot be modeled by a Poisson random variable.

Classical approach: The critical value, with df = 5 – 1 = 4, is $\chi_{0.05}^2 = 9.488$. Since the test statistic is not in the critical region ($\chi_0^2 < \chi_{0.05}^2$), we do not reject H_0.

P-value approach: Using the chi-square table, we find the row that corresponds to 4 degrees of freedom. The value of 1.018 is less than 1.064, which has an area under the chi-square distribution of 0.90 to the right. Therefore, we have P-value > 0.90 [Tech: P-value = 0.907]. Since P-value > α, we do not reject the null hypothesis.

Conclusion: There is not sufficient evidence to conclude that the distribution of rocket hits is different from a Poisson distribution. That is, the rocket hits do appear to be modeled by a Poisson random variable.

31. Answers will vary. One possible answer is that the phrase *Goodness-of-fit* is appropriate, because we are testing to see if the frequency of outcomes from a sample "fits" the theoretical distribution well.

33. If the expected count of a category is less than 1, two (or more) of the categories can be combined so that the goodness-of-fit test can be used. Alternatively, the sample size can be increased.

Section 12.2

1. True

3. (a) $\chi_0^2 = \sum \dfrac{(O_i - E_i)^2}{E_i} = \dfrac{(34-36.26)^2}{36.26} + \dfrac{(43-44.63)^2}{44.63} + \cdots + \dfrac{(17-20.89)^2}{20.89} = 1.701$

(b) Classical Approach:

There are 2 rows and 3 columns, so df $= (2-1)(3-1) = 2$ and the critical value is $\chi_{0.05}^2 = 5.991$. The test statistic, 1.701, is less than the critical value so we do not reject H_0.

P-value Approach:

There are 2 rows and 3 columns, so we find the *P*-value using $(2-1)(3-1) = 2$ degrees of freedom. The

P-value is the area under the chi-square distribution with 2 degrees of freedom to the right of $\chi_0^2 = 1.701$.

Using Table VII, we find the row that corresponds to 2 degrees of freedom. The value of 1.701 is less than 4.605, which has an area under the chi-square distribution of 0.10 to the right. Therefore, we have *P*-value > 0.10 [Tech: *P*-value = 0.427]. Since *P*-value > α, we do not reject H_0.

Conclusion:

There is not sufficient evidence, at the $\alpha = 0.05$ level of significance, to conclude that the two variables are dependent. We conclude that *X* and *Y* are not related.

5. The hypotheses are $H_0 : p_1 = p_2 = p_3$ and H_1 : at least one proportion differs from the others.

The expected counts are calculated as $\dfrac{(\text{row total}) \cdot (\text{column total})}{(\text{table total})} = \dfrac{229 \cdot 120}{363} \approx 75.702$ (for the first cell) and

so on. The observed and expected counts are shown in the following table.

	Category 1	Category 2	Category 3	Total
Success	76	84	69	229
	(75.702)	(78.857)	(74.441)	
Failure	44	41	49	134
	(44.298)	(46.143)	(43.559)	
Total	120	125	118	363

Since none of the expected counts is less than 5, the requirements of the test are satisfied.

The test statistic is $\chi_0^2 = \sum \dfrac{(O_i - E_i)^2}{E_i} = \dfrac{(76-75.702)^2}{75.702} + \cdots + \dfrac{(49-43.559)^2}{43.559} = 1.989$.

Classical Approach:

df $= (2-1)(3-1) = 2$ so the critical value is $\chi_{0.01}^2 = 9.210$. The test statistic, 1.989, is less than the critical value so we do not reject H_0.

P-value Approach:

Using Table VII, we find the row that corresponds to 2 degrees of freedom. The value of 1.989 is less than 4.605, which has an area under the chi-square distribution of 0.10 to the right. Therefore, we have *P*-value > 0.10 [Tech: *P*-value = 0.370]. Since *P*-value > α, we do not reject H_0.

Conclusion:

There is not sufficient evidence, at the $\alpha = 0.01$ level of significance, to conclude that at least one of the proportions is different from the others.

7. (a) The expected counts are calculated as $\dfrac{(\text{row total}) \cdot (\text{column total})}{(\text{table total})} = \dfrac{199 \cdot 150}{380} \approx 78.553$ (for the first cell)

and so on, giving the following table of observed and expected counts:

Sexual Activity	Family Structure				Total
	Both Parents	**One Parent**	**Parent/ Stepparent**	**Nonparental Guardian**	
Had intercourse	64	59	44	32	199
	(78.553)	(52.368)	(41.895)	(26.184)	
Did not have	86	41	36	18	181
	(71.447)	(47.632)	(38.105)	(23.816)	
Total	150	100	80	50	380

(b) All expected frequencies are greater than 5 so all requirements for a chi-square test are satisfied. That is, all expected frequencies are greater than or equal to 1, and no more than 20% of the expected frequencies are less than 5.

(c) $\chi_0^2 = \sum \dfrac{(O_i - E_i)^2}{E_i} = \dfrac{(64 - 78.553)^2}{78.553} + \cdots + \dfrac{(18 - 23.816)^2}{23.816} \approx 10.358$ [Tech: 10.357]

(d) H_0 : the row and column variables are independent

H_1 : the row and column variables are dependent

df $= (2-1)(4-1) = 3$ so the critical value is $\chi_{0.05}^2 = 7.815$.

Classical Approach:
The test statistic is 10.358 which is greater than the critical value so we reject H_0.

P-value Approach:
Using Table VII, we find the row that corresponds to 2 degrees of freedom. The value of 10.358 is greater than 9.348, which has an area under the chi-square distribution of 0.025 to the right. Therefore, we have P-value < 0.025 [Tech: P-value = 0.016]. Since P-value < α, we reject H_0.

Conclusion:
There is sufficient evidence, at the $\alpha = 0.05$ level of significance, to conclude that family structure and sexual activity are dependent.

(e) The biggest difference between observed and expected occurs under the family structure in which both parents are present. Fewer females were sexually active than was expected when both parents were present. This means that having both parents present seems to have an impact on whether the child is sexually active.

(f) The conditional frequencies and bar chart show that sexual activity varies by family structure.

Sexual Intercourse	Both Parents	One Parent	Parent/Stepparent	Nonparent Guardian
Yes	$\dfrac{64}{150} \approx 0.427$	$\dfrac{59}{100} = 0.590$	$\dfrac{44}{80} = 0.550$	$\dfrac{32}{50} = 0.640$
No	$\dfrac{86}{150} \approx 0.573$	$\dfrac{41}{100} = 0.410$	$\dfrac{36}{80} = 0.450$	$\dfrac{18}{50} = 0.360$

9. (a) The expected counts are calculated as $\dfrac{\text{(row total)}\cdot\text{(column total)}}{\text{(table total)}} = \dfrac{634\cdot551}{1996} \approx 175.017$ (for the first

cell) and so on, giving the following table:

Happiness	Health				Total
	Excellent	**Good**	**Fair**	**Poor**	
Very Happy	271	261	82	20	634
	(175.017)	(295.718)	(128.642)	(34.622)	
Pretty Happy	247	567	231	53	1098
	(303.105)	(512.143)	(222.791)	(59.961)	
Not Too Happy	33	103	92	36	264
	(72.878)	(123.138)	(53.567)	(14.417)	
Total	551	931	405	109	1996

All expected frequencies are greater than 5 so the requirements for a chi-square test are satisfied.

$$\chi_0^2 = \sum \frac{(O_i - E_i)^2}{E_i} = \frac{(271-175.017)^2}{175.017} + \cdots + \frac{(36-14.417)^2}{14.417} \approx 182.173 \quad [\text{Tech: } 182.174]$$

$df = (3-1)(4-1) = 6$ so the critical value is $\chi_{0.05}^2 = 12.592$.

H_0 : the row and column variables are independent

H_1 : the row and column variables are dependent

Classical Approach:
The test statistic is 182.173 which is greater than the critical value so we reject H_0.

P-value Approach:
Using Table VII, we find the row that corresponds to 6 degrees of freedom. The value of 182.173 is greater than 18.548, which has an area under the chi-square distribution of 0.005 to the right. Therefore, we have P-value < 0.005 [Tech: P-value < 0.001]. Since P-value < α, we reject H_0.

Conclusion:
There is sufficient evidence, at the $\alpha = 0.05$ level of significance, to conclude that health and happiness are related.

(b) The conditional frequencies and bar chart show that the level of happiness varies by health status.

Happiness	Health			
	Excellent	**Good**	**Fair**	**Poor**
Very Happy	$\dfrac{271}{551} \approx 0.492$	$\dfrac{261}{931} \approx 0.280$	$\dfrac{82}{405} \approx 0.202$	$\dfrac{20}{109} \approx 0.183$
Pretty Happy	$\dfrac{247}{551} \approx 0.448$	$\dfrac{567}{931} \approx 0.609$	$\dfrac{231}{405} \approx 0.570$	$\dfrac{53}{109} \approx 0.486$
Not Too Happy	$\dfrac{33}{551} \approx 0.60$	$\dfrac{103}{931} \approx 0.111$	$\dfrac{92}{405} \approx 0.227$	$\dfrac{36}{109} \approx 0.330$

Health and Happiness

(c)   The proportion of individuals who are "very happy" is much higher for individuals in "excellent" health than any other health category. Further, the proportion of individuals who are "not too happy" is much lower for individuals in "excellent" health compared to the other health categories. The level of happiness seems to decline as health status declines.

11. (a)   The expected counts are calculated as $\dfrac{(\text{row total}) \cdot (\text{column total})}{(\text{table total})} = \dfrac{474 \cdot 393}{1054} \approx 176.738$ (for the first

cell) and so on, giving the following table:

Years of Education	Smoking Status			Total
	Current	**Former**	**Never**	
< 12	178	88	208	474
	(176.738)	(96.688)	(200.573)	
12	137	69	143	349
	(130.130)	(71.191)	(147.679)	
13–15	44	25	44	113
	(42.134)	(23.050)	(47.816)	
16 or more	34	33	51	118
	(43.998)	(24.070)	(49.932)	
Total	393	215	446	1054

All expected frequencies are greater than 5 so the requirements for a chi-square test are satisfied.

$$\chi_0^2 = \sum \frac{(O_i - E_i)^2}{E_i} = \frac{(178 - 176.738)^2}{176.738} + \cdots + \frac{(51 - 49.932)^2}{49.932} \approx 7.803$$

df $= (4-1)(3-1) = 6$ so the critical value is $\chi_{0.05}^2 = 12.592$.

H_0 : the row and column variables are independent

H_1 : the row and column variables are dependent

Classical Approach:
The test statistic is 7.803 which is less than the critical value so we do not reject H_0 .

P-value Approach:
Using Table VII, we find the row that corresponds to 6 degrees of freedom. The value of 7.803 is less than 10.645, which has an area under the chi-square distribution of 0.10 to the right. Therefore, we have P-value > 0.10 [Tech: P-value = 0.253]. Since P-value > α , we do not reject H_0 .

Conclusion:
There is not sufficient evidence, at the $\alpha = 0.05$ level of significance, to conclude that smoking status and years of education are associated. That is, it does not appear that years of education plays a role in determining smoking status.

(b)   The conditional frequencies and bar chart show that the distribution of smoking status is similar at all levels of education. This supports the result from part (a).

Years of Education	Current	Former	Never
< 12	$\dfrac{178}{474} \approx 0.376$	$\dfrac{88}{474} \approx 0.186$	$\dfrac{208}{474} \approx 0.439$
12	$\dfrac{137}{349} \approx 0.393$	$\dfrac{69}{349} \approx 0.198$	$\dfrac{143}{349} \approx 0.410$
13–15	$\dfrac{44}{113} \approx 0.389$	$\dfrac{25}{113} \approx 0.221$	$\dfrac{44}{113} \approx 0.389$
16 or more	$\dfrac{34}{118} \approx 0.288$	$\dfrac{33}{118} \approx 0.280$	$\dfrac{51}{118} \approx 0.432$

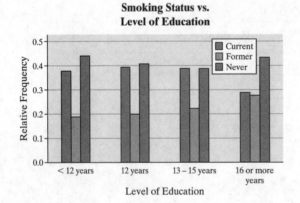

13. (a) The expected counts are calculated as $\dfrac{\text{(row total)} \cdot \text{(column total)}}{\text{(table total)}} = \dfrac{1028 \cdot 385}{1437} \approx 275.421$ (for the first cell) and so on, giving the following table:

Opinion	Highest Degree					Total
	Less Than High School	High School	Junior College	Bachelor	Graduate	
Male	302	551	29	100	46	1028
	(275.421)	(537.250)	(37.915)	(122.330)	(55.084)	
Female	83	200	24	71	31	409
	(109.579)	(213.750)	(15.085)	(48.670)	(21.916)	
Total	385	751	53	171	77	1437

All expected frequencies are greater than 5 so all requirements for a chi-square test are satisfied.

$$\chi_0^2 = \sum \frac{(O_i - E_i)^2}{E_i} = \frac{(302 - 275.421)^2}{275.421} + \cdots + \frac{(31 - 21.916)^2}{21.916} \approx 37.198$$

$\text{df} = (2-1)(5-1) = 4$ so the critical value is $\chi_{0.05}^2 = 9.488$.

H_0: $p_{\text{LTHS}} = p_{\text{HS}} = p_{\text{JC}} = p_{\text{B}} = p_{\text{G}}$

H_1: at least one of the proportions is different from the others

Classical Approach:
The test statistic is 37.198 which is greater than the critical value so we reject H_0.

P-value Approach:
Using Table VII, we find the row that corresponds to 4 degrees of freedom. The value of 37.198 is greater than 14.860, which has an area under the chi-square distribution of 0.005 to the right. Therefore, we have P-value < 0.005 [Tech: P-value < 0.001]. Since P-value < α, we reject H_0.

Conclusion:
There is sufficient evidence, at the $\alpha = 0.05$ level of significance, to conclude that at least one of the proportions is different from the others. That is, the evidence suggests that the proportion of individuals who feel everyone has an equal opportunity to obtain a quality education in the United States is different for at least one level of education.

(b) The conditional frequencies and bar chart show that the distribution of opinion varies by education. Higher proportions of individuals with either a "high school" education or "less than a high school" education seem to believe that everyone has an equal opportunity to obtain a quality education. Individuals whose highest degree is from a junior college appear to agree least with the statement.

Opinion	Less than High School	High School	Junior College	Bachelor	Graduate
Yes	$\frac{302}{385} \approx 0.784$	0.734	0.547	0.585	0.597
No	$\frac{83}{385} \approx 0.216$	0.266	0.453	0.415	0.403

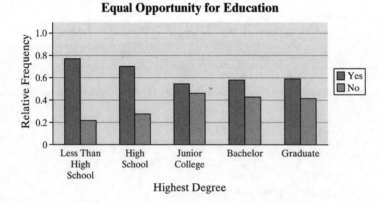

Equal Opportunity for Education

15. (a) The expected counts are calculated as $\dfrac{(\text{row total}) \cdot (\text{column total})}{(\text{table total})} = \dfrac{426 \cdot 499}{1466} \approx 145.003$ (for the first

cell) and so on, giving the following table:

Reaction	Political Party			Total
	Democrat	Independent	Republican	
Positive	220	144	62	426
	(145.003)	(160.985)	(120.012)	
Negative	279	410	351	1040
	(353.997)	(393.015)	(292.988)	
Total	499	554	413	1466

All expected frequencies are greater than 5 so all requirements for a chi-square test are satisfied.

$\chi_0^2 = \sum \dfrac{(O_i - E_i)^2}{E_i} = \dfrac{(220-145.003)^2}{145.003} + \cdots + \dfrac{(351-292.988)^2}{292.988} \approx 96.733$ and df $= (2-1)(3-1) = 2$

$H_0 : p_{\text{Democrats}} = p_{\text{Independents}} = p_{\text{Republicans}}$

H_1 : at least one of the proportions is not equal to the rest

Classical Approach:

With 2 degrees of freedom, the critical value is $\chi_{0.05}^2 = 5.991$. The test statistic is 96.733 which is greater than the critical value so we reject H_0.

P-value Approach:

Using Table VII, we find the row that corresponds to 2 degrees of freedom. The value of 96.733 is greater than 10.597, which has an area under the chi-square distribution of 0.005 to the right. Therefore, we have P-value < 0.005 [Tech: P-value < 0.001]. Since P-value < α, we reject H_0.

Conclusion:
There is sufficient evidence, at the $\alpha = 0.05$ level of significance, to conclude that a different proportion of individuals within each political affiliation react positively to the word "socialism."

(b) Dividing the observed counts within each party by the number of respondents identified with each party, we get:

Reaction	Political Party		
	Democrat	Independent	Republican
Positive	0.4409	0.2599	0.1501
Negative	0.5591	0.7401	0.8499
Total	1	1	1

(c) Independents and Republicans are far more likely to react negatively to the word socialism than Democrats are. However, it is important to note that a majority of Democrats in the sample did have a negative reaction, so the word socialism has a negative connotation among all groups.

17. (a) Since there were no individuals who gave "career" as a reason for dropping, we have omitted that category from the analysis.

Gender	Drop Reason			Total
	Personal	Work	Course	
Female	5	3	13	21
	(4.62)	(6.72)	(9.66)	
Male	6	13	10	29
	(6.38)	(9.28)	(13.34)	
Total	11	16	23	50

(b) The expected counts are calculated as $\dfrac{(\text{row total}) \cdot (\text{column total})}{(\text{table total})} = \dfrac{21 \cdot 11}{50} = 4.62$ (for the first cell) and so on (included in the table from part (a)). All expected frequencies are greater than 1 and only one (out of six) expected frequency is less than 5. All requirements for a chi-square test are satisfied.

$$\chi_0^2 = \sum \frac{(O_i - E_i)^2}{E_i} = \frac{(5-4.62)^2}{4.62} + \cdots + \frac{(10-13.34)^2}{13.34} \approx 5.595$$

$\text{df} = (2-1)(3-1) = 2$ so the critical value is $\chi_{0.10}^2 = 4.605$.

Classical Approach:
The test statistic is 5.595 which is greater than the critical value so we reject H_0.

P-value Approach:
Using Table VII, we find the row that corresponds to 2 degrees of freedom. The value of 5.595 is greater than 4.605, which has an area under the chi-square distribution of 0.10 to the right. Therefore, we have *P*-value < 0.10 [Tech: *P*-value = 0.061]. Since *P*-value < α, we reject H_0.

Conclusion:
There is sufficient evidence, at the $\alpha = 0.1$ level of significance, to conclude that gender and drop reason are dependent. Females are more likely to drop because of the course, while males are more likely to drop because of work.

(c) The conditional frequencies and bar chart show that the distribution of genders varies by reason for dropping. This supports the results from part (b).

	Personal	Work	Course
Female	$\frac{5}{11} \approx 0.455$	$\frac{3}{16} \approx 0.188$	$\frac{13}{23} \approx 0.565$
Male	$\frac{6}{11} \approx 0.545$	$\frac{13}{16} \approx 0.813$	$\frac{10}{23} \approx 0.435$

19. The counts are summarized in the following table. The expected counts are calculated as
$$\frac{(\text{row total}) \cdot (\text{column total})}{(\text{table total})} = \frac{51 \cdot 80}{199} \approx 20.503 \text{ (for the first cell) and so on, giving the following table:}$$

Deficit	Political Party			Total
	Democrat	**Independent**	**Republican**	
Yes	25	14	12	51
	(20.503)	(14.352)	(16.146)	
No	55	42	51	148
	(59.497)	(41.648)	(46.854)	
Total	80	56	63	199

All expected frequencies are greater than 5 so all requirements for a chi-square test are satisfied.

$$\chi_0^2 = \sum \frac{(O_i - E_i)^2}{E_i} = \frac{(25 - 20.503)^2}{20.503} + \cdots + \frac{(51 - 46.854)^2}{46.854} \approx 2.769 \text{ and df} = (2-1)(3-1) = 2$$

$H_0 : p_{\text{Democrats}} = p_{\text{Independents}} = p_{\text{Republicans}}$

$H_1 :$ at least one of the proportions is not equal to the rest

Classical Approach:
With 2 degrees of freedom, the critical value is $\chi_{0.05}^2 = 5.991$. The test statistic is 2.769 which is less than the critical value so we do not reject H_0.

P-value Approach:
Using Table VII, we find the row that corresponds to 2 degrees of freedom. The value of 2.769 is between 0.211 and 4.605, which correspond to areas to the right of 0.90 and 0.10, respectively. Therefore, we have $0.10 < P\text{-value} < 0.90$ [Tech: $P\text{-value} = 0.2504$]. Since $P\text{-value} > \alpha$, we do not reject H_0.

Conclusion:
There is not sufficient evidence, at the $\alpha = 0.05$ level of significance, to conclude that a different proportion of individuals within each political affiliation are willing to pay higher taxes to reduce the federal deficit.

21. (a) The population being studied is healthy adult women aged 45 years or older. The sample consists of the 39,876 women in the study.

(b) The response variable is whether or not the subject had a cardiovascular event (such as a heart attack or stroke). This is a qualitative variable because the values serve to simply classify the subjects.

(c) There are two treatments: 100 mg of aspirin and a placebo

(d) Because the subjects are all randomly assigned to the treatments, this is a completely randomized design.

(e) Randomization controls for any other explanatory variables because individuals affected by these lurking variables should be equally dispersed between the two treatment groups.

(f) $H_0 : p_1 = p_2$ vs. $H_1 : p_1 \neq p_2$

$$\hat{p} = \frac{x_1 + x_2}{n_1 + n_2} = \frac{477 + 522}{19,934 + 19,942} = \frac{999}{39,876} \approx 0.025, \ \hat{p}_1 = \frac{477}{19,934} \approx 0.024, \ \hat{p}_2 = \frac{522}{19,942} \approx 0.026$$

The test statistic is $z_0 = \dfrac{\hat{p}_1 - \hat{p}_2}{\sqrt{\hat{p}(1-\hat{p})}\sqrt{\dfrac{1}{n_1} + \dfrac{1}{n_2}}} = \dfrac{0.024 - 0.026}{\sqrt{0.025(0.975)}\sqrt{\dfrac{1}{19,934} + \dfrac{1}{19,942}}} \approx -1.28$ [Tech: -1.44]

Classical Approach:
The critical values for $\alpha = 0.05$ are $-z_{0.025} = -1.96$ and $z_{0.025} = 1.96$.
The test statistic is -1.28, which is not in the critical region so we do not reject H_0.

P-value Approach:
$$\begin{aligned} P - \text{value} &= 2 \cdot P(Z > 1.28) \\ &= 2(0.1003) \\ &= 0.2006 \ [\text{Tech: } 0.1511] \end{aligned}$$

Therefore, we have *P*-value $> \alpha = 0.05$ and we do not reject H_0.

Conclusion:
There is not sufficient evidence, at the $\alpha = 0.05$ level of significance, to conclude that a difference exists between the proportions of cardiovascular events in the aspirin group versus the placebo group.

(g) The expected counts are calculated as $\dfrac{(\text{row total}) \cdot (\text{column total})}{(\text{table total})} = \dfrac{999 \cdot 19,934}{39,876} \approx 499.400$ (for the first cell) and so on.

	aspirin	placebo	Total
cardiovascular event	477	522	999
	(499.400)	(499.600)	
no event	19,457	19,420	38,877
	(19,434.600)	(19,442.400)	
Total	19,934	19,942	39,876

All expected frequencies are greater than 1 and only one (out of six) expected frequency is less than 5. All requirements for a chi-square test are satisfied.

$$\chi_0^2 = \sum \frac{(O_i - E_i)^2}{E_i} = \frac{(477 - 499.400)^2}{499.400} + \cdots + \frac{(19,420 - 19,442.400)^2}{19,442.400} \approx 2.061$$

df $= (2-1)(2-1) = 1$ so the critical value is $\chi_{0.05}^2 = 3.841$.

Classical Approach:
The test statistic is 2.061 which is less than the critical value so we do not reject H_0.

P-value Approach:
Using Table VII, we find the row that corresponds to 1 degree of freedom. The value of 2.061 is less than 2.706, which has an area under the chi-square distribution of 0.10 to the right. Therefore, we have P-value > 0.10 [Tech: P-value = 0.151]. Since P-value > α, we do not reject H_0.

Conclusion:
There is not sufficient evidence, at the $\alpha = 0.05$ level of significance, to conclude that a difference exists between the proportions of cardiovascular events in the aspirin group versus the placebo group.

(h) Using technology in part (f), the test statistic to four decimal places is $z_0 = -1.4355$ which gives

$(z_0)^2 \approx 2.061 = \chi_0^2$. Our conclusion is that for comparing two proportions, $z_0^2 = \chi_0^2$.

23. Answers may vary.

Consumer Reports®: Dirty Birds?

(a) Calculate the proportion of incidence for each of the brands shown in the table.

Brand	Present	Absent	Total	Proportion
A	8	192	200	0.040
B	17	183	200	0.085
C	27	173	200	0.135
D	14	186	200	0.070
E	20	180	200	0.100
Total	86	914	1000	1.000

(b) Using Minitab, a 95% confidence interval for the incidence of Salmonella for Brand C is (0.090887, 0.190309). Using a TI-84 Plus, a 95% confidence interval for the incidence of Salmonella for Brand C is (0.08764, 0.18236).

(c) Using Minitab or a TI-84 Plus, the test statistic is $\chi_0^2 = 12.646$ with 4 df. The P-value is 0.013, so there is evidence of different rates of incidence among the five brands.

(d) The sample proportion for Brand A is 0.04 and that for Brand D is 0.07. Our hypotheses would be $H_0 : p_A = p_B$ vs. $H_1 : p_A < p_B$. Using a two-sample test for proportions (not adjusted for multiple comparisons) we obtain a test statistic of $z_0 = -1.32$ and a P-value of 0.094. If this were adjusted for multiple comparisons, the significance level would be even lower, so there is no evidence that Brand A is cleaner than Brand D based on this sample.

(e) Answers will vary. Our tests revealed that there is evidence of different rates of incidence of Salmonella in the 5 brands of chicken studied. Incidence levels ranged from a high of 13.5% in Brand C to a low of 4% in Brand A.

Chapter 12 Review Exercises

1. Using $\alpha = 0.05$, we want to test

    H_0: The wheel is balanced   vs. H_1: The wheel is not balanced

    If the wheel is balanced, each slot is equally likely. Thus, we would expect the proportion of red to be

    $\dfrac{18}{38} = \dfrac{9}{19}$, the proportion of black to be $\dfrac{18}{38} = \dfrac{9}{19}$, and the proportion of green to be $\dfrac{2}{18} = \dfrac{1}{19}$. There are 500

    spins. To determine the expected number of each color, multiply 500 by the given proportions for each color.

    For example, the expected count of red would be $500\left(\dfrac{9}{19}\right) \approx 236.842$. We summarize the observed and

    expected counts in the following table:

	O_i	E_i	$(O_i - E_i)^2$	$\dfrac{(O_i - E_i)^2}{E_i}$
Red	233	236.842	14.7610	0.0623
Black	237	236.842	0.0250	0.0001
Green	30	26.316	13.5719	0.5157
			$\chi_0^2 \approx$	0.578

    Since all the expected cell counts are greater than or equal to 5, the requirements for the goodness-of-fit test are satisfied.

    Classical approach: The critical value, with df = 3 – 1 = 2, is $\chi_{0.05}^2 = 5.991$. Since the test statistic is not in the critical region ($\chi_0^2 < \chi_{0.05}^2$), we do not reject the null hypothesis.

    P-value approach: Using the chi-square table, we find the row that corresponds to 2 degrees of freedom. The value of 0.578 is less than 4.605, which has an area under the chi-square distribution of 0.10 to the right. Therefore, we have P-value > 0.10 [Tech: 0.749]. Since P-value $> \alpha$, we do not reject the null hypothesis.

    Conclusion: There is not enough evidence, at the $\alpha = 0.05$ level of significance, to conclude that the wheel is out of balance.

2. Using $\alpha = 0.05$, we want to test

    H_0: The data follow the expected distribution

    H_1: The data do not follow the expected distribution

    There are 80 series to consider. To determine the expected counts, multiply 80 by the given percentage for each number of games in the series. For example, the expected count for a 4 game series would be $80(0.125) = 10.0$. We summarize the observed and expected counts in the following table:

	O_i	E_i	$(O_i - E_i)^2$	$\dfrac{(O_i - E_i)^2}{E_i}$
4	15	10.0	25.0	2.50
5	17	20.0	9.0	0.45
6	18	25.0	49.0	1.96
7	30	25.0	25.0	1.00
			$\chi_0^2 =$	5.91

    Since all the expected cell counts are greater than or equal to 5, the requirements for the goodness-of-fit test are satisfied.

Classical approach: The critical value, with df $= 4 - 1 = 3$, is $\chi^2_{0.05} = 7.815$. Since the test statistic is not in the critical region ($\chi^2_0 < \chi^2_{0.05}$), we do not reject the null hypothesis.

P-value approach: Using the chi-square table, we find the row that corresponds to 4 degrees of freedom. The value of 5.91 is between 1.064 and 7.779, which correspond to areas to the right of 0.90 and 0.10, respectively. Therefore, we have $0.10 < P\text{-value} < 0.90$ [Tech: $P\text{-value} = 0.1161$]. Since $P\text{-value} > \alpha$, we do not reject H_0.

Conclusion: There is not enough evidence, at the $\alpha = 0.05$ level of significance, to conclude that the teams playing in the World Series have not been evenly matched. That is, the evidence suggests that the teams have been evenly matched. On the other hand, the P-value is suggestive of an issue. In particular, there are fewer six-game series than we would expect. Perhaps the team that is down "goes all out" in game 6, trying to force game 7.

3. (a) The expected counts are calculated as $\dfrac{(\text{row total}) \cdot (\text{column total})}{(\text{table total})} = \dfrac{499 \cdot 325}{1316} \approx 123.233$ (for the first cell) and so on, giving the following table:

	Class			Total
	First	**Second**	**Third**	
Survived	203	118	178	499
	(123.233)	(108.066)	(267.701)	
Did Not Survive	122	167	528	817
	(201.767)	(176.934)	(438.299)	
Total	325	285	706	1316

All expected frequencies are greater than 5 so all requirements for a chi-square test are satisfied. That is, all expected frequencies are greater than or equal to 1, and no more than 20% of the expected frequencies are less than 5.

$$\chi^2_0 = \sum \frac{(O_i - E_i)^2}{E_i} = \frac{(203 - 123.333)^2}{123.333} + \cdots + \frac{(528 - 438.299)^2}{438.299} \approx 132.882 \quad [\text{Tech: } 133.052]$$

H_0 : survival status and social class are independent

H_1 : survival status and social class are dependent

df $= (2-1)(3-1) = 2$ so the critical value is $\chi^2_{0.05} = 5.991$.

Classical Approach:
The test statistic is 132.882 which is greater than the critical value so we reject H_0.

P-value Approach:
Using Table VII, we find the row that corresponds to 2 degrees of freedom. The value of 132.882 is greater than 10.597, which has an area under the chi-square distribution of 0.005 to the right. Therefore, we have $P\text{-value} < 0.005$ [Tech: $P\text{-value} < 0.001$]. Since $P\text{-value} < \alpha$, we reject H_0.

Conclusion:
There is sufficient evidence, at the $\alpha = 0.05$ level of significance, to conclude that survival status and social class are dependent.

(b) The conditional distribution and bar graph support the conclusion that a relationship exists between survival status and social class. Individuals with higher-class tickets survived in greater proportions than those with lower-class tickets.

	Class		
	First	**Second**	**Third**
Survived	$\frac{203}{325} \approx 0.625$	$\frac{118}{285} \approx 0.414$	$\frac{178}{706} \approx 0.252$
Did Not Survive	$\frac{122}{325} \approx 0.375$	$\frac{167}{285} \approx 0.586$	$\frac{528}{706} \approx 0.748$

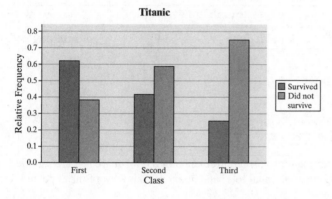

4. The expected counts are calculated as $\frac{(\text{row total}) \cdot (\text{column total})}{(\text{table total})} = \frac{1279 \cdot 40}{4953} \approx 10.329$ (for the first cell) and

so on, giving the following table:

Less than	Class					Total
H.S. degree	**22-27**	**28-32**	**33-36**	**37-42**	**43+**	
Yes	14	34	140	1010	81	1279
	(10.329)	(25.565)	(124.724)	(1043.755)	(74.628)	
No	26	65	343	3032	208	3674
	(29.671)	(73.435)	(358.276)	(2998.245)	(214.372)	
Total	40	99	483	4042	289	4953

All expected frequencies are greater than 5 so all requirements for a chi-square test are satisfied. That is, all expected frequencies are greater than or equal to 1, and no more than 20% of the expected frequencies are less than 5.

$$\chi_0^2 = \sum \frac{(O_i - E_i)^2}{E_i} = \frac{(14 - 10.329)^2}{10.329} + \cdots + \frac{(208 - 214.372)^2}{214.372} \approx 10.238 \quad [\text{Tech: } 10.239]$$

H_0 : gestation period and completing high school are independent

H_1 : gestation period and completing high school are dependent

df $= (2-1)(5-1) = 4$ so the critical value is $\chi_{0.05}^2 = 9.488$.

Classical Approach:
The test statistic is 10.238 which is greater than the critical value so we reject H_0.

Using Table VII, we find the row that corresponds to 4 degrees of freedom. The value of 10.238 is greater than 9.488, which has an area under the chi-square distribution of 0.05 to the right. Therefore, we have *P*-value < 0.05 [Tech: *P*-value < 0.037]. Since *P*-value < α, we reject H_0.

Conclusion:
There is sufficient evidence, at the $\alpha = 0.05$ level of significance, to conclude that gestation period and completing high school are dependent.

5. (a) The expected counts are calculated as $\dfrac{(\text{row total})\cdot(\text{column total})}{(\text{table total})} = \dfrac{2310\cdot1243}{4194} \approx 684.628$ (for the first cell) and so on, giving the following table:

	Funding Level				Total
	High	**Medium**	Low	**No Funding**	
Roosevelt	745	641	513	411	2310
	(684.628)	(619.635)	(523.247)	(482.489)	
Landon	498	484	437	465	1884
	(558.372)	(505.365)	(426.753)	(393.511)	
Total	1243	1125	950	876	4194

All expected frequencies are greater than 5 so all requirements for a chi-square test are satisfied. That is, all expected frequencies are greater than or equal to 1, and no more than 20% of the expected frequencies are less than 5.

$$\chi_0^2 = \sum \frac{(O_i - E_i)^2}{E_i} = \frac{(745 - 684.628)^2}{684.628} + \cdots + \frac{(465 - 393.511)^2}{393.511} \approx 37.518 \text{, and df} = (2-1)(4-1) = 3$$

H_0 : $p_{\text{High}} = p_{\text{Medium}} = p_{\text{Low}} = p_{\text{No Funding}}$

H_1 : at least one proportion is different from the others

Classical Approach:
With 3 degrees of freedom, the critical value is $\chi_{0.05}^2 = 7.815$. The test statistic is 37.518, which is greater than the critical value so we reject H_0.

P-value Approach:
Using Table VII, we find the row that corresponds to 3 degrees of freedom. The value of 37.518 is greater than 12.838, which has an area under the chi-square distribution of 0.005 to the right. Therefore, we have *P*-value < 0.005 [Tech: *P*-value < 0.001]. Since *P*-value < α, we reject H_0.

Conclusion:
There is sufficient evidence, at the $\alpha = 0.05$ level of significance, to conclude that at least one proportion is different from the others. That is, the evidence suggests that the level of funding received by the counties is associated with the candidate.

(b) The conditional distribution and bar graph support the conclusion that the proportion of adults who feel morality is important when deciding how to vote is different for at least one political affiliation. It appears that a higher proportion of Republicans feel that morality is important when deciding how to vote than for Democrats or Independents.

	Funding Level			
	High	**Medium**	**Low**	**No Funding**
Roosevelt	0.5994	0.5698	0.5400	0.4692
Landon	0.4006	0.4302	0.4600	0.5308
Total	1	1	1	1

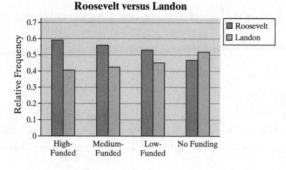

Chapter 12 Test

1. Using $\alpha = 0.01$, we want to test

   H_0 : The dice are fair   vs. H_1 : The dice are loaded

   If the dice are fair, the sum of the two dice will follow the given distribution. There are 400 rolls. To determine the expected number of times for each sum, multiply 400 by the given relative frequency for each

   sum. For example, the expected count for the sum 2 would be $400\left(\dfrac{1}{36}\right) \approx 11.111$. We summarize the

   observed and expected counts in the following table:

Sum	Observed Count (O_i)	Expected Rel. Freq.	Expected Count (E_i)	$(O_i - E_i)^2$	$(O_i - E_i)^2 / E_i$
2	16	1/36	11.111	23.902	2.1512
3	23	2/36	22.222	0.605	0.0272
4	31	3/36	33.333	5.443	0.1633
5	41	4/36	44.444	11.861	0.2669
6	62	5/36	55.556	41.525	0.7474
7	59	6/36	66.667	58.783	0.8817
8	59	5/36	55.556	11.861	0.2135
9	45	4/36	44.444	0.309	0.0070
10	34	3/36	33.333	0.445	0.0133
11	19	2/36	22.222	10.381	0.4672
12	11	1/36	11.111	0.012	0.0011
				$\chi_0^2 \approx$	4.940

Since all the expected cell counts are greater than or equal to 5, the requirements for the goodness-of-fit test are satisfied.

Classical approach: The critical value, with df $= 11 - 1 = 10$, is $\chi^2_{0.01} = 23.209$. Since the test statistic is not in the critical region ($\chi^2_0 < \chi^2_{0.01}$), we do not reject the null hypothesis.

P-value approach: Using the chi-square table, we find the row that corresponds to 10 degrees of freedom. The value of 4.940 is less than 15.987, which has an area under the chi-square distribution of 0.10 to the right. Therefore, we have P-value > 0.10 [Tech: 0.895]. Since P-value $> \alpha$, we do not reject the null hypothesis.

Conclusion: There is not enough evidence, at the $\alpha = 0.01$ level of significance, to conclude that the dice are loaded. The data indicate that the dice are fair.

2. Using $\alpha = 0.1$, we want to test

H_0 : Educational attainment in the U.S. is the same as in 2000 vs.

H_1 : Educational attainment is different today than in 2000.

There are 500 Americans in the sample. To determine the expected number attaining each degree level, multiply 500 by the given relative frequency for each level. For example, the expected count for "not a high school graduate" would be $500(0.158) = 79$. We summarize the observed and expected counts in the following table:

Observed Count (O_i)	Expected Rel. Freq.	Expected Count (E_i)	$(O_i - E_i)^2$	$(O_i - E_i)^2 / E_i$
72	0.158	79	49.00	0.6203
159	0.331	165.5	42.25	0.2553
85	0.176	88	9.00	0.1023
44	0.078	39	25.00	0.6410
92	0.170	85	49.00	0.5765
48	0.087	43.5	20.25	0.4655
			$\chi^2_0 \approx$	2.661

Since all the expected cell counts are greater than or equal to 5, the requirements for the goodness-of-fit test are satisfied.

Classical approach: The critical value, with df $= 6 - 1 = 5$, is $\chi^2_{0.10} = 9.236$. Since the test statistic is not in the critical region ($\chi^2_0 < \chi^2_{0.10}$), we do not reject the null hypothesis.

P-value approach: Using the chi-square table, we find the row that corresponds to 5 degrees of freedom. The value of 2.661 is less than 9.236, which has an area under the chi-square distribution of 0.10 to the right. Therefore, we have P-value > 0.10 [Tech: 0.752]. Since P-value $> \alpha$, we do not reject the null hypothesis.

Conclusion: There is not enough evidence, at the $\alpha = 0.1$ level of significance, to conclude that the distribution of educational attainment has changed since 2000. That is, the data indicate that educational attainment has not changed.

3. The expected counts are calculated as $\dfrac{(\text{row total}) \cdot (\text{column total})}{(\text{table total})} = \dfrac{344 \cdot 550}{1510} \approx 125.298$ (for the first cell) and so on, giving the following table:

	Education			Total
	High School or Less	**Some College**	**College Graduate**	
Enjoy	143	102	99	344
	(125.298)	(105.934)	(112.768)	
Do not Enjoy	407	363	396	1166
	(424.702)	(359.066)	(382.232)	
Total	550	465	495	1510

All expected frequencies are greater than 5 so all requirements for a chi-square test are satisfied. That is, all expected frequencies are greater than or equal to 1, and no more than 20% of the expected frequencies are less than 5.

$$\chi_0^2 = \sum \frac{(O_i - E_i)^2}{E_i} = \frac{(143 - 125.298)^2}{125.298} + \cdots + \frac{(396 - 382.232)^2}{382.232} \approx 5.605$$

$H_0: p_H = p_C = p_G$

$H_1:$ at least one proportion is different from the others

df $= (2-1)(3-1) = 2$ so the critical value is $\chi_{0.05}^2 = 5.991$.

Classical Approach:
The test statistic is 5.605 which is less than the critical value so we do not reject H_0.

P-value Approach:
Using Table VII, we find the row that corresponds to 2 degrees of freedom. The value of 5.605 is less than 5.991, which has an area under the chi-square distribution of 0.05 to the right. Therefore, we have P-value > 0.05 [Tech: P-value = 0.061]. Since P-value > α, we do not reject H_0.

Conclusion:
There is not sufficient evidence, at the $\alpha = 0.05$ level of significance, to conclude that at least one proportion is different from the others. That is, the evidence suggests that the proportions of individuals who enjoy gambling do not differ among the levels of education.

4. The expected counts are calculated as $\dfrac{(\text{row total}) \cdot (\text{column total})}{(\text{table total})} = \dfrac{1997 \cdot 547}{2712} \approx 402.787$ (for the first cell)

and so on, giving the following table:

Religion	Decades				Total
	1970s	**1980s**	**1990s**	**2000s**	
Affiliated	2,395	3,022	2,121	624	8,162
	(1950.23)	(2460.35)	(1809.08)	(1942.34)	
Unaffiliated	327	412	404	2,087	3,230
	(771.77)	(973.65)	(715.92)	(768.66)	
Total	2,722	3,434	2,525	2,711	11,392

All expected frequencies are greater than 5 so all requirements for a chi-square test are satisfied. That is, all expected frequencies are greater than or equal to 1, and no more than 20% of the expected frequencies are less than 5.

$$\chi_0^2 = \sum \frac{(O_i - E_i)^2}{E_i} = \frac{(2395 - 1950.23)^2}{1950.23} + \cdots + \frac{(2087 - 768.66)^2}{768.66} \approx 4155.584 \text{ and df} = (2-1)(4-1) = 3$$

$H_0 : p_{1970s} = p_{1980s} = p_{1990s} = p_{2000s}$

$H_1 :$ at least one of the proportions is not equal to the rest

Classical Approach:
The critical value is $\chi_{0.05}^2 = 7.815$. The test statistic is 4155.584 which is greater than the critical value so we reject H_0.

P-value Approach:
Using Table VII, we find the row that corresponds to 3 degrees of freedom. The value of 4155.584 is greater than 12.838, which has an area under the chi-square distribution of 0.005 to the right. Therefore, we have P-value < 0.005 [Tech: P-value < 0.001]. Since P-value < α, we reject H_0.

Conclusion:
There is sufficient evidence, at the $\alpha = 0.05$ level of significance, to conclude that different proportions of 18-29 year-olds have been affiliated with religion in the past four decades.

5. The expected counts are calculated as $\dfrac{\text{(row total)} \cdot \text{(column total)}}{\text{(table total)}} = \dfrac{3179 \cdot 137}{8874} \approx 49.079$ (for the first cell) and

so on, giving the following table:

	Time Spent in Bars							Total
	Almost Daily	Several Times a Week	Several Times a Month	Once a Month	Several Times a Year	Once a Year	Never	
Smoker	80	409	294	362	433	336	1265	3179
	(49.079)	(271.902)	(241.094)	(298.412)	(360.387)	(323.847)	(1634.280)	
Nonsmoker	57	350	379	471	573	568	3297	5695
	(87.921)	(487.098)	(431.906)	(534.588)	(645.613)	(580.153)	(2927.720)	
Total	137	759	673	833	1006	904	4562	8874

All expected frequencies are greater than 5 so all requirements for a chi-square test are satisfied. That is, all expected frequencies are greater than or equal to 1, and no more than 20% of the expected frequencies are less than 5.

$$\chi_0^2 = \sum \frac{(O_i - E_i)^2}{E_i} = \frac{(80 - 49.079)^2}{49.079} + \cdots + \frac{(3297 - 2927.720)^2}{2927.720} \approx 330.803$$

$H_0 :$ The distribution of bar visits is the same for smokers and nonsmokers

$H_1 :$ The distribution of bar visits is different for smokers and nonsmokers

df $= (2-1)(7-1) = 6$ so the critical value is $\chi_{0.05}^2 = 12.592$.

Classical Approach:
The test statistic is 330.803 which is greater than the critical value so we reject H_0.

P-value Approach:
Using Table VII, we find the row that corresponds to 6 degrees of freedom. The value of 330.803 is greater than 18.548, which has an area under the chi-square distribution of 0.005 to the right. Therefore, we have P-value < 0.005 [Tech: P-value < 0.001]. Since P-value < α, we reject H_0.

Conclusion:

There is sufficient evidence, at the $\alpha = 0.05$ level of significance, to conclude that the distributions of time spent in bars differ between smokers and nonsmokers.

To determine if smokers tend to spend more time in bars than nonsmokers, we can examine the conditional distribution of time spent in bars by smoker status.

	Time Spent in Bars						
	Almost Daily	**Several Times a Week**	**Several Times a Month**	**Once a Month**	**Several Times a Year**	**Once a Year**	**Never**
Smoker	$\frac{80}{137} \approx 0.584$	$\frac{409}{759} \approx 0.539$	$\frac{294}{673} \approx 0.437$	$\frac{362}{833} \approx 0.435$	$\frac{433}{1006} \approx 0.430$	$\frac{336}{904} \approx 0.372$	$\frac{1265}{4562} \approx 0.277$
Nonsmoker	$\frac{57}{137} \approx 0.416$	$\frac{350}{759} \approx 0.461$	$\frac{379}{673} \approx 0.563$	$\frac{471}{833} \approx 0.565$	$\frac{573}{1006} \approx 0.570$	$\frac{568}{904} \approx 0.628$	$\frac{3297}{4562} \approx 0.723$

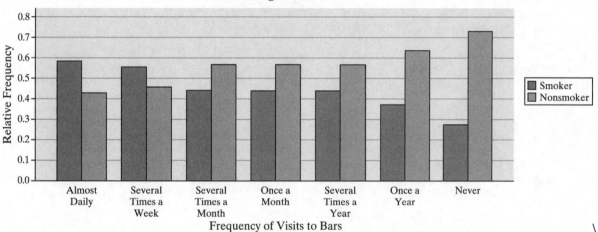

Based on the conditional distribution and the bar graph, it appears that smokers tend to spend more time in bars than nonsmokers.

Case Study: Feeling Lucky? Well, Are You?

Computations for the case study were done using Minitab version 15. The output is provided.

Independence of Game and District

Chi-Square Test: Fantasy 5, Mega Money, Lotto

```
Expected counts are printed below observed counts
Chi-Square contributions are printed below expected counts
```

	Fantasy 5	Mega Money	Lotto	Total
1	87030	27221	270256	384507
	82777.17	28436.62	273293.21	
	218.497	51.966	33.754	
2	186780	56822	934451	1178053
	253612.79	87124.15	837316.06	
	17611.974	10539.218	11268.382	
3	267955	94123	1004651	1366729
	294231.21	101077.89	971419.91	
	2346.586	478.546	1136.795	
4	232451	61019	707934	1001404
	215583.56	74059.89	711760.55	
	1319.722	2296.314	20.572	
5	727390	262840	2746438	3736668
	804434.77	276349.22	2655884.01	
	7378.966	660.393	3087.494	
6	462874	135321	1563491	2161686
	465370.59	159869.77	1536445.64	
	13.394	3769.581	476.067	
7	353532	139421	1492549	1985502
	427441.47	146839.90	1411220.64	
	12779.783	374.830	4686.937	
8	377627	140285	1433077	1950989
	420011.46	144287.45	1386690.09	
	4277.128	111.026	1551.713	
9	470993	153591	1743370	2367954
	509776.23	175124.54	1683053.23	
	2950.587	2647.791	2161.615	
10	588154	185946	1852986	2627086
	565562.51	194288.92	1867234.58	
	902.421	358.251	108.729	
11	1283042	474067	2883453	4640562
	999026.25	343197.66	3298338.09	
	80743.569	49903.558	52186.778	
Total	5037828	1730656	16632656	23401140

```
Chi-Sq = 278452.937, DF = 20, P-Value = 0.000
```

Since the *P*-value is smaller than the significance level, α, we reject the null hypothesis. There is enough evidence at the $\alpha = 0.05$ level to conclude that lottery game and sales district are dependent. We conclude that there is some association between them.

Conditional Distribution Bar Graph (by Sales District)

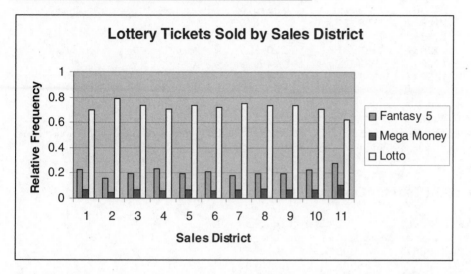

While the overall shape of the distribution is similar for each district, the bar graph tends to support the previous result. There are clear fluctuations in the lottery game as you change the sales district.

Daily Sales Structure

Chi-Square Goodness-of-Fit Test for Observed Counts in Variable: Fantasy 5

Category	Observed	Test Proportion	Expected	Contribution to Chi-Sq
1	12794	0.142857	12434.1	10.41
2	11564	0.142857	12434.1	60.89
3	12562	0.142857	12434.1	1.31
4	11408	0.142857	12434.1	84.68
5	15299	0.142857	12434.1	660.07
6	15684	0.142857	12434.1	849.40
7	7728	0.142857	12434.1	1781.21

N	DF	Chi-Sq	P-Value
87039	6	3447.98	0.000

Since the *P*-value is smaller than the significance level, α, we reject the null hypothesis of equal proportions. It is clear that a highest proportion of tickets are sold on Friday and Saturday, and the lowest on Sunday.

Chi-Square Goodness-of-Fit Test for Observed Counts in Variable: Mega Money

Category	Observed	Test Proportion	Expected	Contribution to Chi-Sq
1	2082	0.142857	3888.71	839.4
2	9983	0.142857	3888.71	9550.8
3	1040	0.142857	3888.71	2086.9
4	1677	0.142857	3888.71	1257.9
5	10439	0.142857	3888.71	11033.5
6	1502	0.142857	3888.71	1464.9
7	498	0.142857	3888.71	2956.5

N	DF	Chi-Sq	P-Value
27221	6	29189.8	0.000

Since the *P*-value is smaller than the significance level, α, we reject the null hypothesis of equal proportions. It is clear that a highest proportion of tickets are sold on Friday and the lowest (by far) on Sunday.

Chi-Square Goodness-of-Fit Test for Observed Counts in Variable: Lotto

Category	Observed	Test Proportion	Expected	Contribution to Chi-Sq
1	13093	0.142857	38608	16862
2	18200	0.142857	38608	10788
3	60746	0.142857	38608	12694
4	18054	0.142857	38608	10942
5	38513	0.142857	38608	0
6	115686	0.142857	38608	153880
7	5964	0.142857	38608	27601

N	DF	Chi-Sq	P-Value
270256	6	232768	0.000

Since the *P*-value is smaller than the significance level, α, we reject the null hypothesis of equal proportions. It is clear that a highest proportion of tickets are sold on Saturday and the lowest (by far) on Sunday. Reports will vary.

Chapter 13

Comparing Three or More Means

Section 13.1

1. analysis of variance.

3. True

5. error; MSE

7.

Source of Variation	Sum of Squares	Degrees of Freedom	Mean Squares	F - Test Statistic
Treatment	387	2	$\dfrac{387}{2}=193.5$	$\dfrac{193.5}{297.852}\approx0.650$
Error	8042	27	$\dfrac{8042}{27}\approx297.852$	
Total	8429	29		

9. $\bar{x}=\dfrac{10\cdot40+10\cdot42+10\cdot44}{10+10+10}=\dfrac{1260}{30}=42$; $\text{MST}=\dfrac{10(40-42)^2+10(42-42)^2+10(44-42)^2}{3-1}=\dfrac{80}{2}=40$;

$\text{MSE}=\dfrac{(10-1)\cdot48+(10-1)\cdot31+(10-1)\cdot25}{30-3}=\dfrac{936}{27}\approx34.667$;

$F_0=\dfrac{40}{34.667}\approx1.154$

11. $\bar{x}_1=\dfrac{28+23+30+27}{4}=\dfrac{108}{4}=27$; $\bar{x}_2=\dfrac{22+25+17+23}{4}=\dfrac{87}{4}=21.75$; $\bar{x}_3=\dfrac{25+24+19+30}{4}=\dfrac{98}{4}=24.5$;

$\bar{x}=\dfrac{28+23+30+...+19+30}{12}=\dfrac{293}{12}\approx24.417$;

$s_1^2=\dfrac{(28-27)^2+(23-27)^2+(30-27)^2+(27-27)^2}{4-1}=\dfrac{26}{3}\approx8.667$;

$s_2^2=\dfrac{(22-21.75)^2+(25-21.75)^2+(17-21.75)^2+(23-21.75)^2}{4-1}=\dfrac{34.75}{3}\approx11.583$;

$s_3^2=\dfrac{(25-24.5)^2+(24-24.5)^2+(19-24.5)^2+(30-24.5)^2}{4-1}=\dfrac{61}{3}\approx20.333$;

$\text{MST}=\dfrac{4(27-24.417)^2+4(21.75-24.417)^2+4(24.5-24.417)^2}{3-1}\approx\dfrac{55.167}{2}\approx27.584$;

$\text{MSE}=\dfrac{(4-1)\cdot8.667+(4-1)\cdot11.583+(4-1)\cdot20.333}{12-3}=\dfrac{121.749}{9}\approx13.528$;

$F_0=\dfrac{27.584}{13.528}\approx2.04$.

13. (a) The hypotheses are $H_0:\mu_{\text{sludge plot}}=\mu_{\text{spring disk}}=\mu_{\text{no till}}$ versus H_1: at least one mean differs from the others.

    (b) (1) Each sample must be a simple random sample.
    (2) The three samples must be independent of each other.
    (3) The samples must come from normally distributed populations.
    (4) The populations must have equal variances.

(c) Since *P*-value = 0.007 is smaller than the $\alpha = 0.05$ level of significance, we reject H_0. There is sufficient evidence to conclude that at least one plot type has a mean number of plants that differs from the other means.

(d) Yes, the boxplots support the results obtained in part (c). The boxplots indicate that significantly more plants are growing in the spring disk plot than in the other plot types.

(e) The needed summary statistics are: $\bar{x}_{\text{sludge plot}} \approx 28.333$, $\bar{x}_{\text{spring disk}} = 33$, $\bar{x}_{\text{no till}} = 28.5$, $\bar{x} \approx 29.944$,

$s^2_{\text{sludge plot}} \approx 7.867$, $s^2_{\text{spring disk}} = 3.2$, $s^2_{\text{no till}} = 6.7$, $n_{\text{sludge plot}} = n_{\text{spring disk}} = n_{\text{no till}} = 6$, $n = 18$, and $k = 3$.

Thus,

$$\text{MST} = \frac{6(28.333 - 29.994)^2 + 6(33 - 29.994)^2 + 6(28.5 - 29.994)^2}{3 - 1} \approx \frac{84.114}{2} \approx 42.057,$$

$$\text{MSE} = \frac{(6-1)\cdot 7.867 + (6-1)\cdot 3.2 + (6-1)\cdot 6.7}{18 - 3} = \frac{88.835}{15} \approx 5.922, \text{ and } F_0 = \frac{42.057}{5.922} \approx 7.10.$$

(f) Because a normal probability plot is roughly linear, it is reasonable to conclude the residuals are normally distributed.

15. (a) The hypotheses are $H_0 : \mu_{\text{I}} = \mu_{\text{II}} = \mu_{\text{III}} = \mu_{\text{IV}}$ versus H_1 : at least one mean differs from the others.

(b) (1) Each sample must be a simple random sample.
(2) The four samples must be independent of each other.
(3) The samples must come from normally distributed populations.
(4) The populations must have equal variances.

(c) Since *P*-value = 0.4204 is larger than the $\alpha = 0.05$ level of significance, we do not reject H_0. There is not sufficient evidence to conclude that at least one of the means for the delivery mechanisms differs from the other means.

(d) Yes, the boxplots support the results obtained in part (c).

(e) Assuming the null hypothesis is true, we would expect this much or more variability in the four sample means about 42 times out of every 100 times the experiment is repeated.

(f) Because a normal probability plot is roughly linear, it is reasonable to conclude the residuals are normally distributed.

17. (a) The hypotheses are $H_0 : \mu_{\text{Financial}} = \mu_{\text{Energy}} = \mu_{\text{Utilities}}$ versus H_1 : at least one mean differs from the others.

(b) (1) Each sample is a simple random sample.
(2) The three samples are independent of each other.
(3) The samples are from normally distributed populations.

(4) $s_{\text{Financial}} \approx 5.12$, $s_{\text{Energy}} \approx 4.87$, and $s_{\text{Utilities}} \approx 4.53$. The largest sample standard deviation is less than twice the smallest sample standard deviation, so the requirement that the populations have equal variances is satisfied.

(c) Using a TI-84 Plus, we obtain the following output:

Since *P*-value ≈ 0.150 is larger than the $\alpha = 0.05$ level of significance, we do not reject H_0. By hand we obtain $F_0 = 2.08 < F_{0.05, 2, 21} \approx F_{0.05, 2, 20} = 3.49$, so we do not reject H_0. There is not sufficient evidence to conclude that at least one mean is different from the others.

(d)

Rate of Return (%)

19. (a) The hypotheses are $H_0 : \mu_{WM} = \mu_{BM} = \mu_{WF} = \mu_{BF}$ versus H_1 : at least one mean differs from the others.

(b) The largest sample standard deviation is less than twice the smallest sample standard deviation ($241.5 < 2 \cdot 157.1$), so the requirement that the populations have equal variances is satisfied.

(c) Using a TI-84 Plus, we obtain the following output:

The level of significance was not specified. We will use $\alpha = 0.05$. Since *P*-value < 0.0001 , the *P*-value is smaller than the $\alpha = 0.05$ level of significance, so we reject H_0 . By hand, we obtain $F_0 = 61.4 > F_{0.05,3,27} \approx F_{0.05,3,25} = 2.99$, so we reject H_0 . There is sufficient evidence to conclude that at least one mean is different from the others. The mean initial price is not the same for the four groups of people.

(d)

Profit on Initial Price (dollars)

(e) The results suggest that the mean initial price offered to blacks and especially black females is higher than that offered to whites, so there does seem to be race discrimination. While gender discrimination against females is apparent among blacks, it is not among whites. In fact, white females in the sample were offered the lowest mean price with the least variability of the four groups. A separate study might be needed to test for gender discrimination specifically.

(f) Because a normal probability plot is roughly linear, it is reasonable to conclude the residuals are normally distributed.

21. (a) The hypotheses are $H_0 : \mu_{Alaska} = \mu_{Florida} = \mu_{Texas}$ versus H_1 : at least one mean differs from the others.

(b) (1) Each sample is a simple random sample.
(2) The three samples are independent of each other.
(3) The samples are from normally distributed populations.

(4) The largest sample standard deviation is less than twice the smallest sample standard deviation ($0.397 < 2 \cdot 0.252$), so the requirement that the populations have equal variances is satisfied.

(c) Using a TI-84 Plus, we obtain the following output:

```
One-way ANOVA          One-way ANOVA
 F=5.810954298        ↑ MS=.583755556
 p=.0135328894          Error
 Factor                   df=15
   df=2                   SS=1.50686667
   SS=1.16751111          MS=.100457778
↓ MS=.583755556         SxP=.31695075
```

Since P-value ≈ 0.014 is less than $\alpha = 0.05$, we reject H_0. By hand we obtain

$F_0 = 5.81 > F_{0.05, 2, 15} = 3.68$, so we reject H_0. There is sufficient evidence to conclude that at least one mean is different from the others.

(d)

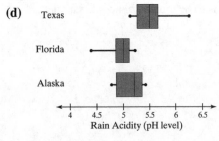

23. **(a)** The hypotheses are $H_0 : \mu_C = \mu_L = \mu_M$ versus $H_1 :$ at least one mean differs from the others.

(b) (1) Each sample is a simple random sample.
(2) The three samples are independent of each other.
(3) The samples are from normally distributed populations.

(4) The largest sample standard deviation is less than twice the smallest sample standard deviation ($310,024 < 2 \cdot 182,988$), so the requirement that the populations have equal variances is satisfied.

(c) Using MINITAB, we obtain the following output:

One-way ANOVA: Rich versus Political philosophy

```
Source                  DF         SS           MS        F      P
Political philosophy     2  4.71605E+11  2.35802E+11  3.10   0.047
Error                  196  1.49246E+13  76145925199
Total                  198  1.53962E+13

S = 275946    R-Sq = 3.06%    R-Sq(adj) = 2.07%
```

Since P-value $= 0.0474$ is less than $\alpha = 0.05$, we reject H_0. By hand we obtain

$F_0 = 3.10 > F_{0.05, 2, 196} \approx F_{0.05, 2, 200} = 3.04$, so we reject H_0. There is sufficient evidence to conclude that at least one mean is different from the others.

(d)

25. **(a)** The hypotheses are $H_0 : \mu_{67-0-301} = \mu_{67-0-400} = \mu_{67-0-353}$ versus $H_1 :$ at least one mean differs from the others.

(b) $s_{67-0-301} \approx 107.8$, $s_{67-0-400} \approx 385.0$, $s_{67-0-353} \approx 195.8$. Since the largest standard deviation is greater than twice the smallest standard deviation (i.e., since $385.0 > 2 \cdot 107.8$), the rule of thumb for meeting the ANOVA requirement of equal variances is violated, and so we cannot perform an ANOVA.

27. (a) $s_{control} = \sqrt{21.6} \approx 4.6$, $s_{RAP} = \sqrt{13.0} \approx 3.6$, and $s_{headache} = \sqrt{8.4} \approx 2.9$

(b) The problem states that children were stratified in two age groups: 4 to 11 years, and 12 to 18 years. That is, the children in the study were separated by a common characteristic, age. Thus, the researchers used stratified sampling.

(c) $H_0 : \mu_1 = \mu_2$ versus $H_1 : \mu_1 \neq \mu_2$. The level of significance is $\alpha = 0.05$. Since the sample size of both groups is 70, we use $n_1 - 1 = 69$ degrees of freedom. The test statistic is

$$t_0 = \frac{(\bar{x}_1 - \bar{x}_2) - (\mu_1 - \mu_2)}{\sqrt{\dfrac{s_1^2}{n_1} + \dfrac{s_2^2}{n_2}}} = \frac{(11.7 - 9.0) - 0}{\sqrt{\dfrac{21.6}{70} + \dfrac{13.0}{70}}} \approx 3.84.$$

Classical approach: Since this is a two-tailed test with 69 degrees of freedom, the critical values are $\pm t_{0.025} = \pm 1.994$. Since the test statistic $t_0 \approx 3.84$ is more than the critical value $t_{0.025} = 1.994$ (i.e., since the test statistic falls within the critical region), we reject H_0.

P-value approach: The *P*-value for this two-tailed test is the area under the *t*-distribution with 69 degrees of freedom to the right of $t_0 = 3.84$ plus the area to the left of -3.84. From the *t*-distribution table in the row corresponding to 70 degrees of freedom (since there is no entry for 69), 3.84 falls above 3.435 whose right-tail area is 0.0005. We must double this value in order to get the total area in both tails: 0.001. So, *P*-value < 0.001. (Using technology, we find *P*-value = 0.000191.) Because the *P*-value is less than the level of significance $\alpha = 0.05$, we reject H_0.

Conclusion: There is sufficient evidence at the $\alpha = 0.05$ level of significance to conclude that the population mean CBCL scores are different.

(d) It is not necessary to check the normality assumption because the samples are all considered to be large samples (sample size greater than 30).

(e) $\bar{x} = \dfrac{11.7 + 9.0 + 12.4}{3} = 11.03$

$$MST = \frac{70(11.7 - 11.03)^2 + 70(9.0 - 11.03)^2 + 70(12.4 - 11.03)^2}{3 - 1} = \frac{451.269}{2} \approx 225.635,$$

$$MSE = \frac{(70-1) \cdot 21.6 + (70-1) \cdot 13.0 + (70-1) \cdot 8.4}{210 - 3} = \frac{2967}{207} \approx 14.333, \text{ and } F_0 = \frac{225.635}{14.333} \approx 15.74.$$

Since $F_0 = 15.74 > F_{0.05,2,207} \approx F_{0.05,2,200} = 3.04$, we reject H_0. There is sufficient evidence to conclude that at least one mean CBCL scores is different from the others.

(f) No; the result from part (e) only indicates that *at least one* mean is different. It does not indicate which pairs are different nor even how many. The result from part (c) indicates that the mean scores for the control group and the RAP group are different, but additional pairs could also be significantly different.

29. The mean square due to treatment estimate of σ^2 is a weighted average of the squared deviations of each sample mean from the grand mean of all the samples. The mean square due to error estimate of σ^2 is the weighted average of the sample variances.

31. Rejecting the statement in the null hypothesis means that there is sufficient evidence to conclude that the mean of one of the populations is different than the other two or that all three population means are different.

Section 13.2

1. **(a)** We look in the table of critical values for Tukey's Test with $\alpha = 0.05$ and where the row corresponding to $\nu = 10$ intersects with the column corresponding to $k = 3$ to find the critical value $q_{0.05,10,3} = 3.877$.

   (b) We look in the table of critical values for Tukey's Test with $\alpha = 0.05$ and where the row corresponding to $\nu = 24$ intersects with the column corresponding to $k = 5$ to find the critical value $q_{0.05,24,5} = 4.166$.

   (c) We look in the table of critical values for Tukey's Test with $\alpha = 0.05$ and where the row corresponding to $\nu = 30$ (since there is no row in the table for $\nu = 32$) intersects with the column corresponding to $k = 8$ to find the critical value $q_{0.05,32,8} \approx q_{0.05,30,8} = 4.602$.

   (d) For this test, $k = 5$ and $\nu = n - k = 65 - 5 = 60$, so we look in the table with $\alpha = 0.05$ and where the row corresponding to $\nu = 60$ intersects with the column corresponding to $k = 5$ to find the critical value $q_{0.05,60,5} = 3.977$.

3. Arrange the sample means in ascending order: $\bar{x}_2 = 9.1$, $\bar{x}_1 = 9.5$, $\bar{x}_3 = 18.1$. Now $n = 5 \cdot 3 = 15$ and $k = 3$, so $\nu = n - k = 12$ and the critical value is $q_{0.05,12,3} = 3.773$.

   We first test $H_0 : \mu_3 = \mu_2$. The test statistic is $q_0 = \dfrac{\bar{x}_3 - \bar{x}_2}{\sqrt{\frac{s^2}{2} \cdot \left(\frac{1}{n_3} + \frac{1}{n_2}\right)}} = \dfrac{18.1 - 9.1}{\sqrt{\frac{26.2}{2} \cdot \left(\frac{1}{5} + \frac{1}{5}\right)}} \approx 3.932$. Since this test

   statistic is greater than $q_{0.05,12,3} = 3.773$, we reject $H_0 : \mu_3 = \mu_2$.

   Next, we test $H_0 : \mu_3 = \mu_1$. The test statistic is $q_0 = \dfrac{\bar{x}_3 - \bar{x}_1}{\sqrt{\frac{s^2}{2} \cdot \left(\frac{1}{n_3} + \frac{1}{n_1}\right)}} = \dfrac{18.1 - 9.5}{\sqrt{\frac{26.2}{2} \cdot \left(\frac{1}{5} + \frac{1}{5}\right)}} \approx 3.757$. Since this test

   statistic is less than $q_{0.05,12,3} = 3.773$, we do not reject $H_0 : \mu_3 = \mu_1$.

   Finally, we test $H_0 : \mu_1 = \mu_2$. The test statistic is $q_0 = \dfrac{\bar{x}_1 - \bar{x}_2}{\sqrt{\frac{s^2}{2} \cdot \left(\frac{1}{n_1} + \frac{1}{n_2}\right)}} = \dfrac{9.5 - 9.1}{\sqrt{\frac{26.2}{2} \cdot \left(\frac{1}{5} + \frac{1}{5}\right)}} \approx 0.175$. Since this test

   statistic is less than $q_{0.05,12,3} = 3.773$, we do not reject $H_0 : \mu_1 = \mu_2$.

   The results of the test are ambiguous since we have that $\mu_1 = \mu_3$ and $\mu_1 = \mu_2$, but $\mu_2 \neq \mu_3$. That is, we have: $\underline{\mu_2 \quad \mu_1 \quad \mu_3}$.

5. **(a)** The hypotheses are $H_0 : \mu_1 = \mu_2 = \mu_3$ versus H_1 : at least one mean differs from the others. Using MINITAB, we obtain the following output:

   One-way ANOVA: Sample 1, Sample 2, Sample 3

Source	DF	SS	MS	F	P
Factor	2	630.3	315.2	9.72	0.002
Error	15	486.5	32.4		
Total	17	1116.8			

   Since P-value ≈ 0.002 is less than the $\alpha = 0.05$ level of significance, we reject H_0. There is sufficient evidence to conclude that at least one mean differs from the others.

(b) Using MINITAB, we obtain the following output:

Tukey 95% Simultaneous Confidence Intervals: All Pairwise Comparisons
```
Individual confidence level = 97.97%

Sample 1 subtracted from:
          Lower   Center   Upper    --------+---------+---------+---------+-
Sample 2  -11.183  -2.650   5.883             (------*------)
Sample 3    2.484  11.017  19.549                      (------*------)
                                     --------+---------+---------+---------+-
                                           -12        0        12        24

Sample 2 subtracted from:
          Lower   Center   Upper    --------+---------+---------+---------+-
Sample 3   5.134  13.667  22.199                       (------*------)
                                     --------+---------+---------+---------+-
                                           -12        0        12        24
```

From the output we have:

$-11.183 < \mu_2 - \mu_1 < 5.883$. This interval contains 0, so we do not reject $H_0 : \mu_1 = \mu_2$.

$2.484 < \mu_3 - \mu_1 < 19.549$. This interval does not contain 0, so we reject $H_0 : \mu_1 = \mu_3$.

$5.134 < \mu_3 - \mu_2 < 22.199$. This interval does not contain 0, so we reject $H_0 : \mu_2 = \mu_3$.

Thus, we have $\mu_1 = \mu_2 \neq \mu_3$. That is, we have: $\underline{\mu_1 \quad \mu_2} \quad \mu_3$.

(c) The boxplots shown to the right support the result of parts (a) and (b).

7. From the provided output, we have:

$1.021 < \mu_{\text{spring disk}} - \mu_{\text{sludge plot}} < 8.313$. This interval does not contain 0, so we reject $H_0 : \mu_{\text{sludge plot}} = \mu_{\text{spring disk}}$

$-3.479 < \mu_{\text{no till}} - \mu_{\text{sludge plot}} < 3.813$. This interval contains 0, so we do not reject $H_0 : \mu_{\text{sludge plot}} = \mu_{\text{no till}}$.

$-8.146 < \mu_{\text{no till}} - \mu_{\text{spring disk}} < -0.854$. This interval does not contain 0, so we reject $H_0 : \mu_{\text{spring disk}} = \mu_{\text{no till}}$.

Thus, we have $\mu_{\text{sludge plot}} = \mu_{\text{no till}} \neq \mu_{\text{spring disk}}$. That is, we have: $\underline{\mu_{\text{sludge plot}} \quad \mu_{\text{no till}}} \quad \underline{\mu_{\text{spring disk}}}$.

From the pairwise comparisons, it appears that the spring disk method has a greater mean number of plants. Therefore, we should recommend spring-disk method of planting.

9. We use MINITAB to obtain the following output:

```
Tukey 95% Simultaneous Confidence Intervals
All Pairwise Comparisons
Individual confidence level = 98.92%


White Male subtracted from:

              Lower   Center   Upper    +---------+---------+---------+---------
Black Male    401.5    724.7  1047.9                            (---*---)
White Female -334.2    -37.6   259.0                  (---*--)
Black Female  928.8   1191.6  1454.5                                  (--*--)
                                        +---------+---------+---------+---------
                                      -1600      -800        0       800
```

```
Black Male subtracted from:
                Lower   Center   Upper    +---------+---------+---------+---------
White Female  -1147.5   -762.3  -377.1              (---*----)
Black Female    107.0    466.9   826.9                              (----*---)
                                          +---------+---------+---------+---------
                                       -1600       -800        0       800


White Female subtracted from:
                Lower   Center   Upper    +---------+---------+---------+---------
Black Female    893.0   1229.2  1565.5                                   (---*----)
                                          +---------+---------+---------+---------
                                       -1600       -800        0       800
```

From the output, we have:

$401.5 < \mu_{BM} - \mu_{WM} < 1047.9$. This interval does not contain 0, so we reject $H_0 : \mu_{BM} = \mu_{WM}$.

$-334.2 < \mu_{WF} - \mu_{WM} < 259.0$. This interval contains 0, so we do not reject $H_0 : \mu_{WF} = \mu_{WM}$.

$928.8 < \mu_{BF} - \mu_{WM} < 1454.5$. This interval does not contain 0, so we reject $H_0 : \mu_{BF} = \mu_{WM}$.

$-1147.5 < \mu_{WF} - \mu_{BM} < -377.1$. This interval does not contain 0, so we reject $H_0 : \mu_{WF} = \mu_{BM}$.

$107.0 < \mu_{BF} - \mu_{BM} < 826.9$. This interval does not contain 0, so we reject $H_0 : \mu_{BF} = \mu_{BM}$.

$893.0 < \mu_{BF} - \mu_{WF} < 1565.5$. This interval does not contain 0, so we reject $H_0 : \mu_{BF} = \mu_{WF}$.

The results of Tukey's test indicate that $\mu_{WM} = \mu_{WF} \neq \mu_{BM} \neq \mu_{BF}$. That is, we have:

$\underline{\mu_{WM} \quad \mu_{WF}} \quad \underline{\mu_{BM}} \quad \underline{\mu_{BF}}$.

11. (a) The hypotheses are $H_0 : \mu_{Placebo} = \mu_{Rit\ 10} = \mu_{Rit\ 17.5} = \mu_{Add\ 7.5} = \mu_{Add\ 12.5}$ versus H_1 : at least one mean differs from the others. Using MINITAB, we obtain the following output:

One-way ANOVA: Placebo, Ritalin 10 m, Ritalin 17.5, Adderall 7.5, Adderall 12.

```
Source  DF     SS    MS    F      P
Factor   4   7297  1824  5.75  0.001
Error   30   9519   317
Total   34  16816
```

Since P-value $= 0.001 < \alpha = 0.05$ level of significance, we reject H_0. There is sufficient evidence to support the hypothesis that at least one of the mean abilities to follow rules is different from the others.

(b) Using MINITAB, we obtain the following output:

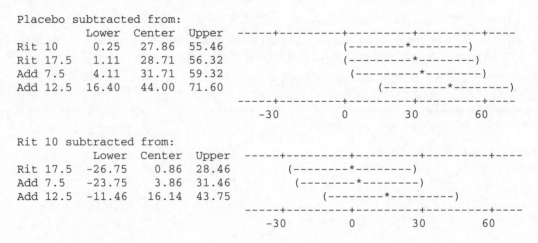

Tukey 95% Simultaneous Confidence Intervals: All Pairwise Comparisons
```
Individual confidence level = 99.31%

Placebo subtracted from:
            Lower   Center   Upper    -----+---------+---------+---------+----
Rit 10       0.25    27.86   55.46                  (--------*--------)
Rit 17.5     1.11    28.71   56.32                  (--------*--------)
Add 7.5      4.11    31.71   59.32                   (--------*--------)
Add 12.5    16.40    44.00   71.60                       (--------*--------)
                                      -----+---------+---------+---------+----
                                         -30        0        30        60


Rit 10 subtracted from:
            Lower   Center   Upper    -----+---------+---------+---------+----
Rit 17.5   -26.75     0.86   28.46         (--------*--------)
Add 7.5    -23.75     3.86   31.46         (--------*--------)
Add 12.5   -11.46    16.14   43.75              (--------*---------)
                                      -----+---------+---------+---------+----
                                         -30        0        30        60
```

```
Rit 17.5 subtracted from:
            Lower   Center   Upper    -----+---------+---------+---------+----
Add 7.5    -24.60     3.00   30.60         (--------*--------)
Add 12.5   -12.32    15.29   42.89              (--------*--------)
                                         -----+---------+---------+---------+----
                                            -30         0        30        60


Add 7.5 subtracted from:
            Lower   Center   Upper    -----+---------+---------+---------+----
Add 12.5   -15.32    12.29   39.89                (--------*--------)
                                         -----+---------+---------+---------+----
                                            -30         0        30        60
```

From the output, we have:

$0.25 < \mu_{\text{Rit 10}} - \mu_{\text{Placebo}} < 55.46$. This interval does not contain 0, so we reject $H_0 : \mu_{\text{Placebo}} = \mu_{\text{Rit 10}}$.

$1.11 < \mu_{\text{Rit 17.5}} - \mu_{\text{Placebo}} < 56.32$. This interval does not contain 0, so we reject $H_0 : \mu_{\text{Placebo}} = \mu_{\text{Rit 17.5}}$.

$4.11 < \mu_{\text{Add 7.5}} - \mu_{\text{Placebo}} < 59.32$. This interval does not contain 0, so we reject $H_0 : \mu_{\text{Placebo}} = \mu_{\text{Add 7.5}}$.

$16.40 < \mu_{\text{Add 12.5}} - \mu_{\text{Placebo}} < 71.60$. This interval does not contain 0, so we reject $H_0 : \mu_{\text{Placebo}} = \mu_{\text{Add 12.5}}$.

$-26.75 < \mu_{\text{Rit 17.5}} - \mu_{\text{Rit 10}} < 28.46$. This interval contains 0, so we do not reject $H_0 : \mu_{\text{Rit 10}} = \mu_{\text{Rit 17.5}}$.

$-23.75 < \mu_{\text{Add 7.5}} - \mu_{\text{Rit 10}} < 31.46$. This interval contains 0, so we do not reject $H_0 : \mu_{\text{Rit 10}} = \mu_{\text{Add 7.5}}$.

$-11.5 < \mu_{\text{Add 12.5}} - \mu_{\text{Rit 10}} < 43.75$. This interval contains 0, so we do not reject $H_0 : \mu_{\text{Rit 10}} = \mu_{\text{Add 12.5}}$.

$-24.60 < \mu_{\text{Add 7.5}} - \mu_{\text{Rit 17.5}} < 30.60$. This interval contains 0, so we do not reject $H_0 : \mu_{\text{Rit 17.5}} = \mu_{\text{Add 7.5}}$.

$-12.32 < \mu_{\text{Add 12.5}} - \mu_{\text{Rit 17.5}} < 42.89$. This interval contains 0, so we do not reject $H_0 : \mu_{\text{Rit 17.5}} = \mu_{\text{Add 12.5}}$.

$-15.32 < \mu_{\text{Add 12.5}} - \mu_{\text{Add 7.5}} < 39.89$. This interval contains 0, so we do not reject $H_0 : \mu_{\text{Add 7.5}} = \mu_{\text{Add 12.5}}$.

Thus, we have that $\mu_{\text{Placebo}} \neq \mu_{\text{Rit 10}} = \mu_{\text{Rit 17.5}} = \mu_{\text{Add 7.5}} = \mu_{\text{Add 12.5}}$. That is, we have:

$\underline{\mu_{\text{Placebo}}} \quad \mu_{\text{Rit 10}} \quad \mu_{\text{Rit 17.5}} \quad \mu_{\text{Add 7.5}} \quad \mu_{\text{Add 12.5}}$. The mean ability to follow rules of a child taking the placebo is less than the mean ability to follow rules of a child taking any of the four drug treatments.

(c) The boxplots shown to the right support the results of parts (a) and (b).

13. (a) The hypotheses are $H_0 : \mu_{\text{nonsmokers}} = \mu_{\text{light smokers}} = \mu_{\text{heavy smokers}}$ versus H_1 : at least one mean differs from the others. Using MINITAB, we obtain the following output:

One-way ANOVA: nonsmoker, light smoker, heavy smoker
```
Source  DF     SS      MS      F      P
Factor   2  1494.9  747.4  19.62  0.000
Error   18   685.7   38.1
Total   20  2180.6
```

Since P-value < 0.001 is less than the $\alpha = 0.05$ level of significance, we reject H_0. There is sufficient evidence to support the hypothesis that at least one of the mean heart rates is different from the others.

(b) Using MINITAB, we obtain the following output:

```
Tukey 95% Simultaneous Confidence Intervals:  All Pairwise Comparisons
Individual confidence level = 98.00%

nonsmoker subtracted from:
         Lower   Center   Upper    ----+---------+---------+---------+-----
light    3.578   12.000   20.422                       (------*------)
heavy   12.150   20.571   28.993                            (------*------)
                                   ----+---------+---------+---------+-----
                                     -12        0         12        24

light subtracted from:
         Lower   Center   Upper    ----+---------+---------+---------+-----
heavy    0.150    8.571   16.993                    (------*------)
                                   ----+---------+---------+---------+-----
                                     -12        0         12        24
```

From the output, we have:

$3.578 < \mu_{\text{light smokers}} - \mu_{\text{nonsmokers}} < 20.422$. This interval does not contain 0, so we reject

$H_0 : \mu_{\text{nonsmokers}} = \mu_{\text{light smokers}}$.

$12.150 < \mu_{\text{heavy smokers}} - \mu_{\text{nonsmokers}} < 28.993$. This interval does not contain 0, so we reject

$H_0 : \mu_{\text{nonsmokers}} = \mu_{\text{heavy smokers}}$.

$0.150 < \mu_{\text{heavy smokers}} - \mu_{\text{light smokers}} < 16.993$. This interval does not contain 0, so we reject

$H_0 : \mu_{\text{light smokers}} = \mu_{\text{heavy smokers}}$.

Thus, we have that $\mu_{\text{nonsmokers}} \neq \mu_{\text{light smokers}} \neq \mu_{\text{heavy smokers}}$. That is, we have:

$\underline{\mu_{\text{nonsmokers}}}$ $\underline{\mu_{\text{light smokers}}}$ $\underline{\mu_{\text{heavy smokers}}}$. It appears that the mean heart rate of nonsmokers is lower than the mean heart rate of smokers and that the mean heat rate of light smokers is lower than the mean heart rate of heavy smokers.

(c) The boxplots shown to the right support the result of parts (a) and (b).

15. (a) The hypotheses are $H_0 : \mu_{\text{skim}} = \mu_{\text{mixed}} = \mu_{\text{whole}}$ versus H_1 : at least one mean differs from the others. Using MINITAB, we obtain the following output:

One-way ANOVA: Skim Milk, Mixed Milk, Whole Milk

```
Source   DF      SS      MS      F       P
Factor    2   117596   58798   70.51   0.000
Error    21    17511     834
Total    23   135107
```

Since *P*-value < 0.001 is less than the $\alpha = 0.05$ level of significance, we reject H_0. There is sufficient evidence to support the hypothesis that at least one of the mean calcium intake levels is different from the others.

(b) Using MINITAB, we obtain the following output:

```
Tukey 95% Simultaneous Confidence Intervals:  All Pairwise Comparisons
Individual confidence level = 98.00%

Skim Milk subtracted from:
            Lower   Center   Upper   ---------+---------+---------+---------+
Mixed Milk  108.65  145.00   181.35                               (---*--)
Whole Milk  -43.10   -6.75    29.60                 (--*---)
                                     ---------+---------+---------+---------+
                                          -100        0       100       200

Mixed Milk subtracted from:
            Lower   Center   Upper   ---------+---------+---------+---------+
Whole Milk  -188.10  -151.75 -115.40 (---*--)
                                     ---------+---------+---------+---------+
                                          -100        0       100       200
```

From the output, we have:

$108.65 < \mu_{\text{mixed}} - \mu_{\text{skim}} < 181.35$. This interval does not contain 0, so we reject $H_0 : \mu_{\text{skim}} = \mu_{\text{mixed}}$.

$-43.10 < \mu_{\text{whole}} - \mu_{\text{skim}} < 29.60$. This interval contains 0, so we do not reject $H_0 : \mu_{\text{skim}} = \mu_{\text{whole}}$.

$-188.10 < \mu_{\text{whole}} - \mu_{\text{mixed}} < -115.40$. This interval does not contain 0, so we reject $H_0 : \mu_{\text{mixed}} = \mu_{\text{whole}}$.

Thus, we have that $\mu_{\text{mixed}} \neq \mu_{\text{skim}} = \mu_{\text{whole}}$. That is, we have: $\underline{\mu_{\text{mixed}}} \quad \underline{\mu_{\text{skim}} \quad \mu_{\text{whole}}}$.

It appears that the mean calcium intake of children who drank mixed milk is different from the mean calcium intakes of children who drank skim or whole milk.

(c) The boxplots shown to the right support the result of parts (a) and (b).

17. (a) This was a cohort study because the researchers identified the participants to be included in the study and then followed them over a period of time.

(b) $\bar{x}_{HS} = \dfrac{2.6 + 4.1 + \ldots + 5.5}{10} = 4.05$, $\bar{x}_S = \dfrac{3.1 + 3.8 + \ldots + 4.1}{10} = 4.2$, $\bar{x}_{NS} = \dfrac{4.7 + 4.2 + \ldots + 5.1}{10} = 4.2$,

$\bar{x}_{OD} = \dfrac{5.0 + 7.3 + \ldots + 6.0}{10} = 5.8$, $\bar{x}_{NR} = \dfrac{4.5 + 5.5 + \ldots + 4.7}{10} = 4.65$

The means do not appear to be equal.

(c) $s_{HS} = \sqrt{\dfrac{(2.6 - 4.05)^2 + \ldots + (5.5 - 4.05)^2}{10 - 1}} \approx 0.991$, $s_S = \sqrt{\dfrac{(3.1 - 4.2)^2 + \ldots + (4.1 - 4.2)^2}{10 - 1}} \approx 0.554$,

$s_{NS} = \sqrt{\dfrac{(4.7 - 4.2)^2 + \ldots + (5.1 - 4.2)^2}{10 - 1}} \approx 1.099$, $s_{OD} = \sqrt{\dfrac{(5.0 - 5.8)^2 + \ldots + (6.0 - 5.8)^2}{10 - 1}} \approx 0.921$,

$s_{NR} = \sqrt{\dfrac{(4.5 - 4.65)^2 + \ldots + (4.7 - 4.65)^2}{10 - 1}} \approx 0.566$

Since the largest sample standard deviation is less than twice the smallest sample standard deviation $(1.099 < 2 \cdot 0.554)$, the population variances can be assumed to be equal.

(d) For 95% confidence, $\alpha/2 = 0.025$. With 9 degrees of freedom, $t_{0.025} = 2.262$. Then:

$$\text{Lower bound} = \bar{x} - t_{0.025} \cdot \frac{s}{\sqrt{n}} = 4.05 - 2.262 \cdot \frac{0.991}{\sqrt{10}} \approx 4.05 - 0.71 = 3.34$$

$$\text{Upper bound} = \bar{x} + t_{0.025} \cdot \frac{s}{\sqrt{n}} = 4.05 + 2.262 \cdot \frac{0.991}{\sqrt{10}} \approx 4.05 + 0.71 = 4.76$$

(e) There are $_5C_2 = \dfrac{5!}{2!(5-2)!} = \dfrac{5!}{2!3!} = \dfrac{5 \cdot 4}{2} = 10$ possible pairwise comparisons among treatment levels.

(f) Assuming all the tests are independent, we obtain $P(\text{Type I error}) = 1 - (1-\alpha)^k$

$$= 1 - (1 - 0.05)^{10}$$
$$= 1 - (0.95)^{10}$$
$$= 1 - 0.5987$$
$$= 0.4013$$

(g) The hypotheses are $H_0 : \mu_{HS} = \mu_S = \mu_{NS} = \mu_{OD} = \mu_{NR}$ versus H_1 : at least one mean differs from the others. Using MINITAB, we obtain the following output:

One-way ANOVA: HS, S, NS, OD, NR

Source	DF	SS	MS	F	P
Factor	4	20.630	5.157	7.03	0.000
Error	45	33.010	0.734		
Total	49	53.640			

Since P-value < 0.001 is less than the $\alpha = 0.05$ level of significance, we reject H_0. There is sufficient evidence to support the hypothesis that at least one mean time to degree completion is different from the others.

Using MINITAB, we obtain the following output:

```
Tukey 95% Simultaneous Confidence Intervals:  All Pairwise Comparisons
Individual confidence level = 99.33%


HS subtracted from:
        Lower   Center   Upper    --------+---------+---------+---------+-
S     -0.9388  0.1500   1.2388            (-------*-------)
NS    -0.9388  0.1500   1.2388            (-------*-------)
OD     0.6612  1.7500   2.8388                      (--------*------)
NR    -0.4888  0.6000   1.6888                 (------*------)
                                 --------+---------+---------+---------+-
                                     -1.5      0.0       1.5       3.0


S subtracted from:
        Lower   Center   Upper    --------+---------+---------+---------+-
NS    -1.0888  0.0000   1.0888          (------*------)
OD     0.5112  1.6000   2.6888                    (-------*------)
NR    -0.6388  0.4500   1.5388               (------*------)
                                 --------+---------+---------+---------+-
                                     -1.5      0.0       1.5       3.0


NS subtracted from:
        Lower   Center   Upper    --------+---------+---------+---------+-
OD     0.5112  1.6000   2.6888                    (-------*------)
NR    -0.6388  0.4500   1.5388               (------*------)
                                 --------+---------+---------+---------+-
                                     -1.5      0.0       1.5       3.0
```

```
OD subtracted from:
       Lower   Center    Upper   --------+---------+---------+---------+-
NR    -2.2388  -1.1500  -0.0612      (------*-------)
                                 --------+---------+---------+---------+-
                                     -1.5       0.0       1.5       3.0
```

From the output, we have:

$-0.9388 < \mu_S - \mu_{HS} < 1.2388$. This interval contains 0, so we do not reject $H_0 : \mu_S = \mu_{HS}$.

$-0.9388 < \mu_{NS} - \mu_{HS} < 1.2388$. This interval contains 0, so we do not reject $H_0 : \mu_{NS} = \mu_{HS}$.

$0.6612 < \mu_{OD} - \mu_{HS} < 2.8388$. This interval does not contain 0, so we reject $H_0 : \mu_{OD} = \mu_{HS}$.

$-0.4888 < \mu_{NR} - \mu_{HS} < 1.6888$. This interval contains 0, so we do not reject $H_0 : \mu_{NR} = \mu_{HS}$.

$-1.0888 < \mu_{NS} - \mu_S < 1.0888$. This interval contains 0, so we do not reject $H_0 : \mu_{NS} = \mu_S$.

$0.5112 < \mu_{OD} - \mu_S < 2.6888$. This interval does not contain 0, so we reject $H_0 : \mu_{OD} = \mu_S$.

$-0.6388 < \mu_{NR} - \mu_S < 1.5388$. This interval contains 0, so we do not reject $H_0 : \mu_{NR} = \mu_S$.

$0.5112 < \mu_{OD} - \mu_{NS} < 2.6888$. This interval does not contain 0, so we reject $H_0 : \mu_{OD} = \mu_{NS}$.

$-0.6388 < \mu_{NR} - \mu_{NS} < 1.5388$. This interval contains 0, so we do not reject $H_0 : \mu_{NR} = \mu_{NS}$.

$-2.2388 < \mu_{NR} - \mu_{OD} < -0.0612$. This interval does not contain 0, so we reject $H_0 : \mu_{NR} = \mu_{OD}$.

Thus, we have that $\mu_{HS} = \mu_S = \mu_{NS} = \mu_{NR} \neq \mu_{OD}$. That is, we have: $\underline{\mu_{HS} \ \mu_S \ \mu_{NS} \ \mu_{NR}} \quad \underline{\mu_{OD}}$.

The mean time to degree completion appears to be the same for all initial institution types except for those with an open-door policy. It appears that students starting at an institution with an open-door policy take longer to complete their bachelor's degree.

Consumer Reports®: Tea Antioxidants: An Unexpected Side Benefit to a Cheap and Tasty Beverage

(a) H_0: μ_i are all equal versus H_1: at least one of the μ_i is different from the others.

(b) The smallest value in Brand 4 and the largest value in Brand 6 are quite different from the other two observations in their respective brands and might be outliers. However, there are too few observations per group to use a boxplot to determine if these observations are outliers.

To perform a one-way ANOVA, the $k = 6$ samples must be independent of each other, the populations must be normally distributed, and the populations must have the same variance. (More sophisticated procedures exist which allow for unequal variances, but they are beyond the scope of this text.)

(c) There is a statistical difference among brands. The *P*-value is 0.001 which would be considered highly significant.

(d) Based on the overlapping confidence intervals, Brands 2, 3, 5, and 6 appear to have similar amounts of ORAC. Brand 4 may have slightly lower ORAC levels, but it's confidence interval overlaps with the confidence intervals of these other brands.

(e) Using MINITAB, we obtain the following output:

Tukey 95% Simultaneous Confidence Intervals All Pairwise Comparisons among Levels of Brand
Individual confidence level = 99.43%

```
Brand = 1 subtracted from:
Brand   Lower   Center   Upper    -----+---------+---------+---------+----
2       74.44   233.13   391.82                         (--------*-------)
3      104.44   263.13   421.82                         (--------*-------)
4      -28.79   129.90   288.59                 (-------*--------)
5       40.61   199.30   357.99                    (-------*-------)
6       85.24   243.93   402.62                     (-------*-------)
                                  -----+---------+---------+---------+----
                                     -200        0       200       400
```

```
Brand = 2 subtracted from:
Brand    Lower   Center   Upper    -----+---------+---------+---------+----
3      -128.69    30.00   188.69            (-------*------)
4      -261.92  -103.23    55.46      (-------*-------)
5      -192.52   -33.83   124.86         (-------*-------)
6      -147.89    10.80   169.49          (-------*------)
                                   -----+---------+---------+---------+----
                                      -200        0       200       400
```

```
Brand = 3 subtracted from:
Brand    Lower   Center   Upper    -----+---------+---------+---------+----
4      -291.92  -133.23    25.46  (-------*--------)
5      -222.52   -63.83    94.86     (-------*-------)
6      -177.89   -19.20   139.49       (-------*-------)
                                   -----+---------+---------+---------+----
                                      -200        0       200       400
```

```
Brand = 4 subtracted from:
Brand   Lower   Center   Upper    -----+---------+---------+---------+----
5      -89.29   69.40   228.09            (------*-------)
6      -44.66  114.03   272.72              (-------*-------)
                                  -----+---------+---------+---------+----
                                     -200        0       200       400
```

```
Brand = 5 subtracted from:
Brand    Lower   Center   Upper    -----+---------+---------+---------+----
6      -114.06    44.63   203.32            (-------*-------)
                                   -----+---------+---------+---------+----
                                      -200        0       200       400
```

Based on the results of our tests, there is sufficient evidence to believe that the mean ORAC levels for the six tested brands are not equal. In particular, Brand 1, with a mean of 445, has the lowest ORAC level. Brands 2, 3, 5, and 6 have similar mean levels of ORAC, ranging from a low of 645 to a high of 709. The mean level of ORAC for Brand 4, with a mean of 576, is somewhat lower, but there is insufficient evidence to conclude that the mean ORAC level of Brand 4 is different from Brands 2, 3, 5, and 6. There was no significant difference between Brands 1 and 4.

(f) Answers will vary.

Section 13.3

1. **(a)** The treatment P-value $= 0.013$ is sufficient evidence against the null hypothesis that the population means are equal. The researcher will conclude that at least one population mean is different from the others.

   (b) The mean square error is MSE $= 17.517$.

   (c) The P-values comparing treatment 3 with treatments 1 and 2 are small (less than 0.05) while the P-value comparing treatments 1 and 2 is large. Therefore, we conclude that means 1 and 2 are the same, but mean 3 is different. That is, we conclude $\mu_1 = \mu_2 \neq \mu_3$ or $\underline{\mu_1 \ \ \mu_2} \ \ \mu_3$.

3. **(a)** The treatment P-value $= 0.108$ is not sufficient evidence against the null hypothesis that the population means are equal. In other words, the researcher does not reject the null hypothesis that population means are equal.

   (b) The mean square error is MSE $= 0.944$.

   (c) We would only use Tukey's test if the null hypothesis from part (a) had been rejected.

5. **(a)** Using MINITAB, we obtain the following output:

   ```
   Two-way ANOVA: Response versus Block, Treatment
   Source      DF      SS      MS      F       P
   Block        4   5.836   1.459   10.13   0.003
   Treatment    2   4.468   2.234   15.51   0.002
   Error        8   1.152   0.144
   Total       14  11.456
   ```

   Since the treatment P-value $= 0.002$, we reject the null hypothesis that the population means are equal. Thus, we conclude that at least one mean is different from the others.

   (b) Using MINITAB, we obtain the following output:

   ```
   Tukey 95.0% Simultaneous Confidence Intervals
   Response Variable Response
   All Pairwise Comparisons among Levels of Treatment
   Treatment = 1   subtracted from:
   Treatment    Lower    Center     Upper   --------+---------+---------+--------
   2           -1.606   -0.920   -0.2344            (---------*---------)
   3           -1.986   -1.300   -0.6144   (--------*---------)
                                           --------+---------+---------+--------
                                             -1.40      -0.70      0.00


   Treatment = 2   subtracted from:
   Treatment    Lower    Center     Upper   --------+---------+---------+--------
   3           -1.066   -0.3800   0.3056                     (---------*--------)
                                           --------+---------+---------+--------
                                             -1.40      -0.70      0.00
   ```

   ```
   Tukey Simultaneous Tests
   Response Variable Response
   All Pairwise Comparisons among Levels of Treatment
   Treatment = 1   subtracted from:
               Difference       SE of               Adjusted
   Treatment    of Means    Difference   T-Value    P-Value
   2             -0.920        0.2400    -3.833      0.0123
   3             -1.300        0.2400    -5.417      0.0016
   Treatment = 2   subtracted from:
               Difference       SE of               Adjusted
   Treatment    of Means    Difference   T-Value    P-Value
   3             -0.3800       0.2400    -1.583      0.3066
   ```

The interval for Treatment 2 minus Treatment 1 does not include 0 and the *P*-value = 0.0123 for this comparison, which is small, so reject the null hypothesis that $\mu_1 = \mu_2$. Likewise, the interval for Treatment 3 minus Treatment 1 does not include 0 and the *P*-value = 0.0016, which is small, so we reject the null hypothesis that $\mu_1 = \mu_3$. Finally, the interval for Treatment 3 minus Treatment 2 does include 0 and the *P*-value = 0.3066, which is not small, so we do not reject the null hypothesis that $\mu_2 = \mu_3$. Thus, the Tukey's tests indicate that the means of Treatments 2 and 3 are equal, but the mean of Treatment 1 is different. That is, we conclude: $\mu_1 \neq \mu_2 = \mu_3$ or $\underline{\mu_1} \quad \underline{\mu_2 \quad \mu_3}$.

(c) The boxplots shown to the right support the results found in parts (a) and (b).

7. (a) The sample standard deviations for the provided data are $s_{87\ \text{Octane}} \approx 1.240$, $s_{89\ \text{Octane}} \approx 1.439$, and $s_{92\ \text{Octane}} \approx 1.225$. Since the largest standard deviation is less than twice the smallest, $1.439 < 2(1.225)$, the requirement for equal population variances is satisfied.

(b) Using MINITAB, we obtain the following output:

```
Two-way ANOVA: Miles per Gallon versus Car and Driver, Octane level
Source          DF      SS       MS       F       P
Car and Driver   5  25.0161  5.00322  92.84   0.000
Octane level     2   1.1411  0.57056  10.59   0.003
Error           10   0.5389  0.05389
Total           17  26.6961
```

Since the treatment (octane level) *P*-value = 0.003, there is sufficient evidence to conclude that the mean number of miles per gallon is different among the three octane levels at the $\alpha = 0.05$ level of significance.

(c) Using MINITAB, we obtain the following output:

```
Tukey 95.0% Simultaneous Confidence Intervals
Response Variable Miles
All Pairwise Comparisons among Levels of Octane level
Octane level = 1  subtracted from:
Octane
level    Lower    Center   Upper    +---------+---------+---------+------
2       -0.3344  0.03333  0.4010    (-----------*----------)
3        0.1823  0.55000  0.9177                   (----------*----------)
                                    +---------+---------+---------+------
                                  -0.35      0.00      0.35      0.70
Octane level = 2  subtracted from:
Octane
level    Lower    Center   Upper    +---------+---------+---------+------
3        0.1490   0.5167   0.8844                   (----------*---------)
                                    +---------+---------+---------+------
                                  -0.35      0.00      0.35      0.70
```

```
Tukey Simultaneous Tests
Response Variable Miles per Gallon
All Pairwise Comparisons among Levels of Octane level
Octane level = 1  subtracted from:
Octane  Difference        SE of                Adjusted
level   of Means    Difference  T-Value    P-Value
2         0.03333       0.1340   0.2487      0.9666
3         0.55000       0.1340   4.1037      0.0055


Octane level = 2  subtracted from:
Octane  Difference        SE of                Adjusted
level   of Means    Difference  T-Value    P-Value
3          0.5167       0.1340    3.855      0.0081
```

The interval for 89 octane minus 87 octane includes 0 and the *P*-value = 0.9666 for this comparison, which is not small, so we do not reject the null hypothesis that $\mu_{87} = \mu_{89}$. The interval for 92 octane minus 87 octane does not include 0 and the *P*-value = 0.0055, which is small, so we reject the null hypothesis that $\mu_{87} = \mu_{92}$. Finally, the interval for 92 octane minus 89 octane does not include 0 and the *P*-value = 0.0081, which is small, so we reject the null hypothesis that $\mu_{89} = \mu_{92}$. We conclude

$$\mu_{87} = \mu_{89} \neq \mu_{92} \text{ or } \underline{\mu_{87} \quad \mu_{89}} \quad \mu_{92}$$

(d) The Tukey's tests indicate that the mean gasoline mileages for 87 octane and 89 octane are equal, but the mean gasoline mileage for 92 octane is different. In fact, since the confidence intervals for 92 octane minus 87 octane and 92 octane minus 87 octane contain all positive values, we can conclude that the mean gasoline mileage for 92 octane is higher than for 87 octane and 89 octane.

9. (a) The sample standard deviations for the provided data are $s_{\text{Jeep}} \approx 448$, $s_{\text{Saturn}} \approx 453$, $s_{\text{Toyota}} \approx 830$, and $s_{\text{Hyundai}} \approx 614$. Since the largest standard deviation is less than twice the smallest, $830 < 2(448)$, the requirement for equal population variances is satisfied.

(b) Using MINITAB, we obtain the following output:

Two-way ANOVA: Cost versus Location, Brand
```
Source     DF      SS        MS       F      P
Location    3  3404468   1134823   10.09   0.003
Brand       3   926488    308829    2.75   0.105
Error       9  1012007    112445
Total      15  5342964
```

Since the treatment (brand) *P*-value = 0.105, there is not sufficient evidence to conclude that the mean cost of repair is different among the car brands at the $\alpha = 0.05$ level of significance.

(c) Since the null hypothesis was not rejected, there is no need to do a post hoc analysis.

11. (a) The sample standard deviations for the provided data are $s_{\text{lift-off}-1} \approx 5.65$, $s_{\text{return}+1} \approx 9.97$, and $s_{+1\text{ month}} \approx 9.00$. Since the largest standard deviation is less than twice the smallest, $9.97 < 2(5.65)$, the requirement for equal population variances is satisfied.

(b) Using MINITAB, we obtain the following output:

Two-way ANOVA: Water Consumption versus Rat, Day
```
Source    DF       SS       MS       F       P
Rat        5   861.78  172.356    8.62   0.002
Day        2   635.44  317.722   15.89   0.001
Error     10   199.89   19.989
Total     17  1697.11
```

Since the treatment (day) *P*-value = 0.001, there is sufficient evidence to conclude that the mean water consumption of rats is different among the experiment levels at the $\alpha = 0.05$ level of significance.

(c) Using MINITAB with Treatment 1 representing Lift-off Minus 1, Treatment 2 representing Return Plus 1, and Treatment 3 representing Return Plus 1 Month, we obtain the following output:

```
Tukey 95.0% Simultaneous Confidence Intervals
Response Variable Water Consumption
All Pairwise Comparisons among Levels of Day
Day = 1   subtracted from:
Day    Lower   Center   Upper    -+---------+---------+---------+-----
2    -0.9152   6.167   13.25     (---------*---------)
3     7.4181  14.500   21.58                    (---------*---------)
                                 -+---------+---------+---------+-----
                                 0.0       7.0      14.0      21.0

Day = 2   subtracted from:
Day  Lower   Center   Upper     -+---------+---------+---------+-----
3    1.251    8.333   15.42      (---------*---------)
                                 -+---------+---------+---------+-----
                                 0.0       7.0      14.0      21.0

Tukey Simultaneous Tests
Response Variable Water Consumption
All Pairwise Comparisons among Levels of Day
Day = 1   subtracted from:
        Difference       SE of                Adjusted
Day     of Means    Difference   T-Value     P-Value
2          6.167         2.581     2.389      0.0885
3         14.500         2.581     5.617      0.0006
Day = 2   subtracted from:
        Difference       SE of                Adjusted
Day     of Means    Difference   T-Value     P-Value
3          8.333         2.581     3.228      0.0225
```

The interval for Return Plus 1 minus Lift-off minus 1 includes 0 and the *P*-value = 0.0885 for this comparison, which is not small, so we do not reject the null hypothesis that $\mu_{\text{lift-off}-1} = \mu_{\text{return}+1}$. The interval for Return Plus 1 Month minus Lift-off Minus 1 does not include 0 and the *P*-value = 0.0006, which is small, so we reject the null hypothesis that $\mu_{\text{lift-off}-1} = \mu_{\text{return}+1\ \text{month}}$. Finally, the interval for Return Plus 1 Month minus Return Plus 1 does not include 0 and the *P*-value = 0.0225, which is small, so we reject the null hypothesis that $\mu_{\text{return}+1} = \mu_{\text{return}+1\ \text{month}}$. Thus, we conclude that

$$\mu_{\text{lift-off}-1} = \mu_{\text{return}+1} \neq \mu_{\text{return}+1\ \text{month}} \quad \text{or} \quad \underline{\mu_{\text{lift-off}-1}} \quad \underline{\mu_{\text{return}+1}} \quad \underline{\mu_{\text{return}+1\ \text{month}}}.$$

13. (a) Using MINITAB, we obtain the following output:

Two-way ANOVA: Wait Time versus Day, Procedure

```
Source       DF       SS        MS        F       P
Day           6  6538.75   1089.79   249.50   0.000
Procedure     1     3.02      3.02     0.69   0.438
Error         6    26.21      4.37
Total        13  6567.97
```

Since the treatment (procedure) *P*-value = 0.438, there is not sufficient evidence to conclude that the mean wait time is different for the two loading procedures at the $\alpha = 0.05$ level of significance.

(b)

Observation	1	2	3	4	5	6	7
Wait time before new procedure	11.6	25.9	20.0	38.2	57.3	32.1	81.8
Wait time after new procedure	10.7	28.3	19.2	35.9	59.2	31.8	75.3
d_i = after − before		−0.9	2.4	−0.8	−2.3	1.9	−0.3	−6.5

Using technology, $\bar{d} \approx -0.93$ minutes and $s_d \approx 2.96$ minutes, where we measure differences as wait time after new procedure minus wait time before new procedure. The hypotheses are $H_0 : \mu_d = 0$ versus $H_1 : \mu_d \neq 0$. The level of significance is $\alpha = 0.05$. The test statistic is

$$t_0 = \frac{\bar{d}}{s_d / \sqrt{n}} = \frac{-0.93}{2.96 / \sqrt{7}} \approx -0.83.$$

Classical approach: Since this is a two-tailed test with 6 degrees of freedom, the critical values are $\pm t_{0.025} = \pm 2.447$. Since the test statistic $t_0 \approx -0.83$ falls between the critical values $\pm t_{0.025} = \pm 2.447$ (i.e., since the test statistic does not fall within the critical region), we do not reject H_0.

P-value approach: The P-value for this two-tailed test is the area under the t-distribution with 6 degrees of freedom to the left of the test statistic $t_0 = -0.83$ plus the area to the right of $t_0 = 0.83$. From the t-distribution table in the row corresponding to 6 degrees of freedom, 0.83 falls between 0.718 and 0.906 whose right-tail areas are 0.25 and 0.20, respectively. We must double these values in order to get the total area in both tails: 0.50 and 0.40. So, $0.40 < P\text{-value} < 0.50$. (Using technology, we find P-value ≈ 0.438.) Because the P-value is greater than the level of significance $\alpha = 0.05$, we do not reject H_0.

Conclusion: There is not sufficient evidence at the $\alpha = 0.05$ level of significance to conclude that the new loading procedure changed the wait time.

(c) The P-values for the two approaches are both 0.438. Yes, it appears that the method presented in this section is a generalization of the matched-pair t-test.

15. Blocking is partitioning the experimental units according to some factor whose level cannot be fixed. Blocking removes variability in the response variable due to the blocked variable.

17. The requirements needed to analyze a randomized complete block design are:

(1) The response variable for each of the k populations is normally distributed.

(2) The response variable for each of the k populations has the same variance.

19. If the experimental units are dependent, then the related experimental units form the blocks of the experiment.

Section 13.4

1. 2; 3

3. interaction effect

5. The plots are relatively parallel, so there is no significant interaction among the factors.

7. The plots cross, so there is significant interaction among the factors.

9. (a) Main effect of factor $A = \dfrac{20+14+1+3}{4} - \dfrac{12+8+6+7}{4} = 1.25$;

Main effect of factor $B = \dfrac{6+7+1+3}{4} - \dfrac{12+8+20+14}{4} = -9.25$.

Interaction effects: If factor B is fixed at low and factor A changes from low to high, the response variable increases by $\dfrac{20+14}{2} - \dfrac{12+8}{2} = 7$ units. On the other hand, if factor B is fixed at high and factor

A changes from low to high, the response variable increases by $\dfrac{1+3}{2} - \dfrac{6+7}{2} = -4.5$ (decreases by 4.5) units.

So the change in the response variable for factor A depends on the level of factor B, meaning there is an interaction effect between factors A and B. If factor A is fixed at low and factor B changes from low to high, the response variable increases by $\dfrac{6+7}{2}-\dfrac{12+8}{2}=-3.5$ (decreases by 3.5) units. On the other hand, if factor A is fixed at high and factor B changes from low to high, the response variable increases by $\dfrac{3+1}{2}-\dfrac{20+14}{2}=-15$ (decreases by 15) units. So the change in the response variable for factor B depends on the level of factor A, meaning there is an interaction effect between factors A and B.

(b)

11. **(a)** Since the interaction P-value is 0.836, which is large, there is no evidence of an interaction effect.

(b) Since the P-value for factor A is 0.003, which is small, there is evidence of a difference in means from factor A. Since the P-value for factor B is < 0.001, which is small, there is also evidence of a difference in means from factor B.

(c) MSE = 45.3

13. **(a)** Since the interaction P-value is < 0.001, which is small, there is evidence of an interaction effect.

(b) In part (a), we found evidence of an interaction effect. Therefore, we do not consider main effects because they could be misleading.

(c) MSE = 31.9

15. Using MINITAB with the provided data, we obtain the following output:

Two-way ANOVA: Response versus Factor A, Factor B

Source	DF	SS	MS	F	P
Factor A	1	30.15	30.150	0.32	0.579
Factor B	2	912.90	456.452	4.84	0.021
Interaction	2	6.80	3.402	0.04	0.965
Error	18	1699.27	94.404		
Total	23	2649.13			

(a) Since the interaction P-value is 0.965, which is large, there is no evidence of an interaction effect.

(b) Since the P-value for factor A is 0.579, which is large, there is no evidence of a difference in the means from factor A. Since the P-value for factor B is 0.021, which is small, there is evidence of a difference in the means from factor B.

(c) The lines in the interaction plot are close to parallel, supporting the conclusion that there is no interaction effect.

(d) The Tukey's test for factor A is unnecessary since there is no evidence for difference in means. For factor B, we use MINITAB to perform Tukey's test. The results are:

Tukey 95.0% Simultaneous Confidence Intervals

```
Response Variable Response
All Pairwise Comparisons among Levels of Factor B
Factor B = 1  subtracted from:
Factor B   Lower  Center   Upper   --+---------+---------+---------+----
2         -18.55  -6.150   6.251   (-------*-------)
3          -3.53   8.875  21.276                (-------*-------)
                                   --+---------+---------+---------+----
                                   -15        0        15        30


Factor B = 2  subtracted from:
Factor B   Lower  Center   Upper   --+---------+---------+---------+----
3          2.624   15.03   27.43                   (-------*-------)
                                   --+---------+---------+---------+----
                                   -15        0        15        30
```

Tukey Simultaneous Tests
Response Variable Response
All Pairwise Comparisons among Levels of Factor B

```
Factor B = 1  subtracted from:
           Difference      SE of            Adjusted
Factor B   of Means    Difference  T-Value  P-Value
2            -6.150       4.858     -1.266   0.4317
3             8.875       4.858      1.827   0.1893
Factor B = 2  subtracted from:
           Difference      SE of            Adjusted
Factor B   of Means    Difference  T-Value  P-Value
3            15.03        4.858      3.093   0.0165
```

The interval for $B_2 - B_1$ includes 0 and the P-value = 0.4317 for this comparison, which is large, so we do not reject the null hypothesis that $B_1 = B_2$. Likewise, the interval for $B_3 - B_1$ includes 0 and the P-value = 0.1893, which is large, so we do not reject the null hypothesis that $B_1 = B_3$. Finally, the interval for $B_3 - B_2$ does not include 0 and the P-value = 0.0165, which is small, so we reject the null hypothesis that $B_2 = B_3$. Thus, the Tukey's test for factor B is ambiguous. It shows $B_1 = B_2$ and $B_1 = B_3$ but $B_2 \neq B_3$. That is, $\underline{B_2 \quad B_1} \quad B_3$.

17. (a) This is a 2×3 factorial design with 2 replications within each cell.

(b) Using MINITAB with the provided data, we obtain the following output:

Two-way ANOVA: Cholesterol versus Gender, Age

Source	DF	SS	MS	F	P
Gender	1	310.08	310.08	2.51	0.164
Age	2	2177.17	1088.58	8.81	0.016
Interaction	2	628.17	314.08	2.54	0.159
Error	6	741.50	123.58		
Total	11	3856.92			

Since the interaction P-value is 0.159, which is large, there is no evidence of an interaction effect.

(c) Since the P-value for gender is 0.164, which is large, there is no evidence of a difference in the means for the two genders. Since the P-value for age is 0.016, which is small, there is evidence of a difference in the means for the three age groups.

(d) The plots are not parallel, so there is significant interaction among the factors. This result contradicts the result found in part (b).

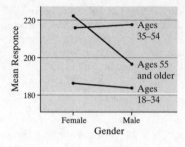

(e) Because a normal probability plot is roughly linear, it is reasonable to conclude the residuals are normally distributed.

(f) The Tukey's test for gender is unnecessary since there is no evidence for difference in means for the two genders. For age, we use MINITAB to perform Tukey's test:

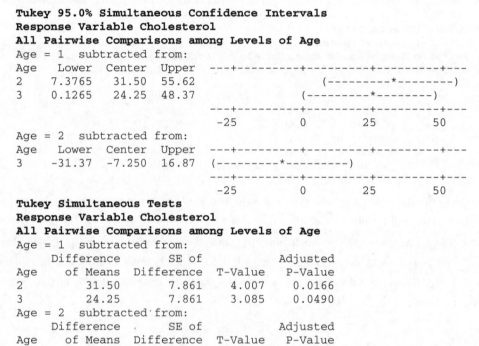

```
Tukey 95.0% Simultaneous Confidence Intervals
Response Variable Cholesterol
All Pairwise Comparisons among Levels of Age
Age = 1  subtracted from:
Age   Lower   Center  Upper   ---+---------+---------+---------+---
2     7.3765  31.50   55.62                   (---------*--------)
3     0.1265  24.25   48.37          (---------*--------)
                              ---+---------+---------+---------+---
                               -25        0         25        50

Age = 2  subtracted from:
Age   Lower   Center  Upper   ---+---------+---------+---------+---
3     -31.37  -7.250  16.87   (----------*----------)
                              ---+---------+---------+---------+---
                               -25        0         25        50
```

```
Tukey Simultaneous Tests
Response Variable Cholesterol
All Pairwise Comparisons among Levels of Age
Age = 1  subtracted from:
      Difference     SE of              Adjusted
Age   of Means    Difference  T-Value  P-Value
2        31.50       7.861     4.007    0.0166
3        24.25       7.861     3.085    0.0490
Age = 2  subtracted from:
      Difference     SE of              Adjusted
Age   of Means    Difference  T-Value  P-Value
3       -7.250       7.861    -0.9223   0.6473
```

The interval for $\mu_{35-54} - \mu_{18-34}$ does not include 0 and the *P*-value = 0.0166 for this comparison, which is small, so we reject the null hypothesis that $\mu_{18-34} = \mu_{35-54}$. Likewise, the interval for $\mu_{55+} - \mu_{18-34}$ does not include 0 and the *P*-value = 0.0490, which is small, so we reject the null hypothesis that $\mu_{18-34} = \mu_{55+}$. Finally, the interval for $\mu_{55+} - \mu_{35-54}$ includes 0 and the *P*-value = 0.6473, which is large, so we do not reject the null hypothesis that $\mu_{35-54} = \mu_{55+}$. Thus, the Tukey's test for age shows $\mu_{18-34} \neq \mu_{35-54} = \mu_{55+}$. That is, $\underline{\mu_{18-34}} \quad \underline{\mu_{35-54} \quad \mu_{55+}}$.

19. (a) Using technology, we find the sample standard deviation of the data within each cell:
$s_{3.75/301} \approx 311$, $s_{3.75/400} \approx 320$, $s_{3.75/353} \approx 257$, $s_{4/301} \approx 298$, $s_{4/400} \approx 250$, $s_{4/353} \approx 168$, $s_{5/301} \approx 236$, $s_{5/400} \approx 222$, and $s_{5/353} \approx 267$. The largest standard deviation (320) is less than twice the smallest standard deviation (168), so the requirement of equal population variances is satisfied. [320 < 2(168).]

(b) Using MINITAB with the provided data, we obtain the following output:

Two-way ANOVA: Strength versus Slump, Mixture

Source	DF	SS	MS	F	P
Slump	2	1234813	617406	8.95	0.002
Mixture	2	1824169	912084	13.23	0.000
Interaction	4	256415	64104	0.93	0.469
Error	18	1241100	68950		
Total	26	4556496			

Since the interaction P-value is 0.469, which is large, there is no evidence of an interaction effect.

(c) Since the P-value for slump is 0.002, which is small, there is sufficient evidence of difference in the means for the three types of slump. Since the P-value for mixture is < 0.001, which is small, there is evidence of a difference in the means for the three mixtures.

(d)

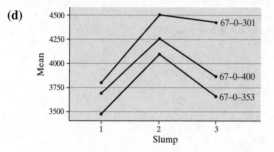

(e) Because a normal probability plot is roughly linear, it is reasonable to conclude the residuals are normally distributed.

(f) We use MINITAB to perform Tukey's test. The results are:

```
Tukey 95.0% Simultaneous Confidence Intervals
Response Variable Strength
All Pairwise Comparisons among Levels of Slump
Slump = 1  subtracted from:
Slump  Lower   Center  Upper   --------+---------+---------+---------
2      -634.3  -318.3   -2.4              (---------*----------)
3      -835.4  -519.4  -203.5   (----------*---------)
                               --------+---------+---------+---------
                                    -600      -300        0

Slump = 2  subtracted from:
Slump  Lower   Center  Upper   --------+---------+---------+---------
3      -517.1  -201.1   114.9                 (---------*----------)
                               --------+---------+---------+---------
                                    -600      -300        0
```

```
Tukey Simultaneous Tests
Response Variable Strength
All Pairwise Comparisons among Levels of Slump
Slump = 1  subtracted from:
        Difference    SE of              Adjusted
Slump   of Means   Difference  T-Value   P-Value
2        -318.3      123.8      -2.572    0.0481
3        -519.4      123.8      -4.196    0.0015
Slump = 2  subtracted from:
        Difference    SE of              Adjusted
Slump   of Means   Difference  T-Value   P-Value
3        -201.1      123.8      -1.625    0.2612
```

```
Tukey 95.0% Simultaneous Confidence Intervals
Response Variable Strength
All Pairwise Comparisons among Levels of Mixture
Mixture = 1  subtracted from:
Mixture   Lower   Center  Upper   ---+---------+---------+---------+---
2        320.691  636.7   952.6                       (------*-----)
3         -2.086  313.9   629.9                 (-----*------)
                                  ---+---------+---------+---------+---
                                  -500        0        500       1000

Mixture = 2  subtracted from:
Mixture   Lower   Center  Upper   ---+---------+---------+---------+---
3        -638.8  -322.8  -6.803   (------*-----)
                                  ---+---------+---------+---------+---
                                  -500        0        500       1000
```

Tukey Simultaneous Tests
Response Variable Strength
All Pairwise Comparisons among Levels of Mixture

```
Mixture = 1  subtracted from:
          Difference    SE of            Adjusted
Mixture   of Means   Difference  T-Value  P-Value
2            636.7      123.8     5.143    0.0002
3            313.9      123.8     2.536    0.0517
Mixture = 2  subtracted from:
          Difference    SE of            Adjusted
Mixture   of Means   Difference  T-Value  P-Value
3           -322.8      123.8    -2.608    0.0448
```

Tukey's test for Slump: The interval for $\mu_4 - \mu_{3.75}$ does not include 0 and the P-value = 0.0481 for this comparison, which is small, so we reject the null hypothesis that $\mu_{3.75} = \mu_4$. Next, the interval for $\mu_5 - \mu_{3.75}$ does not include 0 and the P-value = 0.0015, which is small, so we reject the null hypothesis that $\mu_{3.75} = \mu_5$. Finally, the interval for $\mu_5 - \mu_4$ includes 0 and the P-value = 0.2612, which is large, so we do not reject the null hypothesis that $\mu_4 = \mu_5$. Thus, we have $\mu_{3.75} \neq \mu_4 = \mu_5$ or $\underline{\mu_{3.75}} \quad \underline{\mu_4 \quad \mu_5}$. That is, the Tukey's test indicates that the mean 28-day strength for slump 3.75 is significantly different from the mean 28-day strength for slump 4 and slump 5.

Tukey's test for Mixture: The interval for $\mu_{400} - \mu_{301}$ does not include 0 and the P-value = 0.0002, which is small, so we reject the null hypothesis that $\mu_{301} = \mu_{400}$. Next, the interval for $\mu_{353} - \mu_{301}$ includes 0 and the P-value = 0.0517, which is not small, so we do not reject the null hypothesis that $\mu_{301} = \mu_{353}$. Finally, the interval for $\mu_{353} - \mu_{400}$ does not include 0 and the P-value = 0.0448, which is small, so we reject the null hypothesis that $\mu_{400} = \mu_{353}$. Thus, we have $\mu_{301} = \mu_{353} \neq \mu_{400}$ or $\underline{\mu_{301} \quad \mu_{353}} \quad \underline{\mu_{400}}$. That is, the Tukey's test indicates that the mean 28-day strength of mixture 67-0-400 is significantly different from the mean 28-day strength of the other two mixtures.

21. **(a)** Using technology, we find the sample standard deviation the data within each cell:
$s_{\text{specialty / Chicago}} \approx 4.00$, $s_{\text{specialty / Bolingbrook}} \approx 3.20$, $s_{\text{specialty / Peoria}} \approx 3.63$, $s_{\text{general / Chicago}} \approx 2.50$, $s_{\text{general / Bolingbrook}} \approx 3.91$, , and $s_{\text{general / Peoria}} \approx 2.52$. The largest standard deviation (4.00) is less than twice the smallest standard deviation (2.50), so the requirement of equal population variances is satisfied. [4.00 < 2(2.50).]

(b) Using MINITAB with the provided data, we obtain the following output:

Two-way ANOVA: Cost versus Type, Location

Source	DF	SS	MS	F	P
Type	1	0.720	0.7200	0.06	0.804
Location	2	31.126	15.5630	1.39	0.287
Interaction	2	3.573	1.7867	0.16	0.854
Error	12	134.560	11.2134		
Total	17	169.980			

Since the interaction P-value is 0.854, which is large, there is no evidence of an interaction effect.

(c) Since the P-value for type is 0.804, which is large, there is no evidence of a difference in the means for the type of service center. Since the P-value for location is 0.287, which is large, there is no evidence of a difference in the means for the locations.

(d)

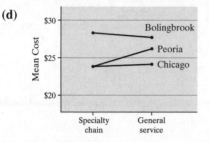

(e) Since there were no significant main effects, there is no need to do Tukey's tests.

23. In a completely randomized design, the researcher examines a single factor fixed at k levels and randomly distributes the experimental units among the levels. In a randomized complete block design, there is a factor whose level cannot be fixed. So the researcher partitions the experimental units according to this factor, forming blocks. The experimental units are then randomly assigned to the blocks, being sure each block has each treatment. In a factorial design, the researcher has two variables A and B, called factor A and factor B. The factors are fixed, so factor A has a levels and factor B has b levels. The experimental units are then uniformly distributed among the cells formed by the factors, and the response variable is measured at each of the ab cells.

24. An interaction effect occurs if changing the level of one factor results in different changes in the value of the response variable for the different levels of the second factor. A significant interaction effect might cause the main effects to be misleading, so the main effects should not be analyzed.

25. An interaction plot graphically displays the mean response for each treatment (combination of factor levels). The plots are useful for illustrating potential interaction between the factors.

Chapter 13 Review Exercises

1. (a) We look in the table of critical values for Tukey's Test with $\alpha = 0.05$ and where the row corresponding to $\nu = 16$ intersects with the column corresponding to $k = 7$ to find the critical value $q_{0.05,16,7} = 4.741$.

(b) We look in the table of critical values for Tukey's Test with $\alpha = 0.05$ and where the row corresponding to $\nu = 30$ intersects with the column corresponding to $k = 6$ to find the critical value $q_{0.05,30,6} = 4.302$.

(c) We look in the table of critical values for Tukey's Test with $\alpha = 0.01$ and where the row corresponding to $\nu = 40$ (since there is no row in the table for $\nu = 42$) intersects with the column corresponding to $k = 4$ to find the critical value $q_{0.01,42,4} \approx q_{0.01,40,4} = 4.696$.

(d) For this test, $k = 6$ and $\nu = n - k = 46 - 6 = 40$, so we look in the table with $\alpha = 0.05$ and where the row corresponding to $\nu = 40$ intersects with the column corresponding to $k = 6$ to find the critical value $q_{0.05,40,6} = 4.232$.

2. Arrange the sample means in ascending order: $\bar{x}_4 = 47$, $\bar{x}_2 = 64$, $\bar{x}_3 = 70$, $\bar{x}_1 = 78$. Now $n = 31 \cdot 4 = 124$ and $k = 4$, so $\nu = n - k = 120$ and the critical value is $q_{0.05,120,4} = 3.685$. We first test $H_0 : \mu_1 = \mu_4$. The

test statistic is $q_0 = \dfrac{\bar{x}_1 - \bar{x}_4}{\sqrt{\dfrac{s^2}{2} \cdot \left(\dfrac{1}{n_1} + \dfrac{1}{n_4}\right)}} = \dfrac{78 - 47}{\sqrt{\dfrac{373}{2} \cdot \left(\dfrac{1}{31} + \dfrac{1}{31}\right)}} \approx 8.937$. Since this test statistic is greater than

$q_{0.05,120,4} = 3.685$, we reject $H_0 : \mu_1 = \mu_4$.

Second, we test $H_0 : \mu_1 = \mu_2$. The test statistic is $q_0 = \dfrac{\bar{x}_1 - \bar{x}_2}{\sqrt{\frac{s^2}{2} \cdot \left(\frac{1}{n_1} + \frac{1}{n_2} \right)}} = \dfrac{78 - 64}{\sqrt{\frac{373}{2} \cdot \left(\frac{1}{31} + \frac{1}{31} \right)}} \approx 4.036$. Since this

test statistic is greater than $q_{0.05,120,4} = 3.685$, we reject $H_0 : \mu_1 = \mu_2$.

Third, we test $H_0 : \mu_1 = \mu_3$. The test statistic is $q_0 = \dfrac{\bar{x}_1 - \bar{x}_3}{\sqrt{\frac{s^2}{2} \cdot \left(\frac{1}{n_1} + \frac{1}{n_3} \right)}} = \dfrac{78 - 70}{\sqrt{\frac{373}{2} \cdot \left(\frac{1}{31} + \frac{1}{31} \right)}} \approx 2.306$. Since this test

statistic is less than $q_{0.05,120,4} = 3.685$, we do not reject $H_0 : \mu_1 = \mu_3$.

Fourth, we test $H_0 : \mu_3 = \mu_4$. The test statistic is $q_0 = \dfrac{\bar{x}_3 - \bar{x}_4}{\sqrt{\frac{s^2}{2} \cdot \left(\frac{1}{n_3} + \frac{1}{n_4} \right)}} = \dfrac{70 - 47}{\sqrt{\frac{373}{2} \cdot \left(\frac{1}{31} + \frac{1}{31} \right)}} \approx 6.631$. Since this test

statistic is greater than $q_{0.05,120,4} = 3.685$, we reject $H_0 : \mu_3 = \mu_4$.

Fifth, we test $H_0 : \mu_3 = \mu_2$. The test statistic is $q_0 = \dfrac{\bar{x}_3 - \bar{x}_2}{\sqrt{\frac{s^2}{2} \cdot \left(\frac{1}{n_3} + \frac{1}{n_2} \right)}} = \dfrac{70 - 64}{\sqrt{\frac{373}{2} \cdot \left(\frac{1}{31} + \frac{1}{31} \right)}} \approx 1.730$. Since this test

statistic is less than $q_{0.05,120,4} = 3.685$, we do not reject $H_0 : \mu_3 = \mu_2$.

Finally, we test $H_0 : \mu_2 = \mu_4$. The test statistic is $q_0 = \dfrac{\bar{x}_2 - \bar{x}_4}{\sqrt{\frac{s^2}{2} \cdot \left(\frac{1}{n_2} + \frac{1}{n_4} \right)}} = \dfrac{64 - 47}{\sqrt{\frac{373}{2} \cdot \left(\frac{1}{31} + \frac{1}{31} \right)}} \approx 4.901$. Since this

test statistic is greater than $q_{0.05,120,4} = 3.685$, we reject $H_0 : \mu_2 = \mu_4$.

The results of the test are ambiguous since we have that $\mu_1 \neq \mu_2$, but $\mu_1 = \mu_3 \neq \mu_4$ and $\mu_2 = \mu_3 \neq \mu_4$.

3. **(a)** The hypotheses are: $H_0 : \mu_{\text{organic}} = \mu_{\text{mineral}} = \mu_{\text{surface}}$ versus H_1 : At least one of the means is different from the others.

(b) The largest standard deviation (6.246) is less than twice the smallest standard deviation (3.927), so the requirement of equal population variances is satisfied.
[6.246 < 2(3.927).]

(c) Since P-value = 0.0002 is smaller than the $\alpha = 0.01$ level of significance, the researcher will reject the null hypothesis and conclude that the mean concentration of dissolved carbon is different in at least one of the collection areas.

(d) Results as extreme as these would only occur in about 2 out of 10,000 similar samples if the null hypothesis were true.

(e) Yes, the boxplots support the conclusion that at least one of the means is different. The plot for water collected from mineral soil appears to differ from the other two.

(f) From the MINITAB output we have: $-7.588 < \mu_{\text{mineral}} - \mu_{\text{organic}} < -2.092$;

$-4.606 < \mu_{\text{surface}} - \mu_{\text{organic}} < 0.964$; and $0.527 < \mu_{\text{surface}} - \mu_{\text{mineral}} < 5.510$.

The first interval does not contain 0, so we reject $H_0 : \mu_{\text{organic}} = \mu_{\text{mineral}}$ ..

The second interval contains 0, so we do not reject $H_0 : \mu_{\text{organic}} = \mu_{\text{surface}}$.

The third interval does not contain 0, so we reject $H_0 : \mu_{\text{mineral}} = \mu_{\text{surface}}$. Thus, we have

$\mu_{\text{organic}} = \mu_{\text{surface}} \neq \mu_{\text{mineral}}$ or $\underline{\mu_{\text{organic}} \quad \mu_{\text{surface}}} \quad \underline{\mu_{\text{mineral}}}$. Thus, the Tukey's test indicates that the mean dissolved carbon is equal in the surface water and in the ground water taken from organic soil, but it differs from the mean dissolved carbon concentration in ground water taken from mineral soil.

4. From the MINITAB output we have: $-6.914 < \mu_{\text{select structural}} - \mu_{\text{no. 1}} < 13.064$;

$1.936 < \mu_{\text{select structural}} - \mu_{\text{no. 2}} < 21.914$; $-2.989 < \mu_{\text{select structural}} - \mu_{\text{below grade}} < 16.989$;

$-1.139 < \mu_{\text{no. 1}} - \mu_{\text{no. 2}} < 18.839$; $-6.064 < \mu_{\text{no. 1}} - \mu_{\text{below grade}} < 13.914$; and

$-14.914 < \mu_{\text{no. 2}} - \mu_{\text{below grade}} < 5.064$. The first interval contains 0, so we do not reject $H_0 : \mu_1 = \mu_2$ .. The second interval does not contain 0, so we reject $H_0 : \mu_1 = \mu_3$. The third interval contains 0, so we do not reject $H_0 : \mu_1 = \mu_4$. The fourth interval contains 0, so we do not reject $H_0 : \mu_2 = \mu_3$. The fifth interval contains 0, so we do not reject $H_0 : \mu_2 = \mu_4$. The sixth interval contains 0, so we do not reject $H_0 : \mu_3 = \mu_4$. Thus, Tukey's test is ambiguous. It indicates that $\mu_{\text{select structural}} \neq \mu_{\text{no. 2}}$, but that

$\mu_{\text{select structural}} = \mu_{\text{no. 1}} = \mu_{\text{no. 2}} = \mu_{\text{below grade}}$ or $\underline{\mu_{\text{select structural}} \quad \mu_{\text{no. 1}} \quad \mu_{\text{below grade}} \quad \mu_{\text{no. 2}}}$.

5. (a) The hypotheses are $H_0 : \mu_{\text{front}} = \mu_{\text{middle}} = \mu_{\text{back}}$ versus H_1 : at least one mean differs from the others. Using MINITAB, we obtain the following output:

One-way ANOVA: Front, Middle, Back

Source	DF	SS	MS	F	P
Factor	2	2.350	1.175	7.50	0.003
Error	27	4.229	0.157		
Total	29	6.580			

The *P*-value is 0.003, so we reject the null hypothesis and conclude that at least one of the means is different.

(b) Using MINITAB, we obtain the following output:

```
Tukey 95% Simultaneous Confidence Intervals
All Pairwise Comparisons
Individual confidence level = 98.04%
Front subtracted from:
           Lower    Center   Upper     --+---------+---------+---------+-------
Middle   -0.9779   -0.5386  -0.0993       (--------*--------)
Back     -1.0760   -0.6367  -0.1974    (--------*--------)
                                       --+---------+---------+---------+-------
                                      -1.00     -0.50      0.00      0.50

Middle subtracted from:
           Lower    Center   Upper     --+---------+---------+---------+-------
Back     -0.5374   -0.0981   0.3412                   (--------*--------)
                                       --+---------+---------+---------+-------
                                      -1.00     -0.50      0.00      0.50
```

From the output we have: $-0.9779 < \mu_{\text{middle}} - \mu_{\text{front}} < -0.0993$; $-1.0760 < \mu_{\text{back}} - \mu_{\text{front}} < -0.1974$; and $-0.5374 < \mu_{\text{back}} - \mu_{\text{middle}} < 0.3412$.

The first interval does not contain 0, so we reject $H_0 : \mu_{\text{front}} = \mu_{\text{middle}}$. Likewise, the second interval does not contain 0, so we reject $H_0 : \mu_{\text{front}} = \mu_{\text{back}}$. Finally, the third interval contains 0, so we do not reject $H_0 : \mu_{\text{middle}} = \mu_{\text{back}}$. Thus, we have $\mu_{\text{front}} \neq \mu_{\text{middle}} = \mu_{\text{back}}$ or $\underline{\mu_{\text{front}} \quad \mu_{\text{middle}} \quad \mu_{\text{back}}}$. That is, there is evidence that students who choose seats in the front of the mathematics classroom have a different mean GPA from that of students choosing seats in the middle or the back of the classroom.

(c) The boxplots shown to the right support the result of parts (a) and (b), indicating that GPAs of students who sit in front are generally higher than students who sit in the middle and back.

6. (a) Using MINITAB, we obtain the following output:

Two-way ANOVA: Distance versus Person, Catapult

Source	DF	SS	MS	F	P
Person	3	45.688	15.2292	16.01	0.001
Catapult	3	105.688	35.2292	37.03	0.000
Error	9	8.563	0.9514		
Total	15	159.938			

Since the treatment (catapult) *P*-value < 0.001, we reject the null hypothesis that the population means are equal. Thus, we conclude that at least one mean is different from the others.

(b) Using MINITAB , we obtain the following output:

```
Tukey 95.0% Simultaneous Confidence Intervals
Response Variable Distance
All Pairwise Comparisons among Levels of Catapult
Catapult = 1  subtracted from:
Catapult  Lower  Center  Upper   ------+---------+---------+---------+
2        -1.656  0.5000  2.656   (-------*------)
3         1.844  4.0000  6.156                (------*--------)
4         4.094  6.2500  8.406                       (------*------)
                                 ------+---------+---------+---------+
                                     0.0       3.0       6.0       9.0

Catapult = 2  subtracted from:
Catapult  Lower  Center  Upper   ------+---------+---------+---------+
3         1.344  3.500   5.656           (--------*-------)
4         3.594  5.750   7.906                    (------*------)
                                 ------+---------+---------+---------+
                                     0.0       3.0       6.0       9.0

Catapult = 3  subtracted from:
Catapult   Lower   Center  Upper  ------+---------+---------+---------+
4         0.09438  2.250   4.406        (------*--------)
                                  ------+---------+---------+---------+
                                      0.0       3.0       6.0       9.0
```

```
Tukey Simultaneous Tests
Response Variable Distance
All Pairwise Comparisons among Levels of Catapult
Catapult = 1  subtracted from:
          Difference    SE of              Adjusted
Catapult  of Means   Difference  T-Value   P-Value
2          0.5000     0.6897     0.7249    0.8847
3          4.0000     0.6897     5.7996    0.0012
4          6.2500     0.6897     9.0618    0.0000
Catapult = 2  subtracted from:
          Difference    SE of              Adjusted
Catapult  of Means   Difference  T-Value   P-Value
3          3.500      0.6897     5.075     0.0031
4          5.750      0.6897     8.337     0.0001
Catapult = 3  subtracted from:
          Difference    SE of              Adjusted
Catapult  of Means   Difference  T-Value   P-Value
4          2.250      0.6897     3.262     0.0405
```

The interval for $\mu_2 - \mu_1$ includes 0 and the *P*-value = 0.8847 for this comparison, which is large, so we do not reject the null hypothesis that $\mu_1 = \mu_2$. The interval for $\mu_3 - \mu_1$ does not include 0 and the *P*-value = 0.0012, which is small, so we reject the null hypothesis that $\mu_1 = \mu_3$. The interval for $\mu_4 - \mu_1$ does not include 0 and the *P*-value < 0.0001, which is small, so we reject the null hypothesis that $\mu_4 = \mu_1$. The interval for $\mu_3 - \mu_2$ does not include 0 and the *P*-value = 0.0031, which is small, so we reject the null hypothesis that $\mu_2 = \mu_3$. The interval for $\mu_4 - \mu_2$ does not include 0 and the *P*-value = 0.0001, which is small, so we reject the null hypothesis that $\mu_2 = \mu_4$. Finally, the interval for $\mu_4 - \mu_3$

does not include 0 and the *P*-value = 0.0405, which is small, so we reject the null hypothesis that $\mu_3 = \mu_4$. Thus, the Tukey's tests indicate that the means of Mark IV V2 and Balloon Ballista are equal, but the means of Hydrolaunch and Waterworks II are different. That is, we conclude: $\mu_1 = \mu_2 \neq \mu_3 \neq \mu_4$ or $\underline{\mu_1 \quad \mu_2} \quad \underline{\mu_3} \quad \underline{\mu_4}$.

(c) The boxplots shown to the right support the results found in parts (a) and (b).

7. **(a)** Using technology, we find the sample standard deviation the data within each cell:
$s_{M/\text{one 8 hr}} \approx 3.42$, $s_{M/\text{two 4 hr}} \approx 2.92$, $s_{M/\text{two 2 hr}} \approx 2.39$, $s_{F/\text{one 8 hr}} \approx 4.72$, $s_{F/\text{two 4 hr}} \approx 2.70$, and $s_{F/\text{two 2 hr}} \approx 3.87$. The largest standard deviation (4.72) is less than twice the smallest standard deviation (2.39), so the requirement of equal population variances is satisfied. [4.72 < 2(2.39).]

(b) Using MINITAB with the provided data, we obtain the following output:

Two-way ANOVA: Score versus Gender, Program

Source	DF	SS	MS	F	P
Gender	1	20.83	20.833	1.77	0.196
Program	2	890.60	445.300	37.90	0.000
Interaction	2	24.07	12.033	1.02	0.374
Error	24	282.00	11.750		
Total	29	1217.50			

Since the interaction *P*-value is 0.374, which is large, there is no evidence of an interaction effect.

(c) Since the *P*-value for gender is 0.196, which is large, there is no evidence of a significant difference in the means for gender. Since we have the program *P*-value program < 0.001, which is small, there is evidence of a significant difference in the means for type of driving program. The null hypothesis $H_0 : \mu_{8 \text{ hour}} = \mu_{4 \text{ hour}} = \mu_{2 \text{ hour}}$ is rejected.

(d) The lines in the interaction plot shown to the right are close to parallel, supporting the conclusion that there is no interaction effect.

(e) The Tukey's test for gender is unnecessary since there is no evidence of a difference in means. For type of program, with Program 1 = One 8-hour session, Program 2 = Two 4-hour session, and Program 3 = Two 2-hour sessions, we use MINITAB to perform Tukey's test. The results are:

```
Tukey 95.0% Simultaneous Confidence Intervals
Response Variable Score
All Pairwise Comparisons among Levels of Program
Program = 1  subtracted from:
Program   Lower  Center   Upper   ---+---------+---------+---------+---
2         -6.03   -2.20   1.626                       (-------*------)
3        -16.33  -12.50  -8.674    (-------*-------)
                                   ---+---------+---------+---------+---
                                 -15.0      -10.0      -5.0       0.0

Program = 2  subtracted from:
Program   Lower  Center   Upper   ---+---------+---------+---------+---
3        -14.13  -10.30  -6.474        (------*-------)
                                   ---+---------+---------+---------+---
                                 -15.0      -10.0      -5.0       0.0
```

```
Tukey Simultaneous Tests
Response Variable Score
All Pairwise Comparisons among Levels of Program
Program = 1  subtracted from:
          Difference     SE of              Adjusted
Program   of Means   Difference  T-Value   P-Value
2           -2.20       1.533     -1.435    0.3394
3          -12.50       1.533     -8.154    0.0000
Program = 2  subtracted from:
          Difference     SE of              Adjusted
Program   of Means   Difference  T-Value   P-Value
3          -10.30       1.533     -6.719    0.0000
```

The interval for $\mu_{\text{two 4 hr}} - \mu_{\text{one 8 hr}}$ includes 0 and the P-value $= 0.3394$ for this comparison, which is large, so we do not reject the null hypothesis that $\mu_{\text{one 8 hr}} = \mu_{\text{two 4 hr}}$. The interval for $\mu_{\text{two 2 hr}} - \mu_{\text{one 8 hr}}$ does not include 0 and the P-value < 0.0001, which is small, so we reject the null hypothesis that $\mu_{\text{one 8 hr}} = \mu_{\text{two 2 hr}}$. Finally, the interval for $\mu_{\text{two 2 hr}} - \mu_{\text{two 4 hr}}$ does not include 0 and the P-value < 0.0001, which is small, so we reject the null hypothesis that $\mu_{\text{two 4 hr}} = \mu_{\text{two 2 hr}}$. Thus, we have $\mu_{\text{one 8 hr}} = \mu_{\text{two 4 hr}} \neq \mu_{\text{two 2 hr}}$ or $\underline{\mu_{\text{one 8 hr}} \quad \mu_{\text{two 4 hr}}} \quad \mu_{\text{two 2 hr}}$. That is, the Tukey's test indicates that the mean score after taking one 8-hour defensive driving class is equal to the mean score after taking two 4-hour classes, but that the mean score after taking two 2-hour classes is different.

Chapter 13 Test

1. (a) The hypotheses are $H_0 : \mu_{\text{nonsmokers}} = \mu_{\text{light smokers}} = \mu_{\text{heavy smokers}}$ versus H_1 : at least one mean differs from the others. Using MINITAB, we obtain the following output:

One-way ANOVA: Machine I, Machine II, Machine III, Machine IV
```
Source   DF    SS     MS     F      P
Factor    3  3.637  1.212  5.81  0.004
Error    24  5.011  0.209
Total    27  8.649
```

Since P-value $= 0.004$ is less than the $\alpha = 0.05$ level of significance, we reject H_0. There is sufficient evidence to support the hypothesis that at least one of the means is different from the others.

   (b) Using MINITAB, we obtain the following output:

```
Tukey 95% Simultaneous Confidence Intervals   All Pairwise Comparisons
Individual confidence level = 98.90%

Machine I subtracted from:
             Lower    Center    Upper
Machine II   -1.4307  -0.7571  -0.0836
Machine III  -0.4736   0.2000   0.8736
Machine IV   -0.7450  -0.0714   0.6022


             ------+---------+---------+---------+---
Machine II      (------*------)
Machine III              (-------*------)
Machine IV         (------*------)
             ------+---------+---------+---------+---
                  -1.0       0.0       1.0       2.0

Machine II subtracted from:
             Lower   Center   Upper  ------+---------+---------+---------+---
Machine III  0.2836  0.9571  1.6307                         (------*-----)
Machine IV   0.0121  0.6857  1.3593                     (------*------)
                                     ------+---------+---------+---------+---
                                          -1.0       0.0       1.0       2.0

Machine III subtracted from:
             Lower    Center   Upper  ------+---------+---------+---------+--
Machine IV   -0.9450  -0.2714  0.4022              (-----*------)
                                      ------+---------+---------+---------+--
                                           -1.0       0.0       1.0       2.0
```

From the output, we have:

$-1.4307 < \mu_{II} - \mu_I < -0.0836$. This interval does not contain 0, so we reject $H_0 : \mu_I = \mu_{II}$.

$-0.4736 < \mu_{III} - \mu_I < 0.8736$. This interval contains 0, so we do not reject $H_0 : \mu_I = \mu_{III}$.

$-0.7450 < \mu_{IV} - \mu_I < 0.6022$. This interval contains 0, so we do not reject $H_0 : \mu_I = \mu_{IV}$.

$0.2836 < \mu_{III} - \mu_{II} < 1.6307$. This interval does not contain 0, so we reject $H_0 : \mu_{II} = \mu_{III}$.

$0.0121 < \mu_{IV} - \mu_{II} < 1.3593$. This interval does not contain 0, so we reject $H_0 : \mu_{II} = \mu_{IV}$.

$-0.9450 < \mu_{IV} - \mu_{III} < 0.4022$. This interval contains 0, so we do not reject $H_0 : \mu_{III} = \mu_{IV}$.

Thus, we have that $\mu_I = \mu_{III} = \mu_{IV} \neq \mu_{II}$. That is, we have: <u>$\mu_I$ μ_{III} μ_{IV}</u> <u>μ_{II}</u> . It appears that the mean amount dispensed by Machine II is less than the mean amount dispensed by the other machines.

(c) The boxplots shown to the right support the result of parts (a) and (b).

2. (a) Michael use a completely randomized design for his experiment.

(b) The response variable is the height after 38 days; this is a quantitative variable.

(c) The explanatory variable is the color of the light bulb; this is a qualitative variable.

(d) Other factors identified that might impact plant growth included type of soil, amount of water, amount of light, and temperature. Michael kept these consistent for all three treatments.

(e) Michael randomly assigned the seeds to each treatment group.

(f) The hypotheses are $H_0 : \mu_B = \mu_R = \mu_G$ versus H_1 : at least one mean differs from the others

(g) The largest standard deviation (0.950) is less than twice the smallest standard deviation (0.595) $[0.950 < 2(0.595)]$.

(h) We assume the level of significance is $\alpha = 0.05$. Since P-value $= 0.000$ (i.e., P-value < 0.0005), it is smaller than the $\alpha = 0.05$ level of significance, so we reject H_0. There is sufficient evidence to conclude that at least one of the mean height is different for at least one group.

(i) Yes, it appears that red light promotes plant growth more than the others.

(j) From the output, we have:

$1.1460 < \mu_R - \mu_B < 2.0740$. This interval does not contain 0, so we reject $H_0 : \mu_R = \mu_B$.

$-0.9973 < \mu_G - \mu_B < -0.0694$. This interval does not contain 0, so we reject $H_0 : \mu_G = \mu_B$.

$-2.6073 < \mu_G - \mu_R < -1.6794$. This interval does not contain 0, so we reject $H_0 : \mu_G = \mu_R$.

Thus, we have that $\mu_G \neq \mu_B \neq \mu_R$. That is, we have: $\underline{\mu_G} \quad \underline{\mu_B} \quad \underline{\mu_R}$. Tukey's test indicates that all three means are different.

3. (a) Using MINITAB, we obtain the following output:

Two-way ANOVA: Tensile Strength versus Thickness, Type
```
Source      DF        SS        MS      F      P
Thickness    3  97239906  32413302  28.27  0.001
Type         2  23628163  11814081  10.30  0.011
Error        6   6880388   1146731
Total       11 127748456
```

Since the treatment (type) P-value $= 0.011$, we reject the null hypothesis that the population means are equal. Thus, we conclude that at least one mean is different from the others.

(b) Using MINITAB , we obtain the following output:

```
Tukey 95.0% Simultaneous Confidence Intervals
Response Variable Tensile Strength
All Pairwise Comparisons among Levels of Type
Type = 1  subtracted from:
Type  Lower   Center  Upper   ---------+---------+---------+-------
2     -4246   -1922   401              (-----------*------------)
3     -5753   -3429   -1105   (-----------*-----------)
                              ---------+---------+---------+-------
                                 -4000      -2000        0

Type = 2  subtracted from:
Type  Lower   Center  Upper   ---------+---------+---------+-------
3     -3830   -1506   817.5            (-----------*------------)
                              ---------+---------+---------+-------
                                 -4000      -2000        0
```

```
Tukey Simultaneous Tests
Response Variable Tensile Strength
All Pairwise Comparisons among Levels of Type
Type = 1  subtracted from:
        Difference      SE of                Adjusted
Type    of Means   Difference  T-Value   P-Value
2           -1922       757.2   -2.539    0.0972
3           -3429       757.2   -4.528    0.0095
Type = 2  subtracted from:
        Difference      SE of                Adjusted
Type    of Means   Difference.  T-Value   P-Value
3           -1506       757.2   -1.989    0.1956
```

The interval for $\mu_2 - \mu_1$ includes 0 and the P-value = 0.0972 for this comparison, which is large, so we do not reject the null hypothesis that $\mu_1 = \mu_2$. The interval for $\mu_3 - \mu_1$ does not include 0 and the P-value = 0.0095, which is small, so we reject the null hypothesis that $\mu_1 = \mu_3$. Finally, the interval for $\mu_3 - \mu_2$ includes 0 and the P-value = 0.1956, which is large, so we do not reject the null hypothesis that $\mu_2 = \mu_3$. Thus, the Tukey's test is ambiguous. It indicates that $\mu_3 \neq \mu_1 = \mu_2$, but $\mu_2 = \mu_3$. That is: $\underline{\mu_1 \quad \mu_2 \quad \mu_3}$.

(c)

4. (a) Using technology, we find the sample standard deviation the data within each cell:
$s_{M/20\text{-}34} \approx 0.46$, $s_{M/35\text{-}49} \approx 0.48$, $s_{M/50\text{-}64} \approx 0.53$, $s_{F/20\text{-}34} \approx 0.31$, $s_{F/35\text{-}49} \approx 0.28$, and $s_{F/50\text{-}64} \approx 0.32$. The largest standard deviation (0.53) is less than twice the smallest standard deviation (0.28), so the requirement of equal population variances is satisfied. [$0.53 < 2(0.28) = 0.56$.]

(b) Using MINITAB with the provided data, we obtain the following output:

Two-way ANOVA: Learning Time versus Gender, Age

Source	DF	SS	MS	F	P
Gender	1	0.05042	0.050417	0.30	0.589
Age	2	0.51083	0.255417	1.53	0.243
Interaction	2	0.27083	0.135417	0.81	0.459
Error	18	2.99750	0.166528		
Total	23	3.82958			

Since the interaction P-value is 0.459, which is larger than $\alpha = 0.05$, there is no evidence of an interaction effect.

(c) Since the P-value for gender is 0.589, which is large, there is no evidence of a difference in the means for gender. Since the P-value for age is 0.243, which is also large, there is no evidence of a difference in the means for the three age groups.

(d) The lines in the interaction plot shown to the right are not very parallel, contradicting the conclusion that there is no interaction effect.

(e) The Tukey's test is unnecessary since there is no evidence of a difference in means.

5. (a)

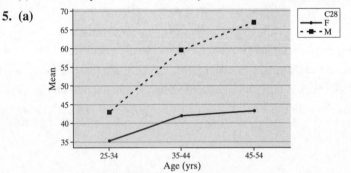

(b) The plot indicates a significant interaction effect since the graphs for the two levels of gender are not parallel.

(c) Since there is a significant interaction effect, any conclusion about main effects would be confounded.

Case Study: Hat Size and Intelligence

Does it appear that at least one of the head-circumference means is different from the others? Explain.

Yes. From the information provided, $k = 8$ and $n = 1672$ data. Using an $\alpha = 0.05$ significance level, the critical value is $F_{\alpha, k-1, n} = F_{0.05, 7, 1672} \approx F_{0.05, 7, 1000} = 2.02$. From the results of the provided ANOVA table, we have $F_0 = 1998.6275$. Therefore, we reject the null hypothesis and conclude that at least one of the head-circumference means is different from the others.

Examine Epstein's table closely. Does it appear that the ANOVA procedure's requirement of equal variances is satisfied? Explain.

From Epstein's table, it does not appear that the ANOVA procedure's requirement of equal variance is satisfied. The largest standard deviation (2.5) is more than two times the smallest standard deviation (0.3). That is, $2.5 > 2(0.3) = 0.6$, so the requirement of equal variances is not satisfied.

Calculate the correct standard deviations by multiplying the tabled value by the square root of its sample size, n.

Vocational Status	n	Mean	Corrected S.D.
Professional	25	569.9	$1.9 \cdot \sqrt{25} = 9.5$
Semiprofessional	61	566.5	$1.5 \cdot \sqrt{61} \approx 11.7$
Clerical	107	566.2	$1.1 \cdot \sqrt{107} \approx 11.4$
Trades	194	565.7	$0.8 \cdot \sqrt{194} \approx 11.1$
Public service	25	564.1	$2.5 \cdot \sqrt{25} = 12.5$
Skill trades	351	562.9	$0.6 \cdot \sqrt{351} \approx 11.2$
Personal services	262	562.7	$0.7 \cdot \sqrt{262} \approx 11.3$
Laborers	647	560.7	$0.3 \cdot \sqrt{647} \approx 7.6$

If the ANOVA table was recalculated with the correct standard deviations, predict the effect on the F-test statistic. Would you be more or less likely to reject the null hypothesis? Explain.

Recalculating with the correct standard deviation will lower the *F*-test statistic, making it less likely to reject the null hypothesis. The corrected standard deviations are larger than those in the original table. These larger standard deviations will make the sum-of-squares error larger than it was originally, which will make the *F*-test statistic smaller.

Using the equation for MSE, calculate a revised sum-of-squares error, mean-squares error, and F-test statistic.

The revised sum of squares for error = $(25-1) \cdot 9.5^2 + (61-1) \cdot 11.7^2 + (107-1) \cdot 11.4^2$

$+ (194-1) \cdot 11.1^2 + (25-1) \cdot 12.5^2 + (351-1) \cdot 11.2^2 + (262-1) \cdot 11.3^2 + (647-1) \cdot 7.6^2 = 166,228.74.$

Therefore, the revised mean-squares error is $\text{MSE} = \dfrac{166,228.74}{1664} \approx 99.90$, and the revised *F*-test statistic is

$F_0 = \dfrac{1123.6284}{99.90} \approx 11.25$.

Retest the claim that all the head circumference population means are equal. Use an $\alpha = 0.05$ significance level.

Comparing $F \approx 11.25$ with the critical value $F_{0.05, 7, 1672} \approx F_{0.05, 7, 1000} = 2.02$, we reject the null hypothesis at $\alpha = 0.05$, and we conclude that at least one of the head-circumference means is different from the others.

From the corrected values for the standard deviation, does it appear that the ANOVA procedure's requirement of equal variances is satisfied?

Yes. The largest corrected standard deviation (12.5) is no more than two times the smallest corrected standard deviation (7.6). That is, 12.5 < 2(7.6) = 15.2, so the requirement of equal variances is satisfied.

Did any of your conclusions change in light of the new information? Explain.

The overall result concerning whether or not to reject the null hypothesis did not change. However, the strength of the rejection did change. Also, the new information allowed us to meet the requirement of equal variances so that we could perform the ANOVA.

If the null hypothesis of equal populations means was rejected, use Tukey's test to determine which of the pairwise means differ using a family wise error rate of $\alpha = 0.05$. Which of these categories appear to have significantly different means?

Arrange the sample means in ascending order are $\bar{x}_1 = 560.7$, $\bar{x}_2 = 562.7$, $\bar{x}_3 = 562.9$, $\bar{x}_4 = 564.1$, $\bar{x}_5 = 565.7$, $\bar{x}_6 = 566.2$, $\bar{x}_7 = 566.5$, and $\bar{x}_8 = 569.9$. Now, $n = 1672$ and $k = 8$, so $\nu = n - k = 1664$ and the critical value is $q_{\alpha, \nu, k} = q_{0.05, 1664, 8} \approx q_{0.05, 120, 8} = 4.363$.

Test 1: $H_0 : \mu_1 = \mu_8$. The test statistic is $q_0 = \dfrac{\overline{x}_8 - \overline{x}_1}{\sqrt{\frac{s^2}{2} \cdot \left(\frac{1}{n_1} + \frac{1}{n_8}\right)}} = \dfrac{569.9 - 560.7}{\sqrt{\frac{75.9}{2} \cdot \left(\frac{1}{647} + \frac{1}{25}\right)}} \approx 7.327$. Since this test statistic is

greater than the critical value $q_{\alpha, \nu, k} \approx 4.363$, we reject $H_0 : \mu_1 = \mu_8$.

Test 2: $H_0 : \mu_1 = \mu_7$. The test statistic is $q_0 = \dfrac{\overline{x}_7 - \overline{x}_1}{\sqrt{\frac{s^2}{2} \cdot \left(\frac{1}{n_1} + \frac{1}{n_7}\right)}} = \dfrac{566.5 - 560.7}{\sqrt{\frac{75.9}{2} \cdot \left(\frac{1}{647} + \frac{1}{61}\right)}} \approx 7.029$. Since this test statistic is

greater than the critical value $q_{\alpha, \nu, k} \approx 4.363$, we reject $H_0 : \mu_1 = \mu_7$.

Test 3: $H_0 : \mu_1 = \mu_6$. The test statistic is $q_0 = \dfrac{\overline{x}_6 - \overline{x}_1}{\sqrt{\frac{s^2}{2} \cdot \left(\frac{1}{n_1} + \frac{1}{n_6}\right)}} = \dfrac{566.2 - 560.7}{\sqrt{\frac{75.9}{2} \cdot \left(\frac{1}{647} + \frac{1}{107}\right)}} \approx 8.555$. Since this test statistic is

greater than the critical value $q_{\alpha, \nu, k} \approx 4.363$, we reject $H_0 : \mu_1 = \mu_6$.

Test 4: $H_0 : \mu_1 = \mu_5$. The test statistic is $q_0 = \dfrac{\overline{x}_5 - \overline{x}_1}{\sqrt{\frac{s^2}{2} \cdot \left(\frac{1}{n_1} + \frac{1}{n_5}\right)}} = \dfrac{565.7 - 560.7}{\sqrt{\frac{75.9}{2} \cdot \left(\frac{1}{647} + \frac{1}{194}\right)}} \approx 9.916$. Since this test statistic is

greater than the critical value $q_{\alpha, \nu, k} \approx 4.363$, we reject $H_0 : \mu_1 = \mu_5$.

Test 5: $H_0 : \mu_1 = \mu_4$. The test statistic is $q_0 = \dfrac{\overline{x}_4 - \overline{x}_1}{\sqrt{\frac{s^2}{2} \cdot \left(\frac{1}{n_1} + \frac{1}{n_4}\right)}} = \dfrac{564.1 - 560.7}{\sqrt{\frac{75.9}{2} \cdot \left(\frac{1}{647} + \frac{1}{25}\right)}} \approx 2.708$. Since this test statistic is

less than the critical value $q_{\alpha, \nu, k} \approx 4.363$, we do not reject $H_0 : \mu_1 = \mu_4$.

Test 6: $H_0 : \mu_1 = \mu_3$. The test statistic is $q_0 = \dfrac{\overline{x}_3 - \overline{x}_1}{\sqrt{\frac{s^2}{2} \cdot \left(\frac{1}{n_1} + \frac{1}{n_3}\right)}} = \dfrac{562.9 - 560.7}{\sqrt{\frac{75.9}{2} \cdot \left(\frac{1}{647} + \frac{1}{351}\right)}} \approx 5.387$. Since this test statistic is

greater than the critical value $q_{\alpha, \nu, k} \approx 4.363$, we reject $H_0 : \mu_1 = \mu_3$.

Test 7: $H_0 : \mu_1 = \mu_2$. The test statistic is $q_0 = \dfrac{\overline{x}_2 - \overline{x}_1}{\sqrt{\frac{s^2}{2} \cdot \left(\frac{1}{n_1} + \frac{1}{n_2}\right)}} = \dfrac{562.7 - 560.7}{\sqrt{\frac{75.9}{2} \cdot \left(\frac{1}{647} + \frac{1}{262}\right)}} \approx 4.433$. Since this test statistic is

greater than the critical value $q_{\alpha, \nu, k} \approx 4.363$, we reject $H_0 : \mu_1 = \mu_2$.

Test 8: $H_0 : \mu_2 = \mu_8$. The test statistic is $q_0 = \dfrac{\overline{x}_8 - \overline{x}_2}{\sqrt{\frac{s^2}{2} \cdot \left(\frac{1}{n_2} + \frac{1}{n_8}\right)}} = \dfrac{569.9 - 562.7}{\sqrt{\frac{75.9}{2} \cdot \left(\frac{1}{262} + \frac{1}{25}\right)}} \approx 5.583$. Since this test statistic is

greater than the critical value $q_{\alpha, \nu, k} \approx 4.363$, we reject $H_0 : \mu_2 = \mu_8$.

Test 9: $H_0 : \mu_2 = \mu_7$. The test statistic is $q_0 = \dfrac{\overline{x}_7 - \overline{x}_2}{\sqrt{\frac{s^2}{2} \cdot \left(\frac{1}{n_2} + \frac{1}{n_7}\right)}} = \dfrac{566.5 - 562.7}{\sqrt{\frac{75.9}{2} \cdot \left(\frac{1}{262} + \frac{1}{61}\right)}} \approx 4.339$. Since this test statistic is

less than the critical value $q_{\alpha, \nu, k} \approx 4.363$, we do not reject $H_0 : \mu_2 = \mu_8$.

Test 10: $H_0 : \mu_2 = \mu_6$. The test statistic is $q_0 = \dfrac{\overline{x}_6 - \overline{x}_2}{\sqrt{\frac{s^2}{2} \cdot \left(\frac{1}{n_2} + \frac{1}{n_6}\right)}} = \dfrac{566.2 - 562.7}{\sqrt{\frac{75.9}{2} \cdot \left(\frac{1}{262} + \frac{1}{107}\right)}} \approx 4.952$. Since this test statistic is

greater than the critical value $q_{\alpha, \nu, k} \approx 4.363$, we reject $H_0 : \mu_2 = \mu_6$.

Test 11: $H_0 : \mu_2 = \mu_5$. The test statistic is $q_0 = \dfrac{\overline{x}_5 - \overline{x}_2}{\sqrt{\frac{s^2}{2} \cdot \left(\frac{1}{n_2} + \frac{1}{n_5}\right)}} = \dfrac{565.7 - 562.7}{\sqrt{\frac{75.9}{2} \cdot \left(\frac{1}{262} + \frac{1}{194}\right)}} \approx 5.141$. Since this test statistic is

greater than the critical value $q_{\alpha, \nu, k} \approx 4.363$, we reject $H_0 : \mu_2 = \mu_5$.

Test 12: $H_0 : \mu_2 = \mu_4$. The test statistic is $q_0 = \dfrac{\overline{x}_4 - \overline{x}_2}{\sqrt{\dfrac{s^2}{2} \cdot \left(\dfrac{1}{n_2} + \dfrac{1}{n_5} \right)}} = \dfrac{564.1 - 562.7}{\sqrt{\dfrac{75.9}{2} \cdot \left(\dfrac{1}{262} + \dfrac{1}{25} \right)}} \approx 1.086$. Since this test statistic is

less than the critical value $q_{\alpha, \nu, k} \approx 4.363$, we do not reject $H_0 : \mu_2 = \mu_4$.

Test 13: $H_0 : \mu_2 = \mu_3$. The test statistic is $q_0 = \dfrac{\overline{x}_3 - \overline{x}_2}{\sqrt{\dfrac{s^2}{2} \cdot \left(\dfrac{1}{n_2} + \dfrac{1}{n_5} \right)}} = \dfrac{562.9 - 562.7}{\sqrt{\dfrac{75.9}{2} \cdot \left(\dfrac{1}{262} + \dfrac{1}{351} \right)}} \approx 0.398$. Since this test statistic is

less than the critical value $q_{\alpha, \nu, k} \approx 4.363$, we do not reject $H_0 : \mu_2 = \mu_3$.

Test 14: $H_0 : \mu_3 = \mu_8$. The test statistic is $q_0 = \dfrac{\overline{x}_8 - \overline{x}_3}{\sqrt{\dfrac{s^2}{2} \cdot \left(\dfrac{1}{n_3} + \dfrac{1}{n_8} \right)}} = \dfrac{569.9 - 562.9}{\sqrt{\dfrac{75.9}{2} \cdot \left(\dfrac{1}{351} + \dfrac{1}{25} \right)}} \approx 5.489$. Since this test statistic is

greater than the critical value $q_{\alpha, \nu, k} \approx 4.363$, we reject $H_0 : \mu_3 = \mu_8$.

Test 15: $H_0 : \mu_3 = \mu_7$. The test statistic is $q_0 = \dfrac{\overline{x}_7 - \overline{x}_3}{\sqrt{\dfrac{s^2}{2} \cdot \left(\dfrac{1}{n_3} + \dfrac{1}{n_7} \right)}} = \dfrac{566.5 - 562.9}{\sqrt{\dfrac{75.9}{2} \cdot \left(\dfrac{1}{351} + \dfrac{1}{61} \right)}} \approx 4.213$. Since this test statistic is

less than the critical value $q_{\alpha, \nu, k} \approx 4.363$, we do not reject $H_0 : \mu_3 = \mu_7$.

Test 16: $H_0 : \mu_3 = \mu_6$. The test statistic is $q_0 = \dfrac{\overline{x}_6 - \overline{x}_3}{\sqrt{\dfrac{s^2}{2} \cdot \left(\dfrac{1}{n_3} + \dfrac{1}{n_6} \right)}} = \dfrac{566.2 - 562.9}{\sqrt{\dfrac{75.9}{2} \cdot \left(\dfrac{1}{351} + \dfrac{1}{107} \right)}} \approx 4.851$. Since this test statistic is

greater than the critical value $q_{\alpha, \nu, k} \approx 4.363$, we reject $H_0 : \mu_3 = \mu_6$.

Test 17: $H_0 : \mu_3 = \mu_5$. The test statistic is $q_0 = \dfrac{\overline{x}_5 - \overline{x}_3}{\sqrt{\dfrac{s^2}{2} \cdot \left(\dfrac{1}{n_3} + \dfrac{1}{n_5} \right)}} = \dfrac{565.7 - 562.9}{\sqrt{\dfrac{75.9}{2} \cdot \left(\dfrac{1}{351} + \dfrac{1}{194} \right)}} \approx 5.081$. Since this test statistic is

greater than the critical value $q_{\alpha, \nu, k} \approx 4.363$, we reject $H_0 : \mu_3 = \mu_5$.

Test 18: $H_0 : \mu_3 = \mu_4$. The test statistic is $q_0 = \dfrac{\overline{x}_4 - \overline{x}_3}{\sqrt{\dfrac{s^2}{2} \cdot \left(\dfrac{1}{n_3} + \dfrac{1}{n_4} \right)}} = \dfrac{564.1 - 562.9}{\sqrt{\dfrac{75.9}{2} \cdot \left(\dfrac{1}{351} + \dfrac{1}{25} \right)}} \approx 0.941$. Since this test statistic is

less than the critical value $q_{\alpha, \nu, k} \approx 4.363$, we do not reject $H_0 : \mu_3 = \mu_4$.

Test 19: $H_0 : \mu_4 = \mu_8$. The test statistic is $q_0 = \dfrac{\overline{x}_8 - \overline{x}_4}{\sqrt{\dfrac{s^2}{2} \cdot \left(\dfrac{1}{n_4} + \dfrac{1}{n_8} \right)}} = \dfrac{569.9 - 564.1}{\sqrt{\dfrac{75.9}{2} \cdot \left(\dfrac{1}{25} + \dfrac{1}{25} \right)}} \approx 3.329$. Since this test statistic is

less than the critical value $q_{\alpha, \nu, k} \approx 4.363$, we do not reject $H_0 : \mu_4 = \mu_8$.

Test 20: $H_0 : \mu_4 = \mu_7$. The test statistic is $q_0 = \dfrac{\overline{x}_7 - \overline{x}_4}{\sqrt{\dfrac{s^2}{2} \cdot \left(\dfrac{1}{n_4} + \dfrac{1}{n_7} \right)}} = \dfrac{566.5 - 564.1}{\sqrt{\dfrac{75.9}{2} \cdot \left(\dfrac{1}{25} + \dfrac{1}{61} \right)}} \approx 1.641$. Since this test statistic is

less than the critical value $q_{\alpha, \nu, k} \approx 4.363$, we do not reject $H_0 : \mu_4 = \mu_7$.

Test 21: $H_0 : \mu_4 = \mu_6$. The test statistic is $q_0 = \dfrac{\overline{x}_6 - \overline{x}_4}{\sqrt{\dfrac{s^2}{2} \cdot \left(\dfrac{1}{n_4} + \dfrac{1}{n_6} \right)}} = \dfrac{566.2 - 564.1}{\sqrt{\dfrac{75.9}{2} \cdot \left(\dfrac{1}{25} + \dfrac{1}{107} \right)}} \approx 1.535$. Since this test statistic is

less than the critical value $q_{\alpha, \nu, k} \approx 4.363$, we do not reject $H_0 : \mu_4 = \mu_6$.

Test 22: $H_0 : \mu_4 = \mu_5$. The test statistic is $q_0 = \dfrac{\overline{x}_5 - \overline{x}_4}{\sqrt{\dfrac{s^2}{2} \cdot \left(\dfrac{1}{n_4} + \dfrac{1}{n_5} \right)}} = \dfrac{565.7 - 564.1}{\sqrt{\dfrac{75.9}{2} \cdot \left(\dfrac{1}{25} + \dfrac{1}{194} \right)}} \approx 1.222$. Since this test statistic is

less than the critical value $q_{\alpha, \nu, k} \approx 4.363$, we do not reject $H_0 : \mu_4 = \mu_5$.

Test 23: $H_0 : \mu_5 = \mu_8$. The test statistic is $q_0 = \dfrac{\overline{x}_8 - \overline{x}_5}{\sqrt{\frac{s^2}{2} \cdot \left(\frac{1}{n_5} + \frac{1}{n_8} \right)}} = \dfrac{569.9 - 565.7}{\sqrt{\frac{75.9}{2} \cdot \left(\frac{1}{194} + \frac{1}{25} \right)}} \approx 3.208$. Since this test statistic is

less than the critical value $q_{\alpha,\nu,k} \approx 4.363$, we do not reject $H_0 : \mu_5 = \mu_8$.

Test 24: $H_0 : \mu_5 = \mu_7$. The test statistic is $q_0 = \dfrac{\overline{x}_7 - \overline{x}_5}{\sqrt{\frac{s^2}{2} \cdot \left(\frac{1}{n_5} + \frac{1}{n_7} \right)}} = \dfrac{566.5 - 565.7}{\sqrt{\frac{75.9}{2} \cdot \left(\frac{1}{194} + \frac{1}{61} \right)}} \approx 0.885$. Since this test statistic is

less than the critical value $q_{\alpha,\nu,k} \approx 4.363$, we do not reject $H_0 : \mu_5 = \mu_7$.

Test 25: $H_0 : \mu_5 = \mu_6$. The test statistic is $q_0 = \dfrac{\overline{x}_6 - \overline{x}_5}{\sqrt{\frac{s^2}{2} \cdot \left(\frac{1}{n_5} + \frac{1}{n_6} \right)}} = \dfrac{566.2 - 565.7}{\sqrt{\frac{75.9}{2} \cdot \left(\frac{1}{194} + \frac{1}{107} \right)}} \approx 0.674$. Since this test statistic is

less than the critical value $q_{\alpha,\nu,k} \approx 4.363$, we do not reject $H_0 : \mu_5 = \mu_6$.

Test 26: $H_0 : \mu_6 = \mu_8$. The test statistic is $q_0 = \dfrac{\overline{x}_8 - \overline{x}_6}{\sqrt{\frac{s^2}{2} \cdot \left(\frac{1}{n_6} + \frac{1}{n_8} \right)}} = \dfrac{569.9 - 566.2}{\sqrt{\frac{75.9}{2} \cdot \left(\frac{1}{107} + \frac{1}{25} \right)}} \approx 2.704$. Since this test statistic is

less than the critical value $q_{\alpha,\nu,k} \approx 4.363$, we do not reject $H_0 : \mu_6 = \mu_8$.

Test 27: $H_0 : \mu_6 = \mu_7$. The test statistic is $q_0 = \dfrac{\overline{x}_7 - \overline{x}_6}{\sqrt{\frac{s^2}{2} \cdot \left(\frac{1}{n_6} + \frac{1}{n_7} \right)}} = \dfrac{566.5 - 566.2}{\sqrt{\frac{75.9}{2} \cdot \left(\frac{1}{107} + \frac{1}{61} \right)}} \approx 0.304$. Since this test statistic is

less than the critical value $q_{\alpha,\nu,k} \approx 4.363$, we do not reject $H_0 : \mu_6 = \mu_7$.

Test 28: $H_0 : \mu_7 = \mu_8$. The test statistic is $q_0 = \dfrac{\overline{x}_8 - \overline{x}_7}{\sqrt{\frac{s^2}{2} \cdot \left(\frac{1}{n_7} + \frac{1}{n_8} \right)}} = \dfrac{569.9 - 566.5}{\sqrt{\frac{75.9}{2} \cdot \left(\frac{1}{61} + \frac{1}{25} \right)}} \approx 2.324$. Since this test statistic is

less than the critical value $q_{\alpha,\nu,k} \approx 4.363$, we do not reject $H_0 : \mu_7 = \mu_8$.

Thus, the Tukey's test results in the following pairwise differences: $\mu_{\text{laborers}} \neq \mu_{\text{professional}}$; $\mu_{\text{laborers}} \neq \mu_{\text{semiprofessional}}$; $\mu_{\text{laborers}} \neq \mu_{\text{clerical}}$; $\mu_{\text{laborers}} \neq \mu_{\text{trades}}$; $\mu_{\text{laborers}} \neq \mu_{\text{skill trades}}$; $\mu_{\text{laborers}} \neq \mu_{\text{personal services}}$; $\mu_{\text{personal services}} \neq \mu_{\text{professional}}$; $\mu_{\text{personal services}} \neq \mu_{\text{clerical}}$; $\mu_{\text{personal services}} \neq \mu_{\text{trades}}$; $\mu_{\text{skill trades}} \neq \mu_{\text{professional}}$; $\mu_{\text{skill trades}} \neq \mu_{\text{clerical}}$; and $\mu_{\text{skill trades}} \neq \mu_{\text{trades}}$.

The result of the Tukey's test is ambiguous since all other pairs were found to be equal.

Reports will vary.

Chapter 14

Inference on the Least-Squares Regression Model and Multiple Regression

Section 14.1

1. 82.7

3. 0; σ^2

5. **(a)** Using technology, we get:
$\beta_0 \approx b_0 = -2.3256$ and $\beta_1 \approx b_1 = 2.0233$.

(b) We calculate $\hat{y} = 2.0233x - 2.3256$ to generate the following table:

x	y	$\hat{y}$	$y - \hat{y}$	$(y - \hat{y})^2$	$(x - \overline{x})^2$
3	4	3.74	0.26	0.0654	5.76
4	6	5.77	0.23	0.0541	1.96
5	7	7.79	−0.79	0.6252	0.16
7	12	11.84	0.16	0.0265	2.56
8	14	13.86	0.14	0.0195	6.76
			Sum:	0.7907	17.20

$s_e = \sqrt{\dfrac{\Sigma(y - \hat{y})^2}{n - 2}} = \sqrt{\dfrac{0.7907}{5 - 2}} \approx 0.5134$ is the point estimate for σ.

(c) $s_{b_1} = \dfrac{s_e}{\sqrt{\Sigma(x - \overline{x})^2}} = \dfrac{0.5134}{\sqrt{17.20}} \approx 0.1238$

(d) We have the hypotheses $H_0 : \beta_1 = 0$ versus $H_1 : \beta_1 \neq 0$. The test statistic is

$t_0 = \dfrac{b_1}{s_{b_1}} = \dfrac{2.0233}{0.1238} = 16.343$ [Tech: 16.344].

Classical approach: The level of significance is $\alpha = 0.05$. Since this is a two-tailed test with $n - 2 = 5 - 2 = 3$ degrees of freedom, the critical values are $\pm t_{0.025} = \pm 3.182$. Since $t_0 = 16.343 > t_{0.025} = 3.182$ (the test statistic falls in a critical region), we reject H_0.

P-value approach: The P-value for this two-tailed test is the area under the t-distribution with 3 degrees of freedom to the right of $t_0 = 16.343$ [Tech: 16.344], plus the area to the left of $-t_0 = -16.343$ [Tech: –16.344].

From the t-distribution table in the row corresponding to 3 degrees of freedom, 16.343 falls to the right of 12.924, whose right-tail area is 0.0005. We must double this value in order to get the total area in both tails: 0.001. So, P-value < 0.001. [Tech: P-value = 0.0005]. Because the P-value is less than the level of significance $\alpha = 0.05$, we reject H_0.

Conclusion: There is sufficient evidence to conclude that a linear relationship exists between x and y.

7. **(a)** Using technology, we get: $\beta_0 \approx b_0 = 1.2$ and $\beta_1 \approx b_1 = 2.2$.

(b) We calculate $\hat{y} = 2.2x + 1.2$ to generate the following table:

x	y	$\hat{y}$	$y - \hat{y}$	$(y - \hat{y})^2$	$(x - \overline{x})^2$
−2	−4	−3.2	−0.8	0.64	4
−1	0	−1	1.	1	1
0	1	1.2	−0.2	0.04	0
1	4	3.4	0.6	0.36	1
2	5	5.6	−0.6	0.36	4
			Sum	2.40	10

$s_e = \sqrt{\dfrac{\Sigma(y - \hat{y})^2}{n - 2}} = \sqrt{\dfrac{2.4}{5 - 2}} = 0.8944$ is the point estimate for σ.

(c) $s_{b_1} = \dfrac{s_e}{\sqrt{\Sigma(x - \overline{x})^2}} = \dfrac{0.8944}{\sqrt{10}} = 0.2828$

(d) The hypotheses are $H_0 : \beta_1 = 0$ versus $H_1 : \beta_1 \neq 0$. The test statistic is

$t_0 = \dfrac{b_1}{s_{b_1}} = \dfrac{2.2}{0.2828} = 7.778$ [Tech: 7.779].

Classical approach: The level of significance is $\alpha = 0.05$. Since this is a two-tailed test with $n - 2 = 5 - 2 = 3$ degrees of freedom, the critical values are $\pm t_{0.025} = \pm 3.182$. Since

$t_0 = 7.778 > t_{0.025} = 3.182$ (the test statistic falls in a critical region), we reject H_0.

P-value approach: The P-value for this two-tailed test is the area under the t-distribution with 3 degrees of freedom to the right of $t_0 = 7.778$ [Tech: 7.779], plus the area to the left of $-t_0 = -7.778$ [Tech: –7.779]. From the t-distribution table in the row corresponding to 3 degrees of freedom, 7.778 falls between 7.453 and 10.215, whose right-tail areas are 0.0025 and 0.001, respectively. We must double these values in order to get the total area in both tails: 0.005 and 0.002. So, $0.002 < P\text{-value} < 0.005$. [Tech: P-value = 0.0044]. Because the P-value is less than the level of significance $\alpha = 0.05$, we reject H_0.

Conclusion: There is sufficient evidence to conclude that a linear relationship exists between x and y.

9. **(a)** Using technology, we get: $\beta_0 \approx b_0 = 116.6$ and $\beta_1 \approx b_1 = -0.72$.

(b) We calculate $\hat{y} = -0.72x + 116.6$ to generate the following table:

x	y	$\hat{y}$	$y - \hat{y}$	$(y - \hat{y})^2$	$(x - \bar{x})^2$
20	100	102.2	–2.2	4.84	400
30	95	95	0	0	100
40	91	87.8	3.2	10.24	0
50	83	80.6	2.4	5.76	100
60	70	73.4	–3.4	11.56	400
			Sum	32.40	1000

$s_e = \sqrt{\dfrac{\Sigma(y - \hat{y})^2}{n - 2}} = \sqrt{\dfrac{32.4}{5 - 2}} = 3.2863$ is the point estimate for σ.

(c) $s_{b_1} = \dfrac{s_e}{\sqrt{\Sigma(x - \bar{x})^2}} = \dfrac{3.2863}{\sqrt{1000}} = 0.1039$

(d) The hypotheses are $H_0 : \beta_1 = 0$ versus $H_1 : \beta_1 \neq 0$. The test statistic is

$t_0 = \dfrac{b_1}{s_{b_1}} = \dfrac{-0.72}{0.1039} = -6.929$ [Tech: –6.928]

Classical approach: The level of significance is $\alpha = 0.05$. Since this is a two-tailed test with $n - 2 = 5 - 2 = 3$ degrees of freedom, the critical values are $\pm t_{0.025} = \pm 3.182$. Since $t_0 = -6.929 < -t_{0.025} = -3.182$ (the test statistic falls in a critical region), we reject H_0.

P-value approach: The P-value for this two-tailed test is the area under the t-distribution with 3 degrees of freedom to the left of $t_0 = -6.929$ [Tech: –6.928], plus the area to the right of 6.929 [Tech: 6.928]. From the t-distribution table in the row corresponding to 3 degrees of freedom, 6.929 falls between 5.841 and 7.453, whose right-tail areas are 0.005 and 0.0025, respectively. We must double these values in order to get the total area in both tails: 0.01 and 0.005. So, $0.005 < P\text{-value} < 0.01$. [Tech: P-value = 0.0062]. Because the P-value is less than the level of significance $\alpha = 0.05$, we reject H_0.

Conclusion: There is sufficient evidence to conclude that a linear relationship exists between x and y.

11. **(a)** Using technology, we get: $b_0 = 69.0296$ and $b_1 = -0.0479$.

(b) We calculate $\hat{y} = -0.0479x + 69.0296$ to generate the following table:

x	y	$\hat{y}$	$y - \hat{y}$	$(y - \hat{y})^2$	$(x - \bar{x})^2$
5	69.2	68.79	0.410	0.1680	1509.9
15	68.3	68.31	–0.011	0.0001	832.7
25	67.5	67.83	–0.332	0.110	355.6
35	67.1	67.35	–0.253	0.064	78.4
50	66.4	66.63	–0.234	0.055	37.7
72	66.1	65.58	0.520	0.270	792.0
105	63.9	64.00	–0.099	0.010	3738.4
			Sum	0.677	7344.9

$s_e = \sqrt{\dfrac{\sum(y-\hat{y})^2}{n-2}} = \sqrt{\dfrac{0.677}{7-2}} = 0.3680$ is the

point estimate for σ.

(c) $s_{b_1} = \dfrac{s_e}{\sqrt{\sum(x-\overline{x})^2}} = \dfrac{0.3680}{\sqrt{7344.9}} = 0.0043$

(d) The hypotheses are $H_0 : \beta_1 = 0$ versus $H_1 : \beta_1 \neq 0$. The test statistic is

$t_0 = \dfrac{b_1}{s_{b_1}} = \dfrac{-0.0479}{0.0043} = -11.140$

[Tech: -11.157]

Classical approach: The level of significance is $\alpha = 0.05$. Since this is a two-tailed test with $n - 2 = 7 - 2 = 5$ degrees of freedom, the critical values are $\pm t_{0.025} = \pm 2.571$. Since $-t_0 = 11.140$ [Tech: 11.157] $> t_{0.025} = 2.571$ (the test statistic falls in a critical region), we reject H_0.

P-value approach: The *P*-value for this two-tailed test is the area under the *t*-distribution with $n - 2 = 7 - 2 = 5$ degrees of freedom to the left of $t_0 = -11.140$ [Tech: -11.157], plus the area to the right of $-t_0 = 11.140$ [Tech: 11.157]. From the *t*-distribution table in the row corresponding to 5 degrees of freedom, 11.140 falls above 6.869, whose right-tail areas is 0.0005. We must double this value in order to get the total area in both tails: 0.001. So, *P*-value < 0.001. [Tech: *P*-value = 0.0001]. Because the *P*-value is less than the level of significance $\alpha = 0.05$, we reject H_0.

Conclusion: There is sufficient evidence to conclude that a linear relationship exists between commute time and the score on a well-being survey.

(e) For a 95% confidence interval we use $t_{\alpha/2} = t_{0.025} = 2.571$:

Lower bound:

$b_1 - t_{\alpha/2} \cdot s_{b_1} = -0.0479 - 2.571 \cdot 0.0043$

$= -0.0590$ [Tech: -0.0589]

Upper bound:

$b_1 + t_{\alpha/2} \cdot s_{b_1} = -0.0479 + 2.571 \cdot 0.0043$

$= -0.0368$ [Tech: -0.0369]

13. (a) Using technology, we get:

$\beta_0 \approx b_0 = 12.4932$ and $\beta_1 \approx b_1 = 0.1827$.

(b) Using technology, we calculate

$\hat{y} = 0.1827x + 12.4932$ to generate the table:

x	y	$\hat{y}$	$y - \hat{y}$	$(y-\hat{y})^2$	$(x-\overline{x})^2$
27.75	17.5	17.5640	−0.0640	0.0041	1.6782
24.5	17.1	16.9701	0.1299	0.0169	3.8202
25.5	17.1	17.1528	−0.0528	0.0028	0.9112
26	17.3	17.2442	0.0558	0.0031	0.2066
25	16.9	17.0615	−0.1615	0.0261	2.1157
27.75	17.6	17.5640	0.0360	0.0013	1.6782
26.5	17.3	17.3356	−0.0356	0.0013	0.0021
27	17.5	17.4269	0.0731	0.0053	0.2975
26.75	17.3	17.3813	−0.0813	0.0066	0.0873
26.75	17.5	17.3813	0.1187	0.0141	0.0873
27.5	17.5	17.5183	−0.0183	0.0003	1.0930
			Sum	0.0819	11.9773

$s_e = \sqrt{\dfrac{\sum(y-\hat{y})^2}{n-2}} = \sqrt{\dfrac{0.0819}{11-2}} = 0.0954$

(c) A normal probability plot shows that the residuals are approximately normally distributed.

(d) $s_{b_1} = \dfrac{s_e}{\sqrt{\sum(x-\overline{x})^2}} = \dfrac{0.0954}{\sqrt{11.9773}} = 0.0276$

(e) The hypotheses are $H_0 : \beta_1 = 0$ versus $H_1 : \beta_1 \neq 0$. The test statistic is

$t_0 = \dfrac{b_1}{s_{b_1}} = \dfrac{0.1827}{0.0276} = 6.620$ [Tech: 6.630]

Classical approach: The level of significance is $\alpha = 0.01$. Since this is a two-tailed test with $n - 2 = 11 - 2 = 9$ degrees of freedom, the critical values are $\pm t_{0.005} = \pm 3.250$. Since $t_0 = 6.620 > t_{0.005} = 3.250$ (the test statistic falls in a critical region), we reject H_0.

P-value approach: The P-value for this two-tailed test is the area under the t-distribution with 9 degrees of freedom to the right of $t_0 = 6.620$ [Tech: 6.630], plus the area to the left of $-t_0 = -6.620$ [Tech: –9.630]. From the t-distribution table in the row corresponding to 9 degrees of freedom, 6.620 falls to the right of 4.781, whose right-tail area is 0.0005. We must double this value in order to get the total area in both tails: 0.001. So, P-value < 0.001. [Tech: P-value = 0.0001]. Because the P-value is less than the level of significance $\alpha = 0.01$, we reject H_0.

Conclusion: There is sufficient evidence to conclude that a linear relationship exists between height and head circumference.

(f) For a 95% confidence interval we use $t_{\alpha/2} = t_{0.025} = 2.262$:

Lower bound:
$$b_1 - t_{\alpha/2} \cdot s_{b_1} = 0.1827 - 2.262 \cdot 0.0276$$
$$= 0.1203 \quad \text{[Tech: 0.1204]}$$

Upper bound:
$$b_1 + t_{\alpha/2} \cdot s_{b_1} = 0.1827 + 2.262 \cdot 0.0276$$
$$= 0.2451$$

(g) We use the regression equation to obtain a good estimate of the child's head circumference if the child's height is 26.5 inches:
$$\hat{y} = 0.1827x + 12.4932$$
$$= 0.1827 \cdot 26.5 + 12.4932$$
$$\approx 17.34 \text{ inches}$$

15. (a) Using technology, we get:
$$\beta_0 \approx b_0 = 2675.6 \text{ and } \beta_1 \approx b_1 = 0.6764$$

(b) Using technology, we calculate $\hat{y} = 0.6764x + 2675.6$ to generate the table:

x	y	$\hat{y}$	$y - \hat{y}$	$(y - \hat{y})^2$	$(x - \bar{x})^2$
2300	4070	4231.32	–161.32	26025.6	338724
3390	5220	4968.62	251.38	63191.6	258064
2430	4640	4319.26	320.74	102874.9	204304
2890	4620	4630.41	–10.41	108.4	64
3330	4850	4928.04	–78.04	6089.5	200704
2480	4120	4353.08	–233.08	54326.2	161604
3380	5020	4961.86	58.14	3380.7	248004
2660	4890	4474.84	415.16	172361.9	49284
2620	4190	4447.78	–257.78	66449.7	68644
3340	4630	4934.80	–304.80	92902.8	209764
			Sum	587711.3	1739160

$$s_e = \sqrt{\frac{\Sigma(y - \hat{y})^2}{n - 2}} = \sqrt{\frac{587711.3}{10 - 2}} = 271.04$$

(c) A normal probability plot shows that the residuals are normally distributed.

(d) $s_{b_1} = \dfrac{s_e}{\sqrt{\Sigma(x - \bar{x})^2}} = \dfrac{271.04}{\sqrt{1{,}739{,}160}} = 0.2055$

(e) The hypotheses are $H_0 : \beta_1 = 0$ versus $H_1 : \beta_1 \neq 0$. The test statistic is
$$t_0 = \frac{b_1}{s_{b_1}} = \frac{0.6764}{0.2055} = 3.291.$$

Classical approach: The level of significance is $\alpha = 0.05$. Since this is a two-tailed test with $n - 2 = 10 - 2 = 8$ degrees of freedom, the critical values are $\pm t_{0.025} = \pm 2.306$. Since $t_0 = 3.291 > t_{0.025} = 2.306$ (the test statistic falls in a critical region), we reject H_0.

P-value approach: The P-value for this two-tailed test is the area under the t-distribution with 8 degrees of freedom to the right of

$t_0 = 3.291$, plus the area to the left of $-t_0 = -3.291$. From the *t*-distribution table in the row corresponding to 8 degrees of freedom, 3.291 falls between 2.896 and 3.355, whose right-tail areas are 0.01 and 0.005, respectively. We must double these values in order to get the total area in both tails: 0.02 and 0.01. So, $0.01 < P\text{-value} < 0.02$. [Tech: *P*-value = 0.0110]. Because the *P*-value is less than the level of significance $\alpha = 0.05$, we reject H_0.

Conclusion: There is sufficient evidence to conclude that a linear relationship exists between the 7-day and 28-day strength of this type of concrete.

(f) For a 95% confidence interval we use $t_{\alpha/2} = t_{0.025} = 2.306$:

Lower bound:
$$b_1 - t_{\alpha/2} \cdot s_{b_1} = 0.6764 - 2.306 \cdot 0.2055$$
$$= 0.2025$$

Upper bound:
$$b_1 + t_{\alpha/2} \cdot s_{b_1} = 0.6764 + 2.306 \cdot 0.2055$$
$$= 1.1503$$

(g) We use the regression equation to calculate the mean 28-day strength of this concrete if the 7-day strength is 3000 psi:
$$\hat{y} = 0.6764x + 2675.6$$
$$= 0.6764 \cdot 3000 + 2675.6$$
$$\approx 4704.8 \text{ psi}$$

17. (a) Using technology, we get:
$$\beta_0 \approx b_0 = 0.1200 \text{ and } \beta_1 \approx b_1 = 1.0997.$$

(b) Using technology, we calculate $\hat{y} = 1.0997x + 0.1200$ to generate the table:

x	y	$\hat{y}$	$y - \hat{y}$	$(y - \hat{y})^2$	$(x - \bar{x})^2$
1.48	1.82	1.7476	0.0724	0.0052	0.0344
–8.2	–9.57	–8.8976	–0.6724	0.4522	90.1464
–5.39	–3.67	–5.8074	2.1374	4.5685	44.6831
6.88	9.53	7.6860	1.8440	3.4004	31.1973
–4.74	–7.72	–5.0926	–2.6274	6.9033	36.4157
8.76	9.24	9.7534	–0.5134	0.2636	55.7330
3.69	4.97	4.1779	0.7921	0.6274	5.7382
–0.23	1.24	–0.1329	1.3729	1.8849	2.3242
6.53	4.59	7.3011	–2.7111	7.3500	27.4100
2.26	3.27	2.6054	0.6646	0.4418	0.9321
3.2	3.28	3.6391	–0.3591	0.1289	3.6308
			Sum	26.0261	298.2453

$$s_e = \sqrt{\frac{\Sigma(y - \hat{y})^2}{n - 2}} = \sqrt{\frac{26.0261}{11 - 2}} = 1.701$$

$$s_{b_1} = \frac{s_e}{\sqrt{\Sigma(x - \bar{x})^2}} = \frac{1.701}{\sqrt{298.2453}} = 0.0985$$

The hypotheses are $H_0 : \beta_1 = 0$ versus $H_1 : \beta_1 \neq 0$. The test statistic is

$$t_0 = \frac{b_1}{s_{b_1}} = \frac{1.0997}{0.0985} = 11.168.$$

Classical approach: The level of significance is $\alpha = 0.1$. Since this is a two-tailed test with $n - 2 = 11 - 2 = 9$ degrees of freedom, the critical values are $\pm t_{0.05} = \pm 1.833$. Since $t_0 = 11.168 > t_{0.05} = 1.833$ (the test statistic falls in a critical region), we reject H_0.

P-value approach: The *P*-value for this two-tailed test is the area under the *t*-distribution with 9 degrees of freedom to the right of $t_0 = 11.168$, plus the area to the left of $-t_0 = -11.168$. From the *t*-distribution table in the row corresponding to 9 degrees of freedom, 11.168 falls to the right of 4.781, whose right-tail area is 0.0005. We must double this value in order to get the total area in both tails: 0.001. So, $P\text{-value} < 0.001$.

[Tech: *P*-value < 0.0001]. Because the *P*-value is less than the level of significance $\alpha = 0.10$, we reject H_0.

Conclusion: There is sufficient evidence to conclude that a linear relationship exists between the rate of return of the S&P 500 index and the rate of return of UTX.

(c) For a 90% confidence interval we use $t_{\alpha/2} = t_{0.05} = 1.833$:

Lower bound:
$$b_1 - t_{\alpha/2} \cdot s_{b_1} = 1.0997 - 1.833 \cdot 0.0985$$
$$= 0.9192$$

Upper bound:
$$b_1 + t_{\alpha/2} \cdot s_{b_1} = 1.0997 + 1.833 \cdot 0.0985$$
$$= 1.2802$$

(d) We can use the regression equation to calculate the mean rate of return:
$$\hat{y} = 1.0997x + 0.1200$$
$$= 1.0997(3.25) + 0.1200 \approx 3.694\%$$

19. (a) Using technology, we get:
$$\beta_0 \approx b_0 = 33.2350 \text{ and } \beta_1 \approx b_1 = -0.9428.$$

(b) Using technology, we calculate
$\hat{y} = -0.9428x + 33.2350$ to generate the table:

x	y	$\hat{y}$	$y - \hat{y}$	$(y - \hat{y})^2$	$(x - \overline{x})^2$
26.35	5.91	8.391	-2.481	6.156	89.624
12.48	30.39	21.468	8.922	79.596	19.386
19.44	31.72	14.906	16.814	282.706	6.538
13.37	79.76	20.629	59.131	3496.450	12.341
12.21	−8.4	21.723	−30.123	907.390	21.837
11.89	2.69	22.025	−19.335	373.828	24.930
26.21	4.53	8.523	−3.993	15.945	86.993
14.95	10.8	19.140	−8.340	69.548	3.736
17.57	4.01	16.669	−12.659	160.257	0.472
14.36	11.76	19.696	−7.936	62.977	6.366
26.35	5.91	8.391	-2.481	6.156	89.624
12.48	30.39	21.468	8.922	79.596	19.386
			Sum	5454.851	272.224

$$s_e = \sqrt{\frac{\Sigma(y - \hat{y})^2}{n - 2}} = \sqrt{\frac{5454.851}{10 - 2}} = 26.112$$

$$s_{b_1} = \frac{s_e}{\sqrt{\Sigma(x - \overline{x})^2}} = \frac{26.112}{\sqrt{272.224}} = 1.583$$

The hypotheses are $H_0 : \beta_1 = 0$ versus $H_1 : \beta_1 \neq 0$. The test statistic is

$$t_0 = \frac{b_1}{s_{b_1}} = \frac{-0.9428}{1.583} = -0.596.$$

Classical approach: The level of significance is $\alpha = 0.05$. Since this is a two-tailed test with $n - 2 = 10 - 2 = 8$ degrees of freedom, the critical values are $\pm t_{0.025} = \pm 2.306$. Since $t_0 = -0.596$ is between $-t_{0.025} = -2.228$ and $t_{0.025} = 2.228$, (the test statistic does not fall in a critical region), we do not reject H_0.

P-value approach: The *P*-value for this two-tailed test is the area under the *t*-distribution with 8 degrees of freedom to the left of $t_0 = -0.596$, plus the area to the right of $-t_0 = 0.596$. From the *t*-distribution table in the row corresponding to 8 degrees of freedom, 0.596 falls below 0.706, whose right-tail area is 0.25. We must double this value in order to get the total area in both tails: 0.50. So, $0.50 < P$-value. [Tech: *P*-value = 0.5678]. Because the *P*-value is greater than the level of significance $\alpha = 0.05$, we do not reject H_0.

Conclusion: There is not sufficient evidence to conclude that a linear relationship exists between CEO compensation and stock return.

(c) For a 95% confidence interval we use $t_{\alpha/2} = t_{0.025} = 2.306$:

Lower bound:
$$b_1 - t_{\alpha/2} \cdot s_{b_1} = -0.9428 - 2.306 \cdot 1.583$$
$$= -4.5924 \text{ [Tech: } -4.5924]$$

Upper bound:
$$b_1 + t_{\alpha/2} \cdot s_{b_1} = -0.9428 + 2.306 \cdot 1.583$$
$$= 2.7068 \text{ [Tech: 2.7067]}$$

(d) No, the results do not indicate that a linear relation exists. We could use $\bar{y} = 17.317\%$ as an estimate of the stock return.

21. (a) No linear relationship appears to exist between HDL cholesterol and age:

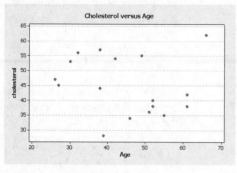

(b) Using technology, we get
$\hat{y} = -0.1298x + 50.7841$.

(c) The residuals appear to be evenly spread about the horizontal axis, so the requirement of constant variance is satisfied. Thus, a linear model seems appropriate.

(d) The five-number summary for age is: 26, 35, 46, 53.5, 66

Lower fence $= 35 - 1.5(53.5 - 35) = 7.25$

Upper fence $= 53.5 + 1.5(53.5 - 35) = 81.25$

The five-number summary for HDL cholesterol is: 28, 37, 44, 54.5, 62

Lower fence $= 37 - 1.5(54.5 - 37) = 10.75$

Upper fence $= 54.5 + 1.5(54.5 - 37) = 80.75$

So, there are no outliers with respect to either variable.

The point corresponding to the 66-year-old man has a large standardized residual so it

may be an outlier or influential. Removing the point from the data set yields the following regression equation:

$\hat{y} = -0.3554x + 59.4005$

Since there seems to be a significant change to the regression equation, we would consider the point $(66, 62)$ an influential observation.

(e) We use the equation from part (b) to generate the table that follows:

x	y	$\hat{y}$	$y - \hat{y}$	$(y - \hat{y})^2$	$(x - \bar{x})^2$
38	57	45.8501	11.1499	124.321	49
42	54	45.3307	8.6693	75.157	9
46	34	44.8113	−10.8113	116.885	1
32	56	46.6291	9.3709	87.813	169
55	35	43.6427	−8.6427	74.697	100
52	40	44.0323	−4.0323	16.259	49
61	42	42.8637	−0.8637	0.746	256
61	38	42.8637	−4.8637	23.655	256
26	47	47.4082	−0.4082	0.167	361
38	44	45.8501	−1.8501	3.423	49
66	62	42.2145	19.7855	391.467	441
30	53	46.8888	6.1112	37.346	225
51	36	44.1621	−8.1621	66.620	36
27	45	47.2784	−2.2784	5.191	324
52	38	44.0323	−6.0323	36.388	49
49	55	44.4218	10.5782	111.898	16
39	28	45.7202	−17.7202	314.007	36
			Sum	1486.041	2426

$$s_e = \sqrt{\frac{\Sigma(y - \hat{y})^2}{n - 2}} = \sqrt{\frac{1486.041}{17 - 2}} = 9.9534$$

$$s_{b_1} = \frac{s_e}{\sqrt{\Sigma(x - \bar{x})^2}} = \frac{9.9534}{\sqrt{2426}} = 0.2021$$

The hypotheses are $H_0 : \beta_1 = 0$ versus $H_1 : \beta_1 \neq 0$. The test statistic is

$$t_0 = \frac{b_1}{s_{b_1}} = \frac{-0.1298}{0.2021} = -0.642 \; [\text{Tech:} -0.643].$$

Classical approach: The level of significance is $\alpha = 0.01$. Since this is a two-tailed test with $n - 2 = 17 - 2 = 15$ degrees of freedom, the critical values are $\pm t_{0.005} = \pm 2.947$.

Since $t_0 = -0.642 > t_{0.005} = -2.947$ (the test statistic does not fall in a critical region), we do not reject H_0.

P-value approach: The P-value for this two-tailed test is the area under the t-distribution with 15 degrees of freedom to the left of $t_0 = -0.642$, plus the area to the right of 0.642. From the t-distribution table in the row corresponding to 15 degrees of freedom, 0.642 falls to the left of 0.691, whose right-tail area is 0.25. We must double this value in order to get the total area in both tails: 0.50. So, P-value > 0.50. [Tech: P-value = 0.5299]. Because the P-value is greater than the level of significance $\alpha = 0.01$, we do not reject H_0.

Conclusion: A linear relationship does not exist between age and HDL cholesterol level.

(f) For a 95% confidence interval we use $t_{\alpha/2} = t_{0.025} = 2.131$:

Lower bound:
$$b_1 - t_{\alpha/2} \cdot s_{b_1} = -0.1298 - 2.131 \cdot 0.2021$$
$$= -0.5605$$

Upper bound:
$$b_1 + t_{\alpha/2} \cdot s_{b_1} = -0.1298 + 2.131 \cdot 0.2021$$
$$= 0.3009 \text{ [Tech: 0.3008]}$$

(g) We have concluded that a linear relationship does not exist between age and HDL cholesterol level, so we do not recommend using the least-squares regression equation to predict the HDL cholesterol levels. A good estimate of the HDL cholesterol level would be $\bar{y} = 44.9$.

23. (a) Using technology, we get
$$\hat{y} = 0.0657 x - 12.4967.$$

(b) The hypotheses are $H_0 : \beta_1 = 0$ versus $H_1 : \beta_1 \neq 0$. For a test at the $\alpha = 0.01$ level of significance with $n - 2 = 9 - 2 = 7$ degrees of freedom, the critical values are

$\pm t_{0.005} = \pm 3.499$. Using technology we find the test statistic to be $t_0 = 17.608$, and we find the P-value to be P-value < 0.0001.

Classical approach: Since $t_0 = 17.608 > t_{0.005} = 3.499$ (the test statistic falls within the critical region), we reject H_0.

P-value approach: Since P-value < 0.0001 $< \alpha = 0.01$ we reject H_0.

Conclusion: A linear relationship exists between distance from the sun and length of the sidereal year.

(c)

(d)

(e) A linear model is not appropriate. The scatter plot seems to show some curvature, and there is a pattern in the residual plot.

(f) The moral is that the inferential procedures may lead us to believe that a linear relation exists between the two variables even though diagnostic tools (such as residual plots) indicate that a linear model is inappropriate.

25. (a)

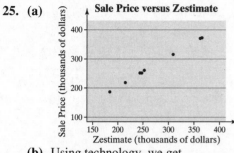

Sale Price versus Zestimate

(b) Using technology, we get
$\hat{y} = 1.0228 x - 0.7590$.

The hypotheses are $H_0 : \beta_1 = 0$ versus
$H_1 : \beta_1 \neq 0$. For a test at the $\alpha = 0.05$ level
of significance with $n - 2 = 8 - 2 = 6$
degrees of freedom, the critical values are
$\pm t_{0.025} = \pm 2.447$. Using technology we find
the test statistic to be $t_0 = 105.426$, and we
find the P-value to be P-value < 0.0001.

Classical approach: Since $t_0 = 105.426 >$
$t_{0.025} = 2.447$ (the test statistic falls within
the critical region), we reject H_0.

P-value approach: Since P-value < 0.0001
$< \alpha = 0.05$ we reject H_0.

Conclusion: A linear relationship exists
between the Zestimate and the sale price.

(c) Using technology, we get
$\hat{y} = 0.5220 x + 115.8094$.

The hypotheses are $H_0 : \beta_1 = 0$ versus
$H_1 : \beta_1 \neq 0$. For a test at the $\alpha = 0.05$ level
of significance with $n - 2 = 9 - 2 = 7$
degrees of freedom, the critical values are
$\pm t_{0.025} = \pm 2.447$. Using technology we find
the test statistic to be $t_0 = 1.441$, and we
find the P-value to be P-value $= 0.1927$.

Classical approach: Since $t_0 = 1.441$ is
between the critical values,
$\pm t_{0.025} = \pm 2.365$ (the test statistic does not
fall within the critical region), we do not
reject H_0.

P-value approach: Using Table VI with 7
degrees of freedom, we find that $t_0 = 1.441$
is between 1.415 and 1.895, which

correspond to areas to the right of 0.10 and
0.05, respectively. Since this is a two-tailed
test, we need to double these values to get
the P-value. $0.10 < P$-value < 0.20 Since
the P-value is greater than the level of
significance $\alpha = 0.05$, we do not reject
H_0.

Conclusion: There is not sufficient evidence
to conclude that there is a linear relation
between the Zestimate and the sale price.

Yes, this observation is influential.

27. The y-coordinates on the least-squares regression
line represent the mean value of the response
variable for any given value of the explanatory
variable.

29. We do not conduct inference on the linear
correlation coefficient because a hypothesis test on
the slope and a hypothesis test on the linear
correlation coefficient yield the same conclusion.
Moreover, the requirements for conducting
inference on the linear correlation coefficient are
very hard to verify.

Section 14.2

1. confidence; mean

3. (a) From Problem 5 in Section 14.1, the least-
squares regression equation is
$\hat{y} = 2.0233 x - 2.3256$. So, the predicted
mean value of y when $x = 7$ is
$\hat{y} = 2.0233(7) - 2.3256 \approx 11.8$.

(b) From Problem 5 in Section 14.1, we have
that $n = 5$, $s_e = 0.5134$, $\bar{x} = 5.4$, and
$\sum (x - \bar{x})^2 = 17.20$. For a 95% confidence
interval with $n - 2 = 3$ degrees of freedom,
we use $t_{0.025} = 3.182$.

Lower Bound:

$$\hat{y} - t_{\alpha/2} \cdot s_e \sqrt{\frac{1}{n} + \frac{(x^* - \bar{x})^2}{\sum (x - \bar{x})^2}}$$

$$= 11.8 - 3.182 \cdot 0.5134 \cdot \sqrt{\frac{1}{5} + \frac{(7 - 5.4)^2}{17.20}}$$

$$\approx 10.8 \ [\text{Tech: } 10.9]$$

Upper Bound:

$$\hat{y} + t_{\alpha/2} \cdot s_e \sqrt{\frac{1}{n} + \frac{(x^* - \overline{x})^2}{\Sigma(x - \overline{x})^2}}$$

$$= 11.8 + 3.182 \cdot 0.5134 \cdot \sqrt{\frac{1}{5} + \frac{(7 - 5.4)^2}{17.20}}$$

$$\approx 12.8$$

(c) The predicted value of y when $x = 7$ is also $\hat{y} = 11.8$.

(d) Lower Bound:

$$\hat{y} - t_{\alpha/2} \cdot s_e \sqrt{1 + \frac{1}{n} + \frac{(x^* - \overline{x})^2}{\Sigma(x - \overline{x})^2}}$$

$$= 11.8 - 3.182 \cdot 0.5134 \cdot \sqrt{1 + \frac{1}{5} + \frac{(7 - 5.4)^2}{17.20}}$$

$$\approx 9.9$$

Upper Bound:

$$\hat{y} + t_{\alpha/2} \cdot s_e \sqrt{1 + \frac{1}{n} + \frac{(x^* - \overline{x})^2}{\Sigma(x - \overline{x})^2}}$$

$$= 11.8 + 3.182 \cdot 0.5134 \cdot \sqrt{1 + \frac{1}{5} + \frac{(7 - 5.4)^2}{17.20}}$$

$$\approx 13.7$$

(e) In (a) and (b) we are predicting the mean response (a point estimate and an interval estimate) for the population of individuals with $x = 7$. In (c) and (d) we are predicting the response (a point estimate and an interval estimate) for a single individual with $x = 7$.

5. (a) From Problem 7 in Section 14.1, the least-squares regression equation is $\hat{y} = 2.2x + 1.2$. The predicted mean value of y when $x = 1.4$ is $\hat{y} = 2.2(1.4) + 1.2 = 4.3$.

(b) From Problem 7 in Section 14.1, we have that $n = 5$, $s_e = 0.8944$, $\overline{x} = 0$, and $\Sigma(x - \overline{x})^2 = 10$. For a 95% confidence interval with $n - 2 = 3$ degrees of freedom, we use $t_{0.025} = 3.182$.

Lower Bound:

$$\hat{y} - t_{\alpha/2} \cdot s_e \sqrt{\frac{1}{n} + \frac{(x^* - \overline{x})^2}{\Sigma(x - \overline{x})^2}}$$

$$= 4.3 - 3.182 \cdot 0.8944 \cdot \sqrt{\frac{1}{5} + \frac{(1.4 - 0)^2}{10}}$$

$$\approx 2.5$$

Upper Bound:

$$\hat{y} + t_{\alpha/2} \cdot s_e \sqrt{\frac{1}{n} + \frac{(x^* - \overline{x})^2}{\Sigma(x - \overline{x})^2}}$$

$$= 4.3 + 3.182 \cdot 0.8944 \cdot \sqrt{\frac{1}{5} + \frac{(1.4 - 0)^2}{10}}$$

$$\approx 6.1$$

(c) The predicted value of y when $x = 1.4$ is also $\hat{y} = 4.3$.

(d) Lower Bound:

$$\hat{y} - t_{\alpha/2} \cdot s_e \sqrt{1 + \frac{1}{n} + \frac{(x^* - \overline{x})^2}{\Sigma(x - \overline{x})^2}}$$

$$= 4.3 - 3.182 \cdot 0.8944 \cdot \sqrt{1 + \frac{1}{5} + \frac{(1.4 - 0)^2}{10}}$$

$$\approx 0.9$$

Upper Bound:

$$\hat{y} + t_{\alpha/2} \cdot s_e \sqrt{1 + \frac{1}{n} + \frac{(x^* - \overline{x})^2}{\Sigma(x - \overline{x})^2}}$$

$$= 4.3 + 3.182 \cdot 0.8944 \cdot \sqrt{1 + \frac{1}{5} + \frac{(1.4 - 0)^2}{10}}$$

$$\approx 7.7$$

7. (a) From Problem 11 in Section 14.1, the least-squares regression equation is $\hat{y} = -0.0479x + 69.0296$. The predicted mean value of y when $x = 20$ is $\hat{y} = -0.0479(20) + 69.0296 = 68.07$.

(b) From Problem 11 in Section 14.1, we have that $n = 7$, $s_e = 0.3680$, $\overline{x} = 43.86$, and $\Sigma(x - \overline{x})^2 = 7344.9$. For a 90% confidence interval with $n - 2 = 7 - 2 = 5$ degrees of freedom, we use $t_{0.05} = 2.015$.

Lower Bound:

$$\hat{y}-t_{\alpha/2}\cdot s_e\sqrt{\frac{1}{n}+\frac{(x^*-\overline{x})^2}{\Sigma(x-\overline{x})^2}}$$

$$=68.07-2.015\cdot0.3680\cdot\sqrt{\frac{1}{7}+\frac{(20-43.86)^2}{7344.9}}$$

$$\approx67.722 \quad \text{[Tech: 67.723]}$$

Upper Bound:

$$\hat{y}+t_{\alpha/2}\cdot s_e\sqrt{\frac{1}{n}+\frac{(x^*-\overline{x})^2}{\Sigma(x-\overline{x})^2}}$$

$$=68.07+2.015\cdot0.3680\cdot\sqrt{\frac{1}{7}+\frac{(20-43.86)^2}{7344.9}}$$

$$\approx68.418 \quad \text{[Tech: 68.420]}$$

(c) The predicted value of y when $x=20$ is also $\hat{y}=68.07$.

(d) Lower Bound:

$$\hat{y}-t_{\alpha/2}\cdot s_e\sqrt{1+\frac{1}{n}+\frac{(x^*-\overline{x})^2}{\Sigma(x-\overline{x})^2}}$$

$$=68.07-2.015\cdot0.3680\cdot\sqrt{1+\frac{1}{7}+\frac{(20-43.86)^2}{7344.9}}$$

$$\approx67.251 \quad \text{[Tech: 67.252]}$$

Upper Bound:

$$\hat{y}+t_{\alpha/2}\cdot s_e\sqrt{1+\frac{1}{n}+\frac{(x^*-\overline{x})^2}{\Sigma(x-\overline{x})^2}}$$

$$=68.07+2.015\cdot0.3680\cdot\sqrt{1+\frac{1}{7}+\frac{(20-43.86)^2}{7344.9}}$$

$$\approx68.889 \quad \text{[Tech: 68.891]}$$

(e) The prediction made in part (a) is an estimate of the mean well-being index composite score for all individuals whose commute time is 20 minutes. The prediction made in part (c) is an estimate of the well-being index composite score of one individual, Jane, whose commute time is 20 minutes.

9. (a) From Problem 13 in Section 14.1, the least-squares regression equation is $\hat{y}=0.1827x+12.4932$. So, the predicted mean value of y when $x=25.75$ is $\hat{y}=0.1827(25.75)+12.4932=17.20$ inches.

(b) From Problem 13 in Section 14.1, we have that $n=11$, $s_e=0.0954$, $\overline{x}=26.4545$, and

$\Sigma(x-\overline{x})^2\approx11.9773$. For a 95% confidence interval with $n-2=9$ degrees of freedom, we use $t_{0.025}=2.262$.

Lower Bound:

$$\hat{y}-t_{\alpha/2}\cdot s_e\sqrt{\frac{1}{n}+\frac{(x^*-\overline{x})^2}{\Sigma(x-\overline{x})^2}}$$

$$=17.20-2.262\cdot0.0954\cdot\sqrt{\frac{1}{11}+\frac{(25.75-26.4545)^2}{11.9773}}$$

$$\approx17.12 \text{ inches}$$

Upper Bound:

$$\hat{y}+t_{\alpha/2}\cdot s_e\sqrt{\frac{1}{n}+\frac{(x^*-\overline{x})^2}{\Sigma(x-\overline{x})^2}}$$

$$=17.20+2.262\cdot0.0954\cdot\sqrt{\frac{1}{11}+\frac{(25.75-26.4545)^2}{11.9773}}$$

$$\approx17.28 \text{ inches}$$

(c) The predicted value of y when $x=25.75$ is also $\hat{y}=17.20$ inches.

(d) Lower Bound:

$$\hat{y}-t_{\alpha/2}\cdot s_e\sqrt{1+\frac{1}{n}+\frac{(x^*-\overline{x})^2}{\Sigma(x-\overline{x})^2}}$$

$$=17.20-2.262\cdot0.0954\cdot\sqrt{1+\frac{1}{11}+\frac{(25.75-26.4545)^2}{11.9773}}$$

$$\approx16.97 \text{ inches}$$

Upper Bound:

$$\hat{y}+t_{\alpha/2}\cdot s_e\sqrt{1+\frac{1}{n}+\frac{(x^*-\overline{x})^2}{\Sigma(x-\overline{x})^2}}$$

$$=17.20+2.262\cdot0.0954\cdot\sqrt{1+\frac{1}{11}+\frac{(25.75-26.4545)^2}{11.9773}}$$

$$\approx17.43 \text{ inches}$$

(e) In (a) and (b) we are predicting the mean head circumference (a point estimate and an interval estimate) for the population of all children who are 25.75 inches tall. In (c) and (d) we are predicting the head circumference (a point estimate and an interval estimate) of a single child who is 25.75 inches tall.

11. (a) From Problem 15 in Section 14.1, the least-squares regression equation is $\hat{y}=0.6764x+2675.6$. So, the predicted mean value of y when $x=2550$ is $\hat{y}=0.6764\cdot2550+2675.6=4400.4$ psi.

(b) From Problem 15 in Section 14.1, we have that $n = 10$, $s_e = 271.04$, $\bar{x} = 2882$, and $\sum(x - \bar{x})^2 = 1{,}739{,}160$. For a 95% confidence interval with $n - 2 = 8$ degrees of freedom, we use $t_{0.025} = 2.306$.

Lower Bound:

$$\hat{y} - t_{\alpha/2} \cdot s_e \sqrt{\frac{1}{n} + \frac{(x^* - \bar{x})^2}{\sum(x - \bar{x})^2}}$$

$$= 4400.4 - 2.306 \cdot 271.04 \cdot \sqrt{\frac{1}{10} + \frac{(2550 - 2882)^2}{1{,}739{,}160}}$$

$$\approx 4147.8 \text{ psi}$$

Upper Bound:

$$\hat{y} + t_{\alpha/2} \cdot s_e \sqrt{\frac{1}{n} + \frac{(x^* - \bar{x})^2}{\sum(x - \bar{x})^2}}$$

$$= 4400.4 + 2.306 \cdot 271.04 \cdot \sqrt{\frac{1}{10} + \frac{(2550 - 2882)^2}{1{,}739{,}160}}$$

$$\approx 4653.0 \text{ [Tech: 4653.1] psi}$$

(c) The predicted value of y when $x = 2550$ is also $\hat{y} = 4400.4$ psi.

(d) Lower Bound:

$$\hat{y} - t_{\alpha/2} \cdot s_e \sqrt{1 + \frac{1}{n} + \frac{(x^* - \bar{x})^2}{\sum(x - \bar{x})^2}}$$

$$= 4400.4 - 2.306 \cdot 271.04 \cdot \sqrt{1 + \frac{1}{10} + \frac{(2550 - 2882)^2}{1{,}739{,}160}}$$

$$\approx 3726.3 \text{ psi}$$

Upper Bound:

$$\hat{y} + t_{\alpha/2} \cdot s_e \sqrt{1 + \frac{1}{n} + \frac{(x^* - \bar{x})^2}{\sum(x - \bar{x})^2}}$$

$$= 4400.4 + 2.306 \cdot 271.04 \cdot \sqrt{1 + \frac{1}{10} + \frac{(2550 - 2882)^2}{1{,}739{,}160}}$$

$$\approx 5074.5 \text{ [Tech: 5074.6] psi}$$

(e) In (a) and (b) we are predicting the mean 28-day strength (a point estimate and an interval estimate) for the population of all concrete that has a 7-day strength of 2550 psi. In (c) and (d) we are predicting the 28-day strength (a point estimate and an interval estimate) for a single batch of concrete with 7-day strength of 2550 psi.

13. (a) From Problem 17 in Section 14.1, the least-squares regression equation is $\hat{y} = 1.0997x + 0.12$. So, the predicted mean value of y when $x = 4.2$ is $\hat{y} = 1.0997 \cdot 4.2 + 0.12 = 4.74\%$.

(b) From Problem 17 in Section 14.1, we have that $n = 11$, $s_e = 3.2309$, $\bar{x} \approx -0.5373$, and $\sum(x - \bar{x})^2 = 125.8797$. For a 90% confidence interval with $n - 2 = 9$ degrees of freedom, we use $t_{0.05} = 1.833$.

Lower Bound:

$$\hat{y} - t_{\alpha/2} \cdot s_e \sqrt{\frac{1}{n} + \frac{(x^* - \bar{x})^2}{\sum(x - \bar{x})^2}}$$

$$= 4.430 - 1.833 \cdot 3.2309 \cdot \sqrt{\frac{1}{11} + \frac{(4.2 - (-0.5373))^2}{125.8797}}$$

$$\approx 1.357\%$$

Upper Bound:

$$\hat{y} + t_{\alpha/2} \cdot s_e \sqrt{\frac{1}{n} + \frac{(x^* - \bar{x})^2}{\sum(x - \bar{x})^2}}$$

$$= 4.430 + 1.833 \cdot 3.2309 \cdot \sqrt{\frac{1}{11} + \frac{(4.2 - (-0.5373))^2}{125.8797}}$$

$$\approx 7.503\%$$

(c) The predicted value of y when $x = 4.2$ is also $\hat{y} = 4.430\%$.

(d) Lower Bound:

$$\hat{y} - t_{\alpha/2} \cdot s_e \sqrt{1 + \frac{1}{n} + \frac{(x^* - \bar{x})^2}{\sum(x - \bar{x})^2}}$$

$$= 4.430 - 1.833 \cdot 3.2309 \cdot \sqrt{1 + \frac{1}{11} + \frac{(4.2 - (-0.5373))^2}{125.8797}}$$

$$\approx -2.242\%$$

Upper Bound:

$$\hat{y} + t_{\alpha/2} \cdot s_e \sqrt{1 + \frac{1}{n} + \frac{(x^* - \bar{x})^2}{\sum(x - \bar{x})^2}}$$

$$= 4.430 + 1.833 \cdot 3.2309 \cdot \sqrt{1 + \frac{1}{11} + \frac{(4.2 - (-0.5373))^2}{125.8797}}$$

$$\approx 11.102\%$$

(e) Although the predicted rates of return in parts (a) and (c) are the same, the intervals are different because the distribution of the mean rate of return, part (a), has less variability than the distribution of the individual rate of return of a particular S&P 500 rate of return, part (c).

15. (a) In Problem 19 of Section 14.1, we showed that no linear relationship exists between CEO compensation and stock return. Therefore, it does not make sense to construct either a confidence interval or prediction interval based on the least-squares regression equation.

(b) Since no linear relation exists between x and y, we use the techniques of Section 9.2 on the y-data to construct the confidence interval. The summary statistics are $n = 10$, $\bar{y} \approx 17.317$ and $s_y \approx 25.159$. For 95% confidence with $n - 1 = 9$, we use $t_{0.025} = 2.262$.

Lower Bound:

$$\bar{y} - t_{0.025} \cdot \frac{s_y}{\sqrt{n}} = 17.317 - 2.262 \cdot \frac{25.159}{\sqrt{10}}$$
$$\approx -0.679 \quad [\text{Tech: } -0.681]$$

Upper Bound:

$$\bar{y} + t_{0.025} \cdot \frac{s_y}{\sqrt{n}} = 17.317 + 2.262 \cdot \frac{25.159}{\sqrt{10}}$$
$$\approx 35.313 \quad [\text{Tech: } 35.315]$$

17. (a)

Price versus Size for 3D Televisions

(b) $n = 9$, $\sum x_i = 450$, $\sum y_i = 20,559$, $\sum x_i^2 = 23,102$, $\sum y_i^2 = 58,880,225$, $\sum x_i y_i = 1,110,828$

$$r = \frac{\sum x_i y_i - \frac{\sum x_i \sum y_i}{n}}{\sqrt{\left(\sum x_i^2 - \frac{(\sum x_i)^2}{n}\right)\left(\sum y_i^2 - \frac{(\sum y_i)^2}{n}\right)}}$$

$$= \frac{1,110,828 - \frac{(450)(20,559)}{9}}{\sqrt{\left(23,102 - \frac{(450)^2}{9}\right)\left(58,880,225 - \frac{(20,559)^2}{9}\right)}}$$

$$\approx 0.979$$

(c) Looking to Table II, the critical value for a correlation coefficient when $n = 9$ is 0.666. Since $0.979 > 0.666$, we conclude that a linear relation exists between size and price.

(d) Using technology, we obtain the regression equation: $\hat{y} = 137.6711x - 4599.2215$

(e) For each inch added to the size of a 3D television, the price will increase by about $137.67, on average.

(f) It is not reasonable to interpret the intercept. It would indicate the price of a 0-inch 3D television, which makes no sense.

(g) We compute the coefficient of determination: $r^2 = (0.979)^2 \approx 0.957$. About 95.7% of the variability in price is explained by the variability in size.

(h) $\beta_0 \approx b_0 = -4599.2215$; $\beta_1 \approx b_1 = 137.6711$

(i) We use the least-squares regression equation $\hat{y} = 137.6711x - 4599.2215$ to complete the table that follows:

x	y	$\hat{y}$	$y - \hat{y}$	$(y - \hat{y})^2$	$(x - \bar{x})^2$
40	1168	907.622	260.378	67796.510	100
40	1170	907.622	262.378	68842.021	100
46	1494	1733.649	−239.649	57431.618	16
46	1500	1733.649	−233.649	54591.831	16
46	1559	1733.649	−174.649	30502.255	16
52	2700	2559.676	140.324	19690.958	4
55	2600	2972.689	−372.689	138896.953	25
60	3890	3661.044	228.956	52420.714	100
65	4478	4349.400	128.600	16538.017	225
			Sum	506710.877	602

$$s_e = \sqrt{\frac{\Sigma(y-\hat{y})^2}{n-2}} = \sqrt{\frac{506710.877}{9-2}}$$
$$= 269.049$$

(j) A normal probability plot shows that the residuals are approximately normally distributed.

(k) $s_{b_1} = \dfrac{s_e}{\sqrt{\Sigma(x-\bar{x})^2}} = \dfrac{269.049}{\sqrt{602}} = 10.966$

(l) The hypotheses are $H_0 : \beta_1 = 0$ versus $H_1 : \beta_1 \neq 0$. The test statistic is

$$t_0 = \frac{b_1}{s_{b_1}} = \frac{137.6711}{10.966} = 12.555 .$$

Classical approach: The level of significance is $\alpha = 0.05$. Since this is a two-tailed test with $n - 2 = 7$ degrees of freedom, the critical values are $\pm t_{0.025} = \pm 2.365$. Since $t_0 = 12.555 > t_{0.025} = 2.365$ (the test statistic falls in a critical region), we reject H_0.

P-value approach: The P-value for this two-tailed test is the area under the t-distribution with 7 degrees of freedom to the right of $t_0 = 12.555$, plus the area to the left of $-t_0 = -12.555$. From the t-distribution table in the row corresponding to 7 degrees of freedom, 12.555 falls above 5.408, whose right-tail area is 0.0005. We must double this value in order to get the total area in both tails: 0.001. So, P-value < 0.001. [Tech: P-value < 0.0001]. Because the P-value is less than the level of significance $\alpha = 0.05$, we reject H_0.

Conclusion: There is sufficient evidence to conclude that a linear relationship exists between the size and price of 3D televisions.

(m) For a 95% confidence interval we use $t_{\alpha/2} = t_{0.025} = 2.365$:

Lower bound:
$$b_1 - t_{\alpha/2} \cdot s_{b_1} = 137.6711 - 2.365 \cdot 10.966$$
$$= 111.7416 \quad [\text{Tech: } 111.7374]$$

Upper bound:
$$b_1 + t_{\alpha/2} \cdot s_{b_1} = 137.6711 + 2.365 \cdot 10.966$$
$$= 163.6007 \quad [\text{Tech: } 163.6048]$$

(n) The predicted mean value of y when $x = 50$ is $\hat{y} = 137.6711(50) - 4599.2215$
$$= 2284.333 \approx \$2284.$$

(o) For a 95% confidence interval with $n - 2 = 7$ degrees of freedom, we use $t_{0.025} = 2.365$.

Lower Bound:

$$\hat{y} - t_{\alpha/2} \cdot s_e \sqrt{\frac{1}{n} + \frac{(x^* - \bar{x})^2}{\Sigma(x-\bar{x})^2}}$$

$$= 2284.3 - 2.365 \cdot 269.049 \cdot \sqrt{\frac{1}{9} + \frac{(50-50)^2}{602}}$$

$$\approx \$2072.3 \quad [\text{Tech: } \$2072.2]$$

Upper Bound:

$$\hat{y} + t_{\alpha/2} \cdot s_e \sqrt{\frac{1}{n} + \frac{(x^* - \bar{x})^2}{\Sigma(x-\bar{x})^2}}$$

$$= 2284.3 + 2.365 \cdot 269.049 \cdot \sqrt{\frac{1}{9} + \frac{(50-50)^2}{602}}$$

$$\approx \$2496.4 \quad [\text{Tech: } \$2496.4]$$

(p) The price of the Sharp 52" 3D television is above the average price for 50-inch 3D televisions. Its price, $2700, is above the upper end of the confidence interval found in part (o).

(q) The predicted value of y when $x = 50$ is also $\hat{y} = \$2284$.

(r) Lower Bound:

$$\hat{y} - t_{\alpha/2} \cdot s_e \sqrt{1 + \frac{1}{n} + \frac{(x^* - \bar{x})^2}{\Sigma(x-\bar{x})^2}}$$

$$= 2284.3 - 2.365 \cdot 269.049 \cdot \sqrt{1 + \frac{1}{9} + \frac{(50-50)^2}{602}}$$

$$\approx \$1613.7 \quad [\text{Tech: } \$1613.6]$$

Upper Bound:

$$\hat{y}+t_{\alpha/2}\cdot s_e\sqrt{1+\frac{1}{n}+\frac{(x^*-\bar{x})^2}{\sum(x-\bar{x})^2}}$$

$$=2284.3+2.365\cdot269.049\cdot\sqrt{1+\frac{1}{9}+\frac{(50-50)^2}{602}}$$

$$\approx \$2954.9 \quad [\text{Tech: } \$2955.1]$$

(s) Answers will vary. Possible lurking variables: brand name, picture quality, and audio quality.

Section 14.3

1. correlation matrix

3. 4.39

5. indicator; dummy

7. (a) The slope coefficient of x_1 is 3. This indicates that $\hat{y}$ will increase 3 units, on average, for every one unit increase in x_1, provided x_2 remains constant. The slope coefficient of x_2 is -4. This indicates that $\hat{y}$ will decrease 4 units, on average, for every one unit increase in x_2, provided x_1 remains constant.

(b) $\hat{y}=5+3x_1-4x_2$
$\hat{y}=5+3(10)-4x_2$
$\hat{y}=35-4x_2$

(c) $\hat{y}=5+3x_1-4x_2$
$\hat{y}=5+3(15)-4x_2$
$\hat{y}=50-4x_2$

(d) $\hat{y}=5+3x_1-4x_2$
$\hat{y}=5+3(20)-4x_2$
$\hat{y}=65-4x_2$

(e) Changing the value of x_1 has the effect of changing the y-intercept of the graph of the regression line.

9. (a) $R^2_{adj}=1-\left(\frac{n-1}{n-k-1}\right)\left(1-R^2\right)$

$$=1-\left(\frac{25-1}{25-3-1}\right)(1-0.653)=0.603$$

(b) $F=\frac{R^2}{1-R^2}\cdot\frac{n-(k+1)}{k}$

$$=\frac{0.653}{1-0.653}\cdot\frac{25-(3+1)}{3}=13.173$$

(c) $R^2_{adj}=1-\left(\frac{25-1}{25-4-1}\right)(1-0.665)=0.598$

Do not add the new variable. Its addition reduces the proportion of variance explained by the model.

11. **(a)** **Correlations: x1, x2, x3, y**

```
          x1        x2        x3
x2   -0.460
x3    0.183   -0.012
y    -0.748    0.788    0.258
```

Because the correlations among the explanatory variables are all between -0.7 and 0.7, multicollinearity does not appear to be a problem.

(b) **Regression Analysis: y versus x1, x2, x3**

```
The regression equation is
y = 7.96 - 0.104 x1 + 0.936 x2 + 0.115 x3

Predictor      Coef   SE Coef       T       P
Constant     7.9647    0.5407   14.73   0.000
x1          -0.10354   0.02080   -4.98   0.003
x2           0.9359    0.1999    4.68   0.003
x3           0.11524   0.03188    3.61   0.011

S = 0.245127    R-Sq = 94.0%    R-Sq(adj) = 91.0%

Analysis of Variance

Source          DF       SS       MS       F       P
Regression       3   5.6485   1.8828   31.33   0.000
Residual Error   6   0.3605   0.0601
Total            9   6.0090
```

The regression equation is $\hat{y} = 7.9647 - 0.1035x_1 + 0.9359x_2 + 0.1152x_3$.

(c) From the output in part (b), our test statistic is $F_0 = 31.33$. Since P-value < 0.001, we reject the null hypothesis and conclude that at least one of the explanatory variables is linearly related to the response variable.

(d) From the output in part (b), conclude the following:

Reject $H_0 : \beta_1 = 0$ since $t_{\beta 1} = -4.98$ and P-value $= 0.003 < \alpha = 0.05$.

Reject $H_0 : \beta_2 = 0$ since $t_{\beta 2} = 4.68$ and P-value $= 0.003 < \alpha = 0.05$

Reject $H_0 : \beta_3 = 0$ since $t_{\beta 3} = 3.61$ and P-value $= 0.011 < \alpha = 0.05$

13. (a) Correlations: x1, x2, x3, x4, y

```
            x1        x2        x3        x4
x2    -0.184
x3     0.182    -0.106
x4    -0.166    -0.558     0.291
y     -0.046     0.932    -0.063    -0.750
```

Because the correlations among the explanatory variables are all between −0.7 and 0.7, multicollinearity does not appear to be a problem.

(b) Regression Analysis: y versus x1, x2, x3, x4

```
The regression equation is
y = 279 + 0.07 x1 + 1.15 x2 + 0.390 x3 - 2.94 x4

Predictor      Coef    SE Coef       T       P
Constant     279.02      61.34    4.55   0.003
x1            0.068      1.159    0.06   0.955
x2           1.1460     0.1546    7.41   0.000
x3           0.3897     0.2639    1.48   0.183
x4          -2.9378     0.8235   -3.57   0.009

S = 13.1177   R-Sq = 95.9%   R-Sq(adj) = 93.6%

Analysis of Variance
Source           DF        SS       MS       F       P
Regression        4   28152.4   7038.1   40.90   0.000
Residual Error    7    1204.5    172.1
Total            11   29356.9
```

The regression equation is $\hat{y} = 279.02 + 0.068x_1 + 1.1460x_2 + 0.3897x_3 - 2.9378x_4$.

$F_0 = 40.90$ and P-value < 0.001 so we reject the null hypothesis and conclude that at least one of the explanatory variables is linearly related to the response variable.

Based on the regression output, we see that x_1 and x_3 have slope coefficients that are not significantly different from zero (P-value > 0.05).

(c) The variable x_1 has the highest P-value. After removing this variable, the regression equation becomes:
$\hat{y} = 282.33 + 1.1426x_2 + 0.3941x_3 - 2.9570x_4$
The new test statistic is $F_0 = 62.29$ and P-value < 0.001, so the model is still significant.

The variable x_3 has a slope coefficient that is not significantly different from zero. Removing this variable gives $\hat{y} = 281.62 + 1.1584x_2 - 2.6267x_4$
The new test statistic is $F_0 = 76.89$ and P-value < 0.001 so the model is still significant. Both of the explanatory variables have slope coefficients that are significantly different from zero.
Reject $H_0 : \beta_2 = 0$ since $t_{\beta 2} = 7.89$ and P-value $< 0.001 < \alpha = 0.05$
Reject $H_0 : \beta_4 = 0$ since $t_{\beta 4} = -3.54$ and P-value $= 0.006 < \alpha = 0.05$

(d)

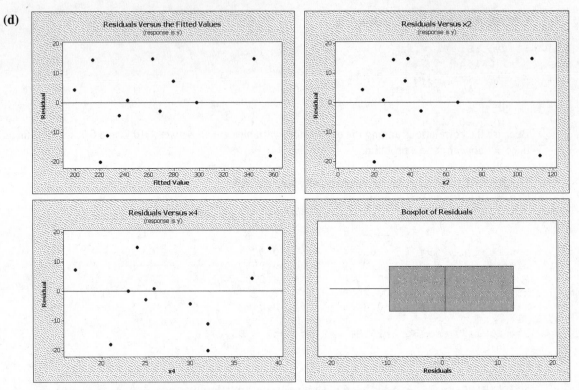

None of the residual plots show any discernible pattern, and the boxplot does not show any outliers. Therefore, the linear model is appropriate.

(e) $\hat{y} = 281.62 + 1.1584x_2 - 2.6267x_4$

$= 281.62 + 1.1584(35.6) - 2.6267(29)$

$= 246.7$

(f)

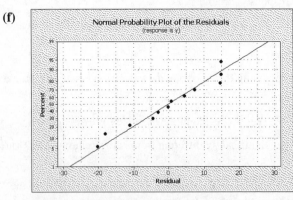

Because the points are roughly linear (i.e. lie within the confidence bounds), it is reasonable to conclude that the residuals are normally distributed. Therefore, it is reasonable to construct confidence and prediction intervals.

(g)   95% Confidence Interval: Lower Bound: 237.45; Upper Bound: 255.91

95% Prediction Interval: Lower Bound: 214.93; Upper Bound: 278.43

15. (a) $y_i = \beta_0 + \beta_1 x_1 + \beta_2 x_2 + \beta_3 x_3 + \varepsilon_i$

(b) $y_i = \beta_0 + \beta_1 x_1 + \beta_2 x_2 + \beta_3 x_1 x_2 + \varepsilon_i$

(c) $y_i = \beta_0 + \beta_1 x_1 + \beta_2 x_2 + \beta_3 x_3 + \beta_4 x_2 x_3 + \beta_5 x_3^2 + \varepsilon_i$

17. (a) Substitute $x_2 = 10$, $x_2 = 20$, and $x_2 = 30$ into the regression equation to get the following equations:

$x_2 = 10$: $\hat{y} = 4 + 0.4x_1 - 1.3x_2 = 4 + 0.4x_1 - 1.3(10) = -9 + 0.4x_1$, or $\hat{y} = -9 + 0.4x_1$

$x_2 = 20$: $\hat{y} = 4 + 0.4x_1 - 1.3x_2 = 4 + 0.4x_1 - 1.3(20) = -22 + 0.4x_1$, or $\hat{y} = -22 + 0.4x_1$

$x_2 = 30$: $\hat{y} = 4 + 0.4x_1 - 1.3x_2 = 4 + 0.4x_1 - 1.3(30) = -35 + 0.4x_1$, or $\hat{y} = -35 + 0.4x_1$

Plotting these three equations simultaneously, we get:

(b) Substitute $x_1 = 40$, $x_1 = 50$, and $x_1 = 60$ into the regression equation to get the following equations:

$x_1 = 40$: $\hat{y} = 4 + 0.4x_1 - 1.3x_2 = 4 + 0.4(40) - 1.3x_2 = 20 - 1.3x_2$, or $\hat{y} = 20 - 1.3x_2$

$x_1 = 50$: $\hat{y} = 4 + 0.4x_1 - 1.3x_2 = 4 + 0.4(50) - 1.3x_2 = 24 - 1.3x_2$, or $\hat{y} = 24 - 1.3x_2$

$x_1 = 60$: $\hat{y} = 4 + 0.4x_1 - 1.3x_2 = 4 + 0.4(60) - 1.3x_2 = 28 - 1.3x_2$, or $\hat{y} = 28 - 1.3x_2$

Plotting these three equations simultaneously, we get:

(c) The lines are parallel, so we know there is not an interaction between x_1 and x_2.

(d) Substitute $x_2 = 10$, $x_2 = 20$, and $x_2 = 30$ into the regression equation to get the following equations:

$x_2 = 10$: $\hat{y} = 4 + 0.4x_1 - 1.3x_2 + 0.05x_1 x_2 = 4 + 0.4x_1 - 1.3(10) + 0.05x_1(10) = -9 + 0.9x_1$, or $\hat{y} = -9 + 0.9x_1$

$x_2 = 20$: $\hat{y} = 4 + 0.4x_1 - 1.3x_2 + 0.05x_1 x_2 = 4 + 0.4x_1 - 1.3(20) + 0.05x_1(20) = -22 + 1.4x_1$, or $\hat{y} = -22 + 1.4x_1$

$x_2 = 30$: $\hat{y} = 4 + 0.4x_1 - 1.3x_2 + 0.05x_1 x_2 = 4 + 0.4x_1 - 1.3(30) + 0.05x_1(30) = -35 + 1.9x_1$, or $\hat{y} = -35 + 1.9x_1$

Plotting these three equations simultaneously, we get the graph on the left (below).

Substitute $x_1 = 40$, $x_1 = 50$, and $x_1 = 60$ into the regression equation to get the following equations:

$x_1 = 40$: $\hat{y} = 4 + 0.4x_1 - 1.3x_2 + 0.05x_1 x_2 = 4 + 0.4(40) - 1.3x_2 + 0.05(40)x_2 = 20 + 0.7x_2$, or $\hat{y} = 20 + 0.7x_2$

$x_1 = 50$: $\hat{y} = 4 + 0.4x_1 - 1.3x_2 + 0.05x_1 x_2 = 4 + 0.4(50) - 1.3x_2 + 0.05(50)x_2 = 24 + 1.2x_2$, or $\hat{y} = 24 + 1.2x_2$

$x_1 = 60$: $\hat{y} = 4 + 0.4x_1 - 1.3x_2 + 0.05x_1 x_2 = 4 + 0.4(60) - 1.3x_2 + 0.05(60)x_2 = 28 + 1.7x_2$, or $\hat{y} = 28 + 1.7x_2$

Plotting these three equations simultaneously, we get the graph on the right (below).

When the interaction is included, the lines are not parallel.

19. (a) Yes, the P-value is very small so we would reject the null hypothesis and conclude that at least one of the explanatory variables is linearly related to the response variable.

(b) $b_1 = -0.87$: A one point increase in the computer confidence variable reduces the computer anxiety measurement by 0.87 point, on average, provided all other explanatory variables remain constant.

$b_2 = -0.51$: A one point increase in the computer knowledge variable reduces the computer anxiety measurement by 0.51 point, on average, provided all other explanatory variables remain constant.

$b_3 = -0.45$: A one point increase in the computer liking variable reduces the computer anxiety measurement by 0.45 point, on average, provided all other explanatory variables remain constant.

$b_4 = 0.33$: A one point increase in the trait anxiety variable increases the computer anxiety measurement by 0.33 point, on average, provided all other explanatory variables remain constant.

Yes, the slope coefficients of the model are all reasonable.

(c) $\hat{y} = 84.04 - 0.87(25) - 0.51(19) - 0.45(20) + 0.33(43) = 57.79$

(d) $R^2 = 0.69$ indicates that 69% of the variance in computer anxiety is explained by this regression model.

(e) This statement means that the requirements of a multiple linear regression model were checked and were met.

21. (a) Substitute $x = 20, 30, 40, 50, 60,$ and 70 into the equation $\hat{y} = -55,961.675 + 4314.374x - 45.66x^2$

x	$\hat{y}$
20	$-55,961.675 + 4314.374(20) - 45.66(20)^2 = \$12,062$
30	$-55,961.675 + 4314.374(30) - 45.66(30)^2 = \$32,376$
40	$-55,961.675 + 4314.374(40) - 45.66(40)^2 = \$43,557$
50	$-55,961.675 + 4314.374(50) - 45.66(50)^2 = \$45,607$
60	$-55,961.675 + 4314.374(60) - 45.66(60)^2 = \$38,525$
70	$-55,961.675 + 4314.374(70) - 45.66(70)^2 = \$22,311$

(b) $\$45,607 > \$43,557$, so the median income is higher when an individual is 50 years old.

(c) The median income decreases from $\$45,607$ to $\$38,525$, a decrease of $\$7,082$.

23. (a) For a female, $x_1 = 0$. Substituting this into the equation with the other values, we get:

$\hat{y} = 9.063 + 0.386x_1 + 1.268x_2 + 0.676x_3 - 0.913x_4 - 0.175x_3x_4$

$= 9.063 + 0.386(0) + 1.268(28) + 0.676(8) - 0.913(18) - 0.175(8)(18)$

≈ 8.4 years

(b) For a male, $x_1 = 1$. Substituting this into the equation with the other values, we get:

$$\hat{y} = 9.063 + 0.386x_1 + 1.268x_2 + 0.676x_3 - 0.913x_4 - 0.175x_3x_4$$

$$= 9.063 + 0.386(1) + 1.268(28) + 0.676(8) - 0.913(18) - 0.175(8)(18)$$

$$\approx 8.7 \text{ years}$$

(c) The interaction term is: $-0.175x_3x_4$. The number of teeth with root development, x_3, and the sum of the normalized heights of seven teeth on the left side of the mouth, x_4, interact.

(d) 86.3% of the variation in the response variable (age) is explained by the least squares regression model.

25. (a)   Regression Analysis: Wind Chill versus Air Temp, Wind Speed

```
The regression equation is
Wind Chill = - 12.5 + 1.33 Air Temp - 0.414 Wind Speed

Predictor        Coef  SE Coef       T       P
Constant     -12.5091   0.9194  -13.61   0.000
Air Temp      1.33408  0.04024   33.16   0.000
Wind Speed   -0.41430  0.03365  -12.31   0.000

S = 1.62942   R-Sq = 99.3%   R-Sq(adj) = 99.1%
```

The regression equation is $\hat{y} = -12.5091 + 1.3341x_1 - 0.4143x_2$.

(b)

There is a pattern in the residual plot, so we conclude that the linear model is not appropriate.

27. (a) Correlations: Slump, 7-day psi, 28-day psi

```
                Slump    7-day psi
7-day psi      -0.460
28-day psi     -0.753       0.737
```

Because the correlation between Slump and 7-day strength is between −0.7 and 0.7, multicollinearity does not appear to be a problem.

(b) Regression Analysis: 28-day psi versus Slump, 7-day psi

```
The regression equation is
28-day psi = 3891 - 296 Slump + 0.552 7-day psi

Predictor     Coef    SE Coef      T       P
Constant     3890.5    828.8     4.69   0.002
Slump        -295.9    109.9    -2.69   0.027
7-day psi    0.5523   0.2171     2.54   0.034

S = 276.505    R-Sq = 76.0%    R-Sq(adj) = 70.1%

Analysis of Variance
Source          DF        SS       MS       F       P
Regression       2   1942115   971058   12.70   0.003
Residual Error   8    611639    76455
Total           10   2553755
```

The regression equation is $\hat{y} = 3890.5 - 295.9x_1 + 0.5523x_2$.

(c)

None of the residual plots show any discernible pattern, and the boxplot does not show any outliers. Therefore, the linear model is appropriate.

(d) $b_1 = -295.9$: A one unit increase in slump of the concrete decreases the 28-day strength by 295.9 pounds per square inch, on average, provided the other variable (7-day strength) remains constant.

$b_2 = 0.5523$: A one pound per square inch increase in the 7-day strength increases the 28-day strength by 0.5523 pounds per square inch, on average, provided the other variable (slump) remains constant.

(e) $R^2 = 0.760$ means that 76.0% of the variance in 28-day strength is explained by the model.
$R^2_{adj} = 0.701$ (or 70.1%) modifies the value of R^2 taking the sample size and number of explanatory variables into account.

(f) From part (b), we have $F_0 = 12.70$ and P-value $= 0.003 < \alpha = 0.05$. Therefore, we reject the null hypothesis and conclude that at least one of the explanatory variables is linearly related to the 28-day strength of the concrete.

(g) From part (b), we have the following:
Reject $H_0 : \beta_1 = 0$ since $t_{\beta 1} = -2.69$ and P-value $= 0.027 < \alpha = 0.05$.
Reject $H_0 : \beta_2 = 0$ since $t_{\beta 2} = 2.54$ and P-value $= 0.034 < \alpha = 0.05$.

(h) $\hat{y} = 3890.5 - 295.9(3.5) + 0.552(2450) = 4207.3$ psi

(i) $\hat{y} = 3890.5 - 295.9(3.5) + 0.552(2450) = 4207.3$ psi (the prediction is the same for an individual as it is for the mean response, given particular values for the explanatory variables).

(j) 95% Confidence Interval: Lower Bound: 3988.6 psi; Upper Bound: 4427.2 psi; We are 95% confident that the mean 28-day strength of all concrete for which slump is 3.5 inches and 7-day strength is 2450 psi is between 3988.6 psi and 4427.2 psi.

95% Prediction Interval: Lower Bound: 3533.6 psi; Upper Bound: 4882.2 psi; We are 95% confident that the mean 28-day strength of a specific sample of concrete for which slump is 3.5 inches and 7-day strength is 2450 psi is between 3533.6 psi and 4882.2 psi.

29. (a) **Correlations: Square Footage, Bedrooms, Baths, Asking Price**

	Square Foota	Bedrooms	Baths
Bedrooms	0.267		
Baths	0.274	0.565	
Asking Price	0.665	0.409	0.626

Because the correlations among the explanatory variables are all between -0.7 and 0.7, multicollinearity does not appear to be a problem.

(b) **Regression Analysis: Asking Price versus Square Footage, Bedrooms, Baths**
```
The regression equation is
Asking Price = 25.6 + 0.0719 Square Footage - 0.8 Bedrooms + 55.3 Baths

Predictor              Coef    SE Coef      T       P
Constant              25.63      98.19    0.26   0.800
Square Footage      0.07193    0.02761    2.61   0.028
Bedrooms              -0.81      27.28   -0.03   0.977
Baths                 55.35      27.46    2.02   0.075

S = 46.2441    R-Sq = 65.6%    R-Sq(adj) = 54.1%

Analysis of Variance

Source           DF       SS      MS      F       P
Regression        3    36675   12225   5.72   0.018
Residual Error    9    19247    2139
Total            12    55921
```

The regression equation is $\hat{y} = 25.63 + 0.0719x_1 - 0.81x_2 + 55.35x_3$.

(c) From the output in (b), we get $F_0 = 5.72$ and $P\text{-value} = 0.018 < \alpha = 0.05$. We reject the null hypothesis and conclude that at least one of the explanatory variables is linearly associated with asking price.

(d) From part (b), we have the following:
Square Footage: Reject $H_0: \beta_1 = 0$ since $t_0 = 2.61$ and $P\text{-value} = 0.028 < \alpha = 0.05$
Bedrooms: Do not reject $H_0: \beta_2 = 0$ since $t_0 = -0.03$ and $P\text{-value} = 0.977 > \alpha = 0.05$
Baths: Do not reject $H_0: \beta_3 = 0$ since $t_0 = 2.02$ and $P\text{-value} = 0.075 > \alpha = 0.05$
We conclude that the square footage is linearly related to asking price for a house.

(e) The variable x_2, bedrooms, has the highest P-value. Removing this variable, we get:

Regression Analysis: Asking Price versus Square Footage, Baths
```
The regression equation is
Asking Price = 24.1 + 0.0718 Square Footage + 54.9 Baths

Predictor              Coef    SE Coef      T       P
Constant              24.09      79.18    0.30   0.767
Square Footage      0.07181    0.02593    2.77   0.020
Baths                 54.92      22.08    2.49   0.032

S = 43.8732    R-Sq = 65.6%    R-Sq(adj) = 58.7%

Analysis of Variance

Source           DF       SS      MS      F       P
Regression        2    36673   18336   9.53   0.005
Residual Error   10    19249    1925
Total            12    55921
```

The new regression equation is: $\hat{y} = 24.09 + 0.0718x_1 + 54.92x_3$

Square Footage: Reject $H_0 : \beta_1 = 0$ since $t_0 = 2.77$ and P-value $= 0.020 < \alpha = 0.05$

Bathrooms: Reject $H_0 : \beta_3 = 0$ since $t_0 = 2.49$ and P-value $= 0.032 < \alpha = 0.05$

We conclude that the square footage and number of bathrooms are both linearly related to asking price.

(f)

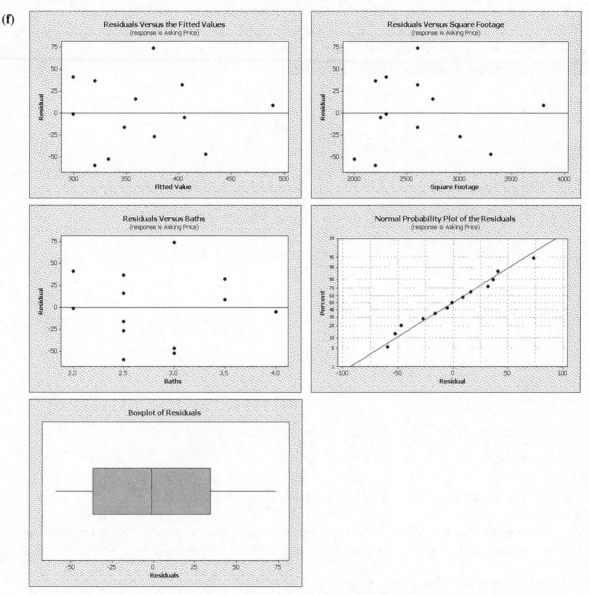

None of the residual plots show any discernible pattern, the boxplot does not show any outliers, and the normal probability plot looks fairly linear. Therefore, the linear model is appropriate.

(g) $b_1 = 0.0718$: If the house increases by 1 square foot, then the asking price will increase by $71.80, on average, assuming that the number of bathrooms remains constant.

$b_3 = 54.9$: For each bathroom added to the house, the asking price will increase by $54,900, on average, assuming that the square footage remains constant.

(h) 95% Confidence Interval: Lower Bound: 365.1; Upper Bound: 429.1; We are 95% confident that the mean asking price for all 2900-square-foot houses with 3 bathrooms is between $365,100 and $429,100.

95% Prediction Interval: Lower Bound: 294.2; Upper Bound: 499.9; We are 95% confident that the asking price of a randomly chosen 2900-square-foot house with 3 bathrooms is between $294,200 and $499,900.

31. (a) **Correlations: Engine, Weight, HP, MPG**

```
          Engine   Weight      HP
Weight    0.764
HP        0.920    0.558
MPG      -0.597   -0.390   -0.735
```

Yes, because the correlation between engine size and weight and the correlation between engine size and horse power are larger than 0.7, there is reason to be concerned about multicollinearity.

(b) **Regression Analysis: MPG versus Engine, Weight, HP**

```
The regression equation is
MPG = 35.8 + 4.80 Engine - 0.00394 Weight - 0.0658 HP

Predictor        Coef    SE Coef       T       P
Constant       35.827      9.212    3.89   0.004
Engine          4.803      3.361    1.43   0.187
Weight      -0.003945   0.003847   -1.03   0.332
HP           -0.06583    0.02636   -2.50   0.034

S = 1.46624    R-Sq = 62.6%    R-Sq(adj) = 50.1%

Analysis of Variance

Source           DF      SS       MS      F      P
Regression        3  32.344   10.781   5.01  0.026
Residual Error    9  19.349    2.150
Total            12  51.692
```

The regression equation is $\hat{y} = 35.827 + 4.803x_1 - 0.00394x_2 - 0.06583x_3$.

(c) From the output in (b), we get $F_0 = 5.01$ and P-value $= 0.026 < \alpha = 0.05$. We reject the null hypothesis and conclude that at least one of the explanatory variables is linearly associated with gasoline mileage.

(d) From part (b), we have the following:

Engine Size: Do not reject $H_0 : \beta_1 = 0$ since $t_0 = 1.43$ and P-value $= 0.187 > \alpha = 0.05$.

Curb Weight: Do not reject $H_0 : \beta_2 = 0$ since $t_0 = -1.03$ and P-value $= 0.332 > \alpha = 0.05$.

Horsepower: Reject $H_0 : \beta_3 = 0$ since $t_0 = -2.50$ and P-value $= 0.034 < \alpha = 0.05$.

We conclude that horsepower is linearly associated with gas mileage, but curb weight and engine size are not linearly associated with gas mileage.

(e) Removing the explanatory variable curb weight, x_2, we obtain:

Regression Analysis: MPG versus Engine, HP

```
The regression equation is
MPG = 26.8 + 2.15 Engine - 0.0503 HP

Predictor       Coef   SE Coef       T      P
Constant      26.805     2.740    9.78  0.000
Engine         2.147     2.147    1.00  0.341
HP          -0.05032   0.02164   -2.32  0.042

S = 1.47003   R-Sq = 58.2%   R-Sq(adj) = 49.8%

Analysis of Variance

Source          DF       SS      MS     F      P
Regression       2   30.082  15.041  6.96  0.013
Residual Error  10   21.610   2.161
Total           12   51.692
```

The new regression equation is $\hat{y} = 26.805 + 2.147x_1 - 0.05032x_3$.

Testing the explanatory variables in this new model, we find:

Engine Size: Do not reject $H_0 : \beta_1 = 0$ since $t_0 = 1.00$ and P-value $= 0.341 > \alpha = 0.05$.

Horsepower: Reject $H_0 : \beta_3 = 0$ since $t_0 = -2.32$ and P-value $= 0.042 < \alpha = 0.05$.

We conclude that horsepower is linearly associated with gas mileage, but engine size is not.

Removing the explanatory variable engine size, x_1, we obtain:

Regression Analysis: MPG versus HP

```
The regression equation is
MPG = 28.7 - 0.0304 HP

Predictor        Coef    SE Coef       T      P
Constant       28.723      1.956   14.68  0.000
HP          -0.030399   0.008457   -3.59  0.004

S = 1.47000   R-Sq = 54.0%   R-Sq(adj) = 49.8%

Analysis of Variance

Source          DF       SS      MS      F      P
Regression       1   27.922  27.922  12.92  0.004
Residual Error  11   23.770   2.161
Total           12   51.692
```

The new regression equation is $\hat{y} = 28.723 - 0.0304x_3$.

(f)

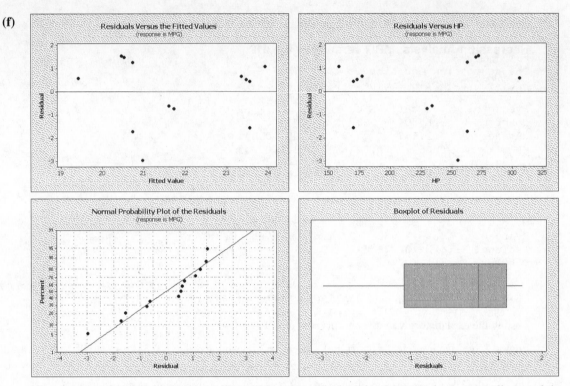

None of the residual plots show any discernible pattern, the boxplot does not show any outliers, and the normal probability plot looks roughly linear. Therefore, the linear model is appropriate.

(g) $b_3 = -0.0304$: If the horsepower is increased by 1, the gas mileage will decrease b 0.0304 miles per gallon, on average.

(h) 95% Confidence Interval: Lower Bound: 21.622; Upper Bound: 23.665; We are 95% confident that the mean gas mileage for all automobiles with 200 horsepower is between 21.622 and 23.665 miles per gallon.

95% Prediction Interval: Lower Bound: 19.251; Upper Bound: 26.036; We are 95% confident that the gas mileage for a randomly chosen automobile with 200 horsepower is between 19.251 and 26.036 miles per gallon.

33. (a) Regression Analysis: IQ versus MRI Count

```
The regression equation is
IQ = 110 + 0.000029 MRI Count


Predictor        Coef      SE Coef       T       P
Constant      109.894        9.559   11.50   0.000
MRI Count   0.00002863   0.00001029    2.78   0.012


S = 3.39589    R-Sq = 30.1%    R-Sq(adj) = 26.2%


Analysis of Variance
Source          DF       SS      MS      F       P
Regression       1    89.22   89.22   7.74   0.012
Residual Error  18   207.58   11.53
Total           19   296.80
```

The regression equation is $\hat{y} = 109.894 + 0.0000286x_1$.

(b) From the result in part (a), we have $t_0 = 2.78$ and P-value $= 0.012 < \alpha = 0.05$. Therefore, we reject the null hypothesis $H_0 : \beta_1 = 0$ and conclude that there is a linear relation between MRI count and IQ.

(c)

(d) Regression Analysis: IQ versus MRI Count, Gender

```
The regression equation is
IQ = 122 + 0.000017 MRI Count - 2.49 Gender

Predictor          Coef      SE Coef        T      P
Constant         121.58        13.36     9.10  0.000
MRI Count     0.00001735   0.00001366     1.27  0.221
Gender           -2.486         2.016    -1.23  0.234

S = 3.34772    R-Sq = 35.8%    R-Sq(adj) = 28.3%

Analysis of Variance
Source           DF      SS      MS      F      P
Regression        2  106.28   53.14   4.74  0.023
Residual Error   17  190.52   11.21
Total            19  296.80
```

The regression equation is $\hat{y} = 121.58 + 0.0000174x_1 - 2.486x_2$

(e) For the explanatory variable gender, $t_0 = -1.23$ and the P-value $= 0.234$. So, we do not reject the null hypothesis and conclude that there is not a significant linear relation between gender and intelligence. For the explanatory variable MRI count, $t_0 = 1.27$ and the P-value $= 0.221$. So, we do not reject the null hypothesis, and we conclude that there is no significant linear relation between MRI count and intelligence.

(f) The variable x_1, MRI count, originally has a slope coefficient that was significantly different from 0. Adding the variable x_2, gender, caused this relationship to change so that the slope coefficient of both variables are not significantly different from 0. This is an example of how two highly correlated predictor variables can adversely affect a model.

35. **(a)** **(i)** Ratio **(ii)** Ordinal **(iii)** Ordinal **(iv)** Ordinal **(v)** Ratio **(vi)** Nominal

(b) **Correlations: Carat, Color, Clarity, Cut, Price**

```
              Carat     Color  Clarity       Cut
Color         0.538
Clarity       0.368     0.318
Cut          -0.123     0.093   -0.147
Price         0.859     0.178    0.025    -0.162
```

Because the correlations among the explanatory variables are all between -0.7 and 0.7, multicollinearity does not appear to be a problem.

(c) **Regression Analysis: Price versus Carat, Color, Clarity, Cut**

```
The regression equation is
Price = 267 + 11319 Carat - 599 Color - 481 Clarity - 146 Cut

Predictor      Coef   SE Coef       T       P
Constant      266.5     649.3    0.41   0.684
Carat       11319.2     573.1   19.75   0.000
Color        -599.47    99.98   -6.00   0.000
Clarity      -481.26    84.72   -5.68   0.000
Cut          -145.6     223.5   -0.65   0.519

S = 1033.20    R-Sq = 92.3%    R-Sq(adj) = 91.5%

Analysis of Variance

Source            DF          SS         MS       F       P
Regression         4   449660719  112415180  105.31   0.000
Residual Error    35    37362264    1067493
Total             39   487022983
```

The regression equation is $\hat{y} = 266.5 + 11{,}319.2x_1 - 599.47x_2 - 481.26x_3 - 145.6x_4$.

(d) From the output in (c), we get $F_0 = 105.31$ and $P\text{-value} = 0.001 < \alpha = 0.05$. We reject the null hypothesis and conclude that at least one of the explanatory variables is linearly associated with price.

(e) From part (b), we have the following:

Carat Weight: Reject $H_0 : \beta_1 = 0$ since $t_0 = 19.75$ and $P\text{-value} < 0.001 < \alpha = 0.05$.

Color Grade: Reject $H_0 : \beta_2 = 0$ since $t_0 = -6.00$ and $P\text{-value} < 0.001 < \alpha = 0.05$.

Clarity Grade: Reject $H_0 : \beta_3 = 0$ since $t_0 = -5.68$ and $P\text{-value} < 0.001 < \alpha = 0.05$.

Cut Grade: Do not reject $H_0 : \beta_4 = 0$ since $t_0 = -0.65$ and $P\text{-value} = 0.519 > \alpha = 0.05$.

We conclude that carat weight, color grade, and clarity grade are all linearly associated with price, but cut grade is not. So, we remove cut grade, x_1, from our model

Regression Analysis: Price versus Carat, Color, Clarity

```
The regression equation is
Price = - 36 + 11382 Carat - 613 Color - 473 Clarity

Predictor      Coef   SE Coef       T       P
Constant      -36.5     449.4   -0.08   0.936
Carat       11381.8     560.4   20.31   0.000
Color        -613.17    96.95   -6.32   0.000
Clarity      -473.36    83.18   -5.69   0.000

S = 1024.90    R-Sq = 92.2%    R-Sq(adj) = 91.6%
```

```
Analysis of Variance

Source           DF        SS         MS        F       P
Regression        3   449207830  149735943  142.55  0.000
Residual Error   36    37815153    1050421
Total            39   487022983
```

The new regression equation is $\hat{y} = -36.5 + 11,381.8x_1 - 613.17x_2 - 473.36x_3$. The P-values for each of the remaining explanatory variables are very small (< 0.001), so we believe that this is our "best" model.

(f)

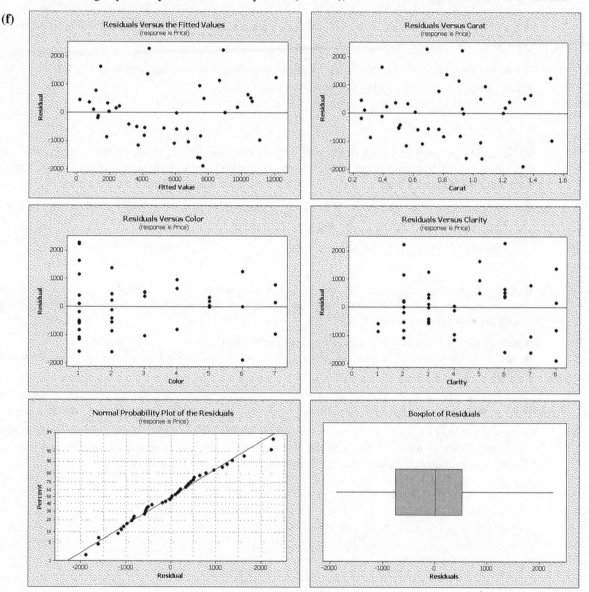

(g) $\hat{y} = -36.5 + 11,381.8x_1 - 613.17x_2 - 473.36x_3$

$b_1 = 11,381.8$: If a diamond increase by 1 carat, then the price will increase by \$11,381.80, on average, assuming that color and clarity remain constant.

$b_2 = -613.17$: If the color rating increases numerically by 1 (say from D to F, which is actually a decrease in color quality), then the price will decrease by $613.17, on average, assuming that carat weight and clarity grade remain constant.

$b_3 = -473.36$: If the clarity rating increases by 1 (say from IF to VVSI, which is actually a decrease in clarity quality), then the price will decrease by $473.36, on average, assuming that carat weight and color grade remain constant.

(h) $R^2 = 92.2\%$; 92.2% of the variation in price is explained by the least-squares regression model. $R^2_{adj} = 91.6\%$ modifies the value of R^2 taking the sample size and number of explanatory variables into account.

(i) We substitute $x_1 = 0.85$, $x_2 = 2$ (for E color grade), and $x_3 = 3$ (for VVS1 clarity grade) into our into our "best" model: $\hat{y} = -36.5 + 11,381.8(0.85) - 613.17(2) -$

$$473.36(3)$$

$$\approx \$6992$$

(j) 95% Confidence Interval: Lower Bound: $6,568; Upper Bound: $7,415; We are 95% confident that the mean price for all diamonds weighing 0.85 carats, having an E color grade, and having a VVS1 clarity grade is between $6,568 and $7,415.

(k) We again substitute $x_1 = 0.85$, $x_2 = 2$ (for E color grade), and $x_3 = 3$ (for VVS1 clarity grade) into our into our "best" model: $\hat{y} = -36.5 + 11,381.8(0.85) - 613.17(2) -$

$$473.36(3)$$

$$\approx \$6992$$

(l) 95% Confidence Interval: Lower Bound: $4,870; Upper Bound: $9,113; We are 95% confident that the price for a randomly chosen diamond weighing 0.85 carats, having an E color grade, and having a VVS1 clarity grade is between $4,870 and $9,113.

(m) Although the predicted prices in parts (i) and (k) are the same, the intervals are different because the distribution of the mean price, part (j), has less variability than the distribution of the individual prices of diamonds, part (l).

37. When the explanatory variables have an additive effect, or do not interact, it means that the effect of any one explanatory variable on the response variable does not depend on the value of the other explanatory variables.

39. Multicollinearity exists when two explanatory variables are highly correlated. We use a correlation matrix to check if multicollinearity exists. A consequence of multicollinearity is that *t*-test statistics might be small and their *P*-values large, falsely indicating that neither explanatory variable is significant.

Chapter 14 Review Exercises

1. The least-squares regression model is
   $y_i = \beta_1 x_i + \beta_0 + \varepsilon_i$. The requirements to
   perform inference on the least-squares regression
   line are (1) for any particular values of the
   explanatory variable x, the mean of the
   corresponding responses in the population
   depends linearly on x, and (2) the response
   variables, y_i, are normally distributed with mean
   $\mu_{y|x} = \beta_1 x + \beta_0$ and the standard deviation σ.
   We verify these requirements by checking to see
   that the residuals are normally distributed, with
   mean 0 and constant variance σ^2 and that the
   residuals are independent. We do this by
   constructing residual plots and a normal
   probability plot of the residuals.

2. (a) Using technology, we get: $\beta_0 \approx b_0 = 3.8589$
   and $\beta_1 \approx b_1 = -0.1049$. Substituting $x = 5$
   into the least-squares regression equation,
   we get: $\hat{y} = -0.1049(5) + 3.8589 \approx 3.334$.
   So, the mean GPA of students who choose
   to sit in the fifth row is 3.334.

(b) We calculate $\hat{y} = -0.1049x + 3.8589$ to
generate the following table:

x	y	$\hat{y}$	$y - \hat{y}$	$(y - \hat{y})^2$	$(x - \bar{x})^2$
1	4	3.754	0.2460	0.0605	21.6960
2	3.35	3.649	-0.2991	0.0895	13.3802
2	3.5	3.649	-0.1491	0.0222	13.3802
2	3.67	3.649	0.0209	0.0004	13.3802
2	3.75	3.649	0.1009	0.0102	13.3802
3	3.37	3.544	-0.1742	0.0303	7.0644
3	3.62	3.544	0.0758	0.0057	7.0644
4	2.35	3.439	-1.0893	1.1866	2.7486
4	2.71	3.439	-0.7293	0.5319	2.7486
4	3.75	3.439	0.3107	0.0965	2.7486
5	3.1	3.334	-0.2344	0.0549	0.4328
5	3.22	3.334	-0.1144	0.0131	0.4328
5	3.36	3.334	0.0256	0.0007	0.4328
5	3.58	3.334	0.2456	0.0603	0.4328
5	3.67	3.334	0.3356	0.1126	0.4328
5	3.69	3.334	0.3556	0.1265	0.4328
5	3.72	3.334	0.3856	0.1487	0.4328
5	3.84	3.334	0.5056	0.2556	0.4328
6	2.35	3.230	-0.8795	0.7735	0.1170
6	2.63	3.230	-0.5995	0.3594	0.1170
6	3.15	3.230	-0.0795	0.0063	0.1170
6	3.69	3.230	0.4605	0.2121	0.1170
6	3.71	3.230	0.4805	0.2309	0.1170
7	2.88	3.125	-0.2446	0.0598	1.8012
7	2.93	3.125	-0.1946	0.0379	1.8012
7	3	3.125	-0.1246	0.0155	1.8012
7	3.21	3.125	0.0854	0.0073	1.8012
7	3.53	3.125	0.4054	0.1643	1.8012
7	3.74	3.125	0.6154	0.3787	1.8012
7	3.75	3.125	0.6254	0.3911	1.8012
7	3.9	3.125	0.7754	0.6012	1.8012
8	2.3	3.020	-0.7197	0.5180	5.4854
8	2.54	3.020	-0.4797	0.2301	5.4854
8	2.61	3.020	-0.4097	0.1679	5.4854
9	2.71	2.915	-0.2048	0.0419	11.1696
9	3.74	2.915	0.8252	0.6810	11.1696
9	3.75	2.915	0.8352	0.6976	11.1696
11	1.71	2.705	-0.9950	0.9900	28.5380
			Sum:	9.3709	194.5526

$$s_e = \sqrt{\frac{\sum(y - \hat{y})^2}{n - 2}} = \sqrt{\frac{9.3709}{38 - 2}} \approx 0.5102$$

(c) A normal probability plot shows that the residuals are approximately normally distributed.

(d) $s_{b_1} = \dfrac{s_e}{\sqrt{\sum(x-\bar{x})^2}} = \dfrac{0.5102}{\sqrt{194.5526}} \approx 0.0366$

(e) We have the hypotheses $H_0 : \beta_1 = 0$ versus $H_1 : \beta_1 \neq 0$. The test statistic is

$$t_0 = \frac{b_1}{s_{b_1}} = \frac{-0.1049}{0.0366} = -2.866 \ [\text{Tech:} -2.868]$$

Classical approach: The level of significance is $\alpha = 0.05$. Since this is a two-tailed test with $n-2 = 38-2 = 36$ degrees of freedom, the critical values are $\pm t_{0.025} = \pm 2.028$. Since $t_0 = -2.866 < -t_{0.025} = -2.028$ (the test statistic falls in a critical region), we reject H_0.

P-value approach: The P-value for this two-tailed test is the area under the t-distribution with 36 degrees of freedom to the left of $t_0 = -2.866$ [Tech: -2.868], plus the area to the right of 2.866 [Tech: 2.868]. From the t-distribution table in the row corresponding to 36 degrees of freedom, 2.866 falls between 2.719 and 2.990, whose right-tail areas are 0.005 and 0.0025, respectively. We must double these values in order to get the total area in both tails: 0.01 and 0.005. So, $0.005 < P\text{-value} < 0.01$. [Tech: $P\text{-value} = 0.0069$]. Because the P-value is less than the level of significance $\alpha = 0.05$, we reject H_0.

Conclusion: There is sufficient evidence to conclude that a linear relationship exists between the row chosen by students on the first day of class and their cumulative GPAs.

(f) For a 95% confidence interval with d.f. = 36, we use $t_{0.025} = 2.028$:

Lower bound:
$$b_1 - t_{\alpha/2} \cdot s_{b_1} = -0.1049 - 2.028 \cdot 0.0366$$
$$= -0.1791$$

Upper bound:
$$b_1 + t_{\alpha/2} \cdot s_{b_1} = -0.1049 + 2.028 \cdot 0.0366$$
$$= -0.0307$$

(g) Lower Bound:
$$\hat{y} - t_{\alpha/2} \cdot s_e \sqrt{\frac{1}{n} + \frac{(x^* - \bar{x})^2}{\sum(x-\bar{x})^2}}$$
$$= 3.334 - 2.028 \cdot 0.5102 \cdot \sqrt{\frac{1}{38} + \frac{(5-5.6579)^2}{194.5526}}$$
$$\approx 3.159$$

Upper Bound:
$$\hat{y} + t_{\alpha/2} \cdot s_e \sqrt{\frac{1}{n} + \frac{(x^* - \bar{x})^2}{\sum(x-\bar{x})^2}}$$
$$= 3.334 + 2.028 \cdot 0.5102 \cdot \sqrt{\frac{1}{38} + \frac{(5-5.6579)^2}{194.5526}}$$
$$\approx 3.509$$

(h) The predicted GPA of a randomly selected student who chose a seat in the fifth row is also $\hat{y} = -0.1049(5) + 3.8589 \approx 3.334$.

(i) Lower Bound:
$$\hat{y} - t_{\alpha/2} \cdot s_e \sqrt{1 + \frac{1}{n} + \frac{(x^* - \bar{x})^2}{\sum(x-\bar{x})^2}}$$
$$= 3.334 - 2.028 \cdot 0.5102 \cdot \sqrt{1 + \frac{1}{38} + \frac{(5-5.6579)^2}{194.5526}}$$
$$\approx 2.285$$

Upper Bound:
$$\hat{y} + t_{\alpha/2} \cdot s_e \sqrt{1 + \frac{1}{n} + \frac{(x^* - \bar{x})^2}{\sum(x-\bar{x})^2}}$$
$$= 3.334 + 2.028 \cdot 0.5102 \cdot \sqrt{1 + \frac{1}{38} + \frac{(5-5.6579)^2}{194.5526}}$$
$$\approx 4.383 \ [\text{Tech: 4.384}]$$

(j) Although the predicted GPAs in parts (a) and (b) are the same, the intervals are different because the distribution of the mean GPA, part (a), has less variability than the distribution of individual GPAs, part (h).

3. (a) Using technology, we get $\beta_0 \approx b_0 = -399.2$ and $\beta_1 \approx b_1 = 2.5315$. Substituting $x = 900$ into the least-squares regression equation, we get $\hat{y} = -399.2 + 2.5315(900) \approx 1879.2$. So, the mean rent of a 900-square-foot apartment in Queens is \$1879.2.

(b) Using technology, we calculate $\hat{y} = -399.2 + 2.5315x$ to generate the table:

x	y	$\hat{y}$	$y - \hat{y}$	$(y - \hat{y})^2$	$(x - \overline{x})^2$
500	650	866.55	−216.55	46,893.9	160,000
588	1215	1089.32	125.678	15,795	97,344
1000	2000	2132.3	−132.3	17,503.3	10,000
688	1655	1342.47	312.528	97,673.8	44,944
825	1250	1689.29	−439.29	19,2974	5625
1259	2700	2787.96	−87.959	7736.7	128,881
650	1200	1246.28	−46.275	2141.38	62,500
560	1250	1018.44	231.56	53620	115,600
1073	2350	2317.1	32.9005	1082.44	29,929
1452	3300	3276.54	23.462	550.465	304,704
1305	3100	2904.41	195.593	38,256.4	164,025
				474,227	1,123,552

$$s_e = \sqrt{\frac{\sum(y - \hat{y})^2}{n - 2}} = \sqrt{\frac{474,227}{11 - 2}} \approx 229.547$$

(c) A normal probability plot shows that the residuals are normally distributed.

(d) $s_{b_1} = \dfrac{s_e}{\sqrt{\sum(x - \overline{x})^2}} = \dfrac{229.547}{\sqrt{1,123,552}} = 0.2166$

(e) We have the hypotheses $H_0 : \beta_1 = 0$ versus $H_1 : \beta_1 \neq 0$. The test statistic is

$$t_0 = \frac{b_1}{s_{b_1}} = \frac{2.5315}{0.2166} = 11.687$$

Classical approach: The level of significance is $\alpha = 0.05$. Since this is a two-tailed test with $n - 2 = 11 - 2 = 9$ degrees of freedom, the critical values are $\pm t_{0.025} = \pm 2.262$. Since $t_0 = 11.687 > t_{0.025} = 2.262$ (the test statistic falls in a critical region), we reject H_0.

P-value approach: The P-value for this two-tailed test is the area under the t-distribution with 9 degrees of freedom to the right of $t_0 = 11.687$, plus the area to the left of $-t_0 = -11.687$. From the t-distribution table in the row corresponding to 9 degrees of freedom, 11.687 falls to the right of 4.781, whose right-tail area is 0.0005. We must double this value in order to get the total area in both tails: 0.001. So, P-value < 0.001. [Tech: P-value < 0.0001]. Because the P-value is less than the level of significance $\alpha = 0.05$, we reject H_0.

Conclusion: There is sufficient evidence to conclude that a linear relationship exists between the square footage of an apartment in Queens, New York, and the monthly rent.

(f) For a 95% confidence interval we use $t_{0.025} = 2.262$:

Lower Bound:
$$b_1 - t_{\alpha/2} \cdot s_{b_1} = 2.5315 - 2.262(0.2166)$$
$$= 2.0416$$

Upper Bound:
$$b_1 + t_{\alpha/2} \cdot s_{b_1} = 2.5315 + 2.262(0.2166)$$
$$= 3.0214$$

(g) We have $n = 11$, $s_e = 229.547$, $\overline{x} = 900$ and $\sum(x - \overline{x})^2 = 1,123,552$. For a 90% confidence interval with $n - 2 = 9$ degrees of freedom, we use $t_{0.05} = 1.833$.

Lower Bound:

$$\hat{y} - t_{\alpha/2} \cdot s_e \sqrt{\frac{1}{n} + \frac{(x^* - \bar{x})^2}{\sum(x - \bar{x})^2}}$$

$$= 1879.2 - 1.833(229.547)\sqrt{\frac{1}{11} + \frac{(900 - 900)^2}{1,123,552}}$$

$$= \$1752.3 \quad [\text{Tech: } \$1752.2]$$

Upper Bound:

$$\hat{y} + t_{\alpha/2} \cdot s_e \sqrt{\frac{1}{n} + \frac{(x^* - \bar{x})^2}{\sum(x - \bar{x})^2}}$$

$$= 1879.2 + 1.833(229.547)\sqrt{\frac{1}{11} + \frac{(900 - 900)^2}{1,123,552}}$$

$$= \$2006.1 \quad [\text{Tech: } \$2006.0]$$

(h) The predicted value of y is also $\hat{y} = \$1879.2$ [Tech: 1879.1].

(i) Lower Bound:

$$\hat{y} - t_{\alpha/2} \cdot s_e \sqrt{1 + \frac{1}{n} + \frac{(x^* - \bar{x})^2}{\sum(x - \bar{x})^2}}$$

$$= 1879.2 - 1.833(229.547)\sqrt{1 + \frac{1}{11} + \frac{(900 - 900)^2}{1,123,552}}$$

$$= \$1439.7 \quad [\text{Tech: } \$1439.6]$$

Upper Bound:

$$\hat{y} + t_{\alpha/2} \cdot s_e \sqrt{1 + \frac{1}{n} + \frac{(x^* - \bar{x})^2}{\sum(x - \bar{x})^2}}$$

$$= 1879.2 + 1.833(229.547)\sqrt{1 + \frac{1}{11} + \frac{(900 - 900)^2}{1,123,552}}$$

$$= \$2318.7 \quad [\text{Tech: } \$2318.6]$$

(j) Although the predicted rents in parts (a) and (h) are the same, the intervals are different because the distribution of the means, in part (a), has less variability than the distribution of the individuals, in part (h).

4. (a) Using technology, we get
$\hat{y} = -2054.6x + 20976.2$.

(b) The hypotheses are $H_0: \beta_1 = 0$, $H_1: \beta_1 \neq 0$. The level of significance is $\alpha = 0.05$. Since this is a two-tailed test with $n - 2 = 12 - 2 = 10$ degrees of freedom, the

critical values are $\pm t_{0.025} = \pm 2.228$. Using technology we obtain the test statistic $t_0 = -10.111$ and P-value < 0.0001.

Classical Approach: Since $t_0 = -10.111 < -t_{0.025} = -2.228$ (the test statistic falls within critical region), we reject H_0.

P-value Approach: Since P-value $< 0.0001 < \alpha = 0.05$, we reject H_0.

Conclusion: There is sufficient evidence to conclude that a linear relationship exists between age of used Chevy Camaros and their prices.

(c)

(d) The plot of residuals shows a pattern which indicates that a linear model is not appropriate. The moral is to perform graphical diagnostic tests along with the inferential procedures before drawing any conclusions regarding the model.

5. (a) Coefficient of x_1: If rainfall during the winter increases by 1 inch while temperature and rainfall during the harvest remain constant, then the wine quality rating increases by 0.00117.
Coefficient of x_2: If temperature increases by 1 degree while rainfall during the winter and rainfall during the harvest remain constant, then the wine quality rating increases by 0.0614.
Coefficient of x_3: If rainfall during the harvest increases by 1 inch while rainfall during the winter and temperature remain constant, then the wine quality rating decreases by 0.00386.

(b) The wine quality rating is predicted to be:
$\hat{y} = 12.15 + 0.00117x_1 + 0.0614x_2 - 0.00386x_3$
$= 12.15 + 0.00117(8) + 0.0614(58) - 0.00386(2.9)$
≈ 15.7

6. **(a)** **Correlations: Midterm, Project, Course Grade**

```
                  Midterm      Project
Project           -0.076
Course Grade       0.704        0.526
```

No; the correlation between midterm and project is between −0.7 and 0.7, multicollinearity does not appear to be a problem.

(b) **Regression Analysis: Course Grade versus Midterm, Project**
```
The regression equation is
Course Grade = - 91.3 + 1.52 Midterm + 0.441 Project

Predictor     Coef    SE Coef       T      P
Constant    -91.29      30.77   -2.97  0.021
Midterm     1.5211     0.3148    4.83  0.002
Project     0.4407     0.1170    3.77  0.007

S = 3.99690   R-Sq = 83.3%   R-Sq(adj) = 78.5%

Analysis of Variance
Source           DF      SS       MS      F      P
Regression        2  558.27   279.14  17.47  0.002
Residual Error    7  111.83    15.98
Total             9  670.10
```

The regression equation is $\hat{y} = -91.29 + 1.5211x_1 + 0.4407x_2$.

(c)

None of the residual plots show any discernible pattern, the boxplot does not show any outliers, and the normal probability plot looks fairly linear. Therefore, the linear model is appropriate.

(d) $b_1 = 1.5211$: This indicates that when midterm grade increases by 1 point, the course grade increases by 1.5211 points, on average, assuming the project grade remains constant.

$b_2 = 0.4407$ **:** This indicates that when the project grade increases by 1 point, the course grade increases by 0.4407 point, on average, assuming the midterm grade remains constant.

(e) $R^2 = 0.833$; 83.3% of the variance in course grade is explained by the regression model. $R^2_{adj} = 0.785$ modifies the value of R^2 taking the sample size and number of explanatory variables into account.

(f) From the output in (b), we get $F_0 = 17.47$ and $P\text{-value} = 0.002 < \alpha = 0.05$. We reject the null hypothesis and conclude that at least one of the explanatory variables is linearly associated with course grade.

(g) From part (b), we get the following:
Midterm: Reject $H_0 : \beta_1 = 0$ since $t_0 = 4.83$ and $P\text{-value} = 0.002 < \alpha = 0.05$
Project: Reject $H_0 : \beta_2 = 0$ since $t_0 = 3.77$ and $P\text{-value} = 0.007 < \alpha = 0.05$
Both explanatory variables have slope coefficients that are significantly different than zero.

(h) $\hat{y} = -91.29 + 1.5211x_1 + 0.4407x_2 = -91.29 + 1.5211(85) + 0.4407(75) \approx 71.1$
The predicted mean final course grade is 71.1.

(i) 95% Confidence Interval: Lower Bound: 69.00; Upper Bound: 82.04; We are 95% confident that the mean course grade of all students who score an 83 on their midterm and a 92 on their project is between 69 and 82.

95% Prediction Interval: Lower Bound: 64.04; Upper Bound: 87.00; We are 95% confident that the final course grade of a randomly selected student who scored an 83 on their midterm and a 92 on their project is between 64 and 87.

7. (a) Correlations: Protein, Fat, Carbohydrates, Calories

	Calories	Protein	Fat
Protein	-0.730		
Fat	-0.064	-0.261	
Carbohydrates	0.801	-0.787	-0.541

Yes; because the correlation between carbohydrate and protein is smaller than -0.7 , multicollinearity may be a problem.

(b) **Regression Analysis: Calories versus Protein, Fat, Carbohydrates**
```
The regression equation is
Calories = 111 - 0.24 Protein + 5.39 Fat + 2.96 Carbohydrates

Predictor          Coef   SE Coef      T      P
Constant         111.24     78.15   1.42  0.192
Protein          -0.239     1.546  -0.15  0.881
Fat               5.388     1.890   2.85  0.021
Carbohydrates    2.9558    0.7944   3.72  0.006

S = 6.65682    R-Sq = 83.5%    R-Sq(adj) = 77.3%

Analysis of Variance
Source          DF       SS      MS      F      P
Regression       3  1794.41  598.14  13.50  0.002
Residual Error   8   354.51   44.31
Total           11  2148.92
```

The regression equation is $\hat{y} = 111.24 - 0.239x_1 + 5.388x_2 + 2.9558x_3$.

(c) From the output in (b), we get $F_0 = 13.50$ and P-value $= 0.002 < \alpha = 0.05$. We reject the null hypothesis and conclude that at least one of the explanatory variables is linearly associated with calories.

(d) From part (b), we get the following:

Protein: Do not reject $H_0: \beta_1 = 0$ since $t_0 = -0.15$ and P-value $= 0.881 > \alpha = 0.05$

Fat: Reject $H_0: \beta_2 = 0$ since $t_0 = 2.85$ and P-value $= 0.021 < \alpha = 0.05$

Carbs: Reject $H_0: \beta_3 = 0$ since $t_0 = 3.72$ and P-value $= 0.006 < \alpha = 0.05$

The explanatory variable x_1, protein, is not significantly different from 0, so we should remove it from the model.

(e) After removing x_2, protein, we get the following:

Regression Analysis: Calories versus Fat, Carbohydrates
```
The regression equation is
Calories = 101 + 5.48 Fat + 3.05 Carbohydrates

Predictor          Coef   SE Coef      T      P
Constant         101.20     41.13   2.46  0.036
Fat               5.481     1.692   3.24  0.010
Carbohydrates    3.0536    0.4543   6.72  0.000

S = 6.28549    R-Sq = 83.5%    R-Sq(adj) = 79.8%

Analysis of Variance
Source          DF       SS      MS      F      P
Regression       2  1793.35  896.68  22.70  0.000
Residual Error   9   355.57   39.51
Total           11  2148.92
```

The new regression equation is $\hat{y} = 101.20 + 5.481x_2 + 3.0536x_3$.

Both explanatory variables have slope coefficients that are significantly different than zero.

(f)

None of the residual plots show any discernible pattern, the boxplot does not show any outliers, and the normal probability plot looks fairly linear. Therefore, the linear model is appropriate.

(g) $b_2 = 5.481$: This indicates that when the fat content increases by 1 gram, the cereal increases by 5.481 calories, on average, assuming carbohydrate content remains constant.

$b_3 = 3.0536$: This indicates that when carbohydrate content increases by 1 gram, the cereal increases by 3.0536 calories, on average, assuming the fat content remains constant.

(h) $R^2 = 0.835$; 83.5% of the variance in calories is explained by the regression model. $R^2_{adj} = 0.798$ modifies the value of R^2 taking the sample size and number of explanatory variables into account.

(i) 95% Confidence Interval:
Lower Bound: 388.08; Upper Bound: 407.81
We are 95% confident that the mean calories in 100 grams of cereal containing 4 grams of fat and 90 grams of carbohydrates will be between 388.08 and 407.81.

95% Prediction Interval:
Lower Bound: 380.64; Upper Bound: 415.25
We are 95% confident that 100 grams of a randomly selected cereal containing 4 grams of fat and 90 grams of carbohydrates contains between 380.64 and 415.25 calories.

Chapter 14 Test

1. The requirements to perform inference on the least-squares regression line are (1) for any particular value of the explanatory variable x, the mean of the corresponding responses in the population depends linearly on x, and (2) the response variables, y_i, are normally distributed with mean $\mu_{y|x} = \beta_1 x + \beta_0$ and standard deviation σ.

2. (a) Using technology, we get:

$\beta_0 \approx b_0 = -0.3091$ and $\beta_1 \approx b_1 = 0.2119$.

We use the regression equation to calculate the mean number of chirps at 80.2°F:
$\hat{y} = 0.2119 \cdot 80.2 - 0.3091 = 16.69$

(b) Using technology, we calculate $\hat{y} = 0.2119 x - 0.3091$ to generate the table:

x	y	$\hat{y}$	$y - \hat{y}$	$(y - \hat{y})^2$	$(x - \bar{x})^2$
88.6	20.0	18.467	1.533	2.349	73.274
93.3	19.8	19.463	0.337	0.113	175.828
80.6	17.1	16.772	0.328	0.108	0.314
69.7	14.7	14.462	0.238	0.057	106.916
69.4	15.4	14.398	1.002	1.003	113.210
79.6	15.0	16.560	−1.560	2.434	0.194
80.6	16.0	16.772	−0.772	0.596	0.314
76.3	14.4	15.861	−1.461	2.134	13.988
71.6	16.0	14.865	1.135	1.289	71.234
84.3	18.4	17.556	0.844	0.712	18.148
75.2	15.5	15.628	−0.128	0.016	23.426
82.0	17.1	17.069	0.031	0.001	3.842
83.3	16.2	17.344	−1.144	1.309	10.628
82.6	17.2	17.196	0.004	0.000	6.554
83.5	17.0	17.387	−0.387	0.149	11.972
			Sum	12.270	629.836

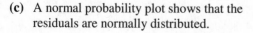

$s_e = \sqrt{\dfrac{\sum(y - \hat{y})^2}{n - 2}} = \sqrt{\dfrac{12.270}{15 - 2}} = 0.9715$

(c) A normal probability plot shows that the residuals are normally distributed.

Normal Probability Plot of the Residuals
(response is Chirps per Second)

(d) $s_{b_1} = \dfrac{s_e}{\sqrt{\sum(x - \bar{x})^2}} = \dfrac{0.9715}{\sqrt{629.836}} = 0.0387$

(e) The hypotheses are $H_0: \beta_1 = 0$ versus $H_1: \beta_1 \neq 0$. The test statistic is

$t_0 = \dfrac{b_1}{s_{b_1}} = \dfrac{0.2119}{0.0387} = 5.475$.

Classical approach: The level of significance is $\alpha = 0.05$. Since this is a two-tailed test with $n - 2 = 15 - 2 = 13$ degrees of freedom, the critical values are $\pm t_{0.025} = \pm 2.160$. Since $t_0 = 5.475 > t_{0.025} = 2.160$ (the test statistic falls in a critical region), we reject H_0.

P-value approach: The P-value for this two-tailed test is the area under the t-distribution with 13 degrees of freedom to the right of $t_0 = 5.475$, plus the area to the left of $-t_0 = -5.475$. From the t-distribution table in the row corresponding to 13 degrees of freedom, 5.475 falls to the right of 4.221, whose right-tail area is 0.0005. We must double this value in order to get the total area in both tails: 0.001. So, P-value < 0.001 [Tech: 0.0001]. Because the P-value is less than the level of significance $\alpha = 0.05$, we reject H_0.

Conclusion: There is sufficient evidence to conclude that a linear relationship exists between temperature and the chirps per second.

(f) For a 95% confidence interval we use $t_{0.025} = 2.160$:

Lower Bound:
$b_1 - t_{\alpha/2} \cdot s_{b_1} = 0.2119 - 2.160 \cdot 0.0387$
$= 0.1283$

Upper Bound:
$$b_1 + t_{\alpha/2} \cdot s_{b_1} = 0.2119 + 2.160 \cdot 0.0387$$
$$= 0.2955$$

(g) We have $n = 15$, $s_e = 0.9715$, $\bar{x} = 80.04$ and $\sum (x - \bar{x})^2 = 629.836$. For a 90% confidence interval with df $= n - 2 = 13$, we use $t_{0.05} = 1.771$:

Lower Bound:
$$\hat{y} - t_{\alpha/2} \cdot s_e \sqrt{\frac{1}{n} + \frac{(x^* - \bar{x})^2}{\sum (x - \bar{x})^2}}$$
$$= 16.69 - 1.771 \cdot 0.9715 \cdot \sqrt{\frac{1}{15} + \frac{(80.2 - 80.04)^2}{629.836}}$$
$$= 16.24 \text{ chirps}$$

Upper Bound:
$$\hat{y} + t_{\alpha/2} \cdot s_e \sqrt{\frac{1}{n} + \frac{(x^* - \bar{x})^2}{\sum (x - \bar{x})^2}}$$
$$= 16.69 + 1.771 \cdot 0.9715 \cdot \sqrt{\frac{1}{15} + \frac{(80.2 - 80.04)^2}{629.836}}$$
$$= 17.13 \text{ chirps}$$

(h) The predicted value of y is also $\hat{y} = 16.69$ chirps.

(i) Lower Bound:
$$\hat{y} - t_{\alpha/2} \cdot s_e \sqrt{1 + \frac{1}{n} + \frac{(x^* - \bar{x})^2}{\sum (x - \bar{x})^2}}$$
$$= 16.69 - 1.771 \cdot 0.9715 \cdot \sqrt{1 + \frac{1}{15} + \frac{(80.2 - 80.04)^2}{629.836}}$$
$$= 14.91 \text{ chirps}$$

Upper Bound:
$$\hat{y} + t_{\alpha/2} \cdot s_e \sqrt{1 + \frac{1}{n} + \frac{(x^* - \bar{x})^2}{\sum (x - \bar{x})^2}}$$
$$= 16.69 + 1.771 \cdot 0.9715 \cdot \sqrt{1 + \frac{1}{15} + \frac{(80.2 - 80.04)^2}{629.836}}$$
$$= 18.47 \quad [\text{Tech: } 18.46] \text{ chirps}$$

(j) Although the predicted numbers of chirps in parts (a) and (h) are the same, the intervals are different because the distribution of the means, part (a), has less variability than the distribution of the individuals, part (h).

3. (a) Using technology we get: $\beta_0 \approx b_0 = 29.705$ and $\beta_1 \approx b_1 = 2.6351$. We use the regression equation to calculate the mean height:
$$\hat{y} = 2.6351(7) + 29.705 = 48.15 \text{ inches.}$$

(b) Using technology, we calculate $\hat{y} = 2.6351 x + 29.705$ to generate the table:

x	y	$\hat{y}$	$y - \hat{y}$	$(y - \hat{y})^2$	$(x - \bar{x})^2$
2	36.1	34.975	1.125	1.266	14.823
2	34.2	34.975	−0.775	0.601	14.823
2	31.1	34.975	−3.875	15.015	14.823
3	36.3	37.610	−1.310	1.716	8.123
3	39.5	37.610	1.890	3.572	8.123
4	41.5	40.245	1.255	1.575	3.423
4	38.6	40.245	−1.645	2.706	3.423
5	45.6	42.880	2.720	7.397	0.722
5	44.8	42.880	1.920	3.686	0.722
5	44.6	42.880	1.720	2.958	0.722
6	49.8	45.515	4.285	18.359	0.023
7	43.2	48.150	−4.950	24.506	1.323
7	47.9	48.150	−0.250	0.063	1.323
8	51.4	50.785	0.615	0.378	4.623
8	48.3	50.785	−2.485	6.177	4.623
8	50.9	50.785	0.115	0.013	4.623
9	52.2	53.420	−1.220	1.490	9.923
9	51.3	53.420	−2.120	4.496	9.923
10	55.6	56.056	−0.456	0.208	17.223
10	59.5	56.056	3.444	11.864	17.223
			Sum	108.045	140.550

$$s_e = \sqrt{\frac{\sum (y - \hat{y})^2}{n - 2}} = \sqrt{\frac{108.045}{20 - 2}} = 2.450$$

(c) $$s_{b_1} = \frac{s_e}{\sqrt{\sum (x - \bar{x})^2}} = \frac{2.450}{\sqrt{140.550}} = 0.2067$$

The hypotheses are $H_0 : \beta_1 = 0$ versus $H_1 : \beta_1 \neq 0$. The test statistic is

$$t_0 = \frac{b_1}{s_{b_1}} = \frac{2.6351}{0.2067} = 12.751.$$

Classical approach: The level of significance is $\alpha = 0.05$. Since this is a two-tailed test with $n - 2 = 20 - 2 = 18$ degrees of freedom, the critical values are $\pm t_{0.025} = \pm 2.101$. Since $t_0 = 12.751 > t_{0.025} = 2.101$ (the test statistic falls in a critical region), we reject H_0.

P-value approach: The P-value for this two-tailed test is the area under the t-distribution with 18 degrees of freedom to the right of $t_0 = 12.751$, plus the area to the left of $-t_0 = -12.751$. From the t-distribution table in the row corresponding to 18 degrees of freedom, 12.751 falls to the right of 3.922, whose right-tail area is 0.0005. We must double this value in order to get the total area in both tails: 0.001. So, P-value < 0.001 [Tech: P-value < 0.0001]. Because the P-value is less than the level of significance $\alpha = 0.05$, we reject H_0.

Conclusion: There is sufficient evidence to conclude that a linear relationship exists between boys' ages and heights.

(d) For a 95% confidence interval we use $t_{0.025} = 2.101$.
Lower Bound:
$$b_1 - t_{\alpha/2} \cdot s_{b_1} = 2.6351 - 2.101 \cdot 0.2067$$
$$= 2.2008$$

Upper Bound:
$$b_1 + t_{\alpha/2} \cdot s_{b_1} = 2.6351 + 2.101 \cdot 0.2067$$
$$= 3.0694$$

(e) We have $n = 20$, $s_e = 2.450$, $\bar{x} = 5.85$ and $\sum (x - \bar{x})^2 = 629.836$. For a 90% confidence interval with $n - 2 = 18$ degrees of freedom, we use $t_{0.05} = 1.734$.

Lower Bound:
$$\hat{y} - t_{\alpha/2} \cdot s_e \sqrt{\frac{1}{n} + \frac{(x^* - \bar{x})^2}{\sum (x - \bar{x})^2}}$$
$$= 48.15 - 1.734 \cdot 2.450 \cdot \sqrt{\frac{1}{20} + \frac{(7 - 5.85)^2}{140.550}}$$
$$= 47.11 \text{ [Tech: 47.12] inches}$$

Upper Bound:
$$\hat{y} + t_{\alpha/2} \cdot s_e \sqrt{\frac{1}{n} + \frac{(x^* - \bar{x})^2}{\sum (x - \bar{x})^2}}$$
$$= 48.15 + 1.734 \cdot 2.450 \cdot \sqrt{\frac{1}{20} + \frac{(7 - 5.85)^2}{140.550}}$$
$$= 49.19 \text{ inches}$$

(f) The predicted value of y is also $\hat{y} = 48.15$ inches.

(g) Lower Bound:
$$\hat{y} - t_{\alpha/2} \cdot s_e \sqrt{1 + \frac{1}{n} + \frac{(x^* - \bar{x})^2}{\sum (x - \bar{x})^2}}$$
$$= 48.15 - 1.734 \cdot 2.450 \cdot \sqrt{1 + \frac{1}{20} + \frac{(7 - 5.85)^2}{140.550}}$$
$$= 43.78 \text{ inches}$$

Upper Bound:
$$\hat{y} + t_{\alpha/2} \cdot s_e \sqrt{1 + \frac{1}{n} + \frac{(x^* - \bar{x})^2}{\sum (x - \bar{x})^2}}$$
$$= 48.15 + 1.734 \cdot 2.450 \cdot \sqrt{1 + \frac{1}{20} + \frac{(7 - 5.85)^2}{140.550}}$$
$$= 52.52 \text{ inches}$$

(h) Although the predicted heights in parts (a) and (h) are the same, the intervals are different because the distribution of the means, part (a), has less variability than the distribution of the individuals, part (h).

4. (a) Using technology we get: $\beta_0 \approx b_0 = 67.388$ and $\beta_1 \approx b_1 = -0.2632$.

(b) Using technology, we calculate $\hat{y} = -0.2632x + 67.388$ to generate the table:

x	y	$\hat{y}$	$y - \hat{y}$	$(y - \hat{y})^2$	$(x - \overline{x})^2$
15	65	63.440	1.560	2.433	598.81
16	60	63.177	–3.177	10.093	550.87
28	58	60.019	–2.019	4.076	131.57
61	60	51.334	8.666	75.101	463.52
53	46	53.439	–7.439	55.344	183.04
43	66	56.071	9.929	98.582	12.46
16	56	63.177	–7.177	51.509	550.87
25	75	60.808	14.192	201.403	209.40
28	46	60.019	–14.019	196.527	131.57
34	45	58.440	–13.440	180.627	29.93
37	58	57.650	0.350	0.122	6.10
41	70	56.597	13.403	179.627	2.34
43	73	56.071	16.929	286.587	12.46
49	45	54.492	–9.492	90.099	90.81
53	60	53.439	6.561	43.042	183.04
61	56	51.334	4.666	21.772	463.52
68	30	49.492	–19.492	379.924	813.93
			Sum	1876.869	4434.24

$$s_e = \sqrt{\frac{\sum(y - \hat{y})^2}{n-2}} = \sqrt{\frac{1876.869}{17-2}} = 11.1859$$

$$s_{b_1} = \frac{s_e}{\sqrt{\sum(x - \overline{x})^2}} = \frac{11.1859}{\sqrt{4434.24}} = 0.1680$$

The hypotheses are $H_0 : \beta_1 = 0$ versus $H_1 : \beta_1 \neq 0$. The test statistic is

$$t_0 = \frac{b_1}{s_{b_1}} = \frac{-0.2632}{0.1680} = -1.567.$$

Classical approach: The level of significance is $\alpha = 0.05$. Since this is a two-tailed test with $n - 2 = 17 - 2 = 15$ degrees of freedom, the critical values are $\pm t_{0.025} = \pm 2.131$. Since $t_0 = -1.567 > -t_{0.025} = -2.131$ (the test statistic does not fall in a critical region), we do not reject H_0.

P-value approach: The P-value for this two-tailed test is the area under the t-distribution with 15 degrees of freedom to the left of $t_0 = -1.567$, plus the area to the right of 1.567. From the t-distribution table in the row corresponding to 15 degrees of freedom, 1.567 falls between 1.341 and 1.753, whose right-tail areas are 0.10 and 0.05, respectively. We must double these values in order to get the total area in both tails: 0.20 and 0.10. So, $0.10 < P\text{-value} < 0.20$ [Tech: P-value = 0.1380]. Because the P-value is larger than $\alpha = 0.05$, we do not reject H_0.

Conclusion: There is not sufficient evidence to support that a linear relationship exists between a woman's age and grip strength.

(c) Since there is no significant linear relationship, the best estimate is the sample mean grip strength, which is $\overline{y} = 57$ psi.

5. (a) Coefficient of x_1: If the rejection rate increases by 1%, while the type of school remains constant, then the average GPA of the school increases by 0.005.
Coefficient of x_2: The average GPA of a private school is 0.2 greater than the average GPA of a public school with the same rejection rate.

(b) The average GPA is predicted to be:
$$\hat{y} = 2.8 + 0.005x_1 + 0.2x_2$$
$$= 2.8 + 0.005(28) + 0.2(1)$$
$$= 3.14$$

(c) The average GPA is predicted to be:
$$\hat{y} = 2.8 + 0.005x_1 + 0.2x_2$$
$$= 2.8 + 0.005(14) + 0.2(0)$$
$$= 2.87$$

6. (a) Correlations: Fat, Protein, Sugar, Carbs, Calories

```
              Fat     Protein     Sugar     Carbs
Protein     0.507
Sugar       0.412     0.247
Carbs       0.576     0.550      0.671
Calories    0.950     0.661      0.542     0.784
```

Because the correlations among the explanatory variables are all between −0.7 and 0.7, multicollinearity does not appear to be a problem.

(b) Regression Analysis: Calories versus Fat, Protein, Sugar, Carbs

```
The regression equation is
Calories = - 11.9 + 9.08 Fat + 4.47 Protein + 1.49 Sugar + 3.52 Carbs


Predictor      Coef    SE Coef      T       P
Constant     -11.90     14.20    -0.84    0.463
Fat           9.0765    0.2421   37.49    0.000
Protein       4.4676    0.5935    7.53    0.005
Sugar         1.488     1.112     1.34    0.273
Carbs         3.5202    0.3035   11.60    0.001


S = 4.39186    R-Sq = 99.9%    R-Sq(adj) = 99.8%


Analysis of Variance

Source            DF     SS      MS       F        P
Regression         4   89342   22336   1157.97   0.000
Residual Error     3      58      19
Total              7   89400
```

The regression equation is $\hat{y} = -11.90 + 9.0765x_1 + 4.4676x_2 + 1.488x_3 + 3.5202x_4$.

(c) From the output in (b), we get $F_0 = 1157.97$ and P-value $< 0.001 < \alpha = 0.05$. We reject the null hypothesis and conclude that at least one of the explanatory variables is linearly associated with calories.

(d) From part (b), we have the following:

Fat: Reject $H_0 : \beta_1 = 0$ since $t_0 = 37.49$ and P-value $< 0.001 < \alpha = 0.05$.

Protein: Reject $H_0 : \beta_2 = 0$ since $t_0 = 7.53$ and P-value $= 0.005 < \alpha = 0.05$.

Sugar: Do not reject $H_0 : \beta_3 = 0$ since $t_0 = 1.34$ and P-value $= 0.273 > \alpha = 0.05$.

Carbohydrates: Reject $H_0 : \beta_4 = 0$ since $t_0 = 11.60$ and P-value $= 0.001 < \alpha = 0.05$.

We conclude that the explanatory variables fat, protein, and carbohydrates all have a significant linear relation with the response variable calories. We also conclude that the explanatory variable sugar is not linearly related to fat. So, we remove x_3, sugar, from the model.

(e) Removing the variable x_3, fat, we obtain:

Regression Analysis: Calories versus Fat, Protein, Carbs

```
The regression equation is
Calories = - 4.4 + 9.11 Fat + 4.29 Protein + 3.77 Carbs


Predictor    Coef   SE Coef      T      P
Constant    -4.42    14.29   -0.31  0.772
Fat        9.1099   0.2635   34.57  0.000
Protein    4.2937   0.6338    6.77  0.002
Carbs      3.7672   0.2637   14.28  0.000


S = 4.80656   R-Sq = 99.9%   R-Sq(adj) = 99.8%

Analysis of Variance

Source          DF      SS     MS        F      P
Regression       3   89308  29769  1288.54  0.000
Residual Error   4      92     23
Total            7   89400
```

The new regression equation is $\hat{y} = -4.42 + 9.1099x_1 + 4.2937x_2 + 3.7672x_4$

Fat: Reject $H_0 : \beta_1 = 0$ since $t_0 = 34.57$ and P-value $< 0.001 < \alpha = 0.05$.

Protein: Reject $H_0 : \beta_2 = 0$ since $t_0 = 6.77$ and P-value $= 0.002 < \alpha = 0.05$.

Carbohydrates: Reject $H_0 : \beta_4 = 0$ since $t_0 = 14.28$ and P-value $< 0.001 < \alpha = 0.05$.

We conclude that the remaining explanatory variables fat, protein, and carbohydrates all have a significant linear relation with the response variable calories. We believe that this is our best model.

(f)

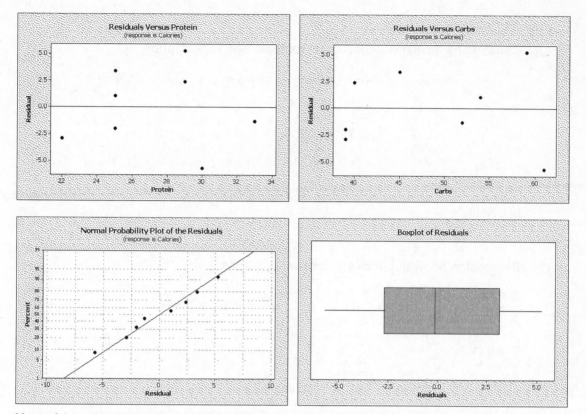

None of the residual plots show any discernible pattern, the boxplot does not show any outliers, and the normal probability plot looks fairly linear. Therefore, the linear model is appropriate.

(g) $b_1 = 9.1099$: If fat is increased by one gram, calories will increase by 9.1099 calories, on average, assuming that protein and carbohydrates remain constant.

$b_2 = 4.2937$: If protein is increased by one gram, calories will increase by 4.2937 calories, on average, assuming that fat and carbohydrates remain constant.

$b_4 = 3.7672$: If carbohydrates are increased by one gram, calories will increase by 3.7672 calories, on average, assuming that fat and protein remain constant.

(h) $R^2 = 99.9\%$; Virtually all (99.9%) of the variation in calories is explained by the least-squares regression model.

$R^2_{adj} = 99.8\%$ modifies the value of R^2 taking the sample size and number of explanatory variables into account.

(i) 95% Confidence Interval: Lower Bound: 656.16; Upper Bound: 668.17; We are 95% confident that the mean number of calories for all fast-food cheeseburgers with 38 grams of fat, 29 grams of protein, and 52 grams of carbohydrates is between 656.16 and 668.17 calories.

95% Prediction Interval: Lower Bound: 647.53; Upper Bound: 676.80; We are 95% confident that the number of calories for a randomly selected fast-food cheeseburger with 38 grams of fat, 29 grams of protein, and 52 grams of carbohydrates is between 647.53 and 676.80 calories.

Chapter 14 Case Study

(a) **Correlations: acerage, beds, baths, footage, age, rooms, price**

	acerage	beds	baths	footage	age	rooms
beds	0.482					
baths	0.543	0.686				
footage	0.692	0.823	0.834			
age	-0.222	-0.446	-0.500	-0.455		
rooms	0.606	0.762	0.668	0.881	-0.267	
price	0.800	0.638	0.736	0.908	-0.259	0.854

Several correlations among the explanatory variables are larger than 0.7. Therefore, multicollinearity may be a problem.

(b) **Regression Analysis: price versus acerage, beds, ...**

```
The regression equation is
price = - 122 + 211 acerage - 60.4 beds + 10.2 baths + 0.161 footage +
        0.789 age + 23.6 rooms
```

Predictor	Coef	SE Coef	T	P
Constant	-122.15	76.69	-1.59	0.126
acerage	210.81	65.18	3.23	0.004
beds	-60.40	24.14	-2.50	0.021
baths	10.18	20.94	0.49	0.632
footage	0.16112	0.04562	3.53	0.002
age	0.7894	0.5347	1.48	0.155
rooms	23.57	13.08	1.80	0.086

$S = 59.7525$ R-Sq = 92.9%   R-Sq(adj) = 90.8%

Analysis of Variance

Source	DF	SS	MS	F	P
Regression	6	976779	162797	45.60	0.000
Residual Error	21	74978	3570		
Total	27	1051757			

The regression equation is:
$$\hat{y} = -122.15 + 210.81x_1 - 60.40x_2 + 10.18x_3 + 0.161x_4 + 0.789x_5 + 23.57x_6$$

(c) From part (b), we have $F_0 = 45.60$ and P-value $< 0.001 < \alpha = 0.05$. We reject the null hypothesis and conclude that at least one of the explanatory variables has a slope coefficient that is significantly different from 0.

(d)

Predictor	Coef	SE Coef	T	P
Constant	-122.15	76.69	-1.59	0.126
acerage	210.81	65.18	3.23	0.004
beds	-60.40	24.14	-2.50	0.021
baths	10.18	20.94	0.49	0.632
footage	0.16112	0.04562	3.53	0.002
age	0.7894	0.5347	1.48	0.155
rooms	23.57	13.08	1.80	0.086

Acres: Reject $H_0 : \beta_1 = 0$ since $t_0 = 3.23$ and P-value $= 0.004 < \alpha = 0.05$

Bedrooms: Reject $H_0 : \beta_2 = 0$ since $t_0 = -2.50$ and P-value $= 0.021 < \alpha = 0.05$

Bathrooms: Do not reject $H_0 : \beta_3 = 0$ since $t_0 = 0.49$ and P-value $= 0.632 > \alpha = 0.05$

Square Footage: Reject $H_0 : \beta_4 = 0$ since $t_0 = 3.53$ and P-value $= 0.002 < \alpha = 0.05$

Age: Do not reject $H_0 : \beta_5 = 0$ since $t_0 = 1.48$ and P-value $= 0.155 > \alpha = 0.05$

Rooms: Reject $H_0 : \beta_6 = 0$ since $t_0 = 1.80$ and P-value $= 0.086 < \alpha = 0.05$

The variables x_3 and x_5 have large P-values so they may need to be removed from the model.

(e) Start by removing x_3, baths, since it has the largest P-value.

Next remove x_5, age, since it has the next highest P-value.

Regression Analysis: price versus acerage, beds, footage, rooms

```
The regression equation is
price = - 67.5 + 220 acerage - 66.4 beds + 0.153 footage + 28.6 rooms

Predictor      Coef   SE Coef       T       P
Constant     -67.47     66.11   -1.02   0.318
acerage      219.66     64.91    3.38   0.003
beds         -66.40     23.91   -2.78   0.011
footage     0.15342   0.03537    4.34   0.000
rooms         28.61     12.11    2.36   0.027


S = 60.0610   R-Sq = 92.1%   R-Sq(adj) = 90.7%

Analysis of Variance
Source           DF       SS      MS      F       P
Regression        4   968788  242197  67.14   0.000
Residual Error   23    82968    3607
Total            27  1051757
```

The regression equation is $\hat{y} = -67.47 + 219.66x_1 - 66.40x_2 + 0.1534x_4 + 28.61x_6$.

(f)

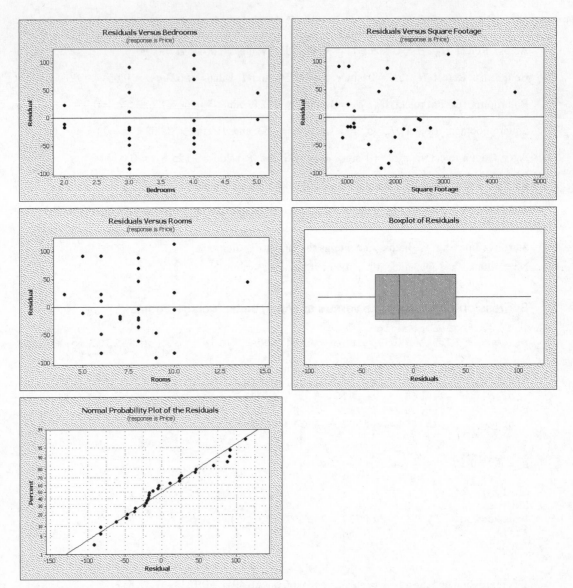

None of the residual plots show any discernible pattern, the boxplot does not show any outliers, and the normal probability plot looks fairly linear. Therefore, the linear model is appropriate.

(g) $R^2 = 0.921$; 92.1% of the variance in price is explained by the regression model. $R^2_{adj} = 0.907$ modifies the value of R^2 taking the sample size and number of explanatory variables into account. It appears that the model fits the data rather well.

(h) $\hat{y} = -67.47 + 219.66x_1 - 66.40x_2 + 0.1534x_4 + 28.61x_6$

$= -67.47 + 219.66(0.18) - 66.40(3) + 0.1534(1176) + 28.61(6)$

$= 124.927$

The predicted price of the house is $124,927. This is quite a bit higher than the actual value of $99,900.

(i)

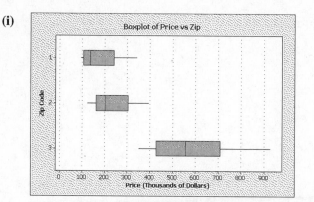

There is reason to believe that price varies by zip code. In particular, it appears that houses in the third zip code are more expensive than those in the first two zip codes.

(j) Let $x_7 = 1$ for the first zip code, $x_7 = 2$ for the second zip code, and $x_7 = 3$ for the third zip code.

Regression Analysis: price versus acerage, beds, footage, rooms, zip

```
The regression equation is
price = - 153 + 69.6 acerage - 44.8 beds + 0.152 footage + 20.0 rooms + 70.7
zip

Predictor       Coef   SE Coef       T       P
Constant     -152.75     55.22   -2.77   0.011
acerage        69.56     62.27    1.12   0.276
beds          -44.83     19.22   -2.33   0.029
footage      0.15217   0.02733    5.57   0.000
rooms         19.993     9.595    2.08   0.049
zip           70.66      17.38    4.07   0.001

S = 46.4040   R-Sq = 95.5%   R-Sq(adj) = 94.5%

Analysis of Variance
Source           DF        SS      MS       F       P
Regression        5   1004384  200877   93.29   0.000
Residual Error   22     47373    2153
Total            27   1051757
```

We proceed to remove variables whose slope coefficient is not significantly different from zero, beginning with x_1, acreage, since it has the largest *P*-value (0.276). After each removal, we compute the regression equation and check the remaining slope coefficients.
Variables removed: rooms (*P*-value = 0.063) and bedrooms (*P*-value = 0.073).

Regression Analysis: price versus footage, zip
```
The regression equation is
price = - 189 + 0.159 footage + 96.1 zip

Predictor       Coef   SE Coef       T      P
Constant     -189.22     27.11   -6.98  0.000
footage      0.15868   0.01275   12.44  0.000
zip            96.13     14.52    6.62  0.000

S = 51.7421    R-Sq = 93.6%    R-Sq(adj) = 93.1%

Analysis of Variance
Source          DF        SS      MS       F      P
Regression       2    984826  492413  183.92  0.000
Residual Error  25     66931    2677
Total           27   1051757
```

The resulting regression equation is $\hat{y} = -189.22 + 0.1587x_4 + 96.13x_7$.

(k) $\hat{y} = -189.22 + 0.1587x_4 + 96.13x_7$

$\qquad = -189.22 + 0.1587(1176) + 96.13(1) = 93.541$

The predicted price of the house is \$93,541. This is below the actual price of \$99,900, but it is closer than the prediction obtained from the first model.

(l)-(n)   Answers will vary.

Chapter 15

Nonparametric Statistics

Section 15.1

1. Parametric procedures are based on some assumptions about the distribution of the population that is being examined, and typically involve the estimation of a population parameter, such as the mean or variance of the population. Nonparametric methods, perhaps better called distribution-free methods, make no assumptions about the underlying population distributions and usually do not involve estimating any population parameters, although they may be used to perform inference about population parameters such as the population median.

3. The power of a test is the probability of making a Type II error, i.e. the error of not rejecting the null hypothesis when the alternative hypothesis is in fact correct.

5. Disadvantages of nonparametric procedures include:
   (1) Nonparametric tests are often misused; that is, they are often used when a more powerful parametric method is appropriate.
   (2) Nonparametric tests are typically less powerful than parametric tests.
   (3) Nonparametric tests are typically less efficient than parametric tests.

Section 15.2

1. run

3. (a) $n = 14$, $n_1 = 6$, $n_2 = 8$ and $r = 6$.

   (b) Since $n_1 \leq 20$ and $n_2 \leq 20$, the critical values are 3 and 12, from Table X.

   (c) The test statistic is $r = 6$, which is not in the critical region, so we do not reject H_0. There is not enough evidence to conclude that the sequence is not random.

5. (a) $n = 19$, $n_1 = 9$, $n_2 = 10$ and $r = 5$.

   (b) Since $n_1 \leq 20$ and $n_2 \leq 20$, the critical values are 5 and 16, from Table X.

(c) The test statistic is $r = 5$, which is in the critical region and so we reject H_0. There is enough evidence to conclude that the sequence is not random.

7. $n = 20$, $n_1 = 9$ (two-seam fastballs), $n_2 = 11$ (split-finger fastballs), and $r = 12$. Since $n_1 \leq 20$ and $n_2 \leq 20$, the critical values are 6 and 16, from Table X. The test statistic is $r = 12$, which is not in the critical region, so we do not reject H_0. There is not enough evidence at the $\alpha = 0.05$ level of significance to conclude that Okajima's pitches are not random.

9. $n = 14$, $n_1 = 8$ (on-time flights), $n_2 = 6$ (late arrivals) and $r = 7$. Since $n_1 \leq 20$ and $n_2 \leq 20$, the critical values are 3 and 12, from Table X. The test statistic is $r = 7$, which is not in the critical region, so we do not reject H_0. There is not enough evidence at the $\alpha = 0.05$ level of significance to conclude that the arrival status of Flight 0426 is not random.

11. $n = 20$, $n_1 = 14$ (accepted bottles), $n_2 = 6$ (rejected bottles) and $r = 4$. Since $n_1 \leq 20$ and $n_2 \leq 20$, the critical values are 5 and 14, from Table X. The test statistic is $r = 4$, which is in the critical region, so we reject H_0. There is enough evidence at the $\alpha = 0.05$ level of significance to conclude that the filling machine is not random in the way that it over- or under-fills bottles.

13. $n = 48$, $n_1 = 26$ (positive changes), $n_2 = 22$ (negative changes) and $r = 28$. Since $n_2 > 20$ the critical values are $\pm z_{0.025} = \pm 1.96$. To compute the test statistic, we calculate
$$\mu_r = \frac{2n_1 n_2}{n} + 1 = \frac{2 \cdot 26 \cdot 22}{48} + 1 \approx 24.83,$$
$$\sigma_r = \sqrt{\frac{2n_1 n_2 (2n_1 n_2 - n)}{n^2 (n-1)}},$$
$$= \sqrt{\frac{2 \cdot 26 \cdot 22 (2 \cdot 26 \cdot 22 - 48)}{48^2 (48-1)}} = 3.403$$

and $z_0 = \dfrac{r - \mu_r}{\sigma_r} = \dfrac{28 - 24.83}{3.403} = 0.93$, which is

not in the critical region so we do not reject H_0. There is not enough evidence at the $\alpha = 0.05$ level of significance to conclude that stock price fluctuations are not random.

15. (a) B A A A A A B B A B B A

    (b) $n = 12$, $n_1 = 7$ (A's), $n_2 = 5$ (B's) and $r = 6$. Since $n_1 \le 20$ and $n_2 \le 20$, the critical values are 3 and 11, from Table X. The test statistic is $r = 6$, which is not in the critical region so we do not reject H_0. There is not enough evidence at the $\alpha = 0.05$ level of significance to conclude that the sequence of residuals is not random. The data appear to support the assumption.

17. Denoting strengths above 75 with A and those below 75 with B, we obtain the sequence
A A B B B A A A A A A B B B A A
A A B B
This gives us $n = 20$, $n_1 = 12$ (above target), $n_2 = 8$ (below target), $r = 6$. Since $n_1 \le 20$ and $n_2 \le 20$, the critical values are 6 and 16, from Table X. The test statistic is $r = 6$, which is in the critical region so we reject H_0. There is enough evidence at the $\alpha = 0.05$ level of significance to conclude that the compression strengths are not random about the target value of 75 psi. It appears the process is no longer in control.

19. A phenomenon is random if individual outcomes are unpredictable (follow no particular pattern), although there may be an overall pattern in the long run. A run is an unbroken sequence of similar events, items, or symbols that is followed by an event, item, or symbol that is mutually exclusive from the first event, item, or symbol.

20. If a sequence of occurrences (of two distinct types) is truly random, then at any point in the sequence the next occurrence is equally likely to be of either type. If there are long runs in the sequence, then the next occurrence is more likely to be of the same type as that preceding it. If there are many short runs in the sequence, then the next occurrence is more likely to be of the different type to that preceding it.

Section 15.3

1. Since $H_1 : M < 8$, this is a left-tailed test and k = the number of plus signs = 8. We have $n = 13 + 8 = 21 \le 25$ and so the critical value, from Table XI, is 6. Since $k > 6$ we do not reject H_0. There is not enough evidence at the 5% level of significance to conclude that the median is less than 8.

3. Since $H_1 : M \ne 100$, this is a two-tailed test and k = the smaller of the number of minus signs and the number of plus signs = 21. We have $n = 21 + 28 = 49 > 25$ and so the critical value, from the normal tables, is $-z_{\alpha/2} = -z_{0.025} = -1.96$. The test statistic is

$$z_0 = \dfrac{(k + 0.5) - \dfrac{n}{2}}{\sqrt{n}/2} = \dfrac{(21 + 0.5) - \dfrac{49}{2}}{\sqrt{49}/2} = -0.86 > -1.96$$

so we do not reject H_0. There is not enough evidence at the 5% level of significance to conclude that the median differs from 100.

5. We first convert the data to plus and minus signs (and zeros) by comparing the data values to $M_0 = 12$. This gives: $+ + + + - + + -$ $+ + - + + + 0$. There are 11 plus signs and 3 minus signs. Since $H_1 : M > 12$, this is a right-tailed test and k = the number of minus signs = 3. We have $n = 11 + 3 = 14 \le 25$ and so the critical value, from Table XI, is 3. Since $k \le 3$ we reject H_0. There is enough evidence at the 5% level of significance to conclude that the median is greater than 12.

7. We record plus signs for values > 5.25 and minus signs for values < 5.25 and count 12 plus signs and 7 minus signs. Since $H_1 : M > 5.25$, this is a right-tailed test and k = the number of minus signs = 5. We have $n = 5 + 10 = 15 \le 25$ and so the critical value, from Table XI, is 3. Since $k > 3$ we do not reject H_0. There is not enough evidence to conclude that the median pH is greater than 5.25. That is, there is not enough evidence at the 5% level of significance to conclude that the acidity of rain in Glacier National Park has decreased.

9. We record plus signs for values $> \$57,283$ and minus signs for values $< \$57,283$ and count 4 plus signs and 8 minus signs. Since

$H_0 : M = \$57,283$ and $H_1 : M < \$57,283$, this is a left-tailed test and k = the number of plus signs = 4. We have $n = 4 + 8 = 12 \leq 25$ and so the critical value, from Table XI, is 2. Since $k > 2$ we do not reject H_0. There is not enough evidence at the 5% level of significance to conclude that the median salary of teachers in the district is less than the median salary of teachers in the state.

11. We record plus signs for values > $800 thousand and minus signs for values < $800 thousand and count 11 plus signs and 3 minus signs. Since $H_1 : M > 800$, this is a right-tailed test and k = the number of minus signs = 3. We have $n = 11 + 3 = 14 \leq 25$ and so the critical value, from Table XI, is 3. Since $k \leq 3$, we reject H_0. There is sufficient evidence at the 5% level of significance to indicate that the median baseball salary is greater than $800,000.

13. P-value $= P(X \leq k) = P(X \leq 5)$ calculated from the binomial formula with $n = 15$ and $p = 0.5$. Using technology we get P-value $= 0.1509$.

15. P-value $= P(X \leq k) = P(X \leq 4)$ calculated from the binomial formula with $n = 12$ and $p = 0.5$. Using technology we get P-value $= 0.1938$.

17. Our hypotheses are $H_0 : p = 0.5$ versus $H_1 : p < 0.5$. It was found that 38 of the 100 were women, so we would have 38 plus signs and 62 minus signs. This is a left-tailed test, so k = number of plus signs = 38. Since $n = 100 > 25$, our test statistic is

$$z_0 = \frac{(k+0.5) - \frac{n}{2}}{\frac{\sqrt{n}}{2}} = \frac{(38+0.5) - \frac{100}{2}}{\frac{\sqrt{100}}{2}} = -2.3.$$

At the $\alpha = 0.01$ level of significance, our critical value is $-z_{0.01} = -2.33$. Since

$z_0 > z_{0.01}$, we do not reject H_0. There is not enough evidence at the $\alpha = 0.01$ level of significance to indicate that less than 50% of gamers are women.

19. Since participants had to agree to be part of the survey, the sample may not be truly random which is required to use the sign test.

21. If the true median $M < M_0$ then we would expect the majority of observations to be less than M_0 (giving minus signs) and few observations to be greater than M_0 (giving plus signs). In this case k, the number of plus signs, will be much smaller than $\frac{1}{2}n$, resulting in a large negative value of the test statistic. If the true median $M > M_0$ then we would expect the majority of observations to be greater than M_0 (giving plus signs) and few observations to be less than M_0 (giving minus signs). In this case k is the number of minus signs, and again it will be much smaller than $\frac{1}{2}n$, resulting in a large negative value of the test statistic. So, in either case, a large negative value of the test statistic is evidence in favor of the alternative hypothesis and against the null hypothesis.

23. Answers will vary. If the true median $M \neq M_0$, then either $M > M_0$ or $M < M_0$. If $M > M_0$ then there will be many observations greater than M_0 (plus signs) and few observations less than M_0 (minus signs). Thus k, the smaller of the number of minus signs and the number of plus signs, will be the number of minus signs, and it will be relatively small. Similarly, if $M < M_0$ then there will be many observations less than M_0 (minus signs) and few observations greater than M_0 (plus signs). Thus k will be the number of plus signs and will be relatively small. In either case a small value of k is evidence that $M \neq M_0$.

Section 15.4

1. If we compute the differences as Population 2 – Population 1, then the researcher believes that $M_D > 0$, where M_D is the median of the differences between the observations in the matched pairs data.

3. Since $H_1 : M_D > 0$, this is a right-tailed test and the test statistic is $T = |T_-| = 16$. The critical value for $n = 12$ and $\alpha = 0.05$ is 17 (from Table XII). Since $T < 17$ we reject H_0. There is enough evidence to conclude that $M_D > 0$.

5. Since $H_1 : M_D < 0$, this is a left-tailed test and the test statistic is $T = T_+ = 33$. The critical value for $n = 15$ and $\alpha = 0.05$ is 30 (from Table XII). Since $T \geq 30$ we do not reject H_0. There is not enough evidence to conclude that $M_D < 0$.

7. Since $H_1 : M_D \neq 0$, this is a two-tailed test and the test statistic is $T = \min(T_+, |T_-|)$
   $= \min(50, 121) = 50$. The critical value for $n = 18$ and $\alpha/2 = 0.025$ is 40 (from Table XII). Since $T \geq 40$ we do not reject H_0. There is not enough evidence to conclude that $M_D \neq 0$.

9. Since $H_1 : M_D > 0$, this is a right-tailed test and $T = |T_-| = 300$. The critical value for $n = 40$ and $\alpha = 0.05$ is
   $-z_\alpha = -z_{0.05} = -1.645$ (since $n > 30$). The test statistic is
   $$z_0 = \frac{T - \dfrac{n(n+1)}{4}}{\sqrt{\dfrac{n(n+1)(2n+1)}{24}}} = \frac{300 - \dfrac{40(40+1)}{4}}{\sqrt{\dfrac{40(40+1)(2 \cdot 40+1)}{24}}}$$
   $= -1.48 > -1.645$
   so we do not reject H_0. There is not enough evidence to conclude that $M_D > 0$.

11. From the computations in the table below we get $T_+ = 12$, $|T_-| = |-33| = 33$ and $n = 9$.

Before	After	After – Before	Absolute Value	Signed Rank
23.5	19.75	–3.75	3.75	–9
18.5	19.25	0.75	0.75	4.5
21.5	21.75	0.25	0.25	1.5
24	22.5	–1.5	1.5	–7.5
25	25	0	0	0
19.75	19.5	–0.25	0.25	–1.5
35	34.25	–0.75	0.75	–4.5
36.5	35	–1.5	1.5	–7.5
52	51.5	–0.5	0.5	–3
30	31	1	1	6

Since $H_1 : M_D < 0$, this is a left-tailed test and the test statistic is $T = T_+ = 12$. The critical value for $n = 9$ and $\alpha = 0.05$ is 8 (from Table XII). Since $T \geq 8$ we do not reject H_0. There is not sufficient evidence to indicate that the median waistline has decreased. The exercise regimen does not appear to be effective in reducing waistline.

13. From the computations in the table below we get $T_+ = 5$, $|T_-| = |-16| = 16$ and $n = 6$.

Blue	Red	Red – Blue	Absolute Value	Signed Rank
0.582	0.408	–0.174	0.174	–4
0.481	0.407	–0.074	0.074	–1
0.841	0.542	–0.299	0.299	–6
0.267	0.402	0.135	0.135	3
0.685	0.456	–0.229	0.229	–5
0.450	0.533	0.083	0.083	2

Since $H_1 : M_D \neq 0$, this is a two-tailed test and the test statistic is $T = \min(T_+, |T_-|)$
$= \min(5, 16) = 5$. The critical value for $n = 6$ and $\alpha/2 = 0.025$ is 0 (from Table XII). Since $T \geq 0$ we do not reject H_0. There is not sufficient evidence to indicate that the median reaction times are different for blue versus red.

15. From the computations in the table below we get $T_+ = 31$, $\left|T_-\right| = \left|-5\right| = 5$ and $n = 8$.

Initial	5 Yrs	5 Yrs – Initial	Absolute Value	Signed Rank
38	52	14	14	8
58	60	2	2	2.5
65	72	7	7	5
74	72	–2	2	–2.5
56	54	–2	2	–2.5
36	48	12	12	7
56	58	2	2	2.5
52	60	8	8	6

Since $H_1 : M_D > 0$, this is a right-tailed test and the test statistic is $T = \left|T_-\right| = 5$. The critical value for $n = 8$ and $\alpha = 0.05$ is 5 (from Table XII). Since $T \geq 5$ we do not reject H_0. There is not enough evidence to indicate that the median clarity of the lake has increased.

17. From the computations in the table below we get $T_+ = 20$, $\left|T_-\right| = \left|-35\right| = 35$ and $n = 10$.

Avis	Hertz	Hertz – Avis	Absolute Value	Signed Rank
67.54	59.14	–8.40	8.40	–10
97.16	97.15	–0.01	0.01	–1
103.43	103.14	–0.29	0.29	–5
38.79	36.63	–2.16	2.16	–9
87.90	87.80	–0.10	0.10	–2
109.86	110.09	0.23	0.23	4
109.35	109.35	0	0	0
82.86	83.03	0.17	0.17	3
112.65	113.32	0.67	0.67	6
116.46	114.53	–1.93	1.93	–8
105.49	107.04	1.55	1.55	7

Since $H_1 : M_D > 0$, this is a right-tailed test and the test statistic is $T = \left|T_-\right| = 35$. The critical value for $n = 10$ and $\alpha = 0.05$ is 10 (from Table XII). Since $T \geq 10$ we do not reject H_0. There is not enough evidence to

indicate that the median car rental cost for Avis is less than that for Hertz.

19. (a) $H_0 : M_D = 0$ and $H_1 : M_D \neq 0$.

(b) Since P-value $< 0.001 < 0.05 = \alpha$ we should reject H_0 and conclude that space flight affects the red blood cell count of rats.

21. Start by computing the difference between the length of stay for completers referred by employers and 107 days (the median for referrals from the criminal justice system). From the computations in the table below we get $T_+ = 8.5$, $\left|T_-\right| = \left|-57.5\right| = 57.5$ and $n = 11$.

Employer – Justice	Absolute Value	Signed Rank
–27	27	–11
1	1	1
–12	12	–7
0	0	0
–18	18	–9
–7	7	–4
–22	22	–10
–5	5	–3
8	8	5.5
2	2	2
–8	8	–5.5
–13	13	–8

Since $H_1 : M_D \neq 0$, this is a two-tailed test and the test statistic is $T = \min\left(T_+, \left|T_-\right|\right)$ $= \min(8.5, 57.5) = 8.5$. The critical value for $n = 11$ and $\alpha / 2 = 0.025$ is 10 (from Table XII). Since $T < 10$ we reject H_0. There is enough evidence at the $\alpha = 0.05$ level of significance to conclude that the median length of stay is different for employer referrals than those referred by the criminal justice system.

23. Her results are suspect because the data were not randomly selected.

Section 15.5

1. Since $H_1 : M_x \neq M_y$, this is a two-tailed test. As $n_1 = 12$, $n_2 = 15$ and $\alpha = 0.05$, the critical values are $w_{0.025} = 50$ (from Table XIII) and $w_{0.975} = n_1 n_2 - w_{0.025} = 12 \cdot 15 - 50 = 130$. The test statistic is

    $T = S - \dfrac{n_1(n_1+1)}{2} = 170 - \dfrac{12 \cdot 13}{2} = 92$ which is

    not in the critical region so we do not reject H_0. There is not enough evidence to indicate that $M_x \neq M_y$.

3. Since $H_1 : M_x < M_y$, this is a left-tailed test. As $n_1 = 18$, $n_2 = 16$ and $\alpha = 0.05$, the critical value is $w_{0.05} = 96$ (from Table XIII). The test statistic is

    $T = S - \dfrac{n_1(n_1+1)}{2} = 210 - \dfrac{18 \cdot 19}{2} = 39$ which

    is less than $w_{0.05} = 96$ and so we reject H_0. There is enough evidence to conclude that $M_x < M_y$.

5. Since $H_1 : M_x > M_y$, this is a right-tailed test. As $n_1 = 15$, $n_2 = 15$ and $\alpha = 0.05$, we first look up $w_{0.05} = 73$ (from Table XIII). Then the critical value is $w_{0.95} = n_1 n_2 - w_{0.05} = 15 \cdot 15 - 73 = 152$. The test statistic is

    $T = S - \dfrac{n_1(n_1+1)}{2} = 250 - \dfrac{15 \cdot 16}{2} = 130$ which

    is not greater than $w_{0.95} = 152$ so we do not reject H_0. There is not enough evidence to indicate that $M_x > M_y$.

7. Since $H_1 : M_x \neq M_y$, this is a two-tailed test. As $n_1 = 22 > 20$, $n_2 = 25 > 20$ and $\alpha = 0.05$, the critical values are $\pm z_{\alpha/2} = \pm z_{0.025} = \pm 1.96$. We first calculate

    $T = S - \dfrac{n_1(n_1+1)}{2} = 590 - \dfrac{22 \cdot 23}{2} = 337$ and

    then the test statistic is given by

    $$z_0 = \frac{T - \dfrac{n_1 n_2}{2}}{\sqrt{\dfrac{n_1 n_2 (n_1 + n_2 + 1)}{12}}} = \frac{337 - \dfrac{22 \cdot 25}{2}}{\sqrt{\dfrac{22 \cdot 25 \cdot (22 + 25 + 1)}{12}}}$$

    $= 1.32$

    which is not in the critical region so we do not reject H_0. There is not enough evidence to indicate that $M_x \neq M_y$.

9. From the computations in the table below we have $n_1 = 15$, $n_2 = 13$ and $S = 193.5$.

Women	Men	Women Ranks	Men Ranks
7	12	3.5	11.5
6	20	2	28
10	13	8.5	14.5
17	14	24.5	17.5
9	9	6.5	6.5
13	13	14.5	14.5
14	10	17.5	8.5
15	18	20.5	26.5
18	15	26.5	20.5
15	8	20.5	5
11	13	10	14.5
12	17	11.5	24.5
16	15	23	20.5
7		3.5	
3		1	
	Sum =	193.5	

    Since $H_0 : M_W \neq M_M$ and $H_1 : M_W \neq M_M$, this is a two-tailed test. The critical values are $w_{0.025} = 55$ (from Table XIII) and $w_{0.975} = n_1 n_2 - w_{0.025} = 15 \cdot 13 - 55 = 140$. The test statistic is

    $T = S - \dfrac{n_1(n_1+1)}{2} = 193.5 - \dfrac{15 \cdot 16}{2} = 73.5$

    which is not in the critical region so we do not reject H_0. There is not enough evidence to indicate that the median sociability score is different for women and men.

11. From the computations in the table below we have $n_1 = 8$, $n_2 = 8$ and $S = 74.5$.

Carpeted	Uncarpeted	Carpeted Ranks	Uncarpeted Ranks
11.8	12.1	10	12
8.2	8.3	4	5
7.1	3.8	2	1
13.0	7.2	13	3
10.8	12.0	8	11
10.1	11.1	6.5	9
14.6	10.1	16	6.5
14.0	13.7	15	14
	Sum =	74.5	

Since $H_1 : M_x > M_y$, this is a right-tailed test. From Table XIII $w_{0.05} = 16$, and so the critical value is $w_{0.95} = n_1 n_2 - w_{0.05} = 8 \cdot 8 - 16 = 48$. The test statistic is

$$T = S - \frac{n_1(n_1 + 1)}{2} = 74.5 - \frac{8 \cdot 9}{2} = 38.5 \text{ which}$$

is not greater than $w_{0.95} = 48$ and so we do not reject H_0. There is not enough evidence to conclude that the median amount of bacteria in carpeted rooms is greater than that in uncarpeted rooms.

13. From the computations in the table below we have $n_1 = 22$, $n_2 = 20$ and $S = 609$.

Lincoln	Clarendon	Lincoln Ranks	Clarendon Ranks
0.11	0.06	12	2
0.09	0.09	7.5	7.5
0.21	0.14	26.5	18
0.09	0.16	7.5	20.5
0.07	0.03	3	1
0.30	0.12	31.5	14
0.41	0.29	36.5	30
0.33	0.10	33.5	10.5
0.20	0.41	24.5	36.5
0.22	0.08	28	4.5
0.34	0.14	35	18
0.46	0.14	39	18
0.19	0.12	22.5	14
0.67	0.08	41	4.5
0.75	0.09	42	7.5
0.19	0.10	22.5	10.5
0.30	0.21	31.5	26.5
0.33	0.16	33.5	20.5
0.20	0.13	24.5	16
0.42	0.12	38	14
0.25		29	
0.47		40	
	Sum =	609	

Since $H_1 : M_x \neq M_y$, this is a two-tailed test. As $n_1 = 22 > 20$ and $\alpha = 0.05$, the critical values are $\pm z_{\alpha/2} = \pm z_{0.025} = \pm 1.96$. We first find $T = S - \frac{n_1(n_1 + 1)}{2} = 609 - \frac{22 \cdot 23}{2} = 356$ and then the test statistic is given by

$$z_0 = \frac{T - \frac{n_1 n_2}{2}}{\sqrt{\frac{n_1 n_2 (n_1 + n_2 + 1)}{12}}} = \frac{356 - \frac{22 \cdot 20}{2}}{\sqrt{\frac{22 \cdot 20 \cdot (22 + 20 + 1)}{12}}}$$

$= 3.43$ which is in the critical region so we reject H_0.

There is enough evidence to conclude that the median calcium level in rainwater in Lincoln County, Nebraska, differs from the median level in Clarendon County, South Carolina.

15. (a) From the computations in the table below we have $n_1 = 15$, $n_2 = 15$ and $S = 262$.

Prof. A	Prof. B	Prof. A's Ranks	Prof. B's Ranks
A	A	3	3
A	A	3	3
B	A	10	3
B	B	10	10
B	B	10	10
C	B	19.5	10
C	B	19.5	10
C	B	19.5	10
C	B	19.5	10
C	C	19.5	19.5
C	C	19.5	19.5
D	C	26.5	19.5
D	C	26.5	19.5
D	D	26.5	26.5
F	F	29.5	29.5
	Sum =	262	

(b) This is a two-tailed test so $H_1 : M_x \neq M_y$.

Assuming that $\alpha = 0.05$, the critical values are $w_{0.025} = 65$ (from Table XIII) and

$$w_{0.975} = n_1 n_2 - w_{0.025} = 15 \cdot 15 - 65 = 160.$$

The test statistic is

$$T = S - \frac{n_1(n_1 + 1)}{2} = 262 - \frac{15 \cdot 16}{2} = 142$$

which is not in the critical region and so we do not reject H_0. There is not enough evidence to support the claim that the median grade differs in the two classes.

17. If the two medians are the same, then the sum of the ranks of the observations from population X should be close to the sum of the ranks of the observations from population Y. The test statistic essentially measures the size of the sum of ranks for the sample from

population X. If this sum is unusually small, then that is evidence that $M_x < M_y$ and if this sum is unusually large then that is evidence that $M_x > M_y$.

19. The Wilcoxon signed-ranks test is used for dependent samples (matched pairs) while the Mann-Whitney test is used for independent samples.

Section 15.6

1. (a) The scatter diagram appears to show a positive linear association.

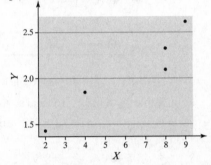

(b) The table shows the data and their ranks:

X	Y	Rank of X	Rank of Y	d_i	d_i^2
2	1.4	1	1	0	0
4	1.8	2	2	0	0
8	2.1	3.5	3	0.5	0.25
8	2.3	3.5	4	−0.5	0.25
9	2.6	5	5	0	0

$$\sum d_i^2 = 0 + 0 + 0.25 + 0.25 + 0 = 0.5$$

$$r_s = 1 - \frac{6 \sum d_i^2}{n(n^2 - 1)} = 1 - \frac{6(0.5)}{5(5^2 - 1)} = 1 - \frac{3}{120} = 0.975$$

(c) With $n = 5$ and $\alpha = 0.05$ (two tails), the critical values are ± 1.000. Since the test statistic, $r_s = 0.975$, is not greater than 1.000, we do not reject H_0. Note that when $n = 5$, this test is too insensitive to detect an association at the $\alpha = 0.05$ level. However at the $\alpha = 0.10$ level of significance, the critical values are ± 0.900 and since $r_s = 0.975 > 0.900$ we would reject H_0 and conclude that there is an association between X and Y.

3. (a) The scatter diagram appears to show a negative linear association.

(b) The table shows the data and their ranks:

X	Y	Rank of X	Rank of Y	d_i	d_i^2
0	5.8	1	6	−5	25
2	5.7	2	5	−3	9
3	5.2	3	4	−1	1
5	2.8	4	3	1	1
6	1.9	5.5	1	4.5	20.25
6	2.2	5.5	2	3.5	12.25

$$\sum d_i^2 = 25 + 9 + 1 + 1 + 20.25 + 12.25 = 68.5$$

$$r_s = 1 - \frac{6 \sum d_i^2}{n(n^2 - 1)} = 1 - \frac{6(68.5)}{6(6^2 - 1)} = 1 - \frac{411}{210} \approx -0.957 \ [\text{Tech: } r_s = -0.968]$$

Note: the text formula and technology values are not identical when there are ties among the ranks and the formula is only approximate.

(c) With $n = 6$ and $\alpha = 0.05$ (two tails), the critical values are ± 0.886. Since the test statistic is less than -0.886, we reject H_0 and conclude that there is an association between X and Y.

5. (a) The table shows the data and their ranks:

% with Bachelor's	Per Capita Income ($)	Rank of X	Rank of Y	d_i	d_i^2
18.8	31,946	1	1	0	0
29.6	42,325	9	9	0	0
24.3	36,751	4	7	−3	9
20.3	35,507	2	3	−1	1
25.4	36,745	6	6	0	0
27.1	38,081	7	8	−1	1
24.1	35,381	3	2	−1	1
28.1	35,667	8	4	4	16
25.3	36,484	5	5	0	0

$$\sum d_i^2 = 0 + 0 + 9 + 1 + 0 + 1 + 1 + 16 + 0 = 28$$

$$r_s = 1 - \frac{6 \sum d_i^2}{n(n^2 - 1)} = 1 - \frac{6(28)}{9(9^2 - 1)} \approx 0.767$$

As we are testing for a positive association, this is a right-tailed test and with $n = 9$ and $\alpha = 0.05$ (one tail). The critical value is 0.600. Since the test statistic is greater than 0.600, we reject H_0. There is enough evidence to conclude that per capita personal income and percentage of the population with at least a bachelor's degree are positively associated.

(b) The scatter diagram appears to show a positive linear association, which supports our conclusion.

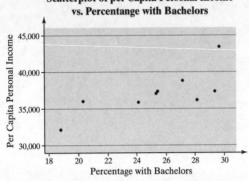

Scatterplot of per Capita Personal Income vs. Percentange with Bachelors

7. (a) The table shows the data and their ranks:

Per Capita Income (\$)	Birth Rate	Rank of X	Rank of Y	d_i	d_i^2
39,817	14.1	6	4.5	1.5	2.25
48,285	13.9	8	3	5	25
66,000	15.1	9	8	1	1
43,874	14.1	7	4.5	2.5	6.25
32,219	12.1	2	1	1	1
34,453	14.5	4	7	−3	9
31,799	14.3	1	6	−5	25
33,786	15.8	3	9	−6	36
37,780	13.1	5	2	3	9

$$\sum d_i^2 = 2.25 + 25 + 1 + 2.25 + 1 + 9 + 25 + 36 + 9 = 110.5$$

$$r_s = 1 - \frac{6\sum d_i^2}{n\left(n^2 - 1\right)} = 1 - \frac{6(114.5)}{9\left(9^2 - 1\right)} = 1 - \frac{519}{720} \approx 0.046$$

As we are testing for a negative association, this is a left-tailed test and with $n = 9$ and $\alpha = 0.05$ (one tail), the critical value is -0.600. Our test statistic is positive so we do not reject H_0. There is not evidence to conclude that per capita personal income and birth rate are negatively associated.

(b) The scatter diagram appears show very little association, which supports our conclusion.

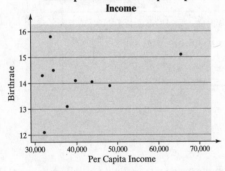

Scatterplot of Birthrate vs. per Capia Income

9. (a) The table shows the data and their ranks:

Family Income ($1000)	Commute Time (min)	Rank of X	Rank of Y	d_i	d_i^2
83	27.1	2.5	5	−2.5	6.25
82	17.0	1	1	0	0
86	26.9	4	4	0	0
90	26.0	5	3	2	4
101	32.1	6	6	0	0
83	22.6	2.5	2	0.5	0.25

$$\sum d_i^2 = 6.25 + 0 + 0 + 4 + 0 + 0.25 = 10.5$$

$$r_s = 1 - \frac{6 \sum d_i^2}{n(n^2 - 1)} = 1 - \frac{6(10.5)}{6(6^2 - 1)} = 1 - \frac{63}{210} = 0.7 \quad [\text{Tech: } 0.696]$$

As we are testing for a positive association, this is a right-tailed test and with $n = 6$ and $\alpha = 0.10$ (one tail), the critical value is 0.657. Since the test statistic is greater than 0.657, we reject H_0. There is enough evidence to conclude that family income cost and commute time are positively associated.

(b) The scatter diagram perhaps shows a positive association, which supports our conclusion.

Scatterplot of Average
Commute Time vs Family Income

11. The table shows the data and their ranks:

X	Y	Rank of X	Rank of Y	$d = Y - X$	d^2
2	5	1	1	0	0
3	8	2	2	0	0
4	11	3	3	0	0
5	14	4	4	0	0
6	17	5	5	0	0
				$\sum d^2 =$	0

$r_s = 1.000$, the correlation coefficient of the ranks. X and Y are perfectly positively associated as we can see from the scatter diagram:

13. The Spearman rank-correlation test can be performed on qualitative data if the requirements are met and if the data can be ranked. For example, if a restaurant price can be ranked as expensive, moderate, or inexpensive, and the quality of the food can be ranked as good, fair, or poor, then the Spearman rank-correlation test is appropriate. Since Spearman's rank correlation test depends only on the ranks of the data values, as long as the data is ordinal, and so can be ranked, we can perform this test to determine if there is an association between the variables.

14. If X and Y are positively associated, then small X-values should be paired with small Y-values, and large with large. This implies that the ranks of the X-values will be close to the ranks of the Y-values, and so the differences between these ranks will be close to 0. This will give a test statistic that is close to 1, and which is therefore large relative to the critical values in Table XIV. A large test statistic (i.e. close in value to 1) is evidence of a positive association between X and Y. If there is a negative association between X and Y, then small X-values should be paired with large Y-values, and large with small. This implies that there will be a large difference between the ranks of the X-values and the ranks of the Y-values, resulting in a relatively large negative value for the test statistic (i.e. close in value to -1). Therefore a relatively large negative test statistic is evidence of a negative association between X and Y.

Section 15.7

1. (a) The tables shows the ranks and the computations based on those ranks:

	X Ranks	Y Ranks	Z Ranks
	6.5	10	4.5
	1.5	12	8
	11	3	1.5
	4.5	6.5	9
R_i	23.5	31.5	23
R_i^2	552.25	992.25	529
n_i	4	4	4

The test statistic is calculated as:

$$H = \frac{12}{N(N+1)}\left[\frac{R_1^2}{n_1} + \cdots + \frac{R_k^2}{n_k}\right] - 3(N+1)$$

$$= \frac{12}{12(12+1)}\left[\frac{552.25}{4} + \frac{992.25}{4} + \frac{529}{4}\right] - 3(12+1) = 0.875$$

(b) Since there are three populations and each sample size is ≤ 5 we use Table XV and get a critical value of 5.6923 (corresponding to $\alpha = 0.049$).

(c) The test statistic is not greater than the critical value so we do not reject H_0. There is not enough evidence to indicate that the distributions are different.

3. (a) H_0: the distributions of the populations are the same.

H_1: at least one population distribution differs from the others.

(b) The calculations are summarized in the table:

	Monday	Tuesday	Wednesday	Thursday	Friday
n_i	8	8	8	8	8
R_i	48	226	144	194.5	207.5
R_i^2	2,304	51,076	20,736	37,830	43,056

$$H = \frac{12}{N(N+1)}\left[\frac{R_1^2}{n_1} + \cdots + \frac{R_k^2}{n_k}\right] - 3(N+1)$$

$$= \frac{12}{40(40+1)}\left[\frac{2,304}{8} + \frac{51,076}{8} + \frac{20,736}{8} + \frac{37,830}{8} + \frac{43,056}{8}\right] - 3(40+1) = 18.770$$

(c) Since we are comparing more than 3 populations, we get the critical value from the χ^2 tables. With df $= k - 1 = 5 - 1 = 4$ we find $\chi^2_{0.05} = 9.488$.

(d) Since the test statistic is greater than the critical value, we reject H_0. There is enough evidence to conclude that the distribution of births varies by day of the week.

(e) Monday births clearly differ from births on other days of the week.

5. The table shows the ranks for the data and the calculations based on those ranks:

	Sludge plot	Spring disk	No till	Spring chisel	Great Lakes bt
	1.5	23.5	18	18	10
	6.5	18	3.5	23.5	23.5
	26.5	26.5	13.5	3.5	6.5
	18	30	23.5	10	18
	10	28.5	1.5	21	13.5
	6.5	28.5	13.5	13.5	6.5
R_i	69	155	73.5	89.5	78
R_i^2	4761	24025	5402.25	8010.25	6084
n_i	6	6	6	6	6

$$H = \frac{12}{N(N+1)}\left[\frac{R_1^2}{n_1} + \cdots + \frac{R_k^2}{n_k}\right] - 3(N+1)$$

$$= \frac{12}{30(30+1)}\left[\frac{4761}{6} + \frac{24{,}025}{6} + \frac{5402.25}{6} + \frac{8010.25}{6} + \frac{6084}{6}\right] - 3(30+1) = 10.833$$

Since we are comparing more than 3 populations, we get the critical value from the χ^2 tables. With df $= k - 1 = 5 - 1 = 4$ we find $\chi_{0.05}^2 = 9.488$. Since the test statistic is greater than the critical value, we reject H_0. There is enough evidence to support the conclusion that the distribution of corn plant yield varies by plot type.

7. The table shows the ranks for the data and the calculations based on those ranks:

	Simple	Go/No-Go	Choice
	7	16	15
	11.5	3	11.5
	9	6	13
	4	17	14
	5	10	8
	1	2	18
R_i	37.5	54	79.5
R_i^2	1406.25	2916	6320.25
n_i	6	6	6

$$H = \frac{12}{N(N+1)}\left[\frac{R_1^2}{n_1} + \cdots + \frac{R_k^2}{n_k}\right] - 3(N+1) = \frac{12}{18(18+1)}\left[\frac{1406.25}{6} + \frac{2916}{6} + \frac{6320.25}{6}\right] - 3(18+1)$$
$$= 5.237$$

Since at least one sample size is greater than 5, we get the critical value from the χ^2 tables. With df $= k - 1$ $= 3 - 1 = 2$ we find $\chi_{0.01}^2 = 9.210$. Since the test statistic is not greater than the critical value, we do not reject H_0. There is not enough evidence to indicate that the distribution of reaction time varies by stimulus.

9. The table shows the ranks for the data and the calculations based on those ranks:

	Large Family	Passenger Vans	Midsize Utility
	12.5	6	6
	20	3.5	18.5
	3.5	2	17
	8.5	8.5	11
	15	1	15
	15	12.5	21
	10	18.5	6
R_i	84.5	52	94.5
R_i^2	7140.25	2704	8930.25
n_i	7	7	7

$$H = \frac{12}{N(N+1)}\left[\frac{R_1^2}{n_1} + \cdots + \frac{R_k^2}{n_k}\right] - 3(N+1) = \frac{12}{21(21+1)}\left[\frac{7140.25}{7} + \frac{2704}{7} + \frac{8930.25}{7}\right] - 3(21+1) \approx 3.664$$

Since at least one sample size is greater than 5, we get the critical value from the χ^2 tables. With df $= k - 1 = 3 - 1 = 2$ we find $\chi^2_{0.01} = 9.210$. Since the test statistic is not greater than the critical value, we do not reject H_0. There is not enough evidence to support the conclusion that the distribution of chest compression varies by vehicle category.

Consumer Reports®: Putting Kids to the Test

Using the Minitab output, we get the following:

(a) H_0: the distributions of preference scores for the eight scooters are the same
H_1: the distributions of preference scores for the eight scooters are not the same.

(b) H = 68.00, or 69.07 (adjusted for ties), but both result in the same conclusion.

(c) *P*-value < 0.001.

(d) Based on the results of this test, there is sufficient evidence to believe that distribution of preference scores for the eight scooters is different. In particular, Scooter 6 (with a median preference score of 1.0) and Scooter 2 (with a median preference score of 2) were the most preferred scooters.

Chapter 15 Review Exercises

1. For the given data, $n = 25$, $n_1 = 11$ (number of E's), $n_2 = 14$ (number of W's) and $r = 11$. Since $n_1 \leq 20$ and $n_2 \leq 20$, the critical values are 8 and 19, from Table X. The test statistic is $r = 11$, which is not in the critical region, so we do not reject H_0. There is not enough evidence to conclude that the sequence is not random. It appears that the winning division occurs randomly.

2. We first convert the data to plus and minus signs (and zeros) by comparing the data values to $M_0 = 15$. This gives: + + + + + − + + + + + − + − + − + − − −. There are 13 plus signs and 7 minus signs. Since $H_1 : M > 15$, this is a right-tailed test and $k =$ the number of minus signs = 7. We have $n = 20 \leq 25$ and so the critical value, from Table XI, is 5. Since $k > 5$ we do not reject H_0. There is not enough evidence to indicate that the median number of hours students talk on the phone per week is greater than 15.

3. From the computations in the table below we get $T_+ = 32$, $|T_-| = |-23| = 23$ and $n = 10$.

Height (inches)	Arm Span (inches)	Arm Span – Height	Absolute Value	Signed Rank
59.5	62	2.5	2.5	6
69	65.5	–3.5	3.5	–9.5
77	76	–1	1	–3
59.5	63	3.5	3.5	9.5
74.5	74	–0.5	0.5	–1.5
63	66	3	3	7.5
61.5	61	–0.5	0.5	–1.5
67.5	69	1.5	1.5	4
73	70	–3	3	–7.5
69	71	2	2	5

$T_+ = 32$ and $T_- = -23$. Since $H_1 : M_D \neq 0$, this is a two-tailed test with test statistic
$T = \min(T_+, |T_-|) = \min(32, 23) = 23$. The critical value for $n = 10$ and $\alpha/2 = 0.025$ is 8 (from Table XII). Since $T > 8$ we do not reject H_0. There is not enough evidence to conclude that an individual's height and arm span are different.

4. From the computations in the table below we have $n_1 = 16$, $n_2 = 12$ and $S = 216$.

Female	Male	Female Ranks	Male Ranks
0.474	0.541	6	10
0.743	0.659	21	17
0.538	1.050	9	28
0.436	0.880	4	25
0.480	0.577	7	13
0.531	0.752	8	22
0.398	0.849	3	24
0.561	0.544	12	11
0.633	0.464	16	5
0.596	0.675	14	18
0.831	0.626	23	15
0.725	0.393	20	2
0.887		26	
0.905		27	
0.711		19	
0.338		1	
	Sum =	216	

Since $H_1 : M_x \neq M_y$, this is a two-tailed test. The critical values are $w_{0.005} = 42$ (from Table XIII) and $w_{0.995} = n_1 n_2 - w_{0.005} = 16 \cdot 12 - 42 = 150$. The test statistic is $T = S - \dfrac{n_1(n_1+1)}{2} = 216 - \dfrac{16 \cdot 17}{2} = 80$ which is not in the critical region so we do not reject H_0. There is not enough evidence to indicate that the median reaction time of males differs from that of females.

5. The table shows the data and their ranks:

Engine Displacement	City mpg	Rank of X	Rank of Y	d_i	d_i^2
3.5	16	9	2.5	6.5	42.25
3.8	17	11.5	4.5	7	49
4.6	15	14.5	1	13.5	182.25
2.2	24	2.5	14	−11.5	132.25
3.5	18	9	7.5	1.5	2.25
2.4	22	5	11.5	−6.5	42.25
2.4	21	5	10	−5	25
2.7	18	7	7.5	−0.5	0.25
3.5	18	9	7.5	1.5	2.25
2.0	24	1	14	−13	169
4.0	17	13	4.5	8.5	72.25
4.6	16	14.5	2.5	12	144
3.8	18	11.5	7.5	4	16
2.2	24	2.5	14	−11.5	132.25
2.4	22	5	11.5	−6.5	42.25

$\sum d_i^2 = 42.25 + 49 + 182.25 + \ldots + 132.25 + 42.25 = 1053.5$

$r_s = 1 - \dfrac{6 \sum d_i^2}{n(n^2-1)} = 1 - \dfrac{6(1053.5)}{15(15^2-1)} = 1 - \dfrac{6321}{3360} \approx -0.881$ [Tech: -0.930]

As we are testing for a negative association, this is a left-tailed test and with $n = 15$ and $\alpha = 0.05$ (one tail). The critical value is -0.446. Since the test statistic is less than -0.446, we reject H_0. There is enough evidence to conclude that engine displacement and fuel economy are negatively associated.

6. (a) H_0: the distributions of the populations are the same.

   H_1: at least one population distribution differs from the others.

   (b) The calculations are summarized in the table:

	Organic	Mineral	Surface
n_i	31	47	44
R_i	2,355.5	2,119.5	3,028.0
R_i^2	5,548,380	4,492,280	9,168,784

$H = \dfrac{12}{N(N+1)} \left[\dfrac{R_1^2}{n_1} + \cdots + \dfrac{R_k^2}{n_k} \right] - 3(N+1)$

$= \dfrac{12}{122(122+1)} \left[\dfrac{5,548,380}{31} + \dfrac{4,492,280}{47} + \dfrac{9,168,784}{44} \right] - 3(122+1) = 17.199$

   (c) Since at least one sample size is greater than 5, we get the critical value from the χ^2 tables. With df $= k - 1 = 3 - 1 = 2$ we find $\chi_{0.05}^2 = 5.991$.

(d) Since the test statistic is greater than the critical value, we reject H_0. There is enough evidence to support the conclusion that the distribution of concentration of dissolved organic carbon varies by collection area.

(e) The distributions do appear to differ.

7. In general, parametric procedures are based on some assumptions about the distribution of the population that is being examined, and typically involve the estimation of a population parameter, such as the mean or variance of the population. Nonparametric methods, perhaps better called distribution-free methods, make no assumptions about the underlying population distributions and usually do not involve estimating any population parameters, although they may be used to perform inference about population parameters such as the population median. Nonparametric tests generally have fewer assumptions, some have simple computations and some can be applied to qualitative data such as rankings. However, they are typically also less powerful and less efficient than parametric tests. The tests are tabulated below with some of the differences.

Nonparametric Test	Parametric Test	Differences
Runs test for randomness	None	
One-sample sign test	z-Test or t-Test	The nonparametric test is a test of the median and the parametric test is of the mean. The parametric test requires that data come from a normal distribution or the sample size be large.
Wilcoxon Matched-Pairs Signed-Ranks Test	Paired t-Test for Dependent Samples	The nonparametric test is a test of the medians and the parametric test is of the means. The parametric test requires that the differences be normally distributed or the sample size be large. The nonparametric test requires that the differences have a symmetric distribution.
Mann-Whitney Test	Welch's t-Test for Independent Samples	The nonparametric test is a test of the medians and the parametric test is of the means. The parametric test requires that the differences be normally distributed or the sample sizes be large. The nonparametric test requires that the distributions of the two populations be the same, except possibly for the medians.
Spearman's Rank-Correlation Test	Hypothesis Test of the Significance of β_1	The nonparametric test is only for an association and so can also be used on ordinal data whereas the parametric test is only applicable to quantitative data. The parametric test requires that the residuals be normally distributed with constant error variance.
Kruskal-Wallis Test	F-test for One-Way ANOVA	The parametric test requires that the data come from populations that are normally distributed with the same variance and tests if the means are all the same. The nonparametric test requires only that the data can be ranked and tests if the distributions are all the same.

Chapter 15 Test

1. For the given data, $n = 25$, $n_1 = 11$ (number of W's), $n_2 = 14$ (number of E's) and $r = 13$. Since $n_1 \leq 20$ and $n_2 \leq 20$, the critical values are 8 and 19, from Table X. The test statistic is $r = 13$, which is not in the critical region and so we do not reject H_0. There is not enough evidence to indicate that the sequence is not random. It appears that the winning division occurs randomly.

2. We first convert the data to plus and minus signs (and zeros) by comparing the data values to $M_0 = 15$. This gives: $- + - - - - - -$ $- - + - + + - - - - - 0$. There are 4 plus signs and 15 minus signs (and one zero). Since $H_1 : M < 15$, this is a left-tailed test and $k =$ the number of plus signs $= 4$. We have $n = 4 + 15 = 19 \leq 25$ and so the critical value, from Table XI, is 5. Since $k \leq 5$ we reject H_0. There is enough evidence to conclude that the median number of hours students study per week is less than 15.

3. From the computations in the table below we get $T_+ = 12$, $|T_-| = |-33| = 33$ and $n = 9$.

Before	After	Before – After	Absolute Value	Signed Rank
136	128	8	8	5
120	111	9	9	6.5
129	129	0	0	0
143	148	–5	5	–3.5
115	110	5	5	3.5
113	112	1	1	1.5
89	98	–9	9	–6.5
122	103	19	19	8.5
102	103	–1	1	–1.5
122	103	19	19	8.5

Since $H_1 : M_D > 0$, this is a right-tailed test with test statistic $T = |T_-| = |-11.5| = 11.5$. The critical value for $n = 9$ and $\alpha = 0.05$ is 8 (from Table XII). Since $T \geq 8$ we do not reject H_0. There is not enough evidence to indicate that the median pulse rate has decreased after the exercise program.

4. From the computations in the table below we have $n_1 = 12$, $n_2 = 14$ and $S = 205$.

Texas	Illinois	Texas Ranks	Illinois Ranks
4.69	4.4	17	6.5
5.22	4.4	25.5	6.5
5.1	4.69	23	17
4.76	4.75	21	20
5.22	4.22	25.5	2
4.25	4.63	3	13
4.46	4.64	9	14
5.14	4.45	24	8
4.93	4.54	22	12
4.11	4.49	1	10
4.65	4.35	15	4
4.71	4.36	19	5
	4.69		17
	4.52		11
Sum =		205	

Since $H_1 : M_H > M_C$, this is a right-tailed test. We first look up $w_{0.05} = 52$ (from Table XIII) and then the critical value is $w_{0.95} = n_1 n_2 - w_{0.05} = 12 \cdot 14 - 52 = 116$. The test statistic is

$$T = S - \frac{n_1(n_1 + 1)}{2} = 205 - \frac{12 \cdot 13}{2} = 127$$ which

is greater than $w_{0.95} = 116$ and so we reject H_0. There is enough evidence to indicate that the median pH of rain near Houston is greater than that of rain near Chicago.

5. The table shows the data and their ranks:

Temp.	Chirps per sec.	Rank of X	Rank of Y	d_i	d_i^2
88.6	20.0	14	15	−1	1
93.3	19.8	15	14	1	1
80.6	17.1	7.5	10.5	−3	9
69.7	14.7	2	2	0	0
69.4	15.4	1	4	−3	9
79.6	15.0	6	3	3	9
80.6	16.0	7.5	6.5	1	1
76.3	14.4	5	1	4	16
71.6	16.0	3	6.5	−3.5	12.25
84.3	18.4	13	13	0	0
75.2	15.5	4	5	−1	1
82.0	17.1	9	10.5	−1.5	2.25
83.3	16.2	11	8	3	9
82.6	17.2	10	12	−2	4
83.5	17.0	12	9	3	9

$$\sum d_i^2 = 1+1+9+\ldots+4+9 = 83.5$$

$$r_s = 1 - \frac{6\sum d_i^2}{n(n^2-1)} = 1 - \frac{6(83.5)}{15(15^2-1)}$$

$$= 1 - \frac{501}{3360} \approx 0.851 \quad [\text{Tech: } 0.850]$$

As we are testing for an association, this is a two-tailed test and with $n = 15$ and $\alpha = 0.05$ (two tails). The critical values are ± 0.521. Since the test statistic is greater than 0.521, we reject H_0. There is enough evidence to conclude that temperature and number of chirps per second are associated.

6. The table shows the ranks for the data and the calculations based on those ranks:

	Large Family	Passenger Vans	Midsize Utility
	18	16.5	4
	6.5	15	1.5
	12	8	13
	3	19	5
	14	21	1.5
	10	11	9
	16.5	20	6.5
R_i	80	110.5	40.5
R_i^2	6400	12,210.25	1640.25
n_i	7	7	7

$$H = \frac{12}{N(N+1)}\left[\frac{R_1^2}{n_1} + \cdots + \frac{R_k^2}{n_k}\right] - 3(N+1)$$

$$= \frac{12}{21(21+1)}\left[\frac{6,400}{7} + \frac{12,210.25}{7} + \frac{1640.25}{7}\right] - 3(21+1)$$

$$= 9.141$$

Since at least one sample size is greater than 5, we get the critical value from the χ^2 tables. With df $= k - 1 = 3 - 1 = 2$ we find $\chi_{0.05}^2 = 5.991$. Since the test statistic is greater than the critical value, we reject H_0. There is enough evidence to conclude that the distribution of femur force varies by vehicle category.

7. An efficiency of 0.713 means the nonparametric test would need a sample size of 1000 to achieve the same results as the corresponding parametric test with a sample size of 713.

Case Study: Evaluating Alabama's 1891 House Bill 504

1. Length of School Year:

1887 B	1887 W	D = W – B	|D|	Signed Ranks
89	61	–28	28	–16.0
85	71	–14	14	–7.0
106	70	–36	36	–18.0
82	65	–17	17	–10.0
131	120	–11	11	–5.0
65	65	0	0	0.0
82	72	–10	10	–3.5
75	60	–15	15	–8.0
106	74	–32	32	–17.0
80	64	–16	16	–9.0
95	71	–24	24	–13.5
83	60	–23	23	–12.0
70	60	–10	10	–3.5
95	70	–25	25	–15.0
59	63	4	4	2.0
100	97	–3	3	–1.0
96	72	–24	24	–13.5
60	60	0	0	0.0
83	71	–12	12	–6.0
79	79	0	0	0.0
78	60	–18	18	–11.0

$T_- = -169$ and $T_+ = 2$. Since $H_1 : M_D \neq 0$, this is a two–tailed test with test statistic $T = T_+ = 2$. The critical value for $n = 18$ and $\alpha = 0.05$ is 40 (from Table XII). Since $T < 40$ we reject H_0. There is enough evidence to conclude that the median school–year length was different for black and white children in 1887.

1915 B	1915 W	D = W – B	|D|	Signed Ranks
93	140	47	47	7.0
91	148	57	57	11.5
86	163	77	77	21.0
84	122	38	38	4.5
91	156	65	65	16.0
56	120	64	64	14.0
72	110	38	38	4.5
108	172	64	64	14.0
94	158	64	64	14.0
102	115	13	13	1.0
89	156	67	67	18.0
91	142	51	51	8.0
101	158	57	57	11.5
93	126	33	33	3.0
65	120	55	55	10.0
121	174	53	53	9.0
109	152	43	43	6.0
80	108	28	28	2.0
80	152	72	72	20.0
86	152	66	66	17.0
81	151	70	70	19.0

$T_- = 0$ and $T_+ = 231$. Since $H_1 : M_D \neq 0$, this is a two–tailed test with test statistic $T = |T_-| = 0$. The critical value for $n = 21$ and $\alpha = 0.05$ is 58 (from Table XII). Since $T < 58$ we reject H_0. There is enough evidence to conclude that the median school–year length was different for black and white children in 1915.

2. Teacher Monthly Salary (dollars):

1887 B	1887 W	D = W – B	|D|	Signed Ranks
22.08	26.50	4.42	4.42	9.0
29.30	23.33	–5.97	5.97	–11.0
27.50	20.00	–7.50	7.50	–13.0
26.04	25.56	–0.48	0.48	–1.0
31.57	38.00	6.43	6.43	12.0
25.93	22.95	–2.98	2.98	–7.0
22.33	24.75	2.42	2.42	5.0
34.33	17.00	–17.33	17.33	–19.5
29.22	19.30	–9.92	9.92	–17.0
36.86	19.53	–17.33	17.33	–19.5
20.75	18.00	–2.75	2.75	–6.0
29.47	26.44	–3.03	3.03	–8.0
20.00	20.00	0.00	0.00	0
34.50	25.18	–9.32	9.32	–16.0
19.93	21.77	1.84	1.84	3.0
22.00	30.00	8.00	8.00	14.0
32.39	20.57	–11.82	11.82	–18.0
14.00	16.00	2.00	2.00	4.0
31.00	36.95	5.95	5.95	10.0
27.75	27.07	–0.68	0.68	–2.0
25.60	16.30	–9.30	9.30	–15.0

$T_- = -153$ and $T_+ = 57$. Since $H_1 : M_D \neq 0$, this is a two–tailed test with test statistic $T = T_+ = 57$. The critical value for $n = 20$ and $\alpha = 0.05$ is 52 (Table XII). Since $T > 52$ we do not reject H_0. There is not enough evidence to conclude that the average monthly pay was different for black and white teachers in 1887.

1915 B	1915 W	D = W – B	|D|	Signed Ranks
24.78	47.93	23.15	23.15	3
27.54	55.72	28.18	28.18	6
25.57	60.48	34.91	34.91	13
25.54	55.30	29.76	29.76	9
30.87	49.76	18.89	18.89	2
22.12	52.58	30.46	30.46	10
30.65	64.23	33.58	33.58	12
22.93	71.73	48.80	48.80	21
24.29	55.89	31.60	31.60	11
25.68	63.55	37.87	37.87	14
32.93	62.28	29.35	29.35	7
27.84	74.58	46.74	46.74	19
28.87	56.65	27.78	27.78	5
22.48	68.66	46.18	46.18	18
30.52	49.27	18.75	18.75	1
31.52	78.96	47.44	47.44	20
26.08	53.65	27.57	27.57	4
20.75	50.22	29.47	29.47	8
29.26	70.71	41.45	41.45	16
26.46	66.68	40.22	40.22	15
19.88	62.63	42.75	42.75	17

$T_- = 0$ and $T_+ = 231$. Since $H_1 : M_D \neq 0$, this is a two–tailed test with test statistic $T = |T_-| = 0$. The critical value for $n = 21$ and $\alpha = 0.05$ is 58 (from Table XII). Since $T < 58$ we reject H_0. There is enough evidence to conclude that the average monthly pay was different for black and white teachers in 1915.

Reports will vary.

Appendix B

Lines

1. (a) Slope $= \dfrac{1-0}{2-0} = \dfrac{1}{2}$

(b) If x increases by 2 units, y will increase by 1 unit.

3. (a) Slope $= \dfrac{1-2}{1-(-2)} = \dfrac{-1}{3} = -\dfrac{1}{3}$

(b) If x increases by 3 units, y will decrease by 1 unit.

5. Slope $= \dfrac{y_2 - y_1}{x_2 - x_1} = \dfrac{0-3}{4-2} = \dfrac{-3}{2} = -\dfrac{3}{2}$

7. Slope $= \dfrac{y_2 - y_1}{x_2 - x_1} = \dfrac{1-3}{2-(-2)} = \dfrac{-2}{4} = -\dfrac{1}{2}$

9. Slope $= \dfrac{y_2 - y_1}{x_2 - x_1} = \dfrac{-1-(-1)}{2-(-3)} = \dfrac{0}{5} = 0$

11. Slope $= \dfrac{y_2 - y_1}{x_2 - x_1} = \dfrac{-2-2}{-1-(-1)}$

$= \dfrac{-4}{0}$ undefined.

13. $P = (1, 2); m = 3 = \dfrac{3}{1}$

15. $P = (2,4); m = -\dfrac{3}{4} = \dfrac{-3}{4}$

17. $P = (-1,3); m = 0$

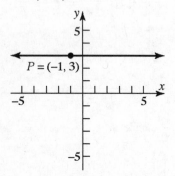

19. $P = (0,3);$ slope undefined

21. Slope $= 3 = \dfrac{3}{1}$

If x increases by 1 unit, then y increases by 3 units.
Answers will vary. Three possible points are:
$x = 1+1 = 2$ and $y = 2+3 = 5$

$(2,5)$

$x = 2+1 = 3$ and $y = 5+3 = 8$

$(3,8)$

$x = 3+1 = 4$ and $y = 8+3 = 11$

$(4,11)$

23. Slope $= -\dfrac{3}{2} = \dfrac{-3}{2}$

If x increases by 2 units, then y decreases by 3 units.
Answers will vary. Three possible points are:
$x = 2+2 = 4$ and $y = -4-3 = -7$

$(4,-7)$

$x = 4+2 = 6$ and $y = -7-3 = -10$

$(6,-10)$

$x = 6+2 = 8$ and $y = -10-3 = -13$

$(8,-13)$

25. Slope $= -2 = \dfrac{-2}{1}$

If x increases by 1 unit, then y decreases by 2 units.
Answers will vary. Three possible points are:
$x = -2+1 = -1$ and $y = -3-2 = -5$

$(-1,-5)$

$x = -1+1 = 0$ and $y = -5-2 = -7$

$(0,-7)$

$x = 0+1 = 1$ and $y = -7-2 = -9$

$(1,-9)$

27. (0, 0) and (2, 1) are points on the line.

Slope $= \dfrac{1-0}{2-0} = \dfrac{1}{2}$

y-intercept is 0; using $y = mx + b$:

$$y = \frac{1}{2}x + 0$$
$$2y = x$$
$$0 = x - 2y$$
$$x - 2y = 0 \text{ or } y = \frac{1}{2}x$$

29. (−1, 3) and (1, 1) are points on the line.

Slope $= \dfrac{1-3}{1-(-1)} = \dfrac{-2}{2} = -1$

Using $y - y_1 = m(x - x_1)$

$$y - 1 = -1(x - 1)$$
$$y - 1 = -x + 1$$
$$y = -x + 2$$
$$x + y = 2 \text{ or } y = -x + 2$$

31. $y = 2x + 3$; Slope = 2; y-intercept = 3

33. $\frac{1}{2}y = x - 1$; $y = 2x - 2$

Slope = 2; y-intercept = –2

35. $y = \frac{1}{2}x + 2$; Slope = $\frac{1}{2}$; y-intercept = 2

37. $x + 2y = 4$; $2y = -x + 4 \rightarrow y = -\frac{1}{2}x + 2$

Slope = $-\frac{1}{2}$; y-intercept = 2

39. $2x - 3y = 6$; $-3y = -2x + 6 \rightarrow y = \frac{2}{3}x - 2$

Slope = $\frac{2}{3}$; y-intercept = –2

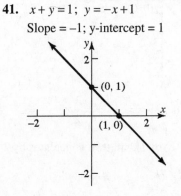

41. $x + y = 1$; $y = -x + 1$

Slope = –1; y-intercept = 1

43. $x = -4$; Slope is undefined
y-intercept - none

45. $y = 5$; Slope = 0; y-intercept = 5

47. $y - x = 0$; $y = x$
Slope = 1; y-intercept = 0

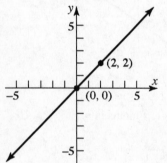

49. $2y - 3x = 0$; $2y = 3x \rightarrow y = \dfrac{3}{2}x$

Slope $= \dfrac{3}{2}$; y-intercept = 0